PROCESS CONTROL AND IDENTIFICATION

PROCESS CONTROL AND IDENTIFICATION

W. Fred Ramirez

University of Colorado at Boulder
Department of Chemical Engineering
Boulder, Colorado

ACADEMIC PRESS, INC.

Harcourt Brace & Company, Publishers

Boston San Diego New York
London Sydney Tokyo Toronto

This book is printed on acid-free paper. ∞

ACADEMIC PRESS, INC.
1250 Sixth Avenue, San Diego, CA 92101-4311

United Kingdom Edition published by
ACADEMIC PRESS LIMITED
24-28 Oval Road, London NW1 7DX

Library of Congress Cataloging-in-Publication Data

Ramirez, W. Fred, 1941-
Process control and identification / W. Fred Ramirez.
p. cm.
Includes bibliographical references and index.
ISBN 0-12-577240-8 (alk. paper)
1. Process control--Mathematical models. I. Title.
TS156.8.R36 1993
629.8'312--dc20 93-29779

Printed in the United States of America

93 94 95 96 CU 9 8 7 6 5 4 3 2 1

Contents

Chapter 3
Dynamic Optimization

Chapter 4
Optimal Control

Chapter 5
The Linear Quadratic Regulator (LQR) Problem

Chapter 6
Model Predictive Control

Chapter 7
Robust Control

Chapter 8
Problems Linear in the Control Variables

Chapter 9
The Discrete Maximum Principle

Chapter 10
State and Parameter Identification

Chapter 11
Sequential Least Squares Parameter Estimation

Chapter 12
The Linear Quadratic Gaussian (LQG) Problem

Appendix

PREFACE

The purpose of this book is to present a time domain approach to modern process control. The time domain approach has several advantages including the fact that process models are naturally developed through conservation laws and mechanistic phenomena in the time domain. This approach also allows for the formulation of precise performance objectives that can be extremized. There is a definite need in the process industries for improved control. New hardware and software tools now allow the control engineer to consider the implementation of more sophisticated control strategies that address critical and difficult process control problems. In general, it is necessary to incorporate process knowledge into the control design in order to improve process operation. Advanced control designs require more engineering analysis but can lead to significant improvements in process behavior and profitability.

The reader will notice that I have tried to include practical examples throughout the book in order to illustrate theoretical concepts. This approach also allows the reader to be aware of computational issues of implementation as well as the interpretation of the results of process testing.

Chapter 1 presents basic time domain system concepts that are needed to mathematically describe an advanced process control problem. The important concepts of observability and controllability are introduced. Observability is used in the design of the measurement system, and controllability is important for the specification of the control variables of the system. The software package MATLAB is introduced. It simplifies many of the control design calculations.

Chapter 2 treats the topic of steady-state optimization. Necessary conditions for extrema of functions are derived using variational principles. These steady-state optimization techniques are used for the determination of optimal setpoints for regulators used in supervisory computer control.

Chapter 3 gives the fundamental mathematical principles of the calculus of variations used for the optimization of dynamic systems. Classical results of the Euler equation for functional extrema and those of constrained optimization given by the Euler–Lagrange equation are developed.

Chapter 4 applies variational calculus to problems that include control variables as well as state variables. Optimal control strategies are developed that extremize precise performance criteria. Necessary conditions for optimization are shown to be conveniently expressed in terms of a mathematical function called the Hamiltonian. Pontryagin's maximum principle is developed for systems that have control constraints. Process applications of optimal control are presented.

Chapter 5 considers optimal regulator control problems. The Kalman linear quadratic regulator (LQR) problem is developed, and this optimal multivariable proportional controller is shown to be easily computable using the Riccati matrix differential equation. The regulator problem with unmeasurable load disturbances is shown to lead to an optimal multivariable proportional-integral feedback structure.

Chapter 6 develops model predictive control concepts. This structure allows for the inclusion of predictive feed-forward control into the optimal control problem. We consider design strategies for completely measurable disturbances as well as systems with both measurable and unmeasurable disturbances.

Chapter 7 discusses robust control. This allows for the inclusion of uncertainty of process parameters in the control design. The concept of robustness refers to the preservation of closed-loop stability under allowable variations in system parameters. General stability results and integrity results are given for the LQR problem.

Chapter 8 considers optimal control problems for systems that are either linear or nonlinear in the state variables but are linear in the controls. The solution of this class of problems leads to bang-bang control strategies. The existence of singular or intermediate control must also be investigated. Both time-optimal control and minimum integral square error problems are discussed.

Chapter 9 develops necessary conditions for optimality of discrete time problems. In implementing optimal control problems using digital computers, the control is usually kept constant over a period of time. Problems that were originally described by differential equations defined over a continuous time domain are transformed to problems that are described by a set of discrete algebraic equations. Necessary conditions for optimality are derived for this class of problems and are applied to several process control situations.

Chapter 10 discusses state and parameter identification. Using uncertainty concepts, an optimal estimate of the state for a linear system is obtained based upon available measurements. The result is the Kalman filter. The Kalman filter is extended for nonlinear systems and discrete-time models. Kalman filtering is also shown to be effective for the estimation of model parameters.

Chapter 11 presents the use of sequential least squares techniques for the recursive estimation of uncertain model parameters. There is a statistical advantage in taking this approach to model parameter identification over that of incorporating model parameter estimation directly into Kalman filtering.

Chapter 12 considers the combination of optimal control with state and parameter estimation. The separation principle is developed, which states that the design of a control problem with measurement and model uncertainty can be treated by first performing a Kalman filter estimate of the states and then developing the optimal control law based upon the estimated states. For linear regulator problems, the problem is known as the linear quadratic Gaussian (LQG) problem. The inclusion of model parameter identification results in adaptive control algorithms.

ACKNOWLEDGMENTS

This book is a result of the author's research and teaching career in the area of optimal process control and identification. I gratefully acknowledge the contributions of my research students to the development of many of the ideas contained in this book. I have been fortunate to have had a group of research students who have stimulated new and creative insights into process control. They deserve credit for many of the novel and important ideas found in this work.

Special thanks are due to Ellen Romig who did the technical word processing and layout of this book. Her talents and personal concern for this project are truly appreciated.

I also thank the University of Colorado for awarding me a faculty fellowship, which provided the time needed to prepare the final manuscript in the excellent academic environment of Cambridge University.

Finally, I want to thank my wife, Marion, who has been my personal inspiration for many years.

Chapter 1

BASIC SYSTEMS CONCEPTS

1.1 Introduction

There have been great advances in the mathematical understanding of the control problem. Sophisticated mathematical techniques of functional optimization, operator theory, and statistical processes have been developed in the control literature. There have also been vast improvements in on-line computational hardware and software. With the introduction of the minicomputer in the late 1960s, the microcomputer in the 1970s, the personal computer in the 1980s, and workstations in the 1990s, a wide range of relatively cheap computer hardware has become available for process applications. Although there have been great advances in computer systems, we have not experienced the same degree of advancement in new methodologies and concepts needed to utilize these computer advances for the control, optimization, and identification of industrial processes. There is still a gap between sophisticated modern control theoretical concepts and every day practice in the process industries. This book aims to bridge that gap and to help stimulate the application of modern control concepts to the industrial environment.

With today's competition in the high-technology industries, the continuing need for material and energy conservation, the need for hazard free operation, and the need for environmentally safe discharges, it is important to develop and apply methodologies and techniques that can lead to improved performance and operation. This is the challenge facing engineers designing and implementing control systems for industrial processes.

In this chapter we present basic time domain system concepts that are needed to describe mathematically a process control problem. The software package MATLAB is introduced. It is an important computational tool that simplifies many of the control design calculations. Also, two important system concepts of observability and controllability are introduced. Observability is important in the design of a measurement system for a process, and controllability is important to the specification and adequacy of the variables that are proposed to control the system.

1.2 The Process Control Problem

In general, the process control problem can be visualized with the aid of the diagram of Figure 1.1. In this figure we define two classes of variables for a process. These are input variables and output variables. The input variables can be further divided into:

> Control variables—these are the input variables that can be directly set or manipulated.
>
> Disturbance variables—these are the input variables that cannot be controlled.

The output variables are classified as:

> State variables—these are the set of output variables that are needed to describe the dynamics of the process.
>
> Measured variables—these are the subset of the output variables that are actually measured.

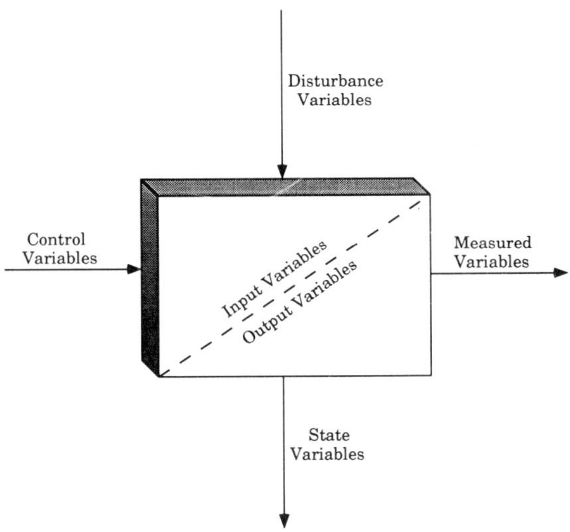

Figure 1.1 The process control problem.

The objective of any process control system is to adjust the control variables to achieve well-defined process objectives in spite of disturbances, using measurements of state and disturbance variables.

1.3 Single Loop Feedback Control

One realization of the general objective of process control is that of single loop feedback control systems. The feedback control concept is shown in Figure 1.2.

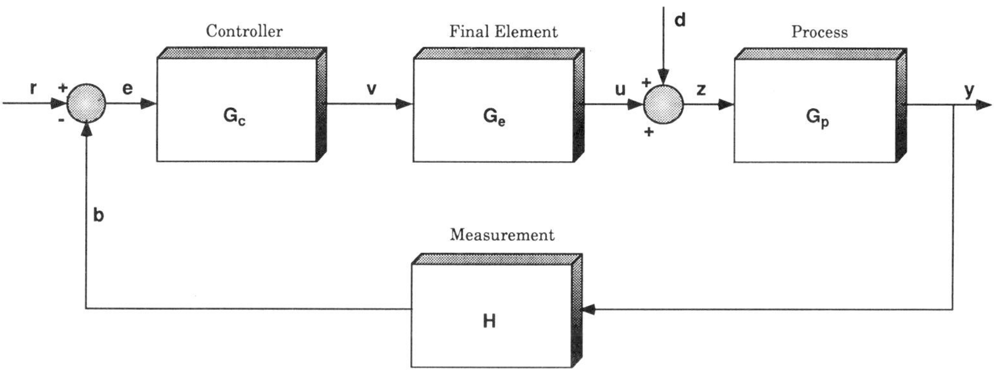

Figure 1.2 Single loop feedback control.

In this figure we define the following variables and transfer functions:

b = the feedback variable
d = the disturbance variable to the process
e = the error
G_c = the controller transfer function
G_e = the final control element transfer function
G_p = the process transfer function
H = the measurement element transfer function
r = the set point or desired value of the output variable
u = the control variable
v = the output variable from the controller
y = the output variable from the process
z = the input variable to the process

The goal of the feedback control problem is to design the controller transfer function, G_c, so as to maintain the output, y, at a desired set point value, r, in spite of the disturbances, d, that affect the process. Most classical feedback control concepts have been developed based upon the study of the response of a second-order linear system. A general second-order linear system is described by the linear differential equation,

$$\tau^2 \frac{d^2y}{dt^2} + 2\xi\tau \frac{dy}{dt} + y = x(t) \tag{1.3.1}$$

where τ = the time constant
ξ = the damping ratio

The transfer function, G_p, for this system is

$$\frac{y(s)}{z(s)} = \frac{1}{\tau^2 s^2 + 2\xi\tau s + 1} \tag{1.3.2}$$

The response of the system to a unit step input is

$$y(s) = \frac{1/\tau^2}{s(s-s_1)(s-s_2)} \tag{1.3.3}$$

with

$$s_1 = -\frac{\xi}{\tau} + \frac{\sqrt{\xi^2 - 1}}{\tau}$$

$$s_2 = -\frac{\xi}{\tau} - \frac{\sqrt{\xi^2 - 1}}{\tau}$$

The roots of the characteristic equation (the denominator of Equation (1.3.2)), s_1 and s_2 can be real or complex, depending upon the value of the damping ratio, ξ. Figure 1.3 shows the effect of different values of the damping constant. When $\xi < 1$, the roots s_1 and s_2 are complex so that the system response is underdamped and oscillatory. When $\xi > 1$, the roots are real so that the system response is overdamped and uniformly monotonic. When $\xi = 1$, the system is critically damped.

The following terms are often used to describe an underdamped system response. These terms are illustrated in Figure 1.4.

Overshoot = (A/B)—The overshoot is the ratio of the maximum peak value to the desired steady-state value of the output variable.

Decay Ratio = (A/C)—this is the ratio of successive peak heights.

Rise Time = (t_r)—The rise time is the time at which the system output first reaches the desired steady-state value.

Response Time = (t_d)—This is the time when the output variable gets to within a specified range, say ±5%, of the steady-state value and stays there.

Control system designers often attempt to make setpoint step responses of

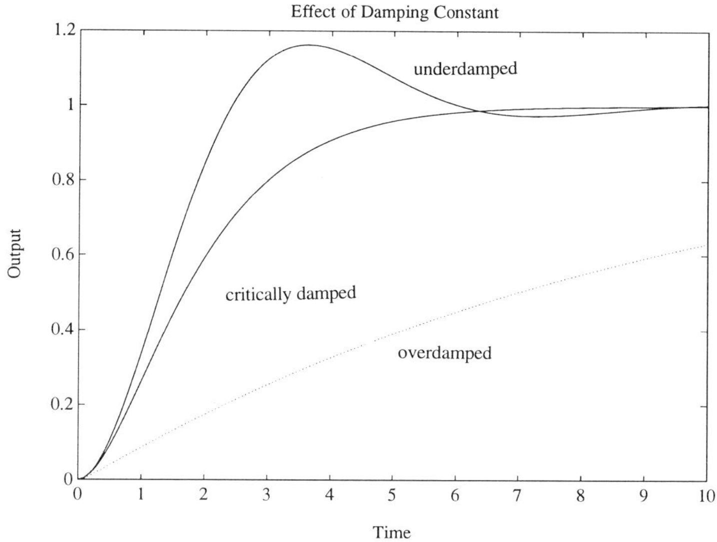

Figure 1.3 Effect of the damping constant on second order systems.

controlled systems approximate the step response of an underdamped second order system. A number of design procedures have been developed to tune proportional-integral-derivative controllers for single loop feedback systems. The PID controller is given by

$$v = -K_p \left[e + K_I \int e \; dt + K_D \frac{de}{dt} \right] \tag{1.3.4}$$

where K_p is the proportional gain constant, K_I is the integral gain constant (1/s), and K_D is the derivative gain constant (s).

Good treatments of both heuristic and analytical procedures for determining tuning constants of K_p, K_I, and K_D are given in the books by Coughanowr and Koppel, (1965), Stephanopoulos (1984), and Seborg *et al.* (1989).

The main drawbacks to classical single loop control designs are the following. Most process control problems involve multivariable systems with interactions between state and control variables. Also, there is a vagueness in the specification of the exact goal or definition of the control system objective.

There are modern engineering and mathematical approaches that can overcome these shortcomings of the classical single loop design procedure. Multivariable models can be formulated for process problems. Dynamic models can be written in terms of the state variables that describe the actual dynamic response of the system. Deterministic models are developed by applying appropriate conservation principles to the system and incorporating operable rate

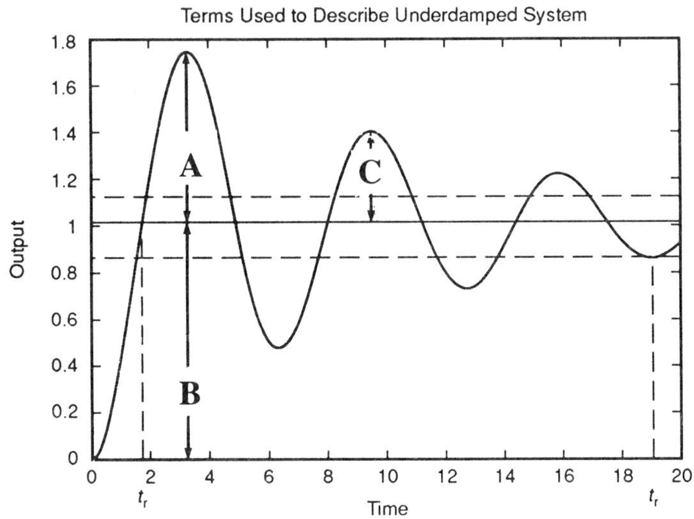

Figure 1.4 Terms used to describe underdamped systems.

phenomena. An excellent treatment of the principles needed for model development is given in the book by Bird, Stewart, and Lightfoot (1960). Modeling and numerical methods are covered in the book by Ramirez (1989).

Dynamic optimization of precise objective criteria is possible through the use of the calculus of variations and Pontryagin's minimum principle. These topics are introduced and discussed in this book.

1.4 State Variable Representation of Systems

The state of a system is defined by the set of variables, called the state variables, $\boldsymbol{x}^T = (x_1, x_2, \cdots, x_n)$, which, if known at a given time, t_o, along with the inputs to the system, allow for their unique determination at any time, $t \geq t_o$. The state variables therefore characterize the dynamic behavior of the system.

Continuous time state variable models are of the form

$$\frac{d\boldsymbol{x}}{dt} = \dot{\boldsymbol{x}} = \boldsymbol{f}(\boldsymbol{x}, \boldsymbol{u}, \boldsymbol{d}, t) \tag{1.4.1}$$

where $\boldsymbol{f}$ = a set of nonlinear time varying functions of order n

$\boldsymbol{x}$ = state variables of order n

$\boldsymbol{u}$ = control variables of order m

$\boldsymbol{d}$ = disturbance variables of order p

The measurement model is usually assumed to be of the form

$$\boldsymbol{y} = \boldsymbol{h}(\boldsymbol{x}, \boldsymbol{u}, \boldsymbol{d}, t) \tag{1.4.2}$$

where $\boldsymbol{h}$ = set of nonlinear time varying functions of order q
$\boldsymbol{y}$ = measured variables of order q

Figure 1.5 gives a schematic description of the continuous time state variable description.

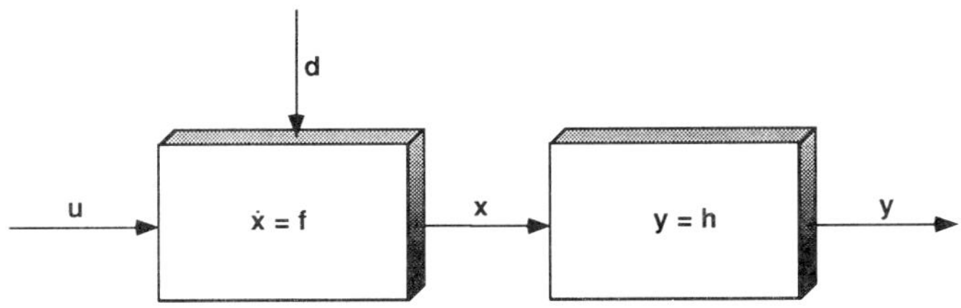

Figure 1.5 Continuous time state variable model.

If the system model is linear, then the state variable description of Equations (1.4.1) and (1.4.2) become

$$\dot{\boldsymbol{x}} = \boldsymbol{A}\boldsymbol{x} + \boldsymbol{B}\boldsymbol{u} + \boldsymbol{W}\boldsymbol{d} \tag{1.4.3}$$

$$\boldsymbol{y} = \boldsymbol{H}\boldsymbol{x} + \boldsymbol{C}\boldsymbol{u} + \boldsymbol{D}\boldsymbol{d} \tag{1.4.4}$$

where $\boldsymbol{A}$, $\boldsymbol{B}$, $\boldsymbol{W}$, $\boldsymbol{H}$, $\boldsymbol{C}$, and $\boldsymbol{D}$ are matrices of appropriate order.

1.5 Transition Matrices

In order to use the state variable representation of processes, it is necessary to seek solutions for the nonlinear differential equations of Equation (1.4.1) or the linear equations of (1.4.3). For the linear model of Equation (1.4.3), we will first consider the solution to the homogeneous problem

$$\dot{\boldsymbol{x}} = \boldsymbol{A}\boldsymbol{x} \tag{1.5.1}$$

The solution to this vector linear differential equation is by analogy to the scalar case (Crowell and Williamson, 1962),

$$\boldsymbol{x} = \exp(\boldsymbol{A}t)\ \boldsymbol{x}(0) \tag{1.5.2}$$

where $\exp(\boldsymbol{A}t)$ is the exponential matrix and is defined by

$$\exp(\boldsymbol{A}t) = \sum_{k=0}^{\infty} \boldsymbol{A}^k \frac{t^k}{k!} \tag{1.5.3}$$

We express the solution to both the linear and nonlinear homogeneous problems in terms of the transition matrix, $\boldsymbol{\phi}(t)$,

$$\boldsymbol{x} = \boldsymbol{\phi}(t)\, \boldsymbol{x}(0) \tag{1.5.4}$$

For the linear homogeneous problem, the transition matrix is the exponential matrix,

$$\boldsymbol{\phi}(t) = \exp(\boldsymbol{A}t) \tag{1.5.5}$$

There are several other ways to calculate this transition or exponential matrix rather than using its infinite series definition of Equation (1.5.3). Petkov *et al.* (1991) present a complete discussion of modern computational methods for the calculation of the exponential matrix as well as other computational issues associated with state space linear control system.

1.5.1 Laplace Transform Method

The transition matrix for linear problems can be computed through the use of Laplace transforms. For the system

$$\dot{\boldsymbol{x}} = \boldsymbol{A}\boldsymbol{x} \tag{1.5.6}$$

we take the Laplace transform of both sides to get

$$s\boldsymbol{x}(s) - \boldsymbol{x}(0) = \boldsymbol{A}\boldsymbol{x}(s) \tag{1.5.7}$$

Rearranging gives

$$\boldsymbol{x}(s) = (s\boldsymbol{I} - \boldsymbol{A})^{-1}\boldsymbol{x}(0) \tag{1.5.8}$$

Taking the inverse transform yields

$$\boldsymbol{x}(t) = \mathcal{L}^{-1}\{(s\boldsymbol{I} - \boldsymbol{A})^{-1}\}\boldsymbol{x}(0) \tag{1.5.9}$$

Therefore, the transition matrix is

$$\boldsymbol{\phi}(t) = \mathcal{L}^{-1}\{s\boldsymbol{I} - \boldsymbol{A})^{-1}\} \tag{1.5.10}$$

Let us consider the following example,

$$\frac{d^2c}{dt^2} - \beta^2 c = 0 \tag{1.5.11}$$

This is a single second-order differential equation. To place it in the state variable form of Equation (1.5.1), we must define an equivalent set of two first-order differential equations. This can be accomplished by defining

$$\frac{dc}{dt} = x_2 \qquad \text{and} \qquad c = x_1 \tag{1.5.12}$$

Equation (1.5.11) is therefore equivalently given by

$$\begin{aligned} \dot{x}_1 &= x_2 \\ \dot{x}_2 &= \beta^2 x_1 \end{aligned} \tag{1.5.13}$$

or

$$\dot{\boldsymbol{x}} = \boldsymbol{A}\boldsymbol{x} \tag{1.5.14}$$

with

$$\boldsymbol{A} = \begin{bmatrix} 0 & 1 \\ \beta^2 & 0 \end{bmatrix}$$

For this example, we have

$$(s\boldsymbol{I} - \boldsymbol{A}) = \begin{bmatrix} s & -1 \\ -\beta^2 & s \end{bmatrix} \tag{1.5.15}$$

The inverse is in general given by

$$(s\boldsymbol{I} - \boldsymbol{A})^{-1} = \frac{adj(s\boldsymbol{I} - \boldsymbol{A})}{det(s\boldsymbol{I} - \boldsymbol{A})} \tag{1.5.16}$$

For this example, the determinant is

$$det(s\boldsymbol{I} - \boldsymbol{A}) = s^2 - \beta^2 \tag{1.5.17}$$

so that the inverse of Equation (1.5.15) is

$$(s\boldsymbol{I} - \boldsymbol{A})^{-1} = \begin{bmatrix} \dfrac{s}{s^2 - \beta^2} & \dfrac{1}{s^2 - \beta^2} \\ \dfrac{\beta^2}{s^2 - \beta^2} & \dfrac{s}{s^2 - \beta^2} \end{bmatrix} \tag{1.5.18}$$

Taking the inverse Laplace transform of each term of Equation (1.5.18) gives the individual terms of the transition matrix

$$\phi_{11} = \mathcal{L}^{-1}\left(\frac{s}{s^2 - \beta^2}\right) = \cosh \beta t = \frac{e^{\beta t} + e^{-\beta t}}{2} \tag{1.5.19}$$

$$\phi_{12} = \mathcal{L}^{-1}\left(\frac{1}{s^2 - \beta^2}\right) = \frac{\sinh \beta t}{\beta} = \frac{e^{\beta t} - e^{-\beta t}}{2\beta} \tag{1.5.20}$$

$$\phi_{21} = \beta \sinh \beta t \tag{1.5.21}$$

$$\phi_{22} = \cosh \beta t \tag{1.5.22}$$

1.5.2 Sylvester's Theorem

We can also compute the exponential transition matrix by the use of Sylvester's theorem. Sylvester's theorem allows us to express any continuous function of

$\boldsymbol{A}$ in terms of the eigenvalues of $\boldsymbol{A}$,

$$f(\boldsymbol{A}t) = \sum_{i=1}^{n} f(\lambda_i t) \prod_{\substack{j=1 \\ j \neq i}}^{n} \left(\frac{\boldsymbol{A} - \lambda_j \boldsymbol{I}}{\lambda_i - \lambda_j} \right) \tag{1.5.23}$$

where λ_i are the distinct eigenvalues of $\boldsymbol{A}$
Π is the product operator

For the example of Equation (1.5.14), we determine the eigenvalues of $\boldsymbol{A}$ from

$$det[\boldsymbol{A} - \lambda \boldsymbol{I}] = 0 \tag{1.5.24}$$

which is

$$det \begin{bmatrix} -\lambda & 1 \\ \beta^2 & -\lambda \end{bmatrix} = 0 \tag{1.5.25}$$

or

$$\lambda^2 - \beta^2 = 0 \tag{1.5.26}$$

so that

$$\begin{aligned} \lambda_1 &= \beta \\ \lambda_2 &= -\beta \end{aligned} \tag{1.5.27}$$

Using Sylvester's theorem, the exponential or transition matrix becomes

$$e^{\boldsymbol{A}t} = e^{\lambda_1 t} \left(\frac{\boldsymbol{A} - \lambda_2 \boldsymbol{I}}{\lambda_1 - \lambda_2} \right) + e^{\lambda_2 t} \left(\frac{\boldsymbol{A} - \lambda_1 \boldsymbol{I}}{\lambda_2 - \lambda_1} \right) \tag{1.5.28}$$

or

$$e^{\boldsymbol{A}t} = \begin{bmatrix} \dfrac{e^{\beta t} + e^{-\beta t}}{2} & \dfrac{e^{\beta t} - e^{-\beta t}}{2\beta} \\ \dfrac{\beta(e^{\beta t} - e^{-\beta t})}{2} & \dfrac{e^{\beta t} + e^{-\beta t}}{2} \end{bmatrix} \tag{1.5.29}$$

1.5.3 Linear Algebra Formulation

Sylvester's theorem can also be expressed by the following linear algebraic formulation which is a use of the Cayley–Hamilton theorem

$$\boldsymbol{f}(\boldsymbol{A}t) = \alpha_0 \boldsymbol{I} + \alpha_1 \boldsymbol{A} + \cdots + \alpha_{n-1} \boldsymbol{A}^{n-1} \tag{1.5.30}$$

where the coefficients α_k are determined from the following set of linear algebraic equations:

$$\begin{aligned} \alpha_0 + \alpha_1 \lambda_1 + \alpha_2 \lambda_1^2 + \cdots + \alpha_{n-1} \lambda_1^{n-1} &= f(\lambda_1 t) \\ \alpha_0 + \alpha_1 \lambda_2 + \alpha_2 \lambda_2^2 + \cdots + \alpha_{n-1} \lambda_2^{n-1} &= f(\lambda_2 t) \\ \vdots \qquad \vdots \qquad \vdots \qquad & \quad \vdots \\ \alpha_0 + \alpha_1 \lambda_n + \alpha_2 \lambda_n^2 + \cdots + \alpha_{n-1} \lambda_n^{n-1} &= f(\lambda_n t) \end{aligned} \tag{1.5.31}$$

where λ_k are the distinct eigenvalues of $\boldsymbol{A}$.

This formulation is particularly convenient for digital computer calculation using linear algebra packages.

Let us consider the example of Equation (1.5.14). Here Equation (1.5.31) becomes

$$\begin{aligned} \alpha_0 + \alpha_1\beta &= e^{\beta t} \\ \alpha_0 + \alpha_1(-\beta) &= e^{-\beta t} \end{aligned} \tag{1.5.32}$$

which gives

$$\alpha_0 = \frac{e^{\beta t} + e^{-\beta t}}{2} \tag{1.5.33}$$

$$\alpha_1 = \frac{e^{\beta t} - e^{-\beta t}}{2\beta} \tag{1.5.34}$$

1.5.4 Multiple Eigenvalues

The linear algebraic method can be easily modified to handle the case when there are multiple eigenvalues for the matrix $\boldsymbol{A}$. Suppose λ_1 is of multiple order 3 and all the other (n–3) eigenvalues are distinct. Then

$$e^{\boldsymbol{A}t} = \alpha_0\boldsymbol{I} + \alpha_1\boldsymbol{A} + \alpha_2\boldsymbol{A}^2 + \cdots + \alpha_{n-1}\boldsymbol{A}^{n-1} \tag{1.5.35}$$

where α_k come from the solution to the following set of linear algebraic equations

$$\begin{aligned} \alpha_0 + \alpha_1\lambda_1 + \alpha_2\lambda_1^2 + \alpha_3\lambda_1^3 + \cdots + \alpha_{n-1}\lambda_1^{n-1} &= e^{\lambda_1 t} \\ \alpha_1 + 2\alpha_2\lambda_1 + 3\alpha_3\lambda_1^2 + \cdots + (n-1)\alpha_{n-1}\lambda_1^{n-2} &= te^{\lambda_1 t} \\ 2\alpha_2 + 3(2)\alpha_3\lambda_1 + \cdots + (n-1)(n-2)\alpha_{n-1}\lambda_1^{n-3} &= t^2e^{\lambda_1 t} \\ \alpha_0 + \alpha_1\lambda_4 + \alpha_2\lambda_4^2 + \alpha_3\lambda_4^3 + \cdots + \alpha_{n-1}\lambda_4^{n-1} &= e^{\lambda_4 t} \\ \vdots \qquad \vdots \qquad \vdots \qquad & \vdots \\ \alpha_0 + \alpha_1\lambda_n + \alpha_2\lambda_n^2 + \alpha_3\lambda_n^3 + \cdots + \alpha_{n-1}\lambda_n^{n-1} &= e^{\lambda_n t} \end{aligned} \tag{1.5.36}$$

The second and third equations are obtained by differentiation with respect to the repeated root. The second equation is the differential of the first equation, and the third equation is the differential of the second equation. In like manner, equations can be generated for any order repeated roots and for as many repeated roots as needed. The equation set of (1.5.36) is a set of n linear independent equations that can be solved for the unknown coefficients α_k. Good references for state variable properties and further background on this topic are those of Ogata, (1967), Kailath (1980), and DeCarlo (1989).

1.6 Nonhomogeneous Differential Equations

We now seek the solution to the linear vector nonhomogeneous differential equation set

$$\dot{\boldsymbol{x}} = \boldsymbol{A}\boldsymbol{x} + \boldsymbol{B}\boldsymbol{u} \tag{1.6.1}$$

From the results of Section 1.5, we know that the solution to the homogeneous problem is

$$\boldsymbol{x}(t) = \boldsymbol{\phi}(t)\boldsymbol{x}(0) \tag{1.6.2}$$

where

$$\boldsymbol{\phi}(t) = e^{\boldsymbol{A}t} \tag{1.6.3}$$

We would like to show that the following property is true:

$$\boldsymbol{A} = -\boldsymbol{\phi}(t)\dot{\boldsymbol{\phi}}(-t) \tag{1.6.4}$$

To do this we differentiate Equation (1.6.2) with respect to t,

$$\dot{\boldsymbol{x}}(t) = \dot{\boldsymbol{\phi}}(t)\boldsymbol{x}(0) \tag{1.6.5}$$

Since, for the homogeneous problem,

$$\dot{\boldsymbol{x}} = \boldsymbol{A}\boldsymbol{x} \tag{1.6.6}$$

then

$$\dot{\boldsymbol{\phi}}(t)\boldsymbol{x}(0) = \boldsymbol{A}\boldsymbol{\phi}(t)\boldsymbol{x}(0) \tag{1.6.7}$$

or

$$\dot{\boldsymbol{\phi}}(t) = \boldsymbol{A}\boldsymbol{\phi}(t) \tag{1.6.8}$$

One of the basic properties of all transition matrices, including the exponential matrix (it is a property of exponential functions), is (Ogata, 1967),

$$\boldsymbol{\phi}^{-1}(t) = \boldsymbol{\phi}(-t) \tag{1.6.9}$$

Then Equation (1.6.8) becomes

$$\boldsymbol{A} = \dot{\boldsymbol{\phi}}(t)\boldsymbol{\phi}(-t) \tag{1.6.10}$$

From the fundamental definition of the inverse, we have

$$\boldsymbol{\phi}(t)\boldsymbol{\phi}(-t) = \boldsymbol{I} \tag{1.6.11}$$

Differentiation of Equation (1.6.11) gives

$$\boldsymbol{\phi}(t)\dot{\boldsymbol{\phi}}(-t) + \dot{\boldsymbol{\phi}}(t)\boldsymbol{\phi}(-t) = \boldsymbol{0} \tag{1.6.12}$$

Therefore, Equation (1.6.10) becomes

$$\boldsymbol{A} = -\boldsymbol{\phi}(t)\dot{\boldsymbol{\phi}}(-t) \tag{1.6.13}$$

which is the property that we wanted to show.

Now we substitute this identity into the original nonhomogeneous Equation (1.6.1) to obtain

$$\dot{\boldsymbol{x}} = -\boldsymbol{\phi}(t)\dot{\boldsymbol{\phi}}(-t)\boldsymbol{x} + \boldsymbol{B}\boldsymbol{u} \tag{1.6.14}$$

Multiplying by $\boldsymbol{\phi}(-t)$ gives

$$\boldsymbol{\phi}(-t)\dot{\boldsymbol{x}}(t) + \dot{\boldsymbol{\phi}}(-t)\boldsymbol{x}(t) = \boldsymbol{\phi}(-t)\boldsymbol{Bu} \tag{1.6.15}$$

which is, due to the chain rule,

$$\frac{d}{dt}\left(\boldsymbol{\phi}(-t)\boldsymbol{x}(t) \right) = \boldsymbol{\phi}(-t)\boldsymbol{Bu} \tag{1.6.16}$$

Integration from t=0 to any time t gives

$$\boldsymbol{\phi}(-t)\boldsymbol{x}(t) - \boldsymbol{\phi}(0)\boldsymbol{x}(0) = \int_0^t \boldsymbol{\phi}(-\tau)\boldsymbol{B}u(\tau)\, d\tau \tag{1.6.17}$$

or

$$\boldsymbol{x}(t) = \boldsymbol{\phi}(t)\boldsymbol{x}(0) + \int_0^t \boldsymbol{\phi}(t-\tau)\boldsymbol{Bu}(\tau)\, d\tau \tag{1.6.18}$$

Equation (1.6.18) shows that the solution to the nonhomogeneous equation is the sum of the homogeneous solution and a particular integral.

Let us now consider the solution to the general linear time varying problem

$$\dot{\boldsymbol{x}} = \boldsymbol{F}(t)\boldsymbol{x}(t) + \boldsymbol{G}(t)\boldsymbol{u}(t) \tag{1.6.19}$$

The general solution is

$$\boldsymbol{x}(t) = \boldsymbol{\phi}(t, t_0)\boldsymbol{x}(t_0) + \int_{t_0}^t \boldsymbol{\phi}(t,\tau)\boldsymbol{G}(\tau)\boldsymbol{u}(\tau)\, d\tau \tag{1.6.20}$$

where the transition matrix $\boldsymbol{\phi}(t, t_0)$ satisfies the following general property for transition matrices,

$$\frac{d}{dt}\boldsymbol{\phi}(t,\tau) = \boldsymbol{F}(t)\boldsymbol{\phi}(t,\tau) \tag{1.6.21}$$

with

$$\boldsymbol{\phi}(\tau,\tau) = \boldsymbol{I} \tag{1.6.22}$$

The form of the transition matrix is given by Equations (1.6.21) and (1.6.22) but analytical expressions or solutions are not possible except for very specific $\boldsymbol{F}(t)$ functions.

1.7 Discrete Time Models

Discrete time models are important for a number of reasons. First they arise because of the existence of serial discrete processes. They also appear when continuous systems are controlled at discrete times using digital computers. In implementing control algorithms via digital computers, the control signals can only be varied at discrete time intervals, and during the control interval the controls usually remain constant.

Let us examine the solution of the linear continuous system for the case that the control variables remain constant over some time interval, T. The system model is

$$\dot{\boldsymbol{x}} = \boldsymbol{A}\boldsymbol{x} + \boldsymbol{B}\boldsymbol{u} \tag{1.7.1}$$

and the solution for a constant value of control vector $\boldsymbol{u}(0)$ applied over the time interval T is

$$\boldsymbol{x}(T) = \boldsymbol{\phi}(T)\boldsymbol{x}(0) + \left[\int_0^T \boldsymbol{\phi}(T-\tau)\boldsymbol{B}\, d\tau\right] \boldsymbol{u}(0) \tag{1.7.2}$$

Defining the particular integral matrix $\boldsymbol{\Delta}(T)$ as

$$\boldsymbol{\Delta}(T) = \int_0^T \boldsymbol{\phi}(T-\tau)\boldsymbol{B}\, d\tau \tag{1.7.3}$$

gives the final state $\boldsymbol{x}(T)$ in terms of the initial state $\boldsymbol{x}(0)$ and initial control $\boldsymbol{u}(0)$:

$$\boldsymbol{x}(T) = \boldsymbol{\phi}(T)\boldsymbol{x}(0) + \boldsymbol{\Delta}(T)\boldsymbol{u}(0) \tag{1.7.4}$$

Since both $\boldsymbol{\phi}(T)$ and $\boldsymbol{\Delta}(T)$ are only functions of the time interval T and not functions of the state $\boldsymbol{x}$ or the control $\boldsymbol{u}$, we have the general recursive relation

$$\boldsymbol{x}_{k+1} = \boldsymbol{\phi}(T)\boldsymbol{x}_k + \boldsymbol{\Delta}(T)\boldsymbol{u}_k \tag{1.7.5}$$

This is a discrete time model for the continuous linear system when the control vector remains constant over a time interval, T. The transition matrices, $\boldsymbol{\phi}(T)$ and $\boldsymbol{\Delta}(T)$, are directly computable from knowledge of the matrices of the original dynamic model of Equation (1.7.1).

1.8 MATLAB Software Package

A special interactive software package, MATLAB, has been developed for use in the solution of matrix and dynamic response problems. MATLAB works on most workstations and mainframes, as well as personal computers using MS-DOS and MacIntosh computers. The Control Toolbox is a good collection of routines that are useful for the design and solution of optimal control and identification problems (Math Works, Inc.; Sherborn, MA). We will use this package often for the solution of various optimal control and identification problems dealing with linear models. Appendix A gives a list of the subroutines that are used in this text and a description of those subroutines. The on-line help and demo facility of MATLAB allows for the efficient development of specific code to solve particular problems.

Let us illustrate the use of MATLAB by solving the exponential matrix at t=2 for the problem

$$\dot{\boldsymbol{x}} = \boldsymbol{A}\boldsymbol{x} + \boldsymbol{B}\boldsymbol{u} \tag{1.8.1}$$

with

$$y = Hx \tag{1.8.2}$$

where

$$A = \begin{bmatrix} 0 & -0.00156 & -0.1711 & 0 & 0. \\ 0 & -0.1419 & 0.1711 & 0 & 0. \\ 0 & -0.00875 & -1.102 & 0 & 0. \\ 0 & -0.00128 & -0.1489 & 0 & 0.00013 \\ 0 & 0.0605 & 0.1489 & 0 & -0.0591 \end{bmatrix}$$

$$B = \begin{bmatrix} 0 & -0.143 & 0 \\ 0 & 0 & 0 \\ 0.392 & 0 & 0 \\ 0 & 0.108 & -0.0592 \\ 0 & -0.0486 & 0 \end{bmatrix}$$

$$H = \begin{bmatrix} 1 & 0 & 0 & 0 & 0 \\ 0 & 0 & 0 & 1 & 0 \\ 0 & 0 & 0 & 0 & 1 \end{bmatrix}$$

We start the on-line development of the appropriate program that is shown in Figure 1.6 by invoking the help command for the exponential matrix routine EXPM. All lines beginning with / are inputted by the operator while those without the / are responses from the program. The % signs indicate a comment line. Everything we need to know about this routine is then displayed. Based upon this information, the A matrix is inputted following the free format MATLAB rules. The value of time T is specified and the product of A times T is computed. Finally we are ready to evaluate the exponential matrix of $\exp(A\,T)$.

Next we use the help facility to learn about the valuation of the eigenvalues and eigenvectors of the original A matrix.

Now we wish to compute the response of the system for a step change in the control vector

$$u = \begin{pmatrix} 0.2 \\ 0.2 \\ 0.2 \end{pmatrix} \tag{1.8.3}$$

over the time interval $0 \le t \le 30$ when the initial state is $x = \mathbf{0}$. To do this we use the time response program STEP. We start the session by using the help facility. In order to transform the actual problem into a form that can use the program STEP, we need to appropriately define the matrices A, B, C, and D.

The plot of the output vector y is given in Figure 1.7. It is easily observed using the PLOT function in MATLAB. The figure shown on the screen is prepared for plotting by creating the meta file "typlot."

```
/%invoke the help command for the exponential matrix
/
/help expm

EXPM      EXPM(X) is the matrix exponential of X. Complex results
          are produced if X has nonpositive eigenvalues. EXPM is
          computed using a Pade approximation. EXPM can be thought
          of as being computed with eigenvalues D and eigenvectors V,
          such that if [V,D] = EIG(X) then EXPM(X) = V*exp(D)/V.

/%type in the A matrix
/
/A=  [0      -0.00156   -0.1711   0    0          ;
      0      -0.1419     0.1711   0    0          ;
      0      -0.00875   -1.102    0    0          ;
      0      -0.00128   -0.1487   0    0.00013    ;
      0       0.0605     0.1489   0   -0.05911    ];
/
/% specify the time t
/
/t = 2;
/
/% compute the product A times t
/
/At = A*t

At =

      0      -0.0031    -0.3422   0    0
      0      -0.2838     0.3422   0    0
      0      -0.0175    -2.2040   0    0
      0      -0.0026    -0.2974   0    0.0003
      0       0.1210     0.2978   0   -0.1182

/% compute the exponential matrix
/
/ex = expm (At)

ex =

      1.0000  -0.0013   -0.1383   0        0
      0        0.7516    0.1144   0        0
      0       -0.0059    0.1097   0        0
      0       -0.0009   -0.1201   1.0000   0.0002
      0        0.0978    0.1206   0        0.8885

/% Now learn about computing eigenvalues and eigenvectors
/
/help eig

EIG  Eigenvalues and eigenvectors.
     EIG(X) is a vector containing the eigenvalues of a square
     matrix X.
     [V,D] = EIG(X) produces a diagonal matrix D of
     eigenvalues and a full matrix V whose columns are the
     corresponding eigenvectors so that X*V = V*D.
```

Figure 1.6 MATLAB example problem.

```
        [V,D] = EIG(X,'nobalance') performs the computation with
        balancing disabled, which gives better results for certain
        ill-conditioned problems.

        Generalized eigenvalues and eigenvectors.
        EIG(A,B) is a vector containing the generalized eigenvalues
        of square matrices A and B.
        [V,D] = EIG(A,B) produces a diagonal matrix D of generalized
        eigenvalues and a full matrix V whose columns are the
        corresponding eigenvectors so that A*V = B*V*D.
/%compute the eigenvalues and eigenvectors of A
/
/[v,d] = eig(A)
v =
        1.0000   0        0         -0.0000    0.1486
        0        0        0          0.8188   -0.1708
        0        0        0         -0.0075    0.9571
        0        1.0000   -0.0022    0.0001    0.1291
        0        0        1.0000    -0.5741   -0.1269

d =
        0        0        0          0         0
        0        0        0          0         0
        0        0        -0.0591    0         0
        0        0        0         -0.1435    0
        0        0        0          0        -1.1004
/% learn about the step response program
/
/help step
 STEP   Step response of continuous-time linear systems.
        Y = STEP(A,B,C,D,iu,T)  calculates the response of the system:

        ẋ = Ax + Bu
        y = Cx + Du

        to a step applied to the iu'th input.  Vector T must be a
        regularly spaced time vector that specifies the time axis for the
        step response.  STEP returns a matrix Y with as many columns as there
        are outputs y, and with LENGTH(T) rows. [Y,X] = STEP(A,B,C,D,iu,T)
        also returns the state time history.

        Y = STEP(NUM,DEN,T) calculates the step response from the transfer
        function description  G(s) = NUM(s)/DEN(s)  where NUM and DEN contain
        the polynomial coefficients in descending powers.
/% define appropriate B so that all initail controls are 0.2 in value
/
/% input original B matrix
```

Figure 1.6 continued.

```
/B = [0        -0.143    0         ;
      0         0        0         ;
      0.392     0        0         ;
      0         0.108   -0.0592    ;
      0        -0.0486   0        ];
/%define an appropriate B so all controls are 0.2 in value
/
/s = [ .2  .2  .2 ]' ;
/
/BB = B*s
BB =

  -0.0286
        0
   0.0784
   0.0098
  -0.0097
/% define the measurement matrix C
/
/C = [ 1 0 0 0 0 ;
       0 0 0 1 0 ;
       0 0 0 0 1 ];
/
/% define the D matrix
/
/D = [ 0  0  0]' ;
/
/% set the time interval
/
/t = [0:0.2:30];
/
/% call the step program
/
/y = step (A,BB,C,D,1,t) ;
/
/help plot
PLOT    Plot vectors or matrices. PLOT(X,Y) plots vector X versus
        vector Y. If X or Y is a matrix, then the vector is plotted
        versus the rows or columns of the matrix, whichever lines
        up. PLOT(X1,Y1,X2,Y2) is another way of producing multiple
        lines on the plot. PLOT(X1,Y1,':',X2,Y2,'+') uses a
        dotted line for the first curve and the point symbol +
        for the second curve. Other line and point types are:
     solid     -       point     .       red         r
     dashed    --      plus      +       green       g
     dotted    :       star      *       blue        b
     dashdot   -.      circle    o       white       w
                       x-mark    x       invisible   i
                                         arbitrary c1, c15, etc.
```

Figure 1.6 continued.

```
    PLOT(Y) plots the columns of Y versus their index. PLOT(Y)
    is equivalent to PLOT(real(Y),imag(Y)) if Y is complex.
    In all other uses of PLOT, the imaginary part is ignored.
    See SEMI, LOGLOG, POLAR, GRID, SHG, CLC, CLG, TITLE, XLABEL
    YLABEL, AXIS, HOLD, MESH, CONTOUR, SUBPLOT.
/% plot the variables/
/plot (t,y)
/title ('Plot of y vector versus time')
/xlabel ('time')
/ylabel ('y vector components')
/
/% save plot file for printing
/
/meta typlot
/
/exit
```

Figure 1.6 continued.

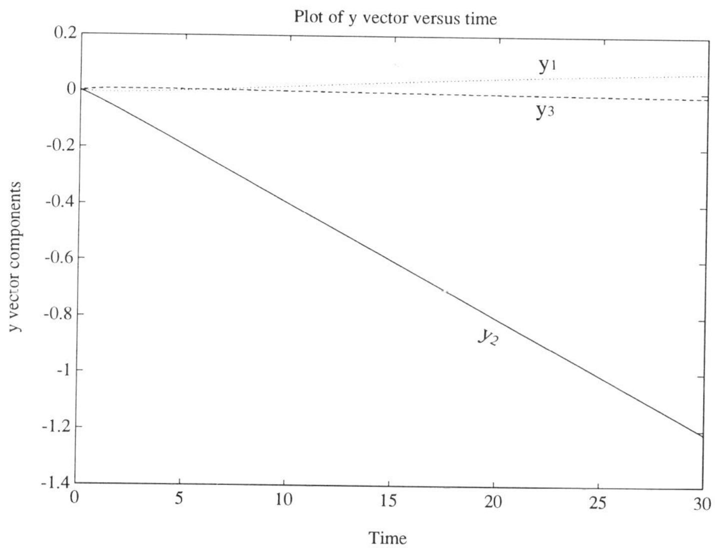

Figure 1.7 Plot of y vector versus time.

1.9 Observability

A system is observable over the time interval $[t_o, t_f]$ if and only if the state of the system at time t_o can be uniquely determined from the measurements of the system over the interval $[t_o, t_f]$.

The observability test is used to design the measurement system for a particular process. It is important that a proposed measurement system involving particular measurements pass the observability test. If the system with the proposed measurement system passes the observability test, then the measurements yield information or knowledge about all of the state variables. The concept of observability was introduced by Kalman (1960).

1.9.1 Discrete Time Systems

Let us consider a linear discrete system

$$\boldsymbol{x}_{k_o+1} = \boldsymbol{\phi}(k_o)\boldsymbol{x}_{k_o} \tag{1.9.1}$$

with the measurement model

$$\boldsymbol{y}_{k_o} = \boldsymbol{H}(k_o)\boldsymbol{x}_{k_o} \tag{1.9.2}$$

The state $\boldsymbol{x}_k$ can be expressed in terms of $\boldsymbol{x}_{k_o}$ by

$$\boldsymbol{x}_k = \psi(k,k_o)\boldsymbol{x}_{k_o} \tag{1.9.3}$$

where

$$\psi(k, k_o) = \boldsymbol{\phi}(k-1)\boldsymbol{\phi}(k-2)\cdots\boldsymbol{\phi}(k_o) \tag{1.9.4}$$

and

$$\boldsymbol{\psi}(k_o, k_o) = \boldsymbol{I} \tag{1.9.5}$$

The measurement model at the kth time increment is

$$\boldsymbol{y}_k = \boldsymbol{H}(k)\boldsymbol{x}_k \tag{1.9.6}$$

or

$$\boldsymbol{y}_k = \boldsymbol{H}(k)\boldsymbol{\psi}(k, k_o)\boldsymbol{x}_{k_o} \tag{1.9.7}$$

A unique representation of $\boldsymbol{x}_{k_o}$ in terms of $\boldsymbol{y}_k$ is possible if and only if the matrix $\boldsymbol{G}(k_o, k_f)$, called the observability Gramian, is nonsingular, where

$$\boldsymbol{G}(k_o, k_f) = \sum_{k=k_o}^{k_f} \boldsymbol{\psi}^T(k, k_o)\boldsymbol{H}^T(k)\boldsymbol{H}(k)\boldsymbol{\psi}(k, k_o) \tag{1.9.8}$$

The proof of this comes from multiplying Equation (1.9.7) by $\boldsymbol{\psi}^T(k, k_o)\boldsymbol{H}^T(k)$ and summing

$$\sum_{k=k_o}^{k_f} \boldsymbol{\psi}^T(k, k_o)\boldsymbol{H}^T(k)\boldsymbol{y}(k) = \boldsymbol{G}(k_o, k_f)\boldsymbol{x}(k_o) \tag{1.9.9}$$

The left-hand side of Equation (1.9.9) is now a vector of length n (the number of state variables) and not m (the number of measurements). Therefore, to have a unique solution for $\boldsymbol{x}(k_o)$, the observability Gramian, $\boldsymbol{G}$, must not be singular; that is, it must have rank n. The condition of rank n means that there are n linearly independent rows to the matrix $\boldsymbol{G}$.

An alternate necessary and sufficient condition for observability is that the rank $\boldsymbol{M}(k_o, k_f)$ is n, where

$$\boldsymbol{M}(k_o, k_f) = \left[\boldsymbol{H}^T(k_o) \vdots \boldsymbol{\psi}^T(k_o+1, k_o)\boldsymbol{H}^T(k_o+1) \vdots \cdots \vdots \boldsymbol{\psi}^T(k_f, k_o)\boldsymbol{H}^T(k_f)\right] \tag{1.9.10}$$

This is shown by forming the augmented form of Equation (1.9.7) as

$$\begin{pmatrix} \boldsymbol{y}(k_o) \\ \cdot \\ \cdot \\ \cdot \\ \boldsymbol{y}(k_f) \end{pmatrix} = \boldsymbol{M}^T(k_o, k_f)\boldsymbol{x}(k_o) \tag{1.9.11}$$

Note that

$$\boldsymbol{G}(k_o, k_f) = \boldsymbol{M}(k_o, k_f)\boldsymbol{M}^T(k_o, k_f) \tag{1.9.12}$$

Therefore, if $\boldsymbol{G}(k_o, k_f)$ is nonsingular, then $\boldsymbol{M}(k_o, k_f)$ must have rank n.

The time invariant case yields a very useful and easily computable result. For time invariant systems

$$\text{rank}\left[\boldsymbol{H}^T \vdots \boldsymbol{\phi}^T\boldsymbol{H}^T \vdots \cdots (\boldsymbol{\phi}^T)^{k_f-k_o}\boldsymbol{H}^T\right] = n \tag{1.9.13}$$

From the dimensions of the matrices, we have for this case

$$\text{rank } \boldsymbol{M}(0, k) = \text{rank } \boldsymbol{M}(0, n) \tag{1.9.14}$$

Therefore, we need only the condition that

$$\text{rank } \boldsymbol{M} = \text{rank}\left[\boldsymbol{H}^T \vdots \boldsymbol{\phi}^T\boldsymbol{H}^T \vdots \cdots \vdots (\boldsymbol{\phi}^T)^{n-1}\boldsymbol{H}^T\right] = n \tag{1.9.15}$$

This is a particularly easy test and one that should always be checked in designing measurement systems. Any measurement system should pass the observability test before being considered for implementation. Nonlinear systems can be linearized about different operating points and checked with Equation (1.9.15). If the linear observability test is valid for the linearized model, it is a good indication that the actual nonlinear system will be observable, especially in the region about the linearization operating condition. If the linear observability test fails for the linearized model, it is a strong indication that the actual measurement system will not yield complete information on the behavior of the system state. Alternative measurement systems should be proposed and investigated.

1.9.2 Continuous Time Systems

Here we consider the linear continuous time model

$$\dot{\boldsymbol{x}}(t) = \boldsymbol{A}(t)\boldsymbol{x}(t) \tag{1.9.16}$$

and the observation measurement model

$$\boldsymbol{y}(t) = \boldsymbol{H}(t)\boldsymbol{x}(t) \tag{1.9.17}$$

The solution of the dynamic model can be given in terms of the transition matrix

$$\boldsymbol{x}(t) = \boldsymbol{\phi}(t, t_o)\boldsymbol{x}(t_o) \tag{1.9.18}$$

where $\boldsymbol{\phi}(t, t_o)$ = the transition matrix.

Therefore, the measurement model becomes

$$\boldsymbol{y}(t) = \boldsymbol{H}(t)\boldsymbol{\phi}(t, t_o)\boldsymbol{x}(t_o) \tag{1.9.19}$$

The observability Gramian is

$$\boldsymbol{G}(t_o, t_f) = \int_{t_o}^{t_f} \boldsymbol{\phi}^T(t, t_o)\boldsymbol{H}^T(t)\boldsymbol{H}(t)\boldsymbol{\phi}(t, t_o)dt \tag{1.9.20}$$

The necessary and sufficient condition for observability is that $\boldsymbol{G}(t_o, t_f)$ is nonsingular. The proof is similar to that of the discrete case by multiplying Equation (1.9.19) by

$$\int_{t_o}^{t_f} \boldsymbol{\phi}^T(t, t_o)\boldsymbol{H}^T(t)dt \tag{1.9.21}$$

For a time invariant problem

$$\dot{\boldsymbol{x}}(t) = \boldsymbol{A}\boldsymbol{x}(t) \tag{1.9.22}$$

$$\boldsymbol{y}(t) = \boldsymbol{H}\boldsymbol{x}(t) \tag{1.9.23}$$

it can be shown that

$$\text{rank } \boldsymbol{M} = \text{rank} \left[\boldsymbol{H}^T \vdots \boldsymbol{A}^T\boldsymbol{H}^T \vdots \cdots \vdots (\boldsymbol{A}^T)^{n-1}\boldsymbol{H}^T\right] = n \tag{1.9.24}$$

This is difficult to prove and requires the notion of duality and the relationship between observability and controllability. A proof of this is given in Lee and Markus (1967).

If a system is observable for all initial times, t_o, and some final times, $t_f > t_o$, then the system is *completely* observable.

If a system is observable for all initial times, t_o, and all final times, $t_f > t_o$, then the system is *totally* observable.

We usually only investigate the linear stationary case since the test is easy to implement. Transition matrices for nonstationary and nonlinear systems are usually difficult to impossible to obtain. The normal practice for nonlinear and linear nonstationary problems is to cast the problem into a linear stationary problem about a specific operating state. If the measurement system passes the stationary test of Equation (1.9.24) about a number of feasible operating states, then the measurement system is accepted, but if the measurement system fails any linear stationary test, then alternative measurement information is sought.

1.10 Controllability

A system is state controllable on $[t_o, t_f]$ if and only if there exists a control function on $[t_o, t_f]$ that transforms any initial state, $\boldsymbol{x}(t_o)$, to any final state vector, $\boldsymbol{x}(t_f)$. The controllability test is used to design the control system. It is important that a proposed control system pass the controllability test. If it does, then the control variables can actually effectively affect all the state variables (Kalman *et al.*, 1963).

1.10.1 Continuous Time Systems

We consider the linear continuous time model

$$\dot{\boldsymbol{x}}(t) = \boldsymbol{A}(t)\boldsymbol{x}(t) + \boldsymbol{B}(t)\boldsymbol{u}(t) \tag{1.10.1}$$

with the measurement model

$$\boldsymbol{y}(t) = \boldsymbol{H}(t)\boldsymbol{x}(t) \tag{1.10.2}$$

The solution to Equation (1.10.1) is

$$\boldsymbol{x}(t) = \boldsymbol{\phi}(t, t_o)\boldsymbol{x}(t_o) + \int_{t_o}^{t} \boldsymbol{\phi}(t, \tau)\boldsymbol{B}(\tau)\boldsymbol{u}(\tau)d\tau \tag{1.10.3}$$

where $\boldsymbol{\phi}(t, \tau)$ is the transition matrix.

We can define the deviation variable

$$\boldsymbol{x}^d(t) = \boldsymbol{x}(t) - \boldsymbol{\phi}(t, t_o)\boldsymbol{x}(t_o) \tag{1.10.4}$$

Therefore the problem of moving $\boldsymbol{x}(t_f)$ from an arbitrary initial condition, $\boldsymbol{x}(t_o)$, is converted into that of moving the deviation variable, $\boldsymbol{x}^d(t)$, from a zero initial condition to the final condition since

$$\boldsymbol{x}^d(t_f) = \int_{t_o}^{t_f} \boldsymbol{\phi}(t_f, \tau)\boldsymbol{B}(\tau)\boldsymbol{u}(\tau)d\tau \tag{1.10.5}$$

For controllability we need to show that every point, $\boldsymbol{x}^d(t_f)$, is within reach of the origin. This can only be true if the scalar product, $\boldsymbol{x}^d(t_f)$, with every fixed length nonzero vector, $\boldsymbol{\xi}$, can be made arbitrarily large by some choice of $\boldsymbol{u}$. If we choose

$$\boldsymbol{u} = k\boldsymbol{B}^T(\tau)\boldsymbol{\phi}^T(t_f, \tau)\boldsymbol{\xi} \tag{1.10.6}$$

then

$$\boldsymbol{\xi} \cdot \boldsymbol{x}^d(t_f) = k \int_{t_o}^{t_f} \boldsymbol{\xi}^T \boldsymbol{\phi}(t_f, \tau)\boldsymbol{B}(\tau)\left(\boldsymbol{B}^T(\tau)\boldsymbol{\phi}^T(t_f, \tau)\boldsymbol{\xi}\right) d\tau \tag{1.10.7}$$

Therefore, for $\boldsymbol{\xi} \cdot \boldsymbol{x}^d(t_f)$ to be nonzero, the controllability Gramian

$$\boldsymbol{W}(t_o, t_f) = \int_{t_o}^{t_f} \boldsymbol{\phi}(t_f, \tau)\boldsymbol{B}(\tau)\boldsymbol{B}^T(\tau)\boldsymbol{\phi}^T(t_f, \tau)d\tau \tag{1.10.8}$$

must be nonsingular and the system is controllable.

If we have a time-invariant system

$$\dot{\boldsymbol{x}}(t) = \boldsymbol{A}\boldsymbol{x}(t) + \boldsymbol{B}\boldsymbol{u}(t) \tag{1.10.9}$$

$$\boldsymbol{y}(t) = \boldsymbol{H}\boldsymbol{x}(t) \tag{1.10.10}$$

then Equation (1.10.7) becomes

$$\boldsymbol{\xi} \cdot \boldsymbol{x}^d(t_f) = k \int_{t_o}^{t_f} \left(\boldsymbol{B}^T \boldsymbol{\phi}^T(t_f, \tau)\boldsymbol{\xi}\right)^T \left(\boldsymbol{B}^T \boldsymbol{\phi}^T(t_f, \tau)\boldsymbol{\xi}\right) d\tau \tag{1.10.11}$$

Therefore, $\boldsymbol{B}^T \boldsymbol{\phi}^T(t_f, \tau)\boldsymbol{\xi}$ must span the space or, equivalently, $\boldsymbol{\phi}(t_f, \tau)\boldsymbol{B}$ and its $(n-1)$ derivatives must not vanish (Lapidus and Luus, 1967). Since

$$\frac{d^i \boldsymbol{\phi}(t_f, \tau)}{d\tau^i} = \boldsymbol{\phi}(t_f, \tau)\, \boldsymbol{A}^i \tag{1.10.12}$$

then

$$\frac{d^o \boldsymbol{\phi}(t_f, \tau)\boldsymbol{B}}{d\tau^o} = \boldsymbol{\phi}(t_f, \tau)\boldsymbol{B} \tag{1.10.13}$$

$$\frac{d \boldsymbol{\phi}(t_f, \tau)\boldsymbol{B}}{d\tau} = \boldsymbol{\phi}(t_f, \tau)\boldsymbol{A}\boldsymbol{B} \tag{1.10.14}$$

$$\frac{d^{n-1} \boldsymbol{\phi}(t_f, \tau)\boldsymbol{B}}{d\tau^{n-1}} = \boldsymbol{\phi}(t_f, \tau)\boldsymbol{A}^{n-1}\boldsymbol{B} \tag{1.10.15}$$

The combined matrix of $\boldsymbol{\phi}(t_f, \tau)\boldsymbol{B}$ and its $(n-1)$ derivatives is

$$\boldsymbol{\phi}(t_f, \tau) \left[\boldsymbol{B} \vdots \boldsymbol{A}\boldsymbol{B} \vdots \boldsymbol{A}^2\boldsymbol{B} \vdots \cdots \vdots \boldsymbol{A}^{n-1}\boldsymbol{B}\right] \tag{1.10.16}$$

Therefore, the necessary and sufficient condition for state controllability of a stationary linear plant is

$$\text{rank } \boldsymbol{\Gamma} = \text{rank} \left[\boldsymbol{B} \vdots \boldsymbol{A}\boldsymbol{B} \vdots \cdots \vdots \boldsymbol{A}^{n-1}\boldsymbol{B}\right] = n \tag{1.10.17}$$

This is a convenient test of controllability and one that should always be performed on any proposed control structure.

The test for output controllability is on the output vector, $\boldsymbol{y}$, and not the state vector, $\boldsymbol{x}$. Since for a linear stationary system, the output vector is related to the state vector by Equation (1.10.10), the necessary and sufficient condition for output controllability of a linear stationary system is

$$\text{rank} = \boldsymbol{\Gamma} = \text{rank} \left[\boldsymbol{H}\boldsymbol{B} \vdots \boldsymbol{H}\boldsymbol{A}\boldsymbol{B} \vdots \cdots \vdots \boldsymbol{H}\boldsymbol{A}^{n-1}\boldsymbol{B}\right] = m \tag{1.10.18}$$

where m is the number of output variables or the order of the output vector.

1.10.2 Discrete Time Systems

If we consider a linear discrete system

$$\boldsymbol{x}_{k_o+1} = \boldsymbol{\phi}(k_o)\boldsymbol{x}_{k_o} + \boldsymbol{\Delta}(k_o)\boldsymbol{u}(k_o) \tag{1.10.19}$$

with the measurement model

$$\boldsymbol{y}_{k_o} = \boldsymbol{H}(k_o)\boldsymbol{x}_{k_o} \tag{1.10.20}$$

then we can define the discrete controllability Gramian as

$$\boldsymbol{W}(k_o, k_f) = \sum_{k=k_o}^{k_f} \boldsymbol{\psi}(k_o, k)\boldsymbol{\Delta}(k)\boldsymbol{\Delta}^T(k)\boldsymbol{\psi}(k_o, k) \tag{1.10.21}$$

where

$$\boldsymbol{\psi}(k_o, k) = \left[\boldsymbol{\phi}(k_o) \vdots \cdots \vdots \boldsymbol{\phi}(k)\right] \tag{1.10.22}$$

For the system to be controllable, the discrete controllability Gramian, $\boldsymbol{W}(k_o, k_f)$, must be nonsingular. This follows from a development that is similar to that presented for the continuous case.

For the time-invariant case, the necessary and sufficient condition for controllability is that

$$\text{rank } \boldsymbol{\Gamma} = \text{rank}\left[\boldsymbol{\Delta} \vdots \boldsymbol{\phi\Delta} \vdots \cdots \vdots \boldsymbol{\phi}^{n-1}\boldsymbol{\Delta}\right] = n \tag{1.10.23}$$

The output controllability test for discrete systems is

$$\text{rank } \boldsymbol{\Gamma} = \text{rank}\left[\boldsymbol{H\Delta} \vdots \boldsymbol{H\phi\Delta} \vdots \cdots \vdots \boldsymbol{H\phi}^{n-1}\boldsymbol{\Delta}\right] = m \tag{1.10.24}$$

1.11 MATLAB Programs for Controllability and Observability

MATLAB has the program CTRB, which can be used to test for the controllability of a linear time-invariant system, and the program OBSV, which is used to test for the observability. We will use these subroutines to test both the controllability and observability of the example given in Section 1.8 (Equations (1.8.1) and (1.8.2)).

The interactive MATLAB program is given in Figure 1.8. It starts by using the help facility to learn about the controllability routine CTRB. We then input the A and B matrix data and compute the controllability matrix which is printed. The rank of that matrix is then computed. The rank is 5, which is

```
/% get help command for controllability
/
/help ctrb
CTRB            Form controllability matrix.
           CTRB(A,B) returns the controllability matrix
           Co = [B AB A**2B ...]
/% input the A and B matrices
/
/A =  [0     -0.00156  -0.1711   0      0         ;
       0     -0.1419    0.1711   0      0         ;
       0     -0.00875  -1.102    0      0         ;
       0     -0.00128  -0.1489   0      0.00013   ;
       0      0.0605    0.1489   0     -0.0591]   ;
/
/B =  [0      -0.143    0        ;
       0       0        0        ;
       0.392   0        0        ;
       0       0.108   -0.0592   ;
       0      -0.0486   0        ];
/
/% evaluate the controllability matrix
/
/co = ctrb (A,B)
/
/co =
/
      Columns 1 through 7
      0         -0.1430    0         -0.0671    0          0     0.0738
      0          0         0          0.0671    0          0    -0.0834
      0.3920     0         0         -0.4320    0          0     0.4755
      0          0.1080   -0.0592    -0.0584   -0.0000     0     0.0642
      0         -0.0486    0          0.0584    0.0029     0    -0.0637
  Columns 8 through 14
      0         0    -0.0812    0          0     0.0894    0
      0         0     0.0932    0          0    -0.1027    0
      0         0    -0.5232    0          0     0.5758    0
      0.0000    0    -0.0707   -0.0000     0     0.0778    0.0000
     -0.0002    0     0.0695    0.0000     0    -0.0764   -0.0000
  Column 15
      0
      0
      0
      0
      0
/determine the rank of the controllabilty matrix
/
/
/help rank
```

Figure 1.8 MATLAB example.

```
RANK                 Rank. K = RANK(X) is the number of singular values of X
                that are larger than MAX(SIZE(X)) * NORM(X) * EPS.
                K = RANK(X,tol) is the number of singular values of X that
                are larger than tol .

/k = rank (co)

k =

     5

/% now get information on observability
/
/help obsv

 OBSV           Form observability matrix.
                OBSV(A,C) returns the observability matrix.
                Ob = [C; CA; CA**2 ...]

/% enter the C or H matrix
/
/H = [1    0    0    0    0    ;
      0    0    0    1    0    ;
      0    0    0    0    1    ];

/% compute the observability matrix
/
/ob = obsv (A,H)

ob =

     1.0000    0          0          0          0
     0         0          0          1.0000     0
     0         0          0          0          1.0000
     0         -0.0016    -0.1711    0          0
     0         -0.0013    -0.1489    0          0.0001
     0         0.0605     0.1489     0          -0.0591
     0         0.0017     0.1883     0          0
     0         0.0015     0.1639     0          -0.0000
     0         -0.0135    -0.1625    0          0.0035
     0         -0.0019    -0.2072    0          0
     0         -0.0016    -0.1804    0          0.0000
     0         0.0035     0.1773     0          -0.0002
     0         0.0021     0.2280     0          0
     0         0.0018     0.1985     0          -0.0000
     0         -0.0021    -0.1948    0          0.0000

/% compute the rank of ob
/
/rk = rank(ob)

rk =

     5

/exit
```

Figure 1.8 continued.

the order of the system. Therefore, the system is controllable. Next we use the help facility to learn about the observability routine OBSV. We input the H measurement matrix and compute the observability matrix. The rank of that matrix is 5, so that the system is observable with this particular measurement system. You should note that the observability matrix of MATLAB is the transpose of that given in Section 1.9 (Equation (1.9.15) or Equation (1.9.24)).

References

Bird, R., Stewart, W., and Lightfoot, E., *Transport Phenomena*, Wiley and Sons, New York (1960).

Coughanowr, D. R. and Koppel, L. B., *Process Systems Analysis and Control*, McGraw–Hill, Inc., New York, NY (1965).

Crowell, R. H. and Williamson, R. E., *Calculus of Vector Functions*, Prentice–Hall, Englewood Cliffs, NJ (1962).

DeCarlo, R. A., *Linear Systems: A State Variable Approach with Numerical Implementation*, Prentice–Hall, Englewood Cliffs, NJ (1989).

Hu, Y. C. and Ramirez, W. F., "Application of Modern Control Theory to Distillation Columns," *AIChE Journal* **18**(3), 479–485 (1972).

Kailath, T., *Linear Systems*, Prentice–Hall, Englewood Cliffs, NJ (1980).

Kalman, R. E., "Contributions to the Theory of Optimal Control," *Bol. Soc. Matem. Mex.*, 102–119 (1960).

Kalman, R. E., Ho, Y., and Narendra, K. S., *Controllability of Linear Dynamical Systems, Contributions to Differential Equations, Vol. 1*, Wiley and Sons, New York (1963).

Lapidus, L. and Luus, R., *Optimal Control of Engineering Processes*, Blaisdell, Waltham, MA (1967).

Lee, E. B. and Markus, L., *Foundations of Optimal Control*, Wiley and Sons, New York (1967).

Ogata, K., *State Space Analysis of Control Systems*, Prentice–Hall, Englewood Cliffs, NJ (1967).

Petkov, P. H., Christov, N. D., and Konstantinov, M. M., *Computational Methods for Linear Control Systems*, Prentice–Hall, Englewood Cliffs, NJ (1991).

Ramirez, W. F., *Computational Methods for Process Simulation*, Butterworths, Boston, MA (1989).

Seborg, D. E., Edgar, T. F., and Mellichamp, D. A., *Process Dynamics and Control*, Wiley and Sons, New York (1989).

Stephanopoulos, G., *Chemical Process Control*, Prentice–Hall, Englewood Cliffs, NJ (1984).

Exercises

1. Solve for the transition matrix when

$$A = \begin{bmatrix} 0 & 1 \\ 0 & -1 \end{bmatrix}$$

using the Laplace transform method.

2. Solve for the transition matrix when

$$A = \begin{bmatrix} 0 & 1 \\ 0 & 0 \end{bmatrix}$$

using the eigenvalues and α's determined by linear algebra.

3. Solve for the transition matrix when

$$A = \begin{bmatrix} 0 & 1 \\ -2 & -3 \end{bmatrix}$$

using Sylvester's theorem.

4. For

$$\begin{bmatrix} -88.824 & 48.024 & 0 & 0 & 0 \\ 40.8 & -88.824 & 48.024 & 0 & 0 \\ 0 & 40.8 & -88.824 & 48.024 & 0 \\ 0 & 0 & 40.8 & -88.824 & 48.024 \\ 0 & 0 & 0 & 40.8 & -88.824 \end{bmatrix} = A$$

$$b = \begin{bmatrix} 0 \\ 0 \\ 0 \\ 0 \\ -20.01 \end{bmatrix}$$

(a) Find the eigenvalues of A.

(b) Find the solution to $Ax = b$.

5. For

$$\begin{aligned} \dot{\boldsymbol{x}} &= \boldsymbol{A}\boldsymbol{x} + \boldsymbol{B}\boldsymbol{u} \\ \boldsymbol{y} &= \boldsymbol{H}\boldsymbol{x} \end{aligned}$$

where

$$\boldsymbol{A} = \begin{bmatrix} 0 & -0.01156 & -0.1711 & 0 & 0 \\ 0 & -0.1419 & 0.1711 & 0 & 0 \\ 0 & -0.00875 & -1.102 & 0 & 0 \\ 0 & -0.00128 & -0.1489 & 0 & 0.00013 \\ 0 & 0.0605 & 0.1489 & 0 & -0.0591 \end{bmatrix}$$

$$\boldsymbol{B} = \begin{bmatrix} 0 & -0.143 & 0 \\ 0 & 0 & 0 \\ 0.392 & 0 & 0 \\ 0 & 0.108 & -0.0592 \\ 0 & -0.0486 & 0 \end{bmatrix}$$

$$\boldsymbol{H} = \begin{bmatrix} 1 & 0 & 0 & 0 & 0 \\ 0 & 0 & 0 & 1 & 0 \\ 0 & 0 & 0 & 0 & 1 \end{bmatrix}$$

(a) Compute and plot the transient output response of the system for the control variables each equal to a constant value of 0.2 from $0 \leq t \leq 30$.

$$\text{initial state } \boldsymbol{x} = \boldsymbol{0}$$

(b) Compute and plot the transient state response of part (a).

6. Determine the state and output controllability and observability of linear time invariant systems. Do by hand.

$$\begin{aligned} \dot{\boldsymbol{x}} &= \boldsymbol{A}\boldsymbol{x} + \boldsymbol{B}\boldsymbol{u} \\ \boldsymbol{y} &= \boldsymbol{H}\boldsymbol{x} \end{aligned}$$

when (a) $\boldsymbol{A} = \begin{bmatrix} 0 & 1 \\ 0 & 0 \end{bmatrix} \quad \boldsymbol{B} = \begin{bmatrix} 0 \\ 1 \end{bmatrix} \quad \boldsymbol{H} = \begin{bmatrix} 1 & 0 \end{bmatrix}$

(b) $\boldsymbol{A} = \begin{bmatrix} 0 & 1 \\ 0 & 0 \end{bmatrix} \quad \boldsymbol{B} = \begin{bmatrix} 0 \\ 1 \end{bmatrix} \quad \boldsymbol{H} = \begin{bmatrix} 0 & 1 \end{bmatrix}$

(c) $\boldsymbol{A} = \begin{bmatrix} -2 & 0 & 1 \\ 0 & -1 & 0 \\ -3 & -4 & -2 \end{bmatrix} \quad \boldsymbol{B} = \begin{bmatrix} 0 & 1 \\ 0 & 0 \\ 1 & 0 \end{bmatrix}$

$$\boldsymbol{H} = \begin{bmatrix} 1 & 0 & 0 \end{bmatrix}$$

7. Using MATLAB, compute state controllability and observability when

$$\boldsymbol{A} = \begin{bmatrix} 0 & -.00156 & -.1711 & 0 & 0 \\ 0 & -.1419 & .1711 & 0 & 0 \\ 0 & -.00875 & -1.102 & 0 & 0 \\ 0 & -.00128 & -0.1489 & 0 & .00013 \\ 0 & .0605 & .1489 & 0 & -.0591 \end{bmatrix}$$

$$\boldsymbol{B} = \begin{bmatrix} 0 & -.143 & 0 \\ 0 & 0 & 0 \\ .392 & 0 & 0 \\ 0 & .108 & -.0592 \\ 0 & -.0486 & 0 \end{bmatrix} \qquad \boldsymbol{H} = \begin{bmatrix} 1 & 0 & 0 & 0 & 0 \\ 0 & 0 & 0 & 1 & 0 \\ 0 & 0 & 0 & 0 & 1 \end{bmatrix}$$

8. For the system

$$\dot{\boldsymbol{x}} = \begin{bmatrix} 1 & 1 \\ -2 & -1 \end{bmatrix} \boldsymbol{x} + \begin{pmatrix} 0 \\ 1 \end{pmatrix} u$$

   (a) Is it controllable?
   (b) With the measurement $y = x_1$, is it observable?
   (c) With the measurement $y = x_2$, is it observable?

9. Hu and Ramirez (1972) developed a model of a distillation tower consisting of five trays, a total condenser, and a reboiler. The state variables are the temperature of each tray, the condenser, and the reboiler. The controls on the system are the reflux ratio of the condenser and the reboiler duty. The linearized model is given by the following state matrix $\boldsymbol{A}$ and control matrix $\boldsymbol{B}$:

$$\boldsymbol{A} = \begin{bmatrix} -7.6397 & 5.0629 & 0 & 0 & 0 & 0 & 0 \\ 4.5985 & -9.9349 & 6.1175 & 0 & 0 & 0 & 0 \\ -0.0263 & 6.3535 & -12.9721 & 8.1940 & 0 & 0 & 0 \\ -0.0236 & -0.0055 & 6.6077 & -19.2419 & 12.0775 & 0 & 0 \\ -0.0245 & -0.0057 & 0 & 11.2851 & -23.7816 & 15.4901 & 0 \\ -0.0199 & -0.0040 & 0 & 0 & 11.3416 & -27.0859 & 18.2897 \\ -0.0002 & -0.0001 & 0 & 0 & 0 & 0.1986 & -0.3470 \end{bmatrix}$$

$$B = \begin{bmatrix} 0 & 0 \\ 0.0595 & 0 \\ 0.0692 & 0 \\ 0.0622 & -0.0003 \\ 0.0645 & -0.0003 \\ 0.0524 & -0.0003 \\ 0.0006 & 0 \end{bmatrix}$$

Is this system controllable? Suppose only the reboiler duty is controlled. Is the system then controllable? Is the system controllable with only reflux ratio as the control variable?

Various temperature measurements can be made. Suppose all temperatures are measured. Is the system then observable? Is the system observable if the condenser and reboiler temperatures are used? Suppose the only measurement is the reboiler temperature. Is the system observable? Is the system observable when the third tray temperature is the only measured state?

10. The kinetics of an isothermal batch reactor conducting the reaction $X \longrightarrow Y \longrightarrow Z$ is given by

$$\begin{aligned} \dot{x} &= -k_1 x \qquad & x(0) &= x_o \\ \dot{y} &= k_1 x - k_2 y \qquad & y(0) &= 0 \\ & & z(0) &= 0 \end{aligned}$$

Develop the state transition matrix using Sylvester's theorem and Laplace transforms.

11. The dynamics of a control process are characterized by

$$\frac{d^2 x}{dt^2} + (a - 2b\cos 2t)x = a$$

Determine the state transition matrix for this system.

12. Using MATLAB, develop the state variable equations for the system described by the following transfer function

$$\frac{Y(s)}{M(s)} = \frac{(s+1)(s+2)}{s(s+4)(s+3)}$$

13. Determine the transition matrix from the response of unit step changes.

(a) Let $\boldsymbol{v} = \begin{pmatrix} \boldsymbol{m} \\ \boldsymbol{x} \end{pmatrix}$ and $\dot{\boldsymbol{v}} = \boldsymbol{A}\,\boldsymbol{v}$ where $\boldsymbol{m}$ is of j dimension and $\boldsymbol{x}$ of r dimension.

The solution is

$$\boldsymbol{v}(\lambda) = \boldsymbol{\phi}(\lambda)\boldsymbol{v}(0^+)$$

where

$$\boldsymbol{\phi}(\lambda) = \begin{bmatrix} a_{11}(\lambda)\cdots\cdots a_{in}(\lambda) \\ \vdots \\ a_{ni}(\lambda)\cdots\cdots a_{nn}(\lambda) \end{bmatrix}$$

(b) Write out the solution for

$$m_i(\lambda) \qquad \text{and} \qquad x_k(\lambda)$$

(c) Show that $a_{i1}(\lambda)$ is the unit step response measured at m_i with $m_1(0^+) = 1$ and all other initial conditions equal to zero.

(d) Show that $a_{k_1 z}(\lambda)$ is the unit step response measured at x_k with $m_z(0^+) = 1$ and all other I.C.'s zero.

(e) Show that $x_{k+j,j+3}(\lambda)$ is the unit step response measured at x_k with $x_3(0^+) = 1$ and all other I.C.'s zero.

(f) What is the transition matrix for the system

$$\dot{x}(t) + a\,x(t) = a\,m(t)$$

assuming $m(t)$ to be a unit step function?

14. Is the following system controllable?

$$\begin{pmatrix} \dot{x}_1 \\ \dot{x}_2 \end{pmatrix} = \begin{pmatrix} 0 & 1 \\ 1 & 2 \end{pmatrix} \begin{pmatrix} x_1 \\ x_2 \end{pmatrix} + \begin{bmatrix} -1 & 1 \\ 0 & 2 \end{bmatrix} \begin{pmatrix} u_1 \\ u_2 \end{pmatrix}$$

15. What conditions on b_1, b_2, and b_3 ensure the controllability of the system

$$\begin{pmatrix} \dot{x}_1 \\ \dot{x}_2 \\ \dot{x}_3 \end{pmatrix} = \begin{bmatrix} \lambda & 1 & 0 \\ 0 & \lambda & 1 \\ 0 & 0 & \lambda \end{bmatrix} \begin{pmatrix} x_1 \\ x_2 \\ x_3 \end{pmatrix} + \begin{pmatrix} b_1 \\ b_2 \\ b_3 \end{pmatrix} u \quad ?$$

Chapter 2

STEADY-STATE OPTIMIZATION

2.1 Supervisory Computer Control

This chapter treats the topic of steady-state optimization. Necessary conditions for extrema of functions are derived using variational principles. Variational principles will also play an important role in the solution of dynamic optimization problems treated in later chapters. Steady-state optimization techniques are used for the determination of optimal setpoints for supervisory computer control.

The first common application of computer control was associated with supervisory control. In supervisory computer control, conventional plant feedback control systems are responsible for the actual control action. However, a supervisory computer program performs a steady-state optimization that prescribes optimal set-point values for classical feedback controllers.

Figure 2.1 gives the general scheme for implementing a supervisory computer control system. Information is fed into a steady-state model concerning the current measurements of the state, controls, and disturbance variables, as well as management directions and market conditions. The result of the steady-state optimization is a set of optimal set-point values for classical feedback control loops.

Although actual feedback control loops have traditionally employed analog equipment, it is now common to install remote microcomputer feedforward-feedback control loops for approximately the same cost. To effectively utilize supervisory control, an adequate steady-state process model must be developed. It is this model of the process that is actually optimized in order to compute desired set-point control values.

To illustrate the supervisory computer control concept, let us consider the following specific example. We want to find the optimum distillate flow D for the operation of a binary distillation column which is subject to feed

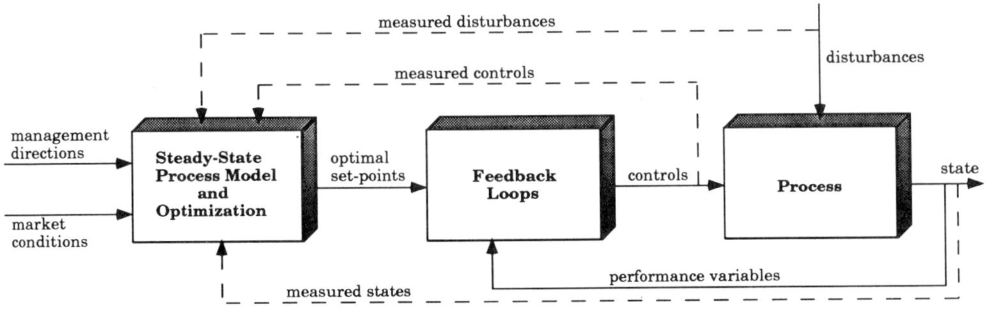

Figure 2.1 Supervisory control system.

composition changes. The column is shown in Figure 2.2.

We first need to develop a mathematical model for the process. This is accomplished by material balances on each tray, an energy balance on the reboiler, and a material balance on the condenser. Appropriate vapor-liquid equilibrium relations must also be used (Ramirez, 1989). Let us consider the problem of separating propylene (species A) and propane (species B).

The component material balance for each internal tray is

$$\begin{aligned} &\text{Rate of } A \text{ In} &&- \text{Rate of } A \text{ Out} = 0 \\ &x_{n+1}L_{n+1} + y_{n-1}V_{n-1} &&- x_nL_n - y_nV_n = 0 \end{aligned} \tag{2.1.1}$$

where x_n = liquid mole fraction of species A on tray n
y_n = vapor mole fraction of species A on tray n

The vapor-liquid equilibrium relation for each tray can be expressed as

$$y_n = f(x_n) \tag{2.1.2}$$

where f is a nonlinear function that describes the vapor-liquid equilibrium condition.

The overall material balance for each internal tray is

$$L_{n+1} + V_{n-1} - L_n - V_n = 0 \tag{2.1.3}$$

For binary distillation, the condition of equal molar overflow is usually valid and is given by

$$V_{n-1} = V_n \tag{2.1.4}$$

The material component balance for the feed tray is also needed

$$x_{F+1}L_{F+1} + y_{F-1}V_{F-1} + Fx^F - x_FL_F - y_FV_F = 0 \tag{2.1.5}$$

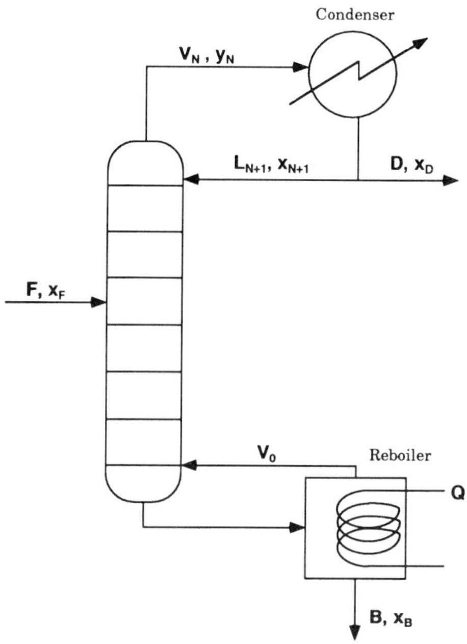

Figure 2.2 Binary distillation column with condenser and reboiler.

and the overall material balance for the feed tray is

$$L_{F+1} + V_{F-1} + F - L_F - V_F = 0 \tag{2.1.6}$$

In addition, the condenser and reboiler must be modeled. The material balances around the total condenser (a condenser where all the vapor is condensed to a saturated liquid) are

$$\begin{aligned} x_D &= x_{N+1} = y_N \\ V_N &= L_{N+1} + D \end{aligned} \tag{2.1.7}$$

The energy balance for the reboiler is

$$V_o = Q/\lambda \tag{2.1.8}$$

where Q = reboiler duty (cal/min)
λ = heat of vaporization (cal/mole)

The mathematical model for the distillation tower are the algebraic equations of (2.1.1) to (2.1.8). For the problem of separating propylene from propane, it is of economic importance to have the distillate essentially pure propylene and free of propane. The bottoms stream may contain both propane and propylene. The question is what is the best distillate product rate, D, that will yield the most profitable distillation column performance and how does that production rate change with changes in feed composition? The economic data for the operation of this column are expressed in Table 2.1.

Table 2.1
Economic Data for Propylene-Propane Separation

Species	Distillate Value	Bottoms Value
Propylene	0.1 ($/gal)	0.05 ($/gal)
Propane	0.0 ($/gal)	0.05 ($/gal)

We define the economic performance index which should be maximized as the value of the products produced and is given mathematically as

$$J = 0.05\ Bx^B_{propylene} + 0.05\ Bx^B_{propane} + 0.1\ Dx^D_{propylene} \tag{2.1.9}$$

Figure 2.3 gives the performance index of an 85 tray column as a function of distillate production rate for various feed compositions of propylene. The results were obtained by running the steady-state model at different distillate rates for various feed compositions. This figure clearly shows that the distillate control set point should vary as there are changes in the feed composition. Implementing such a supervisory control scheme would significantly improve the economic operation of this distillation column.

2.2 Unconstrained Extrema

We will now develop the mathematics necessary for the optimization of algebraic functions. We start with unconstrained extrema.

We let A be a subset of a normed linear vector space having the norm $||\cdot||$, and we let f be a real valued function defined on A, i.e., f is a mapping of R^n to R where R^n is the Euclidean n-space and R is the first-order Euclidean space. The function f has a relative maximum (minimum) at $\boldsymbol{\alpha}$, an element of A, if and only if there exists a positive real number δ such that for any $\boldsymbol{x}$, an element of A satisfying $||\boldsymbol{x} - \boldsymbol{\alpha}|| < \delta$, then $f(\boldsymbol{x}) \leq f(\boldsymbol{\alpha})((f(\boldsymbol{x}) \geq f(\boldsymbol{\alpha}))$. Figure 2.4 shows two values of $\boldsymbol{\alpha}$ that satisfy the maximum condition. In this case x is defined on R, the real line. The necessary condition values are α_1 and α_2.

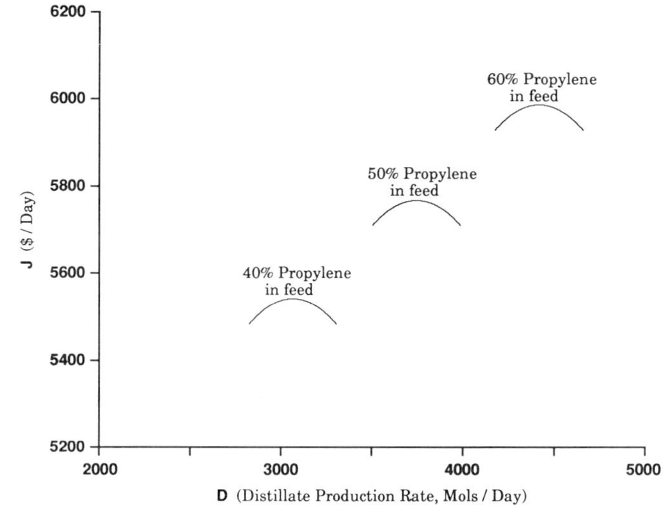

Figure 2.3 Performance as a function of distillate flow for various feed compositions (85 tray column).

If this definition holds for all $\boldsymbol{x}$ in A, then f has an absolute maximum (minimum).

A necessary condition for the vector $\boldsymbol{\alpha}$ to be an interior maximum or minimum is given by calculus as (Crowell and Williamson, 1962)

$$\left.\frac{\partial f(\boldsymbol{x})}{\partial \boldsymbol{x}}\right|_{\boldsymbol{x}=\boldsymbol{\alpha}} = \mathbf{0} \tag{2.2.1}$$

Assuming that the appropriate partial derivatives exist, then a sufficient condition that $\boldsymbol{\alpha}$ be a relative maximum (minimum) is that the matrix

$$\left[\frac{\partial}{\partial \boldsymbol{x}}\left(\frac{\partial f}{\partial \boldsymbol{x}}\right)\right]_{\boldsymbol{x}=\boldsymbol{\alpha}} = \boldsymbol{Q} \tag{2.2.2}$$

is negative (positive) definite. Since $\boldsymbol{Q}$ is a symmetric matrix, we can use Sylvester's criteria (Ogata, 1967) to determine if it is positive or negative definite. To be positive definite, the determinant of the principal minors of

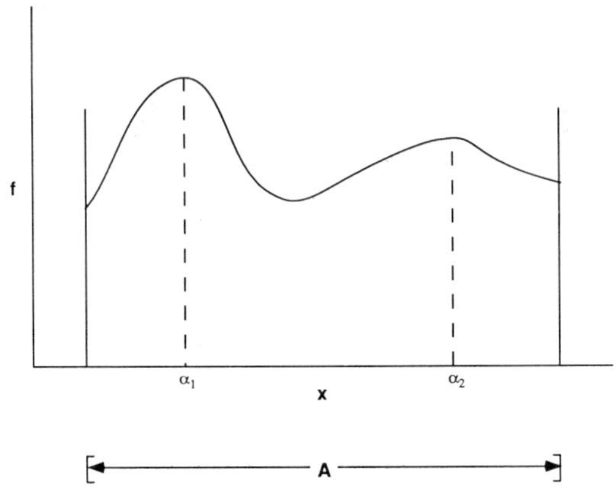

Figure 2.4 Values of α that satisfy the definition of a maximum.

the matrix must be positive. This is

$$
\begin{aligned}
q_{11} &> 0 \\
\begin{vmatrix} q_{11} & q_{12} \\ q_{12} & q_{22} \end{vmatrix} &> 0 \\
&\text{etc.}
\end{aligned}
\tag{2.2.3}
$$

To be negative definite, the determinant of the principal minors must alternate,

$$
\begin{aligned}
q_{11} &< 0 \\
\begin{vmatrix} q_{11} & q_{12} \\ q_{12} & q_{22} \end{vmatrix} &> 0 \\
&\text{etc.}
\end{aligned}
\tag{2.2.4}
$$

The necessary condition of Equation (2.2.1) can be derived using a variational approach. The *increment*, Δf, of the performance measure f is defined as

$$\Delta f = f(\boldsymbol{x} + \delta \boldsymbol{x}) - f(\boldsymbol{x}) \tag{2.2.5}$$

where $\delta \boldsymbol{x}$ are small variations in the variables $\boldsymbol{x}$.

We expand $f(\boldsymbol{x}+\delta\boldsymbol{x})$ in a Taylor series about $f(\boldsymbol{x})$ to obtain

$$f(\boldsymbol{x}+\delta\boldsymbol{x}) = f(\boldsymbol{x}) + \left(\frac{\partial f}{\partial \boldsymbol{x}}\right)^T \delta\boldsymbol{x} + O\left(\delta\boldsymbol{x}^n\right) \qquad n \geq 2 \tag{2.2.6}$$

The increment therefore becomes

$$\Delta f = \left(\frac{\partial f}{\partial \boldsymbol{x}}\right)^T \delta\boldsymbol{x} + O\left(\delta\boldsymbol{x}^n\right) \qquad n \geq 2 \tag{2.2.7}$$

The linear portion of the increment is called the first variation of the function f, δf, and is given as

$$\delta f = \left(\frac{\partial f}{\partial \boldsymbol{x}}\right)^T \delta\boldsymbol{x} \tag{2.2.8}$$

The necessary condition for an extremal in f is that the first variation is zero. Since in general $\delta\boldsymbol{x}$ is not zero, then the necessary condition is

$$\frac{\partial f}{\partial \boldsymbol{x}} = \mathbf{0} \tag{2.2.9}$$

2.3 Extrema with Equality Constraints

We will now consider the optimization of functions that are subject to equality constraints. We consider extremizing the objective function J, subject to the equality constraints $\boldsymbol{f}(\boldsymbol{x}) = \mathbf{0}$ where $f : R^n \rightarrow R^m$. This means that $\boldsymbol{x}$ is a vector of order n and there are m constraining relationships.

For simplicity, we will assume that there are two state variables, x_1 and x_2 (n=2), and that there is one constraining algebraic relation f (m=1). For this case, we will develop the Lagrange multiplier, λ. First we differentiate the objective function J with respect to the state variable, x_1,

$$\frac{dJ}{dx_1} = \frac{\partial J}{\partial x_1} + \frac{\partial J}{\partial x_2}\frac{dx_2}{dx_1} \tag{2.3.1}$$

and we differentiate the constraining relation with respect to the same variable

$$\frac{df}{dx_1} = \frac{\partial f}{\partial x_1} + \frac{\partial f}{\partial x_2}\frac{dx_2}{dx_1} \tag{2.3.2}$$

The term dx_1/dx_2 appears since dx_1 and dx_2 are not independent due to the constraint relation. Since $f(\boldsymbol{x})$=0 for all solutions, then

$$\frac{df}{dx_1} = 0 \tag{2.3.3}$$

Also due to the necessary condition for an extremal

$$\frac{dJ}{dx_1} = 0 \tag{2.3.4}$$

Equating Equation (2.3.1) to zero gives

$$\frac{dx_2}{dx_1} = -\frac{\partial J/\partial x_1}{\partial J/\partial x_2} \tag{2.3.5}$$

and equating Equation (2.3.2) to zero gives

$$\frac{dx_2}{dx_1} = -\frac{\partial f/\partial x_1}{\partial f/\partial x_2} \tag{2.3.6}$$

Equating Equations (2.3.5) and (2.3.6) yields

$$\frac{\partial J/\partial x_1}{\partial J/\partial x_2} = \frac{\partial f/\partial x_1}{\partial f/\partial x_2} \tag{2.3.7}$$

or

$$\frac{\partial J}{\partial x_1} \Big/ \frac{\partial f}{\partial x_1} = \frac{\partial J}{\partial x_2} \Big/ \frac{\partial f}{\partial x_2} = -\lambda \tag{2.3.8}$$

where λ is a constant called the Lagrange multiplier. The only way the partials with respect to x_1 and x_2 of the same functional form can always be equal is that they are both equal to a constant, λ.

Equation (2.3.8) can be expressed as the two following equations:

$$\frac{\partial J}{\partial x_1} + \lambda \frac{\partial f}{\partial x_1} = 0 \tag{2.3.9}$$

and

$$\frac{\partial J}{\partial x_2} + \lambda \frac{\partial f}{\partial x_2} = 0 \tag{2.3.10}$$

We define the Lagrangian, L, for this problem to be

$$L = J + \lambda f \tag{2.3.11}$$

The necessary conditions for constrained extrema of (2.3.9) and (2.3.10) therefore become

$$\frac{\partial L}{\partial \boldsymbol{x}} = \mathbf{0} \tag{2.3.12}$$

and the definition of the constraining relation can be expressed as

$$\frac{\partial L}{\partial \lambda} = 0 \tag{2.3.13}$$

For a general nth-order system with m constraints, we define the Lagrangian as

$$L = J + \boldsymbol{\lambda}^T \boldsymbol{f} \tag{2.3.14}$$

where the Lagrange multiplier vector, $\boldsymbol{\lambda} \in R^m$, and the state vector, $\boldsymbol{x} \in R^n$.

The necessary conditions for an extremum are

$$\frac{\partial L}{\partial \boldsymbol{x}} = \mathbf{0} \qquad (n \text{ equations}) \tag{2.3.15}$$

and

$$\frac{\partial L}{\partial \boldsymbol{\lambda}} = \mathbf{0} \qquad (m \text{ equations}) \tag{2.3.16}$$

2.4 Variational Approach to Constrained Extrema

An alternate approach to obtain the necessary conditions for constrained optimization problems is given by formulating the first variation of the Lagrangian, $L(\boldsymbol{x}, \boldsymbol{u}, \boldsymbol{\lambda})$. The first variation is the linear part of the *increment.* The increment of the Lagrangian is defined as

$$\Delta L = L(\boldsymbol{x} + \delta\boldsymbol{x},\ \boldsymbol{u} + \delta\boldsymbol{u},\ \boldsymbol{\lambda} + \delta\boldsymbol{\lambda}) - L(\boldsymbol{x}, \boldsymbol{u}, \boldsymbol{\lambda}) \tag{2.4.1}$$

where $\delta\boldsymbol{x}, \delta\boldsymbol{u}, \delta\boldsymbol{\lambda}$ are small variations in $\boldsymbol{x}$, $\boldsymbol{u}$, and $\boldsymbol{\lambda}$. We expand $L(\boldsymbol{x}+\delta\boldsymbol{x},\ \boldsymbol{u}+\delta\boldsymbol{u},\ \boldsymbol{\lambda}+\delta\boldsymbol{\lambda})$ in a Taylor series about $L(\boldsymbol{x}, \boldsymbol{u}, \boldsymbol{\lambda})$. The linear terms are defined as the first variation of the Lagrangian. Expanding $L(\boldsymbol{x}+\delta\boldsymbol{x},\ \boldsymbol{u}+\delta\boldsymbol{u},\ \boldsymbol{\lambda}+\delta\boldsymbol{\lambda})$ in a Taylor series gives

$$\begin{aligned} L(\boldsymbol{x} + \delta\boldsymbol{x},\ \boldsymbol{u} + \delta\boldsymbol{u},\ \boldsymbol{\lambda} + \delta\boldsymbol{\lambda}) &= L(\boldsymbol{x}, \boldsymbol{u}, \boldsymbol{\lambda}) + \left(\frac{\partial L}{\partial \boldsymbol{x}}\right)^T \delta\boldsymbol{x} \\ &+ \left(\frac{\partial L}{\partial \boldsymbol{u}}\right)^T \delta\boldsymbol{u} + \left(\frac{\partial L}{\partial \boldsymbol{\lambda}}\right)^T \delta\boldsymbol{\lambda} + O\left(\delta\boldsymbol{x}^n, \delta\boldsymbol{u}^n, \delta\boldsymbol{\lambda}^n\right) \qquad n \geq 2 \end{aligned} \tag{2.4.2}$$

Therefore, the increment of the Lagrangian can be expressed as

$$\Delta L = \left(\frac{\partial L}{\partial \boldsymbol{x}}\right)^T \delta\boldsymbol{x} + \left(\frac{\partial L}{\partial \boldsymbol{u}}\right)^T \delta\boldsymbol{u} + \left(\frac{\partial L}{\partial \boldsymbol{\lambda}}\right)^T \delta\boldsymbol{\lambda} + O\left(\delta\boldsymbol{x}^n, \delta\boldsymbol{u}^n, \delta\boldsymbol{\lambda}^n\right) \qquad n \geq 2 \tag{2.4.3}$$

The first variation of the Lagrangian, being the linear part of the increment, is given by

$$\delta L = \left(\frac{\partial L}{\partial \boldsymbol{x}}\right)^T \delta\boldsymbol{x} + \left(\frac{\partial L}{\partial \boldsymbol{u}}\right)^T \delta\boldsymbol{u} + \left(\frac{\partial L}{\partial \boldsymbol{\lambda}}\right)^T \delta\boldsymbol{\lambda} \tag{2.4.4}$$

The necessary conditions for an extremum of J subject to the constraints, $\boldsymbol{f}(\boldsymbol{x}, \boldsymbol{u}) = \mathbf{0}$, is that the first variation of the Lagrangian be zero,

$$\delta L = 0 \tag{2.4.5}$$

Since the variations, $\delta \boldsymbol{x}$, $\delta \boldsymbol{u}$, and $\delta \boldsymbol{\lambda}$, are independent, we have

$$\begin{aligned} &\frac{\partial L}{\partial \boldsymbol{x}} = \mathbf{0} \\ &\frac{\partial L}{\partial \boldsymbol{u}} = \mathbf{0} \\ &\frac{\partial L}{\partial \boldsymbol{\lambda}} = \mathbf{0} \qquad \text{or} \qquad \boldsymbol{f} = \mathbf{0} \end{aligned} \tag{2.4.6}$$

Now let us consider the second variation, $\delta^2 L$, which we get from the second-order terms of the increment, ΔL. We can neglect the $\delta \boldsymbol{\lambda}$ terms since the constraints must be met. Therefore,

$$\begin{aligned} \delta^2 L = &\frac{1}{2}\delta \boldsymbol{x}^T \left\{ \left[\frac{\partial}{\partial \boldsymbol{x}}\frac{\partial L}{\partial \boldsymbol{x}}\right] \delta \boldsymbol{x} + \left[\frac{\partial}{\partial \boldsymbol{u}}\frac{\partial L}{\partial \boldsymbol{x}}\right] \delta \boldsymbol{u} \right\} \\ &+ \frac{1}{2}\delta \boldsymbol{u}^T \left\{ \left[\frac{\partial}{\partial \boldsymbol{u}}\frac{\partial L}{\partial \boldsymbol{x}}\right]^T \delta \boldsymbol{u} + \left[\frac{\partial}{\partial \boldsymbol{u}}\frac{\partial L}{\partial \boldsymbol{u}}\right] \delta \boldsymbol{u} \right\} \end{aligned} \tag{2.4.7}$$

or

$$\delta^2 L = \frac{1}{2}\left(\delta \boldsymbol{x}^T \;\vdots\; \delta \boldsymbol{u}^T\right) \left[\begin{array}{c|c} \left[\dfrac{\partial}{\partial \boldsymbol{x}}\dfrac{\partial L}{\partial \boldsymbol{x}}\right] & \left[\dfrac{\partial}{\partial \boldsymbol{u}}\dfrac{\partial L}{\partial \boldsymbol{x}}\right] \\ \hline \left[\dfrac{\partial}{\partial \boldsymbol{u}}\dfrac{\partial L}{\partial \boldsymbol{x}}\right]^T & \left[\dfrac{\partial}{\partial \boldsymbol{u}}\dfrac{\partial L}{\partial \boldsymbol{u}}\right] \end{array}\right] \begin{pmatrix} \delta \boldsymbol{x} \\ \hline \delta \boldsymbol{u} \end{pmatrix} \tag{2.4.8}$$

Equation (2.4.8) is a quadratic form which can be expressed as

$$\delta^2 L = \frac{1}{2}\delta \boldsymbol{z}^T \boldsymbol{P} \delta \boldsymbol{z} \tag{2.4.9}$$

with

$$\delta \boldsymbol{z} = \begin{pmatrix} \delta \boldsymbol{x} \\ \hline \delta \boldsymbol{u} \end{pmatrix} \tag{2.4.10}$$

and

$$\boldsymbol{P} = \left[\begin{array}{c|c} \left[\frac{\partial}{\partial \boldsymbol{x}} \frac{\partial L}{\partial \boldsymbol{x}}\right] & \left[\frac{\partial}{\partial \boldsymbol{u}} \frac{\partial L}{\partial \boldsymbol{x}}\right] \\ \hline \left[\frac{\partial}{\partial \boldsymbol{u}} \frac{\partial L}{\partial \boldsymbol{x}}\right]^T & \left[\frac{\partial}{\partial \boldsymbol{u}} \frac{\partial L}{\partial \boldsymbol{u}}\right] \end{array} \right] \tag{2.4.11}$$

A sufficient condition for J to be a maximum is that $\boldsymbol{P}$ be negative definite along $\boldsymbol{f} = \boldsymbol{0}$. A sufficient condition for J to be a minimum is that $\boldsymbol{P}$ be positive definite along $\boldsymbol{f} = \boldsymbol{0}$.

2.5 Supervisory Control Problem

Let us consider the supervisory control problem. It is also sometimes called the static control problem. The state equations are

$$\boldsymbol{f}(\boldsymbol{x}, \boldsymbol{u}) = \boldsymbol{0} \tag{2.5.1}$$

where the state vector, $\boldsymbol{x}$, is an element of the vector space, R^n, and the control vector, $\boldsymbol{u}$, is an element of the vector space, R^m. The set of algebraic relations, $\boldsymbol{f}$, is contained in the vector space, R^n. We want to extremize the scalar performance index

$$J(\boldsymbol{x}, \boldsymbol{u}) \tag{2.5.2}$$

where J is a mapping of $R^{n+m} \to R$. Following the Lagrangian approach for constrained optimization, we first form the Lagrangian as

$$L(\boldsymbol{x}, \boldsymbol{u}, \boldsymbol{\lambda}) = J(\boldsymbol{x}, \boldsymbol{u}) + \boldsymbol{\lambda}^T \boldsymbol{f}(\boldsymbol{x}, \boldsymbol{u}) \tag{2.5.3}$$

where the Lagrange multiplier vector, $\boldsymbol{\lambda}$, is an element of R^n. The necessary conditions for an extremum are

$$\frac{\partial L}{\partial \boldsymbol{x}} = \boldsymbol{0} = \frac{\partial J}{\partial \boldsymbol{x}} + \frac{\partial}{\partial \boldsymbol{x}} \left(\boldsymbol{f}^T \boldsymbol{\lambda}\right) \qquad (n \text{ equations}) \tag{2.5.4}$$

$$\frac{\partial L}{\partial \boldsymbol{u}} = \boldsymbol{0} = \frac{\partial J}{\partial \boldsymbol{u}} + \frac{\partial}{\partial \boldsymbol{u}} \left(\boldsymbol{f}^T \boldsymbol{\lambda}\right) \qquad (m \text{ equations}) \tag{2.5.5}$$

$$\frac{\partial L}{\partial \boldsymbol{\lambda}} = \boldsymbol{0} \quad \text{or} \quad \boldsymbol{f} = \boldsymbol{0} \qquad (n \text{ equations}) \tag{2.5.6}$$

2.5.1 Example

Let us consider finding the optimum value of the coolant flow to a continuous stirred tank reactor shown in Figure 2.5. In this system the reactant undergoes a first-order exothermic reaction. The reaction rate follows the Arrhenius temperature dependency so that the rate model is given by

$$r = k_o C e^{-E/R(T+2.73.15)} \tag{2.5.7}$$

where k_o = frequency constant, E is the activation energy, and R is the gas constant.

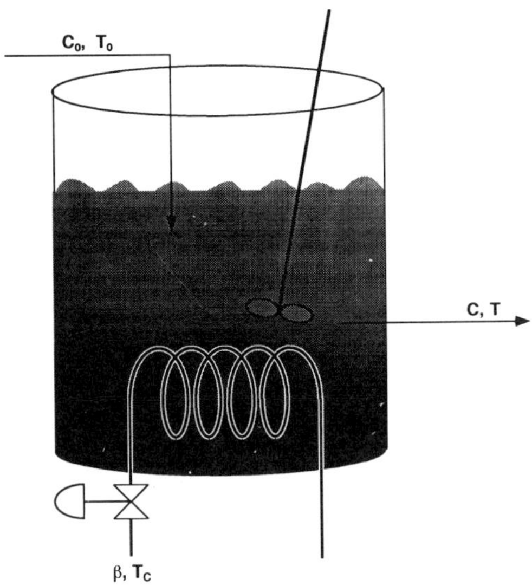

Figure 2.5 Exothermic stirred tank reactor.

The material balance for the stirred tank reactor is

$$V\,\frac{dC}{dt} = FC_o - FC - VCk_o e^{-E/R(T+273.15)} \tag{2.5.8}$$

where V is the reactor volume and F the volumetric flow rate, and the energy balance for the reactor is

$$\rho V C_p\,\frac{dT}{dt} = \rho C_p F T_o - \rho C_p F T - \beta(T - T_c) + \Delta H C V k_o e^{-E/R(T+273.15)} \tag{2.5.9}$$

where ρ is the density, C_p is the heat capacity, and (ΔH) is the heat of reaction.

For this problem the state vector is

$$\boldsymbol{x} = \begin{pmatrix} C \\ T \end{pmatrix} \qquad \begin{matrix} \text{gmole}/\ell \\ {}^\circ\text{C} \end{matrix} \tag{2.5.10}$$

and the control variable is the cooling due to flow of fluid through the cooling valve

$$\boldsymbol{u} = \beta \qquad \text{cal/min °C} \tag{2.5.11}$$

Nyquist and Ramirez (1971) describe a specific reactor system that uses this model. Typical steady-state conditions for their reactor are

$$\begin{aligned} C &= .492 \text{ gmole/liter} \\ T &= 30°\text{C} \\ \beta &= 256 \text{ cal/min°C} \end{aligned} \tag{2.5.12}$$

We can linearize the nonlinear model of Equations (2.5.8) and (2.5.9) about these steady-state conditions to obtain a linearized steady-state model of

$$\begin{aligned} &-0.1048C \quad -0.002034T \quad +0.09884 = 0 \\ &0.03055C \quad -0.0002037T \quad -0.0000346\beta \quad +0.0039409 = 0 \end{aligned} \tag{2.5.13}$$

which can be expressed as

$$\boldsymbol{Ax} + \boldsymbol{Bu} + \boldsymbol{c} = \boldsymbol{0} \tag{2.5.14}$$

A typical performance index might be to find the operating condition that results in minimization of the final reactant concentration (a maximum conversion), the minimization of the reactor temperature, and the minimization of the coolant flow rate. This combined objective can be expressed by the following quadratic performance function

$$J = \frac{1}{2}C^2 + \frac{1}{2}qT^2 + \frac{1}{2}r\beta^2 \tag{2.5.15}$$

Here the parameters q and r give the relative importance of their respective terms as compared to the primary objective of having maximum conversion. The Lagrangian for this problem is

$$L = J + \lambda_1 f_1 + \lambda_2 f_2$$

where

$$f_1 = (a_{11}x_1 + a_{12}x_2 + c_1) \qquad f_2 = (a_{21}x_1 + a_{22}x_2 + b_2u + c_2) \tag{2.5.16}$$

and the necessary conditions for an extremal for the objective function are

$$\frac{\partial L}{\partial x_1} = 0 = x_1 + \lambda_1 a_{11} + \lambda_2 a_{21} \tag{2.5.17}$$

$$\frac{\partial L}{\partial x_2} = 0 = qx_2 + \lambda_1 a_{12} + \lambda_2 a_{22} \tag{2.5.18}$$

$$\frac{\partial L}{\partial u} = 0 = ru + b_2\lambda_2 \tag{2.5.19}$$

$$\frac{\partial L}{\partial \lambda_1} = 0 = a_{11}x_1 + a_{12}x_2 + c_1 \tag{2.5.20}$$

$$\frac{\partial L}{\partial \lambda_2} = 0 = a_{21}x_1 + a_{22}x_2 + b_2u + c_2 \tag{2.5.21}$$

These equations can be expressed in matrix form as

$$\begin{bmatrix} a_{11} & a_{12} & & & \\ a_{21} & a_{22} & b_2 & & \\ & & r & & b_2 \\ 1 & & & a_{11} & a_{21} \\ & q & & a_{12} & a_{22} \end{bmatrix} \begin{bmatrix} x_1 \\ x_2 \\ u \\ \lambda_1 \\ \lambda_2 \end{bmatrix} = \begin{bmatrix} -c_1 \\ -c_2 \\ 0 \\ 0 \\ 0 \end{bmatrix} \tag{2.5.22}$$

We can use MATLAB to solve this set of linear algebraic equations. For the weighting parameters, $q = 5 \times 10^{-4}$ and $r = 5 \times 10^{-6}$, the solution that extremizes the objective function is

$$\begin{aligned} x_1 &= .262 \text{ gmole}/\ell \\ x_2 &= 35.1\ ^\circ\text{C} \\ \beta &= 137.8 \text{ cal/min}\ ^\circ\text{C} \\ \lambda_1 &= 8.29 \\ \lambda_2 &= 19.91 \end{aligned} \tag{2.5.23}$$

2.5.2 Example

We can generalize the previous example. We have the general linear model of

$$\boldsymbol{Ax} + \boldsymbol{Bu} + \boldsymbol{c} = \boldsymbol{0} \tag{2.5.24}$$

with the quadratic performance function of

$$J = \frac{1}{2}\boldsymbol{u}^T\boldsymbol{Ru} + \frac{1}{2}\boldsymbol{x}^T\boldsymbol{Qx} \tag{2.5.25}$$

where $\boldsymbol{Q}$ and $\boldsymbol{R}$ are symmetric positive definite matrices.

The general quadratic form of Equation (2.5.25) can also be written

$$J = \frac{1}{2}\left\| \boldsymbol{u} \right\|_{\boldsymbol{R}}^2 + \frac{1}{2}\left\| \boldsymbol{x} \right\|_{\boldsymbol{Q}}^2 \tag{2.5.26}$$

where $||\cdot||$ denotes the norm of the vector and $||\cdot||^2$ is the quadratic norm.

The Lagrangian is

$$L = \frac{1}{2}\boldsymbol{u}^T\boldsymbol{R}\boldsymbol{u} + \frac{1}{2}\boldsymbol{x}^T\boldsymbol{Q}\boldsymbol{x} + \boldsymbol{\lambda}^T[\boldsymbol{A}\boldsymbol{x} + \boldsymbol{B}\boldsymbol{u} + \boldsymbol{c}] \tag{2.5.27}$$

and the necessary conditions for an extremum are

$$\frac{\partial L}{\partial \boldsymbol{x}} = \boldsymbol{0} = \boldsymbol{Q}\boldsymbol{x} + \boldsymbol{A}^T\boldsymbol{\lambda} \tag{2.5.28}$$

$$\frac{\partial L}{\partial \boldsymbol{u}} = \boldsymbol{0} = \boldsymbol{R}\boldsymbol{u} + \boldsymbol{B}^T\boldsymbol{\lambda} \tag{2.5.29}$$

$$\frac{\partial L}{\partial \boldsymbol{\lambda}} = \boldsymbol{0} = \boldsymbol{A}\boldsymbol{x} + \boldsymbol{B}\boldsymbol{u} + \boldsymbol{c} \tag{2.5.30}$$

We can combine Equations (2.5.28) through (2.5.30) to get the solution for the optimal control as

$$\boldsymbol{u} = -\left[\boldsymbol{R} + \boldsymbol{B}^T\boldsymbol{A}^{T^{-1}}\boldsymbol{Q}\boldsymbol{A}^{-1}\boldsymbol{B}\right]^{-1}\boldsymbol{B}^T\boldsymbol{A}^{T^{-1}}\boldsymbol{Q}\boldsymbol{A}^{-1}\boldsymbol{c} \tag{2.5.31}$$

2.5.3 Example

For the reactor problem of Example 2.5.1, suppose that the performance function is not a function of the control but is only quadratic in the state variables

$$J = \frac{1}{2}c^2 + \frac{1}{2}qT^2 \qquad \text{with} \qquad u_{min} \le u \le u_{max} \tag{2.5.32}$$

Note that the control is bounded by u_{min} and u_{max}. The necessary condition

$$\frac{\partial L}{\partial u} = 0 = \lambda_2 b_2 \tag{2.5.33}$$

becomes

$$\lambda_2 = 0 \tag{2.5.34}$$

A problem when the necessary condition does not involve the control variable is called a singular control problem. Depending upon the value of q, the singular control may or may not fall in the feasible bounded range. (To solve the singular control problem, we solve the set of equations of (2.5.27) except that thc last equation is removed since $\lambda_2 = 0$.) For example, when $q = 10^{-5}$, we obtain a negative value of the control flow, which is not a feasible solution. However, when $q = 10^{-4}$, the singular solution is feasible. The results of this singular control problem are

$$\begin{aligned} x_1 &= .198 \text{ gmole}/\ell & \lambda_1 &= 1.89 \\ x_2 &= 38.4°\text{C} & \lambda_2 &= 0 \\ u &= 62.52 \text{ cal/min}°\text{C} \end{aligned} \tag{2.5.35}$$

References

Crowell, R. H. and Williamson, R. E., *Calculus of Vector Functions*, Prentice–Hall, Englewood Cliffs, NJ (1962).

Nyquist, J. K. and Ramirez, W. F., "Time Optimal Control of an Experimental Continuous Stirred Tank Reactor," *Chem. Eng. Sci.* **26**, 1673 (1971).

Ogata, K., *State Space Analysis of Control Systems*, Prentice–Hall, Englewood Cliffs, NJ, p. 277 (1967).

Ramirez, W. F., *Computational Methods for Process Simulation*, Butterworths, Boston, MA (1989).

Exercises

1. Find u such that

$$J = x^2 + u^2$$

is minimized subject to the equation

$$xu = 1$$

Use the Lagrange multiplier technique.

2. Suppose that observations of a constant vector are taken after being corrupted with noise,

$$\boldsymbol{z} = \boldsymbol{H}\boldsymbol{x} + \boldsymbol{v}$$

where $\boldsymbol{z}$ are the observations of order m, $\boldsymbol{x}$ is a vector of order n, and $\boldsymbol{v}$ is a vector representing measurement noise. Find the best estimate of $\boldsymbol{x}$, denoted as $\hat{\boldsymbol{x}}$, such that

$$J = \frac{1}{2} \left\| \boldsymbol{z} - \boldsymbol{H}\hat{\boldsymbol{x}} \right\|^2_{R^{-1}}$$

is minimized. This is done by setting $\dfrac{\partial J}{\partial \hat{\boldsymbol{x}}} = \mathbf{0}$ and solving for $\hat{\boldsymbol{x}}$. Show that the result is

$$\hat{\boldsymbol{x}} = \left(\boldsymbol{H}^T \boldsymbol{R}^{-1} \boldsymbol{H}\right)^{-1} \boldsymbol{H}^T \boldsymbol{R}^{-1} \boldsymbol{z}$$

Find $\hat{\boldsymbol{x}}_6$ for a set of measurements where

$$\boldsymbol{z} = \begin{pmatrix} 1.01 \\ 2.03 \\ 3.00 \\ 3.05 \\ 1.95 \\ .97 \end{pmatrix} \qquad \boldsymbol{H} = \begin{pmatrix} 1 & 0 \\ 0 & 1 \\ 1 & 1 \\ 1 & 1 \\ 0 & 1 \\ 1 & 0 \end{pmatrix} \qquad \boldsymbol{R} = \boldsymbol{I}$$

Now suppose that an additional measurement,

$$z_7 = 3.0 \qquad \boldsymbol{h} = [1,\ 1]$$

is taken. Compute $\hat{\boldsymbol{x}}_7$.

3. Determine the optimum operating pressure and recycle ratio for a process where the hydrocarbon feed is mixed with the recycle stream and compressed before being passed into a catalytic reactor. The product and unreacted material are separated by distillation, and the unreacted material is recycled. The pressure P and recycle ratio R must be selected to minimize the total annual cost for the required production of 10^7 pounds per year. The feed is brought up to pressure at an annual cost of \$1000 P, mixed with the recycle stream, and fed to the reactor at an annual cost of $4 \times 10^9/PR$. The product is removed in a separator at a cost of \$$10^5$ R per year, and the unreacted material is recycled in a recirculating compressor which consumes 1.5×10^5 R annually.

   Also, determine the annual cost and show that as a minimum.

4. For the process of problem 3, it is necessary to maintain the product of the pressure and recycle ratio equal to 9,000 psi. Determine the optimum values of the pressure and recycle ratio and the minimum cost within this constraint by using Lagrange multipliers.

Chapter 3

DYNAMIC OPTIMIZATION

The fundamental mathematical principles of the calculus of variations are developed in this chapter. The classical results of functional extrema given by the Euler equation and those of constrained optimization given by the Euler–Lagrange equation are presented.

The purpose of an optimal control problem is to determine the control policy that will extremize a specific performance criterion, subject to the constraints imposed by the physical nature of the problem. The necessary conditions of Chapter 2 cannot optimize dynamic systems. This is because for a dynamic system the state, $\boldsymbol{x}$, is a function of another variable, namely time, t. We write this as $\boldsymbol{x}(t)$. Therefore, in the optimal control problem for dynamic systems, we want to extremize a performance index, say, $J(\boldsymbol{x}(t))$, which is a function of the state variables, $\boldsymbol{x}$, which are themselves a function of time, t. We call $J(\boldsymbol{x}(t))$ a *functional* since it is a function of a function. We need to develop methods for extremizing functionals. The technique that we will develop to solve this class of problems is the calculus of variations and its extension, the minimum principle of Pontryagin. The fundamental theorem of the calculus of variations is applied to problems with unconstrained states and controls. Consideration of the effect of control constraints leads to Pontryagin's minimum principle. In this chapter, we will introduce the fundamental concepts that define functionals, properties of functionals, and variational concepts needed for extremizing functionals. Supplementary material can be found in the texts by Kirk (1970), Sage and White (1977), Ray (1981), and Lewis (1986).

3.1 Performance Functionals

Optimization objectives can be expressed as a performance index to be extremized. Performance measures are used in control problems to determine the quality of the control. The performance index is a mathematical expression which, when extremized, indicates that the system is performing in a desirable manner. The performance index should be chosen so that the optimization results in an implementable control law.

A general form for the performance index is

$$J = h(\boldsymbol{x}(t_f), t_f) + \int_{t_o}^{t_f} F(\boldsymbol{x}, \boldsymbol{u}, t)dt \tag{3.1.1}$$

This performance index is a function of the final state, $\boldsymbol{x}(t_f)$, and the integral of the system dynamic behavior from the initial time to the final time. Specific examples follow.

3.1.1 The Minimum Time Problem

$$J = \int_{t_o}^{t_f} dt = (t_f - t_o) \tag{3.1.2}$$

Here we minimize the time, (t_f), it takes to transfer the system from the initial state to a final state or target set.

3.1.2 The Terminal Control Problem

$$J = \left(\boldsymbol{x}(t_f) - \boldsymbol{r}(t_f)\right)^T \boldsymbol{H} \left(\boldsymbol{x}(t_f) - \boldsymbol{r}(t_f)\right) \tag{3.1.3}$$

where $\boldsymbol{r}(t_f)$ is the desired state at the final time. This is a quadratic index in the terminal error. If $\boldsymbol{H}$ is a real symmetric positive definite matrix, the performance index, J, is always positive. The matrix, $\boldsymbol{H}$, is considered as a weighting matrix. For a symmetric $\boldsymbol{H}$ matrix to be positive definite, the principal minors of $\boldsymbol{H}$ must be positive. Diagonal weighting matrices, $\boldsymbol{H}$, are often used.

3.1.3 The Minimum Control Effort Problem

The idea here is to transfer the system from an arbitrary initial state, $\boldsymbol{x}(t_o) = \boldsymbol{x}_o$, to a specified target set, S, using the minimum expenditure of control effort

$$J = \int_{t_o}^{t_f} (\boldsymbol{u}^T \boldsymbol{R} \boldsymbol{u})dt \tag{3.1.4}$$

where $\boldsymbol{R}$ is a real positive definite weighting matrix. This forces the quadratic performance index to always be positive.

3.1.4 The Tracking Problem

Here we want to keep the state, $\boldsymbol{x}(t)$, as close as possible to a desired state, $\boldsymbol{r}(t)$, over the interval, $[t_o, t_f]$. The performance index can be expressed as

$$J = \int_{t_o}^{t_f} \left((\boldsymbol{x} - \boldsymbol{r})^T \boldsymbol{Q} (\boldsymbol{x} - \boldsymbol{r})\right) dt \tag{3.1.5}$$

where $\boldsymbol{Q}$ is a real positive semi-definite matrix. When the controls are not bounded, the index of Equation (3.1.5) is reasonable. Often it is not desirable to allow the controls free access but to place a penalty on the control action. Therefore, we often use the modified performance index

$$J = \int_{t_o}^{t_f} \left((\boldsymbol{x} - \boldsymbol{r})^T \boldsymbol{Q} (\boldsymbol{x} - \boldsymbol{r}) + \boldsymbol{u}^T \boldsymbol{R} \boldsymbol{u} \right) \, dt \tag{3.1.6}$$

When it is especially important that the final state be close to the desired final state, $\boldsymbol{r}(t_f)$, then a terminal term can also be added to give

$$\begin{aligned} J = & \left(\boldsymbol{x}(t_f) - \boldsymbol{r}(t_f) \right)^T \boldsymbol{H} \left(\boldsymbol{x}(t_f) - \boldsymbol{r}(t_f) \right) \\ & + \int_{t_o}^{t_f} \left((\boldsymbol{x} - \boldsymbol{r})^T \boldsymbol{Q} (\boldsymbol{x} - \boldsymbol{r}) + \boldsymbol{u}^T \boldsymbol{R} \boldsymbol{u} \right) \, dt \end{aligned} \tag{3.1.7}$$

3.1.5 The Regulator Problem

A special case of the tracking problem is when $\boldsymbol{r}(t) = \boldsymbol{c}$. Here we want to keep the system at a desired steady state, $\boldsymbol{c}$. This is called a regulator problem and the performance index is

$$J = \left(\boldsymbol{x}(t_f) - \boldsymbol{c} \right)^T \boldsymbol{H} \left(\boldsymbol{x}(t_f) - \boldsymbol{c} \right) + \int_{t_o}^{t_f} \left((\boldsymbol{x} - \boldsymbol{c})^T \boldsymbol{Q} (\boldsymbol{x} - \boldsymbol{c}) + \boldsymbol{u}^T \boldsymbol{R} \boldsymbol{u} \right) \, dt \tag{3.1.8}$$

3.2 Definition of a Functional

A functional, J, is defined as a rule of correspondence that assigns to each function, $\boldsymbol{x}(t)$, contained in a certain set of functions, X, a unique real number. The set of functions, X, is called the domain of the functional, and the set of real numbers associated with the functions in X is called the range of the functional.

3.2.1 Example

An example is to suppose that $x(t)$ is a continuous function of t defined on the interval, $[t_o, t_f]$, and that the functional is

$$J(x) = \int_{t_o}^{t_f} x(t) \, dt \tag{3.2.1}$$

The real number assigned by the functional, J, is the area under the $x(t)$ curve and is the range of this functional.

The domain of a functional is a class of functions so that we can say that a functional is a "function of a function."

3.3 Linearity of Functionals

J is a linear functional of the function, $x(t)$, if and only if it first satisfies the scalar commutative property

$$J(\alpha \boldsymbol{x}) = \alpha J(\boldsymbol{x}) \tag{3.3.1}$$

for all $\boldsymbol{x} \in X$ and for all real numbers, α, such that $\alpha\, \boldsymbol{x} \in X$. To be linear, the functional must also satisfy the associative property under addition

$$J(\boldsymbol{x}_1 + \boldsymbol{x}_2) = J(\boldsymbol{x}_1) + J(\boldsymbol{x}_2) \tag{3.3.2}$$

for all $\boldsymbol{x}_1$, $\boldsymbol{x}_2$, and $\boldsymbol{x}_1 + \boldsymbol{x}_2$ in X.

3.3.1 Example

For example, let us consider the functional

$$J(x) = \int_{t_o}^{t_f} x(t)\; dt \tag{3.3.3}$$

where $x(t)$ is a continuous function of t. Let us first test the scalar commutative property:

$$\alpha J(x) = \alpha \int_{t_o}^{t_f} x(t)\; dt \tag{3.3.4}$$

$$J(\alpha x) = \int_{t_o}^{t_f} \alpha x(t)\; dt \tag{3.3.5}$$

Since the scalar, α, can commute with an integral, the functional defined by Equation (3.3.3) does satisfy the scalar commutative property. Now let's test the associative addition property:

$$J(x_1 + x_2) = \int_{t_o}^{t_f} \Big(x_1(t) + x_2(t) \Big)\; dt \tag{3.3.6}$$

$$J(x_1) + J(x_2) = \int_{t_o}^{t_f} x_1(t)\; dt + \int_{t_o}^{t_f} x_2(t)\; dt \tag{3.3.7}$$

Since integrals are associative under addition, then for this case it is true that

$$J(x_1 + x_2) = J(x_1) + J(x_2) \tag{3.3.8}$$

Therefore, since both the commutative and associative properties hold, the functional defined by Equation (3.3.3) is linear.

3.4 Norm of Functions

The norm of a function is a rule of correspondence that assigns to each function, $\boldsymbol{x}(t) \in X$, defined for $t \in [t_o, t_f]$, a real number. The form of $\boldsymbol{x}$ is denoted by $||\boldsymbol{x}||$, and satisfies the following properties:

1. $||\boldsymbol{x}|| = 0$ if and only if $\boldsymbol{x}(t) = \boldsymbol{0}$ for all $t \in [t_o, t_f]$.

2. $|\alpha| \; ||\boldsymbol{x}|| = ||\alpha \boldsymbol{x}||$ for all real numbers.

3. $||\boldsymbol{x}_1 + \boldsymbol{x}_2|| \leq ||\boldsymbol{x}_1|| + ||\boldsymbol{x}_2||$.

Therefore, a norm has a zero property, is commutative with scalars, and satisfies the triangle inequality. A common norm is the Euclidean norm and is defined as

$$\left|\left|\boldsymbol{x}(t)\right|\right|_E = \left[\sum_{i=1}^{n} x_i^2(t)\right]^{1/2} = (\boldsymbol{x}^T\boldsymbol{x})^{1/2} \tag{3.4.1}$$

Another common norm is the quadratic norm,

$$\left|\left|x(t)\right|\right|^2 = \left(\boldsymbol{x}^T\boldsymbol{x}\right) \tag{3.4.2}$$

To compare the closeness of two functions $\boldsymbol{x}(t)$ and $\boldsymbol{y}(t)$, we compute the distance between $\boldsymbol{x}$ and $\boldsymbol{y}$ as the norm of the difference between the two functions

$$d(\boldsymbol{x}, \boldsymbol{y}) = ||\boldsymbol{x} - \boldsymbol{y}|| \tag{3.4.3}$$

Intuitively speaking, if the distance defined by Equation (3.4.3) is zero, then the functions are identical; if it is small, the functions are close; and if it is large, then the functions are far apart.

3.5 The Increment of a Functional

In order to consider extremal properties of a functional, we introduce the concept of an increment. If $\boldsymbol{x}$ and $\boldsymbol{x} + \boldsymbol{\delta x}$ are both functions for which the functional, J, is defined, then the increment of the functional, ΔJ, is defined as

$$\Delta J = J(\boldsymbol{x} + \boldsymbol{\delta x}) - J(\boldsymbol{x}) \tag{3.5.1}$$

From definition (3.5.1), the increment of a functional depends upon the functions $\boldsymbol{x}$ and $\boldsymbol{\delta x}$. Therefore, the increment can be functionally expressed as $\Delta J(\boldsymbol{x}, \boldsymbol{\delta x})$. The function $\boldsymbol{\delta x}$ is called the variation of the function $\boldsymbol{x}$.

3.5.1 Example

Let's find the increment of the functional

$$J(x) = \int_{t_o}^{t_f} x^2(t)\, dt \tag{3.5.2}$$

where $x(t)$ is a continuous function of t. The increment is

$$\Delta J(x, \delta x) = J(x + \delta x) - J(x) \tag{3.5.3}$$

Upon substitution

$$\Delta J(x, \delta x) = \int_{t_o}^{t_f} [x(t) + \delta x(t)]^2 dt - \int_{t_o}^{t_f} x^2(t)\, dt \tag{3.5.4}$$

and simplifying yields

$$\Delta J(x, \delta x) = \int_{t_o}^{t_f} [2x(t)\delta x(t) + \delta x^2(t)]\, dt \tag{3.5.5}$$

3.6 The First Variation of a Functional

The first variation of a functional, δJ, is the part of the increment of the functional, ΔJ, that is linear in the variation, $\boldsymbol{\delta x}$. The increment of a functional can therefore be written as

$$\Delta J(\boldsymbol{x}, \boldsymbol{\delta x}) = \delta J(\boldsymbol{x}, \boldsymbol{\delta x}) + g(\boldsymbol{x}, \boldsymbol{\delta x})\ ||\boldsymbol{\delta x}|| \tag{3.6.1}$$

where δJ is linear in $\boldsymbol{\delta x}$. If

$$\lim_{||\boldsymbol{\delta x}|| \to 0} g(\boldsymbol{x}, \delta \boldsymbol{x}) = 0 \tag{3.6.2}$$

then the functional, J, is differentiable on $\boldsymbol{x}$ and δJ is the first variation of J evaluated for the function, $\boldsymbol{x}(t)$.

3.6.1 Example

Let us develop the first variation of the following functional:

$$J = \int_{t_o}^{t_f} F\left(x(t), \dot{x}(t), t\right)\, dt \tag{3.6.3}$$

The function, $x(t)$, is continuously differentiable on the interval, $[t_o, t_f]$. Also, the function, F, is continuous in $x(t), \dot{x}(t)$, and t and has continuous partial derivatives with respect to x and $\dot{x}$. In other words, both the functions, $x(t)$ and F, are well behaved.

To compute the first variation of the functional, J, we first form the increment, ΔJ,

$$\Delta J = \int_{t_o}^{t_f} F(x + \delta x,\ \dot{x} + \delta \dot{x},\ t)\ dt - \int_{t_o}^{t_f} F(x, \dot{x}, t)\ dt \tag{3.6.4}$$

or

$$\Delta J = \int_{t_o}^{t_f} \Big(F(x + \delta x,\ \dot{x} + \delta \dot{x},\ t) - F(x, \dot{x}, t) \Big)\ dt \tag{3.6.5}$$

We can expand $F(x + \delta x,\ \dot{x} + \delta \dot{x},\ t)$ in a Taylor series about $F(x, \dot{x}, t)$ as

$$\begin{aligned} F(x + \delta x,\ \dot{x} + \delta \dot{x},\ t) &= F(x, \dot{x}, t) + \frac{\partial F}{\partial x}\ \delta x \\ &+ \frac{\partial F}{\partial \dot{x}}\ \delta \dot{x} + \mathcal{O}(\delta \boldsymbol{x}^n, \delta \dot{\boldsymbol{x}}^n) \qquad n \geq 2 \end{aligned} \tag{3.6.6}$$

The increment therefore becomes

$$\Delta J = \int_{t_o}^{t_f} \left[\frac{\partial F}{\partial x}\ \delta x + \frac{\partial F}{\partial \dot{x}}\ \delta \dot{x} \right]\ dt + \int_{t_o}^{t_f} \mathcal{O}(\delta \boldsymbol{x}^n, \delta \dot{\boldsymbol{x}}^n)\ dt \qquad n \geq 2 \tag{3.6.7}$$

The portion of this expression which is linear in δx and $\delta \dot{x}$ is the first variation of the functional, δJ,

$$\delta J = \int_{t_o}^{t_f} \left(\frac{\partial F}{\partial x}\ \delta x + \frac{\partial F}{\partial \dot{x}}\ \delta \dot{x} \right)\ dt \tag{3.6.8}$$

Because of the smooth behavior of the functions, $x(t)$ and F, the higher order terms in δx and $\delta \dot{x}$ vanish as $||\delta x, \delta \dot{x}||$ goes to zero.

3.6.2 Example

As a specific example of a problem that satisfies the functional of Equation (3.6.3), let us consider

$$F = x^2 + \dot{x}^2 \tag{3.6.9}$$

The appropriate first derivatives are

$$\frac{\partial F}{\partial x} = 2x \qquad \frac{\partial F}{\partial \dot{x}} = 2\dot{x} \tag{3.6.10}$$

and the first variation of the functional is

$$\delta J = \int_{t_o}^{t_f} 2(x\delta x + \dot{x}\delta \dot{x})\ dt \tag{3.6.11}$$

3.7 Extrema of Functionals

A functional, J, with domain, X, has a relative extremum at $\boldsymbol{x}^*$ if there is an $\epsilon > 0$ such that for all functions, $\boldsymbol{x}$, in X which satisfy $||\boldsymbol{x} - \boldsymbol{x}^*|| < \epsilon$, the increment of J has the same sign. If

$$\Delta J = J(\boldsymbol{x}) - J(\boldsymbol{x}^*) \geq 0 \tag{3.7.1}$$

$J(\boldsymbol{x}^*)$ is a relative minimum; if

$$\Delta J = J(\boldsymbol{x}) - J(\boldsymbol{x}^*) \leq 0 \tag{3.7.2}$$

$J(\boldsymbol{x}^*)$ is a relative maximum.

If Equation (3.7.1) is satisfied for arbitrarily large ϵ, then $J(\boldsymbol{x}^*)$ is a global or absolute minimum. Similarly, if Equation (3.7.2) holds for arbitrarily large ϵ, then $J(\boldsymbol{x}^*)$ is a global or absolute maximum. The function, $\boldsymbol{x}^*$, is called an extremal of the functional, and $J(\boldsymbol{x}^*)$ is referred to as an extremum.

3.8 The Fundamental Theorem of Variational Calculus

The fundamental theorem of variational calculus is now given. If $\boldsymbol{x}^*$ is an extremal, then it is a necessary condition that the first variation of J must vanish on $\boldsymbol{x}$. That is,

$$\delta J(\boldsymbol{x}^*, \boldsymbol{\delta x}) = 0 \qquad \text{for all admissible } \boldsymbol{\delta x} \tag{3.8.1}$$

By admissible $\boldsymbol{\delta x}$, we mean that $\boldsymbol{x} + \boldsymbol{\delta x}$ must be a member of the set X. Thus, if X is the set of continuous functions, both $\boldsymbol{x}$ and $\boldsymbol{\delta x}$ are required to be continuous.

This theorem can be easily proven by contradiction. As developed by Kirk (1970), we assume that $\boldsymbol{x}^*$ is an extremal and that $\delta J(\boldsymbol{x}^*, \boldsymbol{\delta x}) \neq 0$. Let us show that these assumptions imply that the increment, ΔJ, can be made to change sign in an arbitrarily small neighborhood of $\boldsymbol{x}^*$. The increment is

$$\Delta J(\boldsymbol{x}^*, \boldsymbol{\delta x}) = \delta J(\boldsymbol{x}^*, \boldsymbol{\delta x}) + g(\boldsymbol{x}^*, \boldsymbol{\delta x})||\boldsymbol{\delta x}|| \tag{3.8.2}$$

where $g(\boldsymbol{x}^*, \boldsymbol{\delta x}) \to 0$ as $||\boldsymbol{\delta x}|| \to 0$. Thus, there is a neighborhood, $||\boldsymbol{\delta x}|| < \epsilon$, where $g(\boldsymbol{x}^*, \boldsymbol{\delta x}||\boldsymbol{\delta x}||$ is small enough so that δJ dominates the expression for ΔJ. Let us select the variation

$$\boldsymbol{\delta x} = \alpha \boldsymbol{\delta x}_1 \tag{3.8.3}$$

Such a variation is shown in Figure 3.1 for a scalar function, x^*. Here $\alpha > 0$ and $||\alpha \boldsymbol{\delta x}_1|| < \epsilon$. Suppose that

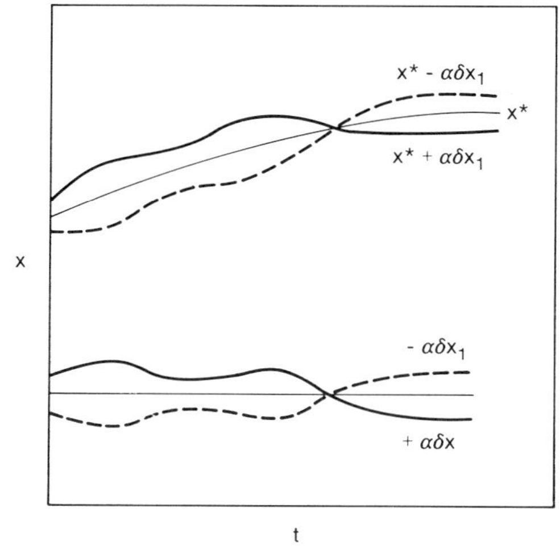

Figure 3.1 An extremal and two specific variations.

$$\delta J(\boldsymbol{x}^*, \alpha \boldsymbol{\delta x}_1) < 0 \tag{3.8.4}$$

Since δJ is a linear functional of $\boldsymbol{\delta x}_1$, the scalar commutative property of Equation (2.2.2) gives

$$\delta J(\boldsymbol{x}^*, \alpha \boldsymbol{\delta x}_1) = \alpha \delta J(\boldsymbol{x}^*, \boldsymbol{\delta x}_1) \tag{3.8.5}$$

The signs of ΔJ and δJ are the same for $||\boldsymbol{\delta x}_1|| < \epsilon$. Thus,

$$\Delta J(\boldsymbol{x}^*, \alpha \boldsymbol{\delta x}_1) < 0 \tag{3.8.6}$$

Next we consider the variation

$$\boldsymbol{\delta x} = -\alpha \boldsymbol{\delta x}_1 \tag{3.8.7}$$

Such a variation is also shown in Figure 3.1. Clearly $||\alpha \boldsymbol{\delta x}_1|| < \epsilon$ implies that $|| - \alpha \boldsymbol{\delta x}_1|| < \epsilon$. Therefore, the sign of the increment, $\Delta J(\boldsymbol{x}^*, -\alpha \boldsymbol{\delta x}_1)$, is the same as the sign of $\delta J(\boldsymbol{x}^*, -\alpha \boldsymbol{\delta x}_1)$. Again, using the scalar commutative property, we obtain

$$\delta J(\boldsymbol{x}^*, -\alpha \boldsymbol{\delta x}_1) = -\alpha \delta J(\boldsymbol{x}^*, \boldsymbol{\delta x}_1) \tag{3.8.8}$$

Since $\delta J(\boldsymbol{x}^*, \alpha \boldsymbol{\delta x}_1) < 0$ from Equation (3.8.4), then from Equation (3.8.8) $\delta J(\boldsymbol{x}^*, -\alpha \boldsymbol{\delta x}_1) > 0$, which implies that

$$\Delta J(\boldsymbol{x}^*, -\alpha \boldsymbol{\delta x}_1) > 0 \tag{3.8.9}$$

Equations (3.8.6) and (3.8.9) mean that $\boldsymbol{x}^*$ cannot be an extremal since it violates the definition of an extremal given in Section 3.7. The assumption that the functions in X are not bounded guarantees that $\alpha\boldsymbol{\delta x}_1$ and $-\alpha\boldsymbol{\delta x}_1$ are both admissible variations.

Therefore, if $\boldsymbol{x}^*$ is an extremal, it is necessary that

$$\delta J(\boldsymbol{x}^*, \boldsymbol{\delta x}) = 0 \tag{3.8.10}$$

3.9 Integral Functionals of a Single Variable

Let us now apply the fundamental theorem to obtain the necessary condition for an extremal of the functional given by Equation (3.6.3).

$$J = \int_{t_o}^{t_f} F\Big(\ x(t), \dot{x}(t), t\ \Big)\ dt \tag{3.9.1}$$

where t_o and t_f are known.

We have previously computed in Section 3.6, Equation (3.6.8), the first variation to be

$$\delta J = \int_{t_o}^{t_f} \left(\frac{\partial F}{\partial x}\ \delta x + \frac{\partial F}{\partial \dot{x}}\ \delta \dot{x}\right)\ dt \tag{3.6.8}$$

The necessary condition for an extremum of J is therefore

$$0 = \int_{t_o}^{t_f} \left(\frac{\partial F}{\partial x}\ \delta x + \frac{\partial F}{\partial \dot{x}}\ \delta \dot{x}\right)\ dt \tag{3.9.2}$$

The variations, δx and $\delta \dot{x}$, are not independent. Any variation in δx will introduce an effect in its derivative. We can eliminate $\delta \dot{x}$ from the problem through the integration by parts

$$\int_{t_o}^{t_f} \frac{\partial F}{\partial \dot{x}}\ \delta \dot{x}\ dt = \frac{\partial F}{\partial \dot{x}}\ \delta x\Big|_{t_o}^{t_f} - \int_{t_o}^{t_f} \frac{d}{dt}\left(\frac{\partial F}{\partial \dot{x}}\right)\ \delta x\ dt \tag{3.9.3}$$

Using Equation (3.9.3) in Equation (3.9.2) gives as the necessary condition for an extremum of J

$$0 = \int_{t_o}^{t_f} \left[\frac{\partial F}{\partial x} - \frac{d}{dt}\left(\frac{\partial F}{\partial \dot{x}}\right)\right]\ \delta x\ dt + \frac{\partial F}{\partial \dot{x}}\ \delta x\Big|_{t_o}^{t_f} \tag{3.9.4}$$

Equation (3.9.4) is satisfied if

$$\frac{\partial F}{\partial \dot{x}}\ \delta x = 0 \qquad \text{at } t = t_o \text{ and } t = t_f \tag{3.9.5}$$

and

$$\int_{t_o}^{t_f} \left[\frac{\partial F}{\partial x} - \frac{d}{dt}\left(\frac{\partial F}{\partial \dot{x}}\right)\right]\ \delta x\ dt = 0 \tag{3.9.6}$$

Since the variation, δx, is free and unconstrained, the integral of Equation (3.9.6) can only be zero if

$$\frac{\partial F}{\partial x} - \frac{d}{dt}\left(\frac{\partial F}{\partial \dot{x}}\right) = 0 \tag{3.9.7}$$

Equation (3.9.7) is a necessary condition for the existence of an extremal of the functional J of Equation (3.9.1) and was first derived by Euler in 1744. Equation (3.9.7) is known as a Euler equation. The conditions of Equation (3.9.5) are known as *transversality conditions* and vary with the specifications of $x(t)$ at the boundaries t_o and t_f. For example, if $x(t_o)$ and $x(t_f)$ are fixed, then the variation, δx, must be zero at the boundaries. For this case, admissible variations are illustrated in Figure 3.2. On the other hand, if $x(t_o)$ and $x(t_f)$ are not specified, then the variations, δx, at the boundaries are free. From Equation (3.9.5), it is necessary for an extremum that

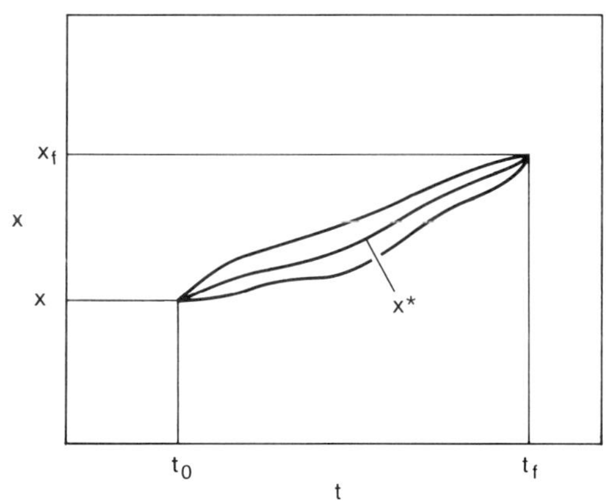

Figure 3.2 Fixed end points.

$$\frac{\partial F}{\partial \dot{x}} = 0 \qquad \text{at } t = t_o \text{ and } t = t_f \tag{3.9.8}$$

This latter case is illustrated in Figure 3.3.

3.9.1 Example

Let us now consider a specific numerical example by computing the necessary conditions for an extremal for the functional

$$J = \int_0^1 (x^2 + \dot{x}^2)\, dt \tag{3.9.9}$$

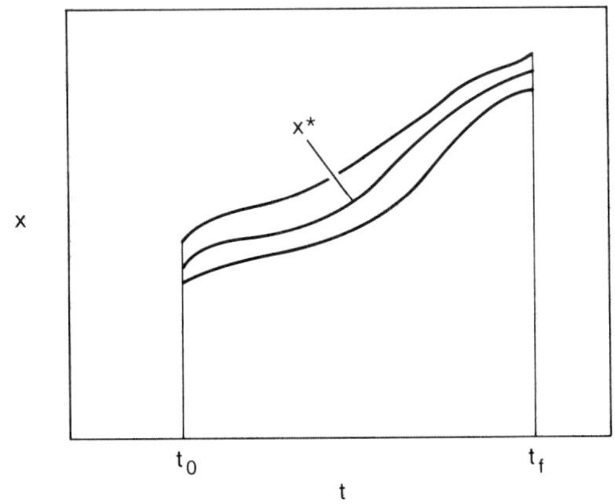

Figure 3.3 Free end points.

$$\text{with} \qquad x(0) = 0$$
$$x(1) = 1$$

The Euler equation of Equation (3.9.7) becomes,

$$2x - \frac{d}{dt}(2\dot{x}) = 0 \tag{3.9.10}$$

or

$$\frac{d^2x}{dt^2} - x = 0 \tag{3.9.11}$$

which has an analytical solution

$$x(t) = A \sinh t + B \cosh t \tag{3.9.12}$$

The transversality conditions of Equation (3.9.5) state that for this problem, the variation, δx, is zero at the boundaries, $t = 0$ and $t = 1$. The specified end point conditions become the boundary conditions for Equation (3.9.12). Using these boundary conditions, we have for the extremum of the function, J, the following state trajectory,

$$x(t) = \frac{\sinh t}{\sinh 1} \tag{3.9.13}$$

3.9.2 Fixed Final State/Free Final Time

Often for functionals of the form of Equation (3.9.1), the final time is not specified. Let us first consider the case when the final time is free and the final state is specified. This is illustrated in Figure 3.4. For this problem, the increment, ΔJ, is

$$\Delta J = \int_{t_o}^{t_f+\delta t_f} F(x+\delta x,\ \dot{x}+\delta \dot{x},\ t)\ dt - \int_{t_o}^{t_f} F(x,\dot{x},t)\ dt \tag{3.9.14}$$

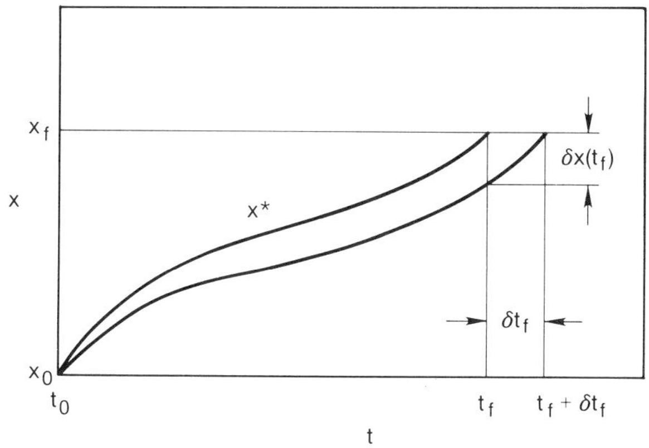

Figure 3.4 Fixed final state/free final time.

which can be written as

$$\begin{aligned}\Delta J = &\int_{t_o}^{t_f} F(x+\delta x,\ \dot{x}+\delta \dot{x},\ t)\ dt - \int_{t_o}^{t_f} F(x,\dot{x},t)\ dt \\ &+ \int_{t_f}^{t_f+\delta t_f} F(x+\delta x,\ \dot{x}+\delta \dot{x},\ t)\ dt\end{aligned} \tag{3.9.15}$$

The first two terms on the right-hand side are exactly those of the previous problem for a specified final time. It is the last term that is new. The last term can be evaluated, correct to the first order in δt_f, as

$$\begin{aligned}&\int_{t_f}^{t_f+\delta t_f} F(x+\delta x,\ \dot{x}+\delta \dot{x},\ t)\ dt \\ &\quad = F(x_f+\delta x_f,\ \dot{x}_f+\delta \dot{x}_f,\ t_f)\ \delta t_f\end{aligned} \tag{3.9.16}$$

We now expand $F(x_f+\delta x_f,\ \dot{x}_f+\delta \dot{x}_f,\ t_f)$ in a Taylor series about $F(x_f, \dot{x}_f, t_f)$ to get

$$\begin{aligned}F(x_f+\delta x_f,\ \dot{x}_f+\delta \dot{x}_f,\ t_f) &= F(x_f,\dot{x}_f,t_f) + \frac{\partial F}{\partial x}\ (t_f)\ \delta x(t_f) \\ &+\frac{\partial F}{\partial \dot{x}}\ (t_f)\ \delta \dot{x}(t_f) + \mathcal{O}\Big(\delta \boldsymbol{x}^n(t_f), \delta \dot{\boldsymbol{x}}^n(t_f)\Big) \qquad n \geq 2\end{aligned} \tag{3.9.17}$$

Substituting this expression into Equation (3.9.16) gives

$$\int_{t_f}^{t_f+\delta t_f} F(x+\delta x,\ \dot{x}+\delta\dot{x},\ t)\ dt = F(x_f, \dot{x}_f, t_f)\ \delta t_f + \frac{\partial F}{\partial x}\ (t_f)\ \delta x(t_f)\ \delta t_f$$
$$+\frac{\partial F}{\partial \dot{x}}\ (t_f)\ \delta\dot{x}(t_f)\ \delta t_f + \mathcal{O}\Big(\delta \boldsymbol{x}^n(t_f), \delta\dot{\boldsymbol{x}}^n(t_f)\Big)\ \delta t_f \qquad n \geq 2 \tag{3.9.18}$$

The first variation, which is the linear part of the increment, ΔJ, therefore becomes

$$\delta J = \int_{t_o}^{t_f} \left(\frac{\partial F}{\partial x} - \frac{d}{dt}\frac{\partial F}{\partial \dot{x}}\right)\ \delta x\ dt + \frac{\partial F}{\partial \dot{x}}\ (t_f)\ \delta x(t_f)$$
$$-\frac{\partial F}{\partial \dot{x}}\ (t_o)\ \delta x(t_o) + F(t_f)\ \delta t_f \tag{3.9.19}$$

The variations, $\delta x(t_f)$ and δt_f, are not independent but are related by the first order relation (see Figure 3.4)

$$\delta x(t_f) = -\dot{x}(t_f)\ \delta t_f \tag{3.9.20}$$

The necessary condition for an extremum is that $\delta J = 0$, which is

$$\delta J = 0 = \int_{t_o}^{t_f} \left(\frac{\partial F}{\partial x} - \frac{d}{dt}\frac{\partial F}{\partial \dot{x}}\right)\ \delta x\ dt$$
$$+\left(F(t_f) - \frac{\partial F}{\partial \dot{x}}\ (t_f)\ \dot{x}(t_f)\right)\ \delta t_f - \frac{\partial F}{\partial \dot{x}}\ (t_o)\ \delta x(t_o) \tag{3.9.21}$$

For Equation (3.9.21) to hold, it is necessary that the following relations be satisfied:

Euler Equation

$$\frac{\partial F}{\partial x} - \frac{d}{dt}\left(\frac{\partial F}{\partial \dot{x}}\right) = 0 \tag{3.9.22}$$

Final Condition

$$x(t_f) = x_f \tag{3.9.23}$$

Initial Transversality Condition

$$\frac{\partial F}{\partial \dot{x}}\ (t_o)\ \delta x(t_o) = 0 \tag{3.9.24}$$

Final Time Transversality Condition

$$F(t_f) - \frac{\partial F}{\partial \dot{x}}\ (t_f)\ \dot{x}(t_f) = 0 \tag{3.9.25}$$

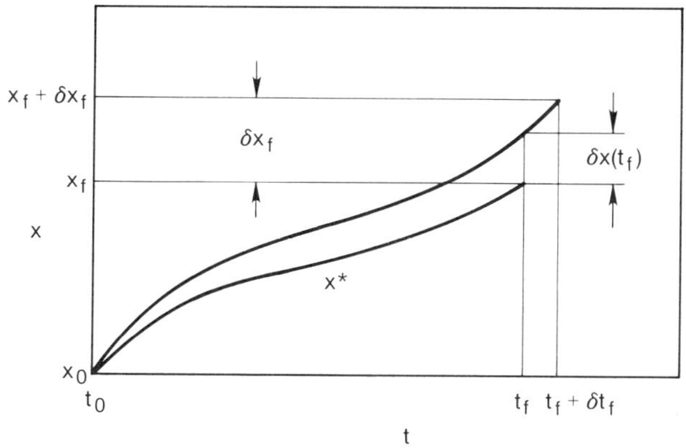

Figure 3.5 Final time free/final state free.

3.9.3 Final Time Free/Final State Free

Another common occurrence is the fact that the final time is free and the final state is also free. This is illustrated in Figure 3.5. The analysis of this problem is the same as the previous case except the variations, $\delta x(t_f)$, δt_f, and δx_f, are related to the first order by

$$\delta x(t_f) = \delta x_f - \dot{x}(t_f)\,\delta t_f \tag{3.9.26}$$

The necessary conditions for this case are as follows:

Euler Equation

$$\frac{\partial F}{\partial x} - \frac{d}{dt}\left(\frac{\partial F}{\partial \dot{x}}\right) = 0 \tag{3.9.27}$$

Initial Transversality Condition

$$\frac{\partial F}{\partial \dot{x}}\,(t_o)\,\delta x(t_o) = 0 \tag{3.9.28}$$

Final Transversality Condition

$$\frac{\partial F}{\partial \dot{x}}\,(t_f)\,\delta x_f + \left(F(t_f) - \frac{\partial F}{\partial \dot{x}}\,(t_f)\,\dot{x}(t_f)\right)\,\delta t_f = 0 \tag{3.9.29}$$

If the variations, δt_f and δx_f, are independent, then

$$\frac{\partial F}{\partial \dot{x}}\,(t_f)\,\delta x_f = 0 \tag{3.9.30}$$

and

$$\left(F(t_f) - \frac{\partial F}{\partial \dot{x}}\,(t_f)\,\dot{x}(t_f)\right)\,\delta t_f = 0 \tag{3.9.31}$$

3.10 Piecewise Smooth Extremals

For the previous development, we have assumed that the admissible response curves were continuous and have smooth first derivatives. This is due to the fact that we have expanded functions in Taylor series as functions of δx and $\delta \dot{x}$. This is quite restrictive. Let us allow for piecewise smooth first derivatives. We want to find the extrema of the functional

$$J(x) = \int_{t_o}^{t_f} F(x, \dot{x}, t)\, dt \tag{3.10.1}$$

where F has continuous first and second derivatives and x is piecewise continuous. For simplicity, let us assume that

$$\begin{aligned} x(t_o) &= x_o \\ x(t_f) &= x_f \\ t_f &= \text{known} \end{aligned} \tag{3.10.2}$$

We also assume that there is a discontinuity of $\dot{x}$ at time, t_1, as shown in Figure 3.6.

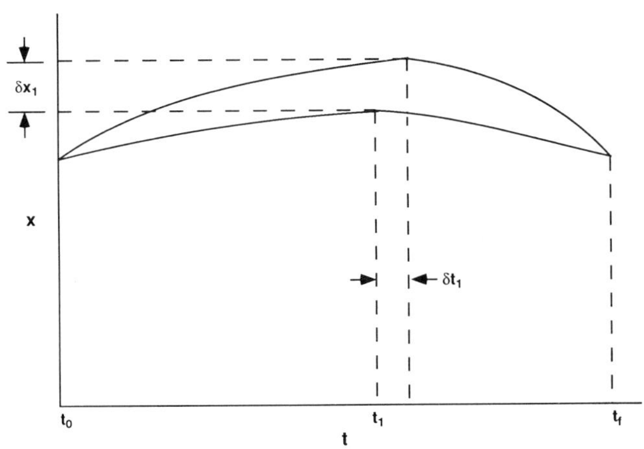

Figure 3.6 Piecewise smooth function.

We can write the performance functional as the sum of the two integrals

$$J(x) = \int_{t_o}^{t_1} F(x, \dot{x}, t)\, dt + \int_{t_1}^{t_f} F(x, \dot{x}, t)\, dt \tag{3.10.3}$$

Since the time, t_1, is not fixed, the variations, δx_1 and δt_1, are free and not fixed. Using t_1 as a free end point for the first integral and a free initial condition for the second integral gives the following first variation for this problem,

$$\begin{aligned}\delta J = &\int_{t_o}^{t_1} \left(\frac{\partial F}{\partial x} - \frac{d}{dt}\frac{\partial F}{\partial \dot{x}}\right) \delta x \, dt + \frac{\partial F}{\partial \dot{x}}\,(t_1^-)\delta x_1 \\ &+ \left(F(t_1^-) - \frac{\partial F}{\partial \dot{x}}\,(t_1^-)\dot{x}(t_1^-)\right)\,\delta t_1 \\ &+ \int_{t_1}^{t_f} \left(\frac{\partial F}{\partial x} - \frac{d}{dt}\frac{\partial F}{\partial \dot{x}}\right) \delta x \, dt - \frac{\partial F}{\partial \dot{x}}\,(t_1^+)\,\delta x_1 \\ &- \left(F(t_1^+) - \frac{\partial F}{\partial \dot{x}}\,(t_1^+)\dot{x}(t_1^-)\right)\,\delta t_1 \end{aligned} \tag{3.10.4}$$

The necessary conditions for an extremal are, therefore, as follows:

The Euler Equation

$$\frac{\partial F}{\partial x} - \frac{d}{dt}\frac{\partial F}{\partial \dot{x}} = 0 \tag{3.10.5}$$

The Natural Boundary Conditions

$$\begin{aligned} x(t_o) &= x_o \\ x(t_f) &= x_f \end{aligned} \tag{3.10.6}$$

The Weierstrass-Erdmann Corner Conditions

$$\begin{aligned}&\left(\frac{\partial F}{\partial \dot{x}}\,(t_1^-) - \frac{\partial F}{\partial \dot{x}}\,(t_1^+)\right)\,\delta x_1 + \left[\left(F(t_1^-) - \frac{\partial F}{\partial \dot{x}}\,(t_1^-)\dot{x}(t_1^-)\right)\right. \\ &\left. - \left(F(t_1^+) - \frac{\partial F}{\partial \dot{x}}\,(t_1^+)\dot{x}(t_1^+)\right)\right]\,\delta t_1 = 0\end{aligned} \tag{3.10.7}$$

These conditions were first derived by Weierstrass in 1927. If δx_1 and δt_1 are independent, then the corner conditions become

$$\frac{\partial F}{\partial \dot{x}}\,(t_1^-) = \frac{\partial F}{\partial \dot{x}}\,(t_1^+) \tag{3.10.8}$$

$$F(t_1^-) - \frac{\partial F}{\partial \dot{x}}\,(t_1^-)\dot{x}(t_1^-) = F(t_1^+) - \frac{\partial F}{\partial \dot{x}}\,(t_1^+)\dot{x}(t_1^+) \tag{3.10.9}$$

If δx_1 and δt_1 are related by

$$x(t_1) = \theta(t_1) \tag{3.10.10}$$

then

$$\delta x_1 = \frac{d\theta}{dt}\,(t_1)\,\delta t_1 \tag{3.10.11}$$

This case yields one corner condition plus the defining θ relation.

3.11 Constrained Extrema

We now consider the development of the necessary conditions for an extremum of the functional

$$J(\boldsymbol{x}(t)) = \int_{t_o}^{t_f} F(\boldsymbol{x}, \dot{\boldsymbol{x}}, t)\, dt \tag{3.11.1}$$

where $\boldsymbol{x}$ is an nth order vector which is a member of R^n. We also introduce the m equality constraints

$$\boldsymbol{\Lambda}(\boldsymbol{x}, \dot{\boldsymbol{x}}, t) = \mathbf{0} \qquad\qquad \epsilon\ R^m \tag{3.11.2}$$

The necessary conditions for an extremal will be given in terms of the Lagrangian, L, which is defined as

$$L = F + \boldsymbol{\lambda}^T \boldsymbol{\Lambda} \tag{3.11.3}$$

where $\boldsymbol{\lambda}(t)$ are called the dynamic Lagrange multipliers. An augmented performance functional, which has the same extrema as Equation (3.11.1), is

$$J_A(\boldsymbol{x}, \boldsymbol{\lambda}) = \int_{t_o}^{t_f} L(\boldsymbol{x}, \dot{\boldsymbol{x}}, \boldsymbol{\lambda}, t)\, dt \tag{3.11.4}$$

By introducing the variations, $\boldsymbol{\delta x}$, $\boldsymbol{\delta \dot{x}}$, $\boldsymbol{\delta\lambda}$, and δt_f, to the increment, ΔJ_A, we get the first variation

$$\begin{aligned}\delta J_A = \int_{t_o}^{t_f} &\left\{ \left(\frac{\partial L}{\partial \boldsymbol{x}} - \frac{d}{dt}\frac{\partial L}{\partial \dot{\boldsymbol{x}}} \right)^T \boldsymbol{\delta x} + \left(\frac{\partial L}{\partial \boldsymbol{\lambda}} \right)^T \boldsymbol{\delta\lambda} \right\} dt + \left(\frac{\partial L}{\partial \dot{\boldsymbol{x}}}(t_f) \right)^T \boldsymbol{\delta x}_f \\ &- \left(\frac{\partial L}{\partial \dot{\boldsymbol{x}}}(t_o) \right)^T \boldsymbol{\delta x}(t_o) + \left(L(t_f) - \left(\frac{\partial L}{\partial \dot{\boldsymbol{x}}}(t_f) \right)^T \dot{\boldsymbol{x}}(t_f) \right) \delta t_f \end{aligned} \tag{3.11.5}$$

The fundamental theorem of the calculus of variations states that a necessary condition for an extremal is that $\delta J_A = 0$. For unbounded admissible states, $\boldsymbol{x}(t)$, we have the following necessary conditions for an extremal:

The Euler–Lagrange Equations

$$\frac{\partial L}{\partial \boldsymbol{x}} - \frac{d}{dt}\left(\frac{\partial L}{\partial \dot{\boldsymbol{x}}} \right) = \mathbf{0} \tag{3.11.6}$$

The Initial Transversality Conditions

$$\left(\frac{\partial L}{\partial \dot{\boldsymbol{x}}}(t_o) \right)^T \boldsymbol{\delta x}(t_o) = 0 \tag{3.11.7}$$

The Final Transversality Conditions

$$\left(\frac{\partial L}{\partial \dot{\boldsymbol{x}}}(t_f) \right)^T \boldsymbol{\delta x}_f + \left(L(t_f) - \left(\frac{\partial L}{\partial \dot{\boldsymbol{x}}}(t_f) \right)^T \dot{\boldsymbol{x}}(t_f) \right) \delta t_f = 0 \tag{3.11.8}$$

The Constraint Equations

$$\frac{\partial L}{\partial \boldsymbol{\lambda}} = \mathbf{0} \tag{3.11.9}$$

or

$$\boldsymbol{\Lambda}(\boldsymbol{x}, \dot{\boldsymbol{x}}, t) = \mathbf{0} \tag{3.11.10}$$

3.11.1 Example

As a specific example, let us consider finding the extremals for the functional

$$J(\boldsymbol{x}, t) = \int_0^1 (\dot{x}_1^2 + t^2)\, dt \tag{3.11.11}$$

subject to the equality constraint

$$x_1^2 - \dot{x}_2 = 0 \tag{3.11.12}$$

with the boundary conditions

$$x_1(0) = 0 \tag{3.11.13}$$

$$x_1(1) = 0 \tag{3.11.14}$$

$$x_2(0) = 0 \tag{3.11.15}$$

$$x_2(1) = 2 \tag{3.11.16}$$

The Lagrangian is

$$L = \dot{x}_1^2 + t^2 + \lambda_1(x_1^2 - \dot{x}_2) \tag{3.11.17}$$

The Euler–Lagrange equations for this problem are

$$\frac{d}{dt}\, \dot{x}_1 - \lambda_1 x_1 = 0 \tag{3.11.18}$$

$$\frac{d}{dt}\, \lambda_1 = 0 \tag{3.11.19}$$

Equation (3.11.19) has the solution

$$\lambda_1 = -c^2 \tag{3.11.20}$$

Substituting Equation (3.11.20) into Equation (3.11.18) gives

$$\frac{d^2 x_1}{dt^2} + c^2 x_1^2 = 0 \tag{3.11.21}$$

The analytical solution to the differential equation of (3.11.21) is

$$x_1(t) = A\, \sin(\mathrm{ct}) + B\, \cos(\mathrm{ct}) \tag{3.11.22}$$

Using boundary conditions of Equations (3.11.13) and (3.11.14) gives

$$x_1(t) = A \, \sin(n\pi t) \qquad n = 1, 2 \cdots \tag{3.11.23}$$

The integrated form of the constraint equation (3.11.12) is

$$x_2(1) - x_2(0) = A^2 \int_0^1 \sin^2(2\pi t) \, dt \tag{3.11.24}$$

Using boundary conditions (3.11.15) and (3.11.16) yields

$$A = \pm 2 \tag{3.11.25}$$

The extrema of the functional are therefore

$$x(t) = \pm 2 \, \sin(n\pi t) \qquad n = 1, 2, \cdots \tag{3.11.26}$$

References

Keller, J. B., "Optimal Velocity in a Race," *Am. Math. Monthly* **81**, No. 5, 474 (1974).

Kirk, D. E., *Optimal Control Theory*, Prentice–Hall, Englewood Cliffs, NJ (1970).

Lewis, F. L., *Optimal Control*, Wiley, New York (1986).

Ray, W. H., *Advanced Process Control*, McGraw–Hill, New York (1981).

Sage, A. P. and White, C. C., III, *Optimal Systems Control*, Second Edition, Prentice–Hall, Englewood Cliffs, NJ (1977).

Szepe, S. and Levenspiel, O., "Optimal Temperature Policies for Reactors Subject to Catalyst Deactivation—I. Batch Reactors," *Chem. E. Sci.* **23** 881–894 (1968).

Exercises

1. Find the extremal of

$$J(x) = \int_0^{\pi/2} \left[x^2 - \dot{x}^2\right] dt$$

with

$$\begin{aligned} x(0) &= 0 \\ x(\pi/2) &= 1 \end{aligned}$$

2. Find the extremal of

$$J(x) = \int_0^1 \left(x^2 + \dot{x}^2\right) dt$$

with

$$\begin{aligned} x(0) &= 1 \\ x(1) &= \text{free} \end{aligned}$$

3. Find the extremal of

$$J(x) = \int_0^2 \left(x^2 + 2x\dot{x} + \dot{x}^2\right) dt$$

with

$$\begin{aligned} x(0) &= 1 \\ x(2) &= -3 \end{aligned}$$

4. Find the extremal of

$$J(x) = \int_0^{\pi/2} \left(\dot{x}_1^2 + \dot{x}_2^2 + 2x_1x_2\right) dt$$

with

$$x_1(0) = 0 \ , \qquad x_1(\pi/2) = \text{free}$$

$$x_2(0) = 0 \ , \qquad x_2(\pi/2) = 1$$

5. Determine the Weierstrass-Erdman corner conditions and the optimum function $x(t)$ for the minimization of the cost function

$$J = \int_0^1 x^2 (2 - \dot{x})^2 \, dt$$

with $x(0) = 0$ and $x(1) = 1$.

6. What inequalities must the constraints a and b satisfy so that the functional

$$J = \int_0^1 \left(\dot{x} + a)^2(\dot{x} + b\right)^2 dt \qquad x(0) = 0 \qquad x(1) = 1$$

has just one corner?

7. Determine an extremal for the functional

$$J(x) = \int_0^{t_f} \left(1 + \dot{x}^2\right)^{1/2} dt$$

which has $x(0) = 2$ and terminates on the curve $\theta = -4t + 5$

8. Find a curve that is an extremal for the functional

$$J(x) = \int_0^{t_f} \left(1 + \dot{x}^2\right)^{1/2} dt$$

with $x(0) = 5$ and end points that must lie on the circle $x^2(t) + (t-5)^2 - 4 = 0$.

This is an end point constraint of the form $m[x(t_f),\, t_f] = 0$. Draw a picture to help determine the relationship between δx_f and δt_f.

9. Find the increment and the first variation for

$$\begin{aligned} J(x) &= \int_{t_0}^{t_f} \left[x^3(t) - x^2(t)\dot{x}(t)\right] dt \\ J(x) &= \int_{t_0}^{t_f} \left[x_1^2(t) + x_1(t)x_2(t) + x_2^2(t) + 2\dot{x}_1(t)\dot{x}_2(t)\right] dt \\ J(x) &= \int_{t_0}^{t_f} e^{x(t)} dt \end{aligned}$$

10. Find the extremal of the functional

$$J = \int_0^{\pi/4} (x^2 - \dot{x}^2) dt$$

where

$$x(0) = 0$$

$$x(\pi/4) \text{ lies on the line } t = \pi/4$$

11. Find the extremal of the functional

$$J = \int_0^{\pi/2} \left(\ddot{x}^2 - x^2 + t^2\right) dt$$

with

$$x(0) = 0 \qquad \dot{x}(0) = 0$$

$$x(\pi/2) = 0 \qquad \dot{x}(\pi/2) = -1$$

Introduce two state variables (note that in doing this an equality constraint must be satisfied).

12. Keller (1974) presents an interesting analysis for the optimal running of a race. He considers the distance functional

$$D = \int_0^T v(t)dt$$

and wants to find the minimum value of time T for a given distance D. An equivalent problem is to maximize D for a given value of T. Assume that the runner can be modeled by an equation of motion and an energy balance. The equation of motion is

$$\frac{dv}{dt} + \frac{v}{\tau} = f$$

where v/τ is the resistance force and f the propulsive force which is bounded by

$$f \leq F$$

The energy balance is

$$\frac{dE}{dt} = \sigma - fv$$

where σ is the oxygen supplied and fv the energy used. There is the state constraint that E cannot be negative:

$$E(t) \geq 0$$

Show that extremal control ($f = F$) leads to the solution

$$v(t) = F\tau\left(1 - e^{-t/\tau}\right) \qquad 0 \leq t \leq t_1$$

This solution is valid only for $E(t) \geq 0$ and therefore holds only up to the condition when $E = 0$. Show that when $E(t) = 0$ and stays zero

$$\sigma - \frac{1}{2}\frac{dv^2}{dt} - \frac{v^2}{\tau} = 0 \qquad t_2 \leq t \leq T$$

Verify that the solution to this equation is

$$v^2(t) = \sigma\tau + \left[v^2(t_2) - \sigma\tau\right] e^{2(t_2-t)/\tau} \qquad t_2 \leq t \leq T$$

Now for the augmented functional

$$D = \int_0^{t_1} F\tau\left(1 - e^{-t/\tau}\right) dt + \int_{t_1}^{t_2} v(t)dt$$
$$+ \int_{t_2}^{T} \left\{\sigma\tau + \left[v^2(t_2) - \sigma\tau\right] e^{2(t_2-t)/\tau}\right\}^{1/2} dt$$

maximize D subject to the constraint $E(t_2) = 0$. Using variational calculus, show that this leads to the solution

$$v(t) = \tau/\lambda$$

with the Lagrange multiplier

$$\lambda = 2\int_{t_2}^{T} \left\{\sigma\tau + \left[v^2(t_2) - \sigma\tau\right] e^{2(t_2-t)/\tau}\right\}^{-1/2} e^{2(t_2-t)/\tau} dt$$

Sketch the solution in each time domain. Interpret the results in terms of running short dashes and long distance races.

13. The operation of heterogeneous catalytic reactors yields dynamic optimization problems due to catalyst deactivation. Consider the single irreversible reaction

$$A \rightarrow R$$

carried out in a batch reactor. The system is described by a material balance on the reactant species and a catalyst deactivation law

$$\frac{dC_A}{dt} + ra = 0$$
$$\frac{da}{dt} + \rho = 0$$

where C_A is the reactant concentration and a is the catalyst activity. The reaction rate r is a function of temperature and reactant concentration:

$$r = k_0 \exp\left(-\frac{E}{RT}\right) C_A$$

and the catalyst deactivation function ρ is given as

$$\rho = K_0 \exp\left(-\frac{\epsilon}{RT}\right) a^m$$

We wish to determine the optimum operational strategy (C_A, a, T) as a function of time so that the reaction time is minimized for a specified conversion and final catalyst activity. It is convenient to work with the reciprocal temperature rather than the temperature:

$$y = \frac{1}{T}$$

Derive the necessary conditions for optimality.

Show that as a result of the Euler–Lagrange equations that the Lagrangian satisfies

$$\frac{d}{dt}\left(-L + \left(\frac{\partial L}{\partial \dot{\boldsymbol{x}}}\right)^T \dot{\boldsymbol{x}}\right) + \frac{\partial F}{\partial t} = 0$$

and that for this problem we have

$$-\lambda_1 r - \lambda_2 \rho = C$$

where C is a constant of integration.

Show that using the Euler–Lagrange relations that

$$\lambda_1 = \frac{C\,\dfrac{\partial \rho}{\partial y}}{\rho\,\dfrac{\partial r}{\partial y} - r\,\dfrac{\partial \rho}{\partial y}}$$

After differentiation show that

$$\dot{r} = -r\,\frac{\partial r}{\partial C_A}$$

which reduces to

$$\frac{d}{dt}\,(ka) = 0$$

Show that if $m = 0$, this relation results in the optimal temperature profile

$$y = \left(\frac{R}{\epsilon}\right) \ln \left[C_1 - K_0\,\frac{\epsilon}{E}\,t\right]$$

A more complete discussion of this problem is given by Szepe and Levenspiel (1968).

Chapter 4

OPTIMAL CONTROL

4.1 An Optimal Control Problem

This chapter applies variational calculus to problems that include control variables as well as state variables. This allows for the computation of control strategies that extremize precise performance criteria expressed as performance functionals. Necessary conditions for optimization are shown to be conveniently expressed in terms of a mathematical function called the Hamiltonian. The consideration of inequality constraints on the control variables results in the well known Pontryagin maximum principle. Necessary conditions for the dynamic optimization of distributed parameter systems are also derived from variational principles. Process applications of optimal control are presented.

In this section we wish to determine the necessary conditions that will determine an optimal control policy, $\boldsymbol{u}(t)$, that causes the system

$$\dot{\boldsymbol{x}} = \boldsymbol{f}(\boldsymbol{x}, \boldsymbol{u}, t) \qquad \boldsymbol{x}(t_o) = \boldsymbol{x}_o \tag{4.1.1}$$

to follow a path, $\boldsymbol{x}(t)$, such that the performance functional

$$J(\boldsymbol{u}) = \int_{t_o}^{t_f} F(\boldsymbol{x}, \boldsymbol{u}, t)\, dt \tag{4.1.2}$$

is extremized. Specific examples were given in Section 3.1 as minimum time problems, minimum control effort problems, tracking problems, and regulator problems. We will assume that the admissible state region, X, and the admissible control region, U, are not bounded. The dimension of the state vector, $\boldsymbol{x}$, is assumed to be n while that of the control vector, $\boldsymbol{u}$, is m. Good references are the books by Lewis (1986) and Bryson and Ho (1975).

The state differential equations of Equation (4.1.1) are a special form of the general equality constraint relations considered in Section 3.11. Namely,

$$\boldsymbol{\Lambda} = \boldsymbol{0} = \boldsymbol{f}(\boldsymbol{x}, \boldsymbol{u}, t) - \dot{\boldsymbol{x}}(t) \tag{4.1.3}$$

Since this is a constrained optimization problem, we introduce the Lagrangian

$$L = F(\boldsymbol{x}, \boldsymbol{u}, t) + \boldsymbol{\lambda}^T(\boldsymbol{f} - \dot{\boldsymbol{x}}) \tag{4.1.4}$$

where $\boldsymbol{\lambda}$ are the dynamic Lagrange multipliers or costate variables and are of dimension n. An augmented functional, with the same extrema of Equation (4.1.2), is therefore

$$J_A = \int_{t_o}^{t_f} L(\boldsymbol{x}, \dot{\boldsymbol{x}}, \boldsymbol{u}, \boldsymbol{\lambda}, t)\, dt \tag{4.1.5}$$

By introducing the variations, $\boldsymbol{\delta x}$, $\boldsymbol{\delta \dot{x}}$, $\boldsymbol{\delta u}$, $\boldsymbol{\delta \lambda}$, and δt_f, we can get the first variation of the functional to be

$$\begin{aligned}\delta J_A = & \int_{t_o}^{t_f} \left[\frac{\partial L}{\partial \boldsymbol{x}} - \frac{d}{dt}\frac{\partial L}{\partial \dot{\boldsymbol{x}}}\right]^T \boldsymbol{\delta x}\, dt + \left[\frac{\partial L}{\partial \dot{\boldsymbol{x}}}(t_f)\right]^T \boldsymbol{\delta x}_f \\ & + \left[L(t_f) - \left(\frac{\partial L}{\partial \dot{\boldsymbol{x}}}(t_f)\right)^T \dot{\boldsymbol{x}}(t_f)\right] \delta t_f \\ & + \int_{t_o}^{t_f} \left(\frac{\partial L}{\partial \boldsymbol{\lambda}}\right)^T \boldsymbol{\delta \lambda}\, dt + \int_{t_o}^{t_f} \left(\frac{\partial L}{\partial \boldsymbol{u}}\right)^T \boldsymbol{\delta u}\, dt\end{aligned} \tag{4.1.6}$$

The fundamental theorem of the calculus of variations states that for $\boldsymbol{x}(t)$ to be an extremum of the functional, J_A, it is necessary that $\delta J_A = 0$. Since the controls and states are unbounded, the variations, $\boldsymbol{\delta x}$, $\boldsymbol{\delta \lambda}$, and $\boldsymbol{\delta u}$, are free and unconstrained. The necessary conditions for an extremal are explained in the following subsections:

The Euler–Lagrange equations

Since the variation, $\boldsymbol{\delta x}$, is free, we have

$$\frac{\partial L}{\partial \boldsymbol{x}} - \frac{d}{dt}\frac{\partial L}{\partial \dot{\boldsymbol{x}}} = 0 \tag{4.1.7}$$

Using Equation (4.1.4) gives

$$\frac{\partial L}{\partial \dot{\boldsymbol{x}}} = -\boldsymbol{\lambda} \tag{4.1.8}$$

Therefore the Euler–Lagrange equations can be written as

$$\dot{\boldsymbol{\lambda}} = -\frac{\partial L}{\partial \boldsymbol{x}} \tag{4.1.9}$$

Using the definition of the Lagrangian, Equation (4.1.9) becomes

$$\dot{\boldsymbol{\lambda}} = -\left[\frac{\partial \boldsymbol{f}}{\partial \boldsymbol{x}}\right]^T \boldsymbol{\lambda} - \frac{\partial F}{\partial \boldsymbol{x}} \tag{4.1.10}$$

Errata

Several errors were noted in this edition. The following are corrections for these errors:

pp. 145-146: The order in which Figure 5.6 is shown is incorrect. The correct order is shown on the revised pages.

p. 156: There should be an additional line in the figure at the bottom of this page which reads "end."

pp. 165-167: The order in which Figure 5.17 is shown is incorrect. The correct order is shown on the revised pages.

p. 196: The figure caption for Figure 6.9 should read "*et al* ."

p. 221: The first 22 lines on the page (Figure 6.23) should not be there. The correct p. 221 is shown on the new pages.

p. 225: Figure 6.26 is missing a line. See the corrected page for where it should be.

p. 324: Both Figure 10.9 and Figure 10.10 are missing a line. See the corrected page for where they should be.

p. 325: There is an extra "C_{KI}" over the "20 minutes of time" in Figure 10.11. See corrected figure on new pages.

```
>> help lqr

 LQR     Linear quadratic regulator design for continuous systems.
         [K,S,E] = LQR(A,B,Q,R)  calculates the optimal feedback gain
         matrix K such that the feedback law  u = -Kx  minimizes the cost
         function:
                 J = Integral {x'Qx + u'Ru} dt

         subject to the constraint equation:
                 .
                 x = Ax + Bu

         Also returned is S, the steady-state solution to the associated
         algebraic Riccati equation and the closed loop eigenvalues E:
                                     -1
                 0 = SA + A'S - SBR  B'S + Q      E = EIG(A-B*K)

         [K,S,E] = LQR(A,B,Q,R,N) includes the cross-term N that relates
         u to x in the cost function.

                 J = Integral {x'Qx + u'Ru + 2*x'Nu}

         The controller can be formed with REG.

         See also: LQRY, LQR2, and REG.

>> %input the a and b matrices
>> a =  [-0.05156  -0.05877;
          0.01503  -0.005887];
>> b = [0  -0.03384]';
>> help eye

EYE     Identity matrix. EYE(N) is the N-by-N identity matrix.
        EYE(M,N) is an M-by-N matrix with 1's on the diagonal and
        zeros elsewhere. EYE(A) is the same size as A.

>> % imput the q and r maatrices
>>
>> q=eye(2)

q =

     1     0
     0     1
>> r=1;
>> % compute the Kalman gain and Riccati  maatrix
>>
>>  [k,p] = lqr(a,b,q,r)

k =

    0.0505   -0.9241


p =

    9.2375   -1.4929
   -1.4929   27.3078

>> % use IMPLUSE to get the response of the linear system
>>
>> help impulse

 IMPULSE Impulse response of continuous-time linear systems.
```

Figure 5.6 MATLAB program for LQR of Nyquist–Ramirez reactor.

```
        IMPULSE(A,B,C,D,IU)  plots the time response of the linear system
                .
                x = Ax + Bu
                y = Cx + Du
        to an impulse applied to the single input IU.  The time vector is
        automatically determined.

        IMPULSE(NUM,DEN) plots the impulse response of the polynomial
        transfer function  G(s) = NUM(s)/DEN(s)  where NUM and DEN contain
        the polynomial coefficients in descending powers of s.

        IMPULSE(A,B,C,D,IU,T) or IMPULSE(NUM,DEN,T) uses the user-supplied
        time vector T which must be regularly spaced.  When invoked with
        left hand arguments,
                [Y,X,T] = IMPULSE(A,B,C,D,...)
                [Y,X,T] = IMPULSE(NUM,DEN,...)
        returns the output and state time history in the matrices Y and X.
        No plot is drawn on the screen.  Y has as many columns as there
        are outputs and length(T) rows.  X has as many columns as there
        are states.

        See also: STEP,INITIAL,LSIM and DIMPULSE.

>> % redefine some matrices to put in proper form so that the immpulse is
>> % applied to both state variables with magnitude unity
>> aa = a-b*k

aa =

   -0.0516   -0.0588
    0.0167   -0.0372

>>
>> bb = [1 1]';
>> c = eye(2)

c =

     1     0
     0     1

>> d =  [0  0]'
>> t = [0:1:100];
>> y = impulse (aa,bb,c,d,1,t);
>> u = -k*y';
>> uu=u';
>> plot (t,y,t,uu)
>> title ('Plot of optimal state and control vs time')
>> xlabel ('Time, min')
>> ylabel ('State vector components and control variable')
>> meta plot1
>>% readjusting the R weighting
>> r=.1;
>> [k,p]=lqr(a,b,q,r)

k =

    0.7054   -3.3588


p =

    8.6073   -2.0845
   -2.0845    9.9255

>> aa
```

Figure 5.6 continued.

```
aa =

   -0.0516   -0.0588
    0.0167   -0.0372

>> aa=a-b*k

aa =

   -0.0516   -0.0588
    0.0389   -0.1195

>> y=impulse (aa,bb,c,d,1,t);
>> u=-k*y';
>> plot (t,y,t,u')
>> meta plot2
>> title('Plot of optimal state and contro vs time')
>> ylabel('State vector components and control variable')
>> xlabel('Time, min')
>> meta plot2
>>% redifining the Q and R weighting matrices
>> q(1,1)=10;
>> r=1;
>> [k,p]=lqr (a,b,q,r)

k =

    1.0172   -1.9623


p =

   78.1773  -30.0600
  -30.0600   57.9868

>> aa=a-b*k

aa =

   -0.0516   -0.0588
    0.0495   -0.0723

>> y=impulse (aa,bb,c,d,1,t);
>> u = -k*y';
>> plot (t,y,t,u')
>> title('Plot of optimal state and control vs time')
>> ylabel('State vector components and control variable')
>> xlabel('Time,min')
>> meta plot3
>> quit

 17423 flop(s).
```

Figure 5.6 continued.

```
% input the a and b matrices
a = [-0.05156  -0.05877;
      0.01503  -0.005887];
b = [0  -0.03384]';
% input the w matrix
w = [0.0125  0;
     0       0.0125];

% define the Johnson transformation

aa= [a   eye(2);
     zeros(2) zeros(2)];
bb= [0;
     0;
     b];
% define the weighting matrices

q = [1  0;
     0  0];

r = 1;

% Johnson transformation weighting matrices

qq = [q      zeros(2);
      zeros(2) zeros(2)];

% calculate the feedback gain using recursive method of section 5.2

z = [aa      -bb*inv(r)*bb';
     -qq     -aa'          ];

% chose the integration time interval dt and calculate transition matrix

dt = 1;

theta = expm(z);

% partition the transition matrix

theta11 = theta(1:4,1:4);
theta12 = theta(1:4,5:8);
theta21 = theta(5:8,1:4);
theta22 = theta(5:8,5:8);

% define the final Raccati matrix p(T)

pt = zeros(4);

% define t max

tmax = 70;
%loop for calcuating feedback gain matrix k

for t = tmax:-1:1,
        pt1 = inv(theta22 - pt*theta12)*(pt*theta11 - theta21);
        kt1 = inv(r)*bb'*pt1;
%       this loop is to transform the  matrix into a vector
        for col = 1:4
                k(t,col) = kt1(1,col);
        end
        pt=pt1;
end
```

Figure 5.11 MATLAB program for the Johnson algorithm control of the Nyquist–Ramirez reactor.

```
>> % input the a, b, and w matrices
>> a = [-0.033181  -0.002511;
        0.05237    0.010216];
>> b = [0   0.0988]';
>> w = [0.030864  0;
        0          0.030864];
>>
>> % check the eigenvalues of a
>> eig(a)

ans =

   -0.0299
    0.0069

>> % the positive eigenvalue shows that the system is unstable
>>
>> % let's look at the response of the uncontrolled system
>>
>> % use a unit impulse on each state
>> % define appropriate matrices
>>
>> bb = [1   1]';
>> c = eye(2);
>> t = [0:1:100];
>> d = [0  0]';
>> y = impulse(a,bb,c,d,1,t);
>> x1 = y(1:101,1);
>> x2 = y(1:101,2);
>> plot (t,y)
>> title('Uncontrolled Response');
>> xlabel('Time');
>> ylabel('State Vector Components');
>> meta uncontroll
>>
>> % compute the Kalman feedback gain matrix for the system
>> % input the weighting matrices
>>
>> q = [1  0;
        0  1];
>> r = 1;
>> [k,p,e] = lqr(a,b,q,r)
k =

    0.4015    1.0985


p =

   19.0538    4.0642
    4.0642   11.1187


e =

   -0.0978
   -0.0337

>> % these results show that a stable optimal proportional feedback system
>> % has been designed
>>
>> % now let's display the results of the controlled system to an unit
>> % impulse disturbance to each state variable
>>
>> % define the closed loop response matrix
```

Figure 5.17 MATLAB solution to regulation of unstable reactor.

```
>>
>> aa = a - b*k;
>>
>> y = impulse(aa,bb,c,d,1,t);
>> u = -k*y';
>> plot (t,y,t,u');
>> title('Optimal State and Control Variables');
>> xlabel ('Time');
>> ylabel ('State Vector Components and Control Variables');
>> meta optimal2_
>>
>> % implement the Johnson algorithm
>>
>> aa = [a          eye(2);
        zeros(2)    zeros(2)];
>> bb = [0;
         0;
         b];
>> q = [1    0;
        0    0];
>> qq = [q          zeros(2);
         zeros(2)   zeros(2)];
>>
>> % integrate the Riccati Equation backwards in time
>>
>> t0 = 0; tf = 200;
>> p0 =[0  0  0  0  0  0  0  0  0  0]';
>> [t,pv] = ode45('xprime',t0,tf,p0);
>> plot (t,pv)
>> title ('Riccati Matrix vs tf - t');
>> xlabel('tf - t'); ylabel('Riccati Gain Components');
>> meta riccati1
>>
>> % preparing for the response of the system to unit step disturbances
>>
>> % steady state values of the Riccati matrix elements
>>
>> pv(65,1:10)

ans =

   1.0e+03 *

  Columns 1 through 7

    0.0128   -0.0005    0.1786   -0.0032    0.0000   -0.0174    0.0006

  Columns 8 through 10

    6.1447   -0.1913    0.0110
>> p(1,1) = pv(65,1);
>> p(1,2) = pv(65,2);
>> p(1,3) = pv(65,3);
>> p(1,4) = pv(65,4);
>> p(2,1) = p(1,2);
>> p(2,2) = pv(65,5);
>> p(2,3) = pv(65,6);
>> p(2,4) = pv(65,7);
>> p(3,1) = p(1,3);
>> p(3,2) = p(2,3);
>> p(3,3) = pv(65,8);
>> p(3,4) = pv(65,9);
```

Figure 5.17 continued.

```
>> p(4,1) = p(1,4);
>> p(4,2) = p(2,4);
>> p(4,3) = p(3,4);
>> p(4,4) = pv(65,10);
>> k = inv(r)*bb'*p;
>> kp = k(1,3:4)

kp =

  -18.8991    1.0890

>> ki = k(1,1:2) -k(1,3:4)*a

ki =

   -1.0000   -0.0000

>> api = [a-b*kp  -b*ki;
          eye(2)   zeros(2)];
>> wi = [w        zeros(2);
         zeros(2) zeros(2)];
>> c = eye(4);
>> d = zeros(4);
>> t = [0:1:100];
>> xpi = step(api,wi,c,d,1,t);
>> x = xpi(1:101,1:2);
>> plot (t,x)
>> title('Johnson Algorithm State Response');
>> ylabel('Normalized Composition and Temperature');
>> xlabel('Time (min)');
>> meta xpi3;
```

Figure 5.17 continued

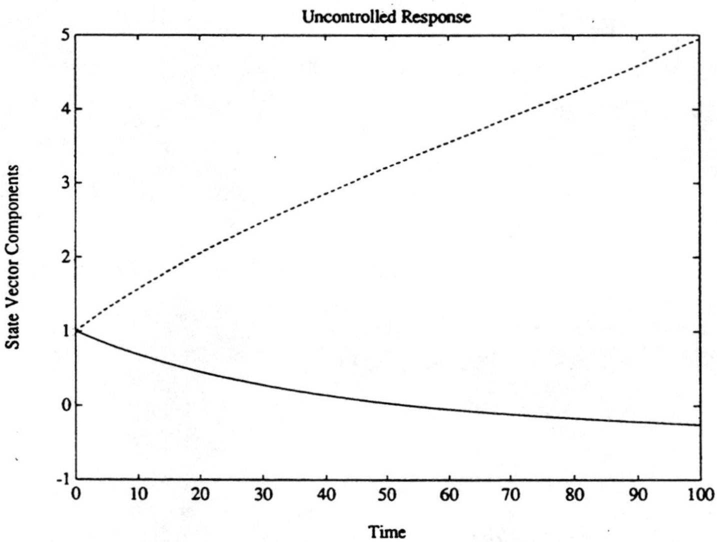

Figure 5.18 Uncontrolled system response. Temperature (- - - -), composition (—).

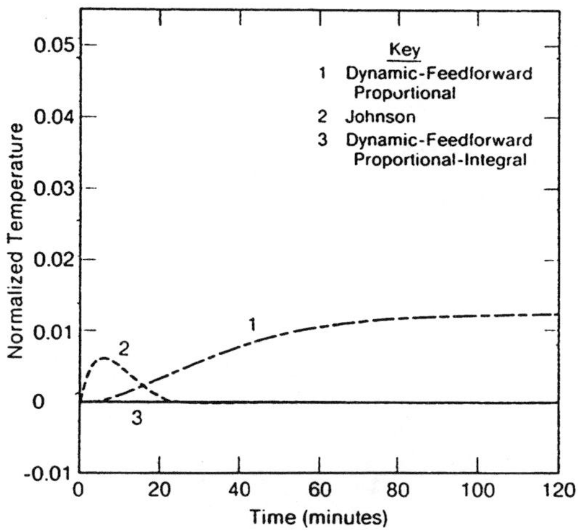

Figure 6.9 Outlet temperature regulated with only inlet temperature measurable. From Ramirez *et al.* (1987).

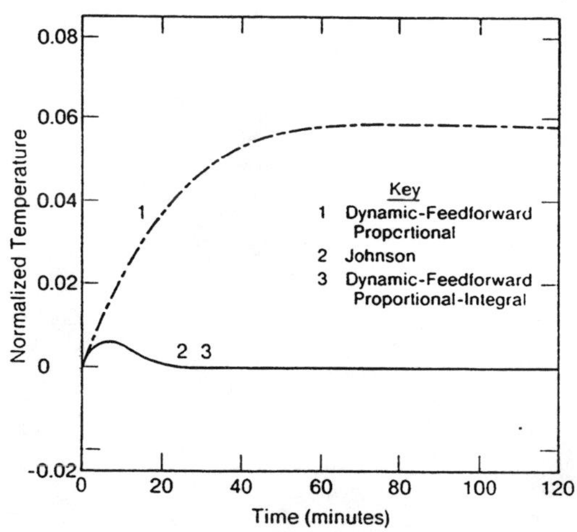

Figure 6.10 Outlet temperature regulated with only inlet concentration measurable. From Ramirez *et al.* (1987).

```
>> y4(12:27,1) = xx41(1:16,1) - xx41(1:16,2);
>> y4(12:27,2) = xx41(1:16,3) - xx41(1:16,4);
>> xx420 = xx41(16,1:16);
>> [t42,xx42] = ode45('xprime',3,50,xx420);
Singularity likely at t = 7.000081
>> size(xx42)

ans =

    18    16

>> y4(28:45,1) = xx42(1:18,1) - xx42(1:18,2);
>> y4(28:45,2) = xx42(1:18,3) - xx42(1:18,4);
>> xx430 = xx42(18,1:16);
>> [t43,xx43] = ode45('xprime',7,50,xx430);
>> size(xx43)

ans =

    36    16

>> y4(46:81,1) = xx43(1:36,1) - xx43(1:36,2);
>> y4(46:81,2) = xx43(1:36,3) - xx43(1:36,4);
>> tp4 = [t3;
          t41;
          t42;
          t43];
>> plot(tt,y3,tp4,y4)
>> plot(tt,y3)
>> meta y1
>> plot(tt,y3,tp4,y4)
>> meta y2
>> diary off
```

Figure 6.23 continued.

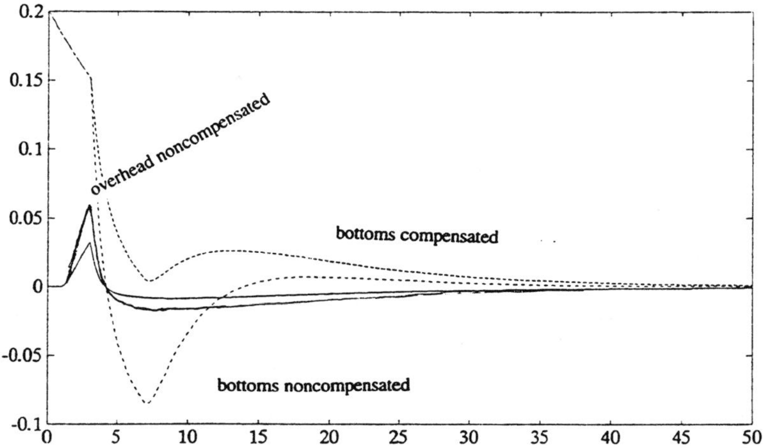

Figure 6.26 Output response to bottoms set point change of −.2 mole fraction.

the general time delayed system given by

$$\dot{\boldsymbol{x}}(t) = \boldsymbol{A}_0\boldsymbol{x}(t) + \sum_{i=1}^{\delta} \boldsymbol{A}_i\boldsymbol{x}(t-\alpha_i) + \boldsymbol{B}_0\boldsymbol{u}(t) + \sum_{j=1}^{\mu} \boldsymbol{B}_j\boldsymbol{u}(t-\beta_j) \tag{6.8.24}$$

For the quadratic performance functional of

$$J = \frac{1}{2}\int_0^\infty \left(\boldsymbol{x}^T(t)\boldsymbol{Q}\boldsymbol{x}(t) + \boldsymbol{u}^T(t)\boldsymbol{R}\boldsymbol{u}(t)\right) dt \tag{6.8.25}$$

the optimal controller takes the form

$$\begin{aligned}\boldsymbol{u} &= -\boldsymbol{R}^{-1}\Big[\left(\boldsymbol{B}_0^T\boldsymbol{P}_0 + \boldsymbol{P}_3(0)\right)\boldsymbol{x} + \int_{-\alpha_\delta}^{0}\left(\boldsymbol{B}_0^T\boldsymbol{P}_1(s) + \boldsymbol{P}_5^T(s,0)\right)\boldsymbol{x}(t+s)ds \\ &\quad + \int_{-\beta_\mu}^{0}\left(\boldsymbol{B}_0^T\boldsymbol{P}_3(s) + \boldsymbol{P}_4(0,s)\right)\boldsymbol{u}(t+s)ds\Big]\end{aligned} \tag{6.8.26}$$

and the Riccati matrices $\boldsymbol{P}_i$ are given by

$$\begin{aligned}\boldsymbol{0} &= -\boldsymbol{P}_0\boldsymbol{A}_0 - \boldsymbol{A}_0^T\boldsymbol{P}_0 - \boldsymbol{P}_1^T(0) - \boldsymbol{P}_1(0) \\ &\quad + \left(\boldsymbol{P}_0\boldsymbol{B}_0 + \boldsymbol{P}_3(0)\right)\boldsymbol{R}^{-1}\left(\boldsymbol{B}_0^T\boldsymbol{P}_0 + \boldsymbol{P}_3^T(0)\right) - \boldsymbol{Q}\end{aligned} \tag{6.8.27}$$

$$\begin{aligned}\boldsymbol{0} &= \frac{d\boldsymbol{P}_1}{ds} - \boldsymbol{A}_0^T\boldsymbol{P}_1(s) - \boldsymbol{P}_2(0,s) \\ &\quad + \left(\boldsymbol{P}_0\boldsymbol{B}_0 + \boldsymbol{P}_3(0)\right)\boldsymbol{R}^{-1}\left(\boldsymbol{B}_0^T\boldsymbol{P}_1(s) + \boldsymbol{P}_5^T(s,0)\right)\end{aligned} \tag{6.8.28}$$

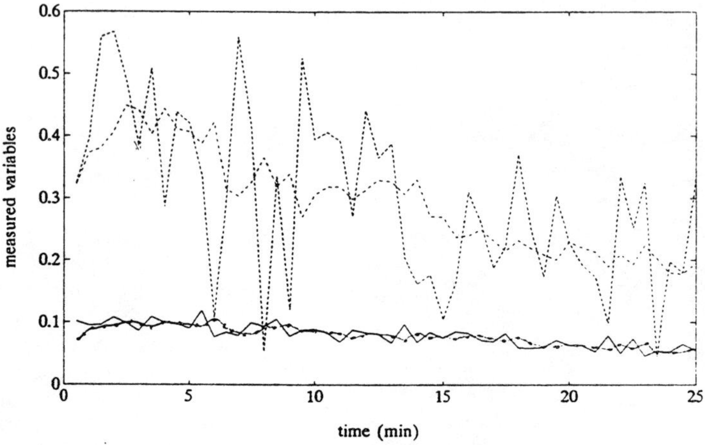

Figure 10.9 Comparison of actual and filtered measurements.

T_m = meas. temperature state (—)
R_m = meas. normalized reaction rate (- - - -)
$\hat{T}$ = filtered temperature state (-·-)
$\hat{R}$ = filtered normalized reaction rate (— · —·)

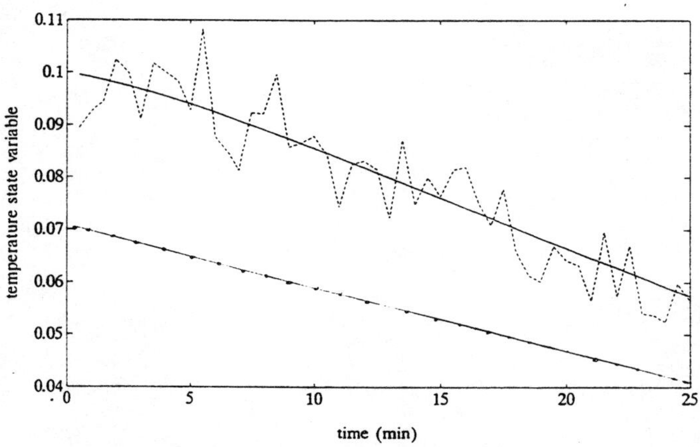

Figure 10.10 Comparison of actual, filtered, and model temperature profiles.

T = actual temperature state (—)
$\hat{T}$ = filtered temperature state (- - - -)
$\overline{T}$ = model temperature state (-·-·)

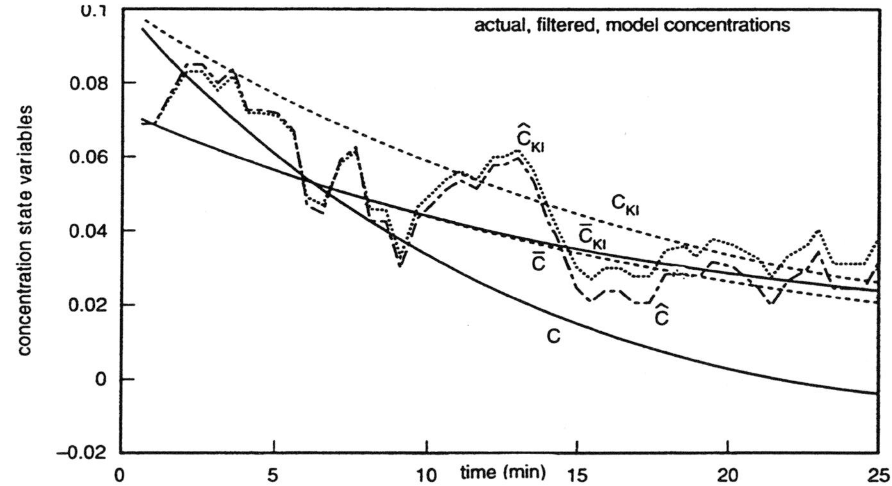

Figure 10.11 Comparison of actual, filtered, and model concentrations.

C	=	actual reactant conc. state	C_{KI}	=	actual catalyst conc. state
$\hat{C}$	=	filtered reactant conc. state	$\hat{C}$	=	filtered catalyst conc. state
$\overline{C}$	=	model reactant conc. state	$\overline{C}$	=	model catalyst conc. state

very weak. Finally there is more discrepancy in the reactant concentration due to the fact that the linearized model did not well represent that state's dynamic response (Figure 10.6). The tracking ability for the reactant concentration can be improved by performing the step by step updating of Equations (10.6.45) to (10.6.47).

10.8 Estimation of Model Parameters

Kalman filtering can be used effectively for the sequential estimation of unknown or uncertain model parameters. This is accomplished by introducing new state variables corresponding to the model parameters to be identified. If the parameters are expected to be constants or slowly varying over the process time domain, then the following state dynamic model is appropriate:

$$\dot{\boldsymbol{a}} = 0 + \boldsymbol{w}_1 \tag{10.8.1}$$

This model simply states that we expect the parameters to be constant and that they have an uncertainty that is characterized by the random variable $\boldsymbol{w}_1$. The random variable vector $\boldsymbol{w}_1$ has an expected value of zero and a covariance

$$\text{cov}\ (\boldsymbol{w}_1) = \boldsymbol{Q}_1 \tag{10.8.2}$$

Equation (4.1.10) shows that the Euler–Lagrange equations are the equations that specify the dynamic Lagrange multipliers or costate variables.

The Constraint Relations

Since the variation, $\boldsymbol{\delta\lambda}$, is free, we have,

$$\frac{\partial L}{\partial \boldsymbol{\lambda}} = \mathbf{0} \tag{4.1.11}$$

which is equivalent to

$$\dot{\boldsymbol{x}} = \boldsymbol{f}(\boldsymbol{x}, \boldsymbol{u}, t) \tag{4.1.12}$$

This states that along the optimal trajectory, the state differential equations must hold.

Optimal Control

Since the variation, $\boldsymbol{\delta u}$, is free, the optimal control policy must be consistent with

$$\frac{\partial L}{\partial \boldsymbol{u}} = \mathbf{0} \tag{4.1.13}$$

or

$$\frac{\partial F}{\partial \boldsymbol{u}} + \left[\frac{\partial \boldsymbol{f}}{\partial \boldsymbol{u}}\right]^T \boldsymbol{\lambda} = \mathbf{0} \tag{4.1.14}$$

Transversality Boundary Conditions

$$-\boldsymbol{\lambda}^T(t_f)\boldsymbol{\delta x}_f + \left[F(t_f) + \boldsymbol{\lambda}^T(t_f)\boldsymbol{f}(t_f)\right]\,\delta t_f = 0 \tag{4.1.15}$$

The transversality conditions specify missing boundary conditions for the costate variables as well as a condition to specify the final time when it is not specified.

The set of necessary conditions can be simplified by the introduction of the Hamiltonian, which is defined as

$$H = F(\boldsymbol{x}, \boldsymbol{u}, t) + \boldsymbol{\lambda}^T \boldsymbol{f}(\boldsymbol{x}, \boldsymbol{u}, t) \tag{4.1.16}$$

Euler–Lagrange Equations

$$\dot{\boldsymbol{\lambda}} = -\frac{\partial H}{\partial \boldsymbol{x}} \tag{4.1.17}$$

Constraint Relations

$$\dot{\boldsymbol{x}} = \boldsymbol{f} = \frac{\partial H}{\partial \boldsymbol{\lambda}} \tag{4.1.18}$$

Optimal Control

$$\frac{\partial H}{\partial \boldsymbol{u}} = \mathbf{0} \tag{4.1.19}$$

Transversality

$$-\boldsymbol{\lambda}^T(t_f)\,\delta\boldsymbol{x}_f + H(t_f)\,\delta t_f = 0 \tag{4.1.20}$$

4.1.1 Minimum Control Effort

Let us apply the necessary conditions of Equations (4.1.17) to (4.1.20) to a specific problem. We will consider the system

$$\dot{\boldsymbol{x}} = \boldsymbol{A}\,\boldsymbol{x} + \boldsymbol{B}\,\boldsymbol{u} \tag{4.1.21}$$

where

$$\boldsymbol{x} = \begin{pmatrix} x_1 \\ x_2 \end{pmatrix} \qquad \boldsymbol{u} = \begin{pmatrix} u_1 \\ u_2 \end{pmatrix}$$

$$\boldsymbol{A} = \begin{bmatrix} 0 & 1 \\ 0 & 0 \end{bmatrix} \qquad \boldsymbol{B} = \begin{bmatrix} 1 & 0 \\ 0 & 1 \end{bmatrix} \tag{4.1.22}$$

We want to determine the control policy, $\boldsymbol{u}(t)$, which minimizes the quadratic minimum control effort objective functional

$$J = \int_0^2 (u_1^2 + u_2^2)\, dt \tag{4.1.23}$$

The initial state is specified

$$x_1(0) = 1 \qquad x_2(0) = 1 \tag{4.1.24}$$

and one final state is specified

$$x_1(2) = 0 \tag{4.1.25}$$

The Hamiltonian is

$$H = u_1^2 + u_2^2 + \lambda_1(x_2 + u_1) + \lambda_2 u_2 \tag{4.1.26}$$

The Euler–Lagrange equations defining the costate variables are

$$\dot{\lambda}_1 = 0 \tag{4.1.27}$$

$$\dot{\lambda}_2 = -\lambda_1 \tag{4.1.28}$$

which have the analytical solutions

$$\lambda_1 = -C_1 \tag{4.1.29}$$

$$\lambda_2 = C_1 t + C_2 \tag{4.1.30}$$

The state constraint relations are

$$\dot{x}_1 = x_2 + u_1 \tag{4.1.31}$$

$$\dot{x}_2 = u_2 \tag{4.1.32}$$

The transversality relation gives

$$\lambda_2(2) = 0 \tag{4.1.33}$$

This relation combined with Equation (4.1.30) yields

$$C_2 = -2C_1 \tag{4.1.34}$$

The optimal control conditions are

$$0 = 2u_1 + \lambda_1 \tag{4.1.35}$$

$$0 = 2u_2 + \lambda_2 \tag{4.1.36}$$

or

$$u_1 = \frac{C_1}{2} \tag{4.1.37}$$

$$u_2 = -\frac{C_1 t}{2} + C_1 \tag{4.1.38}$$

Using Equations (4.1.37) and (4.1.38) for the controls gives the solution of the state equations as

$$x_1 = -\frac{C_1 t^3}{12} + \frac{C_1 t^2}{2} + \left(C_3 + \frac{C_1}{2}\right) t + C_4 \tag{4.1.39}$$

$$x_2 = -\frac{C_1 t^2}{4} + C_1 t + C_3 \tag{4.1.40}$$

Using the specified boundary information allows for the determination of the constants of integration. The final optimal state trajectories are

$$x_1(t) = \frac{t^3}{12} - \frac{t^2}{2} + \frac{t}{2} + 1 \tag{4.1.41}$$

$$x_2(t) = \frac{t^2}{4} - t + 1 \tag{4.1.42}$$

and the optimal controls are

$$u_1 = -\frac{1}{2} \tag{4.1.43}$$

$$u_2(t) = \frac{t}{4} - \frac{1}{2} \tag{4.1.44}$$

4.1.2 Optimal Temperature Profile for a Batch Reactor

We want to obtain the optimal temperature profile for a batch reactor in which the consecutive reactions

$$A \xrightarrow{k_1} B \xrightarrow{k_2} C \tag{4.1.45}$$

are carried out. Both reactions are first order. The material balance relations for species, $A(x_1)$ and $B(x_2)$, are

$$\begin{aligned} \frac{dx_1}{dt} &= f_1 \\ f_1 &= -k_1 x_1 \end{aligned} \tag{4.1.46}$$

$$\begin{aligned} \frac{dx_2}{dt} &= f_2 \\ f_2 &= k_1 x_1 - k_2 x_2 \end{aligned} \tag{4.1.47}$$

with

$$k_1 = k_{10} e^{-E_1/RT} \tag{4.1.48}$$

$$k_2 = k_{20} e^{-E_2/RT} \tag{4.1.49}$$

We desire to maximize the yield of species B, x_2, at the final time, t_f. The objective functional to be maximized is therefore

$$J = \int_{t_o}^{t_f} f_2 \, dt \tag{4.1.50}$$

and the control on the system is the process temperature

$$u = T(t) \tag{4.1.51}$$

The Hamiltonian for this problem is

$$H = \lambda_1 f_1 + (1 + \lambda_2) f_2 \tag{4.1.52}$$

and the necessary Conditions for a maximum of the performance functional are as follows:

Euler–Lagrange Equations

$$\dot{\lambda}_1 = -\lambda_1 \frac{\partial f_1}{\partial x_1} - (1 + \lambda_2) \frac{\partial f_2}{\partial x_1} \tag{4.1.53}$$

$$\dot{\lambda}_2 = -\lambda_1 \frac{\partial f_1}{\partial x_2} - (1 + \lambda_2) \frac{\partial f_2}{\partial x_2} \tag{4.1.54}$$

Optimal Control

$$0 = \lambda_1 \frac{\partial f_1}{\partial u} + (\lambda_2 + 1) \frac{\partial f_2}{\partial u} \tag{4.1.55}$$

Transversality (t_f known and x_f is free)

$$\begin{aligned}\lambda_1(t_f) &= 0 \\ \lambda_2(t_f) &= 0\end{aligned} \tag{4.1.56}$$

And at the final time the Hamiltonian is

$$H(t_f) = f_2(t_f) \tag{4.1.57}$$

In order to make the costate equations homogeneous, we can define two new costate variables as

$$\psi_1 = \lambda_1 \tag{4.1.58}$$

$$\psi_2 = 1 + \lambda_2 \tag{4.1.59}$$

The necessary conditions in terms of these new costate variables are

$$\dot{\psi}_1 = -\psi_1 \frac{\partial f_1}{\partial x_1} - \psi_2 \frac{\partial f_2}{\partial x_1} \tag{4.1.60}$$

$$\dot{\psi}_2 = -\psi_1 \frac{\partial f_1}{\partial x_2} - \psi_2 \frac{\partial f_2}{\partial x_2} \tag{4.1.61}$$

$$0 = \psi_1 \frac{\partial f_1}{\partial u} + \psi_2 \frac{\partial f_2}{\partial u} \tag{4.1.62}$$

$$\psi_1(t_f) = 0 \qquad \psi_2(t_f) = 1 \tag{4.1.63}$$

We can take the time derivative of the optimal control condition of Equation (4.1.62) and using (4.1.60) and (4.1.61), we get

$$\begin{aligned}0 = \psi_1 &\left(-\frac{\partial f_1}{\partial u}\frac{\partial f_1}{\partial x_1} - \frac{\partial f_2}{\partial u}\frac{\partial f_1}{\partial x_2} + \frac{\partial^2 f_1}{\partial u \partial x_1} f_1 + \frac{\partial^2 f_1}{\partial u \partial x_2} f_2 + \frac{\partial^2 f_1}{\partial u^2}\dot{u}\right) \\ +\psi_2 &\left(-\frac{\partial f_1}{\partial u}\frac{\partial f_2}{\partial x_1} - \frac{\partial f_2}{\partial u}\frac{\partial f_2}{\partial x_2} + \frac{\partial^2 f_2}{\partial u \partial x_1} f_1 + \frac{\partial^2 f_2}{\partial u \partial x_2} f_2 + \frac{\partial^2 f_2}{\partial u^2}\dot{u}\right)\end{aligned} \tag{4.1.64}$$

Equations (4.1.64) and (4.1.62) can be solved simultaneously to yield values for the costate variables. These algebraic equations are homogeneous, and a nontrivial solution is possible only when the determinant of the matrix of coefficients is zero. This gives

$$\begin{aligned}&\frac{\partial f_2}{\partial u}\left(-\frac{\partial f_1}{\partial u}\frac{\partial f_1}{\partial x_1} - \frac{\partial f_2}{\partial u}\frac{\partial f_1}{\partial x_2} + \frac{\partial^2 f_1}{\partial u \partial x_1} f_1 + \frac{\partial^2 f_1}{\partial u \partial x_2} f_2 + \frac{\partial^2 f_1}{\partial u^2}\dot{u}\right) \\ &-\frac{\partial f_1}{\partial u}\left(-\frac{\partial f_1}{\partial u}\frac{\partial f_2}{\partial x_1} - \frac{\partial f_2}{\partial u}\frac{\partial f_2}{\partial x_2} + \frac{\partial^2 f_2}{\partial u \partial x_1} f_1 + \frac{\partial^2 f_2}{\partial u \partial x_2} f_2 + \frac{\partial^2 f_2}{\partial u^2}\dot{u}\right) = 0\end{aligned} \tag{4.1.65}$$

Equation (4.1.65) can be solved for $\dot{u}$ to be

$$\dot{u} = \left\{ \frac{\partial f_2}{\partial u} \left(\frac{\partial^2 f_1}{\partial u \partial x_1} f_1 + \frac{\partial^2 f_1}{\partial u \partial x_2} f_2 - \frac{\partial f_1}{\partial u} \frac{\partial f_1}{\partial x_1} - \frac{\partial f_2}{\partial u} \frac{\partial f_1}{\partial x_2} \right) - \frac{\partial f_1}{\partial u} \left(\frac{\partial^2 f_2}{\partial u \partial x_1} f_1 + \frac{\partial^2 f_2}{\partial u \partial x_2} f_2 - \frac{\partial f_1}{\partial u} \frac{\partial f_2}{\partial x_1} - \frac{\partial f_2}{\partial u} \frac{\partial f_2}{\partial x_2} \right) \right\} \Bigg/ \left(\frac{\partial f_1}{\partial u} \frac{\partial^2 f_2}{\partial u^2} - \frac{\partial f_2}{\partial u} \frac{\partial^2 f_1}{\partial u^2} \right) \tag{4.1.66}$$

The transversality conditions are given by Equation (4.1.63), which can be substituted into the optimal control relation of Equation (4.1.62) to give

$$\frac{\partial f_2}{\partial u} = 0 \qquad \text{at } t = t_f \tag{4.1.67}$$

Using the functions of Equations (4.1.46) and (4.1.47) gives for the temperature time derivative

$$\dot{T} = -\frac{RT^2 x_1}{E_2 x_2} k_{10} e^{-E_1/RT} \tag{4.1.68}$$

and the final time condition of

$$T = \frac{(E_1 - E_2)/R}{\ln \left(\dfrac{E_2 k_{20} x_2}{E_1 k_{10} x_1} \right)} \qquad \text{at } t = t_f \tag{4.1.69}$$

The numerical technique for solving this problem is to assume an initial temperature $T(t_o)$ and solve the state differential equations of (4.1.46) and (4.1.47) with the optimal control law of Equation (4.1.68). At the final time, the boundary condition of Equation (4.1.69) is checked. Iteration is continued until the final boundary condition relation is satisfied. This problem was first considered by Bilous and Amundson (1956). Typical results are shown in Figure 4.1.

$k_{10} = 0.535 \text{ x } 10^{11} \text{min}^{-1}$ $\qquad$ $R = 2$ cal/mole°K

$k_{20} = 0.461 \text{ x } 10^{18} \text{min}^{-1}$ $\qquad$ $t_f = 8$ min

$E_1 = 18{,}000$ cal/mole $\qquad$ $x_1(0) = 0.53$ mole/liter

$E_1 = 30{,}000$ cal/mole $\qquad$ $x_2(0) = 0.43$ mole/liter

4.2 Pontryagin's Maximum Principle

In 1962, Pontryagin developed his celebrated maximum principle which allowed for the inclusion of bounds on the control variables (Pontryagin *et al.*,

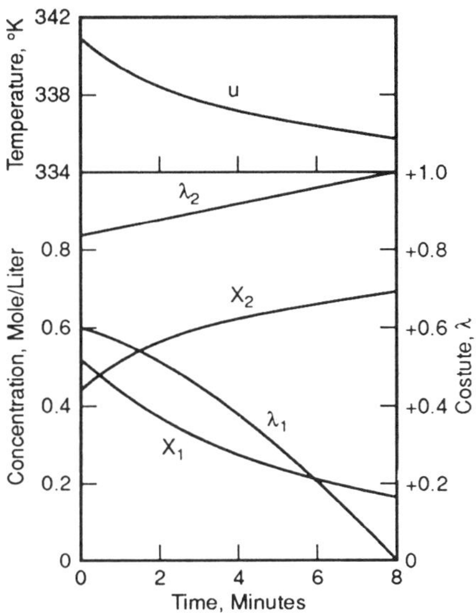

Figure 4.1 Optimal temperature and composition profiles.

1962). The problem that Pontryagin and his colleagues considered was determining the admissible control function that maximizes or minimizes the performance index

$$J = \int_{t_o}^{t_f} F(\boldsymbol{x}, \boldsymbol{u}, t)\, dt \tag{4.2.1}$$

subject to the equality constraints

$$\dot{\boldsymbol{x}} = \boldsymbol{f}(\boldsymbol{x}, \boldsymbol{u}, t) \tag{4.2.2}$$

and the control variable constraint that

$$\boldsymbol{u}(t) \;\; \epsilon\, U \tag{4.2.3}$$

where U is a subset of R^m and $t\, \epsilon [t_o, t_f]$. We assume that the initial state is known, $\boldsymbol{x}(t_o) = \boldsymbol{x}_o$, and that the initial time is known, $t = t_o$.

By definition, the optimal control, $\boldsymbol{u}$, causes the functional, J, to have a relative maximum or minimum. If we consider the case of finding a relative maximum, then, from Section 3.6,

$$\Delta J = J(\boldsymbol{w}) - J(\boldsymbol{u}) \leq 0 \tag{4.2.4}$$

The increment can be expressed in terms of the first variation of the functional as

$$\Delta J = \delta J(\boldsymbol{u}, \boldsymbol{\delta u}) + \boldsymbol{g}(\boldsymbol{u}, \boldsymbol{\delta u}) \, || \, \boldsymbol{\delta u} \, || \tag{4.2.5}$$

As the norm, $||\boldsymbol{\delta u}||$, goes to zero, the function, $\boldsymbol{g}$, which has higher order terms in $\boldsymbol{\delta u}$ goes to zero. Therefore, a necessary condition for a relative maximum is

$$\delta J(\boldsymbol{u}, \boldsymbol{\delta u}) \le 0 \tag{4.2.6}$$

If the optimal control lies on the boundary of the admissible set, then the variation, $\boldsymbol{\delta u}$, is not free. This is illustrated in Figure 4.2. In Section 4.1 we have developed the first variation of the augmented functional for an optimal control problem as Equation (4.1.6). This equation can be written as

$$\delta J_A(\boldsymbol{u}, \boldsymbol{\delta u}) = -\boldsymbol{\lambda}^T(t_f)\, \boldsymbol{\delta x}_f + H(t_f)\, \delta t_f + \int_{t_o}^{t_f} \left(\dot{\boldsymbol{\lambda}} + \frac{\partial H}{\partial \boldsymbol{x}}\right)^T \boldsymbol{\delta x}\, dt$$
$$+ \int_{t_o}^{t_f} \left(\frac{\partial H}{\partial \boldsymbol{\lambda}} - \dot{\boldsymbol{x}}\right)^T \boldsymbol{\delta \lambda}\, dt + \int_{t_o}^{t_f} \left(\frac{\partial H}{\partial \boldsymbol{u}}\right)^T \boldsymbol{\delta u}\, dt \tag{4.2.7}$$

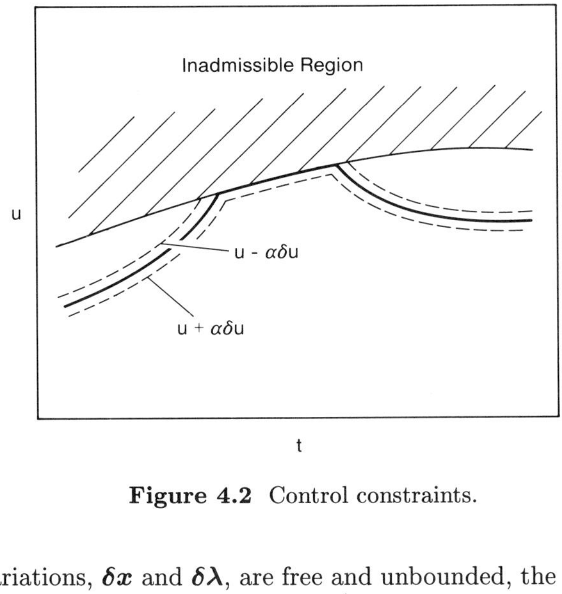

Figure 4.2 Control constraints.

Since the variations, $\boldsymbol{\delta x}$ and $\boldsymbol{\delta \lambda}$, are free and unbounded, the Euler–Lagrange equations, the equality constraint equations, and the transversality conditions are still necessary conditions for an extremum of the functional. If these conditions are satisfied, then the first variation simplifies to

$$\delta J_A(\boldsymbol{u}, \boldsymbol{\delta u}) = \int_{t_o}^{t_f} \left(\frac{\partial H}{\partial \boldsymbol{u}}\right)^T \boldsymbol{\delta u}\, dt \tag{4.2.8}$$

When the control hits the boundary of the admissible control set, the variation, $\boldsymbol{\delta u}$, is not free. The kernel of the integral of Equation (4.2.8) is, to the first order,

$$\left(\frac{\partial H}{\partial \boldsymbol{u}}\right)^T \boldsymbol{\delta u} = H(\boldsymbol{u} + \boldsymbol{\delta u}) - H(\boldsymbol{u}) \tag{4.2.9}$$

or

$$\left(\frac{\partial H}{\partial \boldsymbol{u}}\right)^T \boldsymbol{\delta u} = H(\boldsymbol{w}) - H(\boldsymbol{u}) \tag{4.2.10}$$

Therefore, the first variation is

$$\delta J_A = \int_{t_o}^{t_f} [H(\boldsymbol{w}) - H(\boldsymbol{u})]\, dt \leq 0 \tag{4.2.11}$$

A necessary condition for a maximum is that

$$H(\boldsymbol{x}, \boldsymbol{\lambda}, \boldsymbol{u}) \geq H(\boldsymbol{x}\boldsymbol{\lambda}, \boldsymbol{w}) \tag{4.2.12}$$

Pontryagin's maximum principle states that the optimal control that maximizes the performance functional, J, must maximize the Hamiltonian, H. Likewise, the control that minimizes the performance functional must minimize the Hamiltonian. A necessary condition for a minimum is

$$H(\boldsymbol{x}, \boldsymbol{\lambda}, \boldsymbol{u}) \leq H(\boldsymbol{x}, \boldsymbol{\lambda}, \boldsymbol{w}) \tag{4.2.13}$$

The necessary conditions for an extremum of the functional, J, can be summarized as follows:

Euler–Lagrange Equations Defining the Costate Variables

$$\dot{\boldsymbol{\lambda}} = -\frac{\partial H}{\partial \boldsymbol{x}} \tag{4.2.14}$$

Equality Constraint State Relations

$$\dot{\boldsymbol{x}} = \frac{\partial H}{\partial \boldsymbol{\lambda}} = \boldsymbol{f} \tag{4.2.15}$$

Given Boundary Conditions

$$\boldsymbol{x}(t_o) = \boldsymbol{x}_o \tag{4.2.16}$$

Transversality Boundary Conditions

$$-\boldsymbol{\lambda}^T(t_f)\, \delta \boldsymbol{x}_f + H(t_f)\, \delta t_f = 0 \tag{4.2.17}$$

Optimal Control Conditions

To find a relative maximum in the performance functional,

$$H(\boldsymbol{x}, \boldsymbol{\lambda}, \boldsymbol{u}) \geq H(\boldsymbol{x}, \boldsymbol{\lambda}, \boldsymbol{w}) \tag{4.2.18}$$

To find a relative minimum in the performance functional,

$$H(\boldsymbol{x}, \boldsymbol{\lambda}, \boldsymbol{u}) \leq H(\boldsymbol{x}, \boldsymbol{\lambda}, \boldsymbol{w}) \tag{4.2.19}$$

When the control does not lie on the boundary of the admissible set, U, the variation $\boldsymbol{\delta u}$ is free; therefore,

$$\frac{\partial H}{\partial \boldsymbol{u}} = \mathbf{0} \tag{4.2.20}$$

4.2.1 Numerical Example

To illustrate the use of Pontryagin's maximum principle, let us consider finding the optimal control that minimizes the quadratic functional

$$J = \frac{1}{2}\int_0^2 x^2(t)\,dt \tag{4.2.21}$$

subject to the state equality constraint

$$\dot{x} = u \tag{4.2.22}$$

and the controls bounded by

$$|u(t)| \leq 1 \tag{4.2.23}$$

The specified state boundary conditions are

$$x(0) = 1 \tag{4.2.24}$$

$$x(2) = 1 \tag{4.2.25}$$

The Hamiltonian for this problem is

$$H = \frac{x^2}{2} + \lambda u \tag{4.2.26}$$

The Euler–Lagrange equation is

$$\dot{\lambda} = -\frac{\partial H}{\partial x} = -x \tag{4.2.27}$$

Since the Hamiltonian is linear in the control, u, the following policy will minimize the Hamiltonian with respect to the control,

$$u = +1 \qquad \text{when } \lambda < 0 \tag{4.2.28}$$

$$u = -1 \qquad \text{when } \lambda > 0 \tag{4.2.29}$$

This is illustrated in Figure 4.3, which shows that when H is linear in u and λ is positive, the minimum value of the Hamiltonian is at $u = u_{min}$. When λ is negative, H is a minimum at $u = u_{max}$.

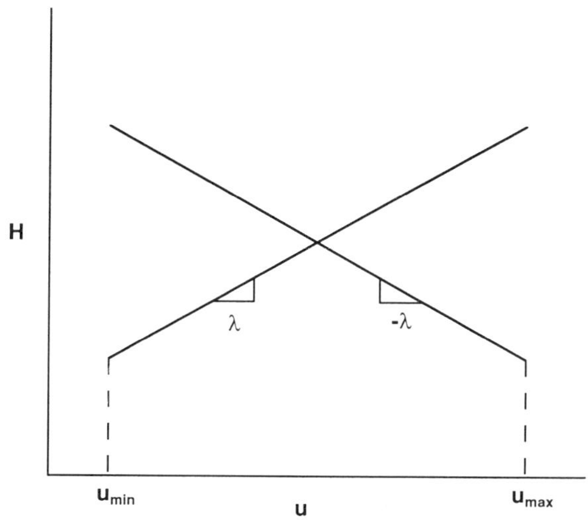

Figure 4.3 Hamiltonian linear in the controls.

The state trajectories under these extremal controls are

$$\text{when } u = +1 \qquad x = t + C_1 \tag{4.2.30}$$

$$\text{when } u = -1 \qquad x = -t + C_2 \tag{4.2.31}$$

The family of curves generated by these policies is illustrated in Figure 4.4. Two possible solutions that satisfy the boundary conditions are also illustrated. One starts with $u = +1$ and then switches to $u = -1$ at $t = 1$. The other starts with $u = -1$ and then switches to $u = +1$ at $t = 1$. Both solutions satisfy the necessary conditions, but the latter policy generates a feasible solution that yields the absolute minimum in the objective functional.

4.2.2 Performance Functionals with Terminal Costs

In this section we want to consider problems where the objective functional has a terminal cost. The performance functional can be written

$$J = h\left(\boldsymbol{x}(t_f), t_f\right) + \int_{t_o}^{t_f} F(\boldsymbol{x}, \boldsymbol{u}, t)\, dt \tag{4.2.32}$$

To develop the necessary conditions for extremizing such a functional, we get extra terms in the transversality conditions since the variations, $\delta \boldsymbol{x}_f$ and δt_f, affect the terminal cost function, $h(\boldsymbol{x}(t_f), t_f)$. The other terms of the necessary

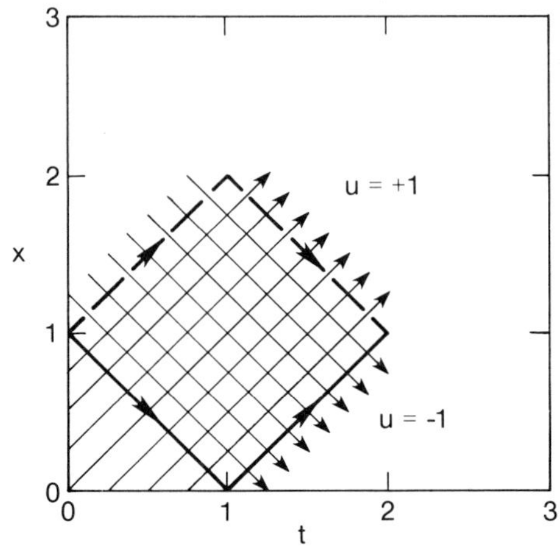

Figure 4.4 Extremal control solution.

conditions remain the same. The necessary conditions for this problem are, therefore, as follows:

Euler–Lagrange Equations Defining the Costate Variables

$$\dot{\boldsymbol{\lambda}} = -\frac{\partial H}{\partial \boldsymbol{x}} \tag{4.2.33}$$

Equality Constraint State Relations

$$\dot{\boldsymbol{x}} = \frac{\partial H}{\partial \boldsymbol{\lambda}} = \boldsymbol{f} \tag{4.2.34}$$

Given Boundary Conditions

$$\boldsymbol{x}(t_o) = \boldsymbol{x}_o \tag{4.2.35}$$

Transversality Boundary Conditions

$$\left[\frac{\partial h}{\partial \boldsymbol{x}}(t_f) - \boldsymbol{\lambda}(t_f)\right]^T \delta \boldsymbol{x}_f + \left(H(t_f) + \frac{\partial h}{\partial t}(t_f)\right) \delta t_f = 0 \tag{4.2.36}$$

Optimal Control Conditions

To find a relative maximum in the performance functional,

$$H(\boldsymbol{x}, \boldsymbol{\lambda}, \boldsymbol{u}) \geq H(\boldsymbol{x}, \boldsymbol{\lambda}, \boldsymbol{w}) \tag{4.2.37}$$

To find a relative minimum in the performance functional,

$$H(\boldsymbol{x}, \boldsymbol{\lambda}, \boldsymbol{u}) \leq H(\boldsymbol{x}, \boldsymbol{\lambda}, \boldsymbol{w}) \tag{4.2.38}$$

When the control does not lie on the boundary of the admissible set, U,

$$\frac{\partial H}{\partial \boldsymbol{u}} = \mathbf{0} \tag{4.2.39}$$

The Kalman problem discussed in Section 5.1 is a good example of a problem with terminal costs.

4.2.3 Problems with a Terminal Manifold or a Target Set

Often the optimal control problem is to hit a terminal manifold or terminal target set. The terminal manifold or target set can be described by the set of algebraic equations

$$\boldsymbol{N}\left(\boldsymbol{x}(t_f), t_f\right) = \mathbf{0} \tag{4.2.40}$$

In order to find the necessary conditions for this class of problems, it is necessary to adjoin the terminal manifold to the performance functional. This introduces new Lagrange multipliers, $\boldsymbol{\nu}$. The modified performance functional is

$$J' = J + \int_{t_o}^{t_f} \boldsymbol{\lambda}^T(\boldsymbol{f} - \dot{\boldsymbol{x}})\, dt + \boldsymbol{\nu}^T \boldsymbol{N}\left(\boldsymbol{x}(t_f), t_f\right) \tag{4.2.41}$$

To find the necessary conditions for optimality, we need to introduce the variation, $\delta\boldsymbol{\nu}$, as well as the standard variations, $\delta\boldsymbol{x}$, $\delta\dot{\boldsymbol{x}}$, $\delta\boldsymbol{\lambda}$, $\delta\boldsymbol{u}$, and δt_f. Since the terminal manifold is a function of $\boldsymbol{x}(t_f)$ and t_f, we get additional terms in the transversality relations of Equation (4.2.36). The new transversality relations are as follows:

Transversality Boundary Conditions

$$\left[\frac{\partial h}{\partial \boldsymbol{x}}(t_f) + \frac{\partial \boldsymbol{N}^T}{\partial \boldsymbol{x}}(t_f)\boldsymbol{\nu} - \boldsymbol{\lambda}(t_f)\right] \delta\boldsymbol{x}_f + \left[H(t_f) + \frac{\partial h}{\partial t}(t_f) + \frac{\partial \boldsymbol{N}^T}{\partial t}(t_f)\boldsymbol{\nu}\right] \delta t_f \tag{4.2.42}$$

Since the variation, $\delta\boldsymbol{\nu}$, is free, we get the additional necessary condition that the terminal manifold equations must hold.

Terminal Manifold Relations

$$\boldsymbol{N}\left(\boldsymbol{x}(t_f), t_f\right) = \mathbf{0} \tag{4.2.43}$$

The time optimal stirred tank reactor problem of Nyquist and Ramirez (1971) discussed in Section 8.1 is a good example of including terminal manifolds.

4.3 State Variable Inequality Constraints

Often control problems have state variable inequality constraints as well as control variable constraints. We will write the state variable inequality constraints as the following set of inequalities:

$$\boldsymbol{g}[\boldsymbol{x}(t), t] \geq \mathbf{0} \qquad\qquad \boldsymbol{g} : R^n \rightarrow R^s \tag{4.3.1}$$

There are a couple of approaches for including these inequality constraints into the variational approach for determining the necessary conditions for extremizing a performance functional. The first is to define a new state variable

$$\dot{x}_{n+1} = f_{n+1} = \sum_{i=1}^{s} g_i^2 U[g_i] \tag{4.3.2}$$

where $U[g_i(\boldsymbol{x}, t)]$ is a step function defined by

$$U[g_i] = \begin{cases} 0 \text{ if } g_i \geq 0 & \text{constraint satisfied} \\ K_i \text{ if } g_i < 0 & \text{constraint violated} \end{cases} \tag{4.3.3}$$

The boundary conditions for this new auxiliary state variable are

$$x_{n+1}(t_o) = 0 \tag{4.3.4}$$

$$x_{n+1}(t_f) = 0 \tag{4.3.5}$$

These boundary conditions force the inequality constraints to be satisfied at both the initial and final times. The differential equation of Equation (4.3.2) requires that the inequality constraints are also satisfied over the entire time domain. We include this new state variable in the optimal control problem by defining a new augmented performance functional

$$J_m = J + \int_{t_o}^{t_f} \left(\boldsymbol{\lambda}^T(\boldsymbol{f} - \dot{\boldsymbol{x}}) + \lambda_{n+1}[f_{n+1} - \dot{x}_{n+1}] \right) dt \tag{4.3.6}$$

There is also a modified Hamiltonian of

$$H_m = F + \boldsymbol{\lambda}^T \boldsymbol{f} + \lambda_{n+1} f_{n+1} \tag{4.3.7}$$

In developing the first variation, δJ_m, we get a defining relation for λ_{n+1} as

$$\dot{\lambda}_{n+1} = -\frac{\partial H_m}{\partial x_{n+1}} = 0 \tag{4.3.8}$$

This term is zero since x_{n+1} does not appear explicitly in H_m.

An alternative approach to including state variable inequality constraints into the optimal control problem is the *penalty function approach*. The penalty function approach is often better computationally. In this approach, we have the original performance functional,

$$J = \int_{t_o}^{t_f} F(\boldsymbol{x}, \boldsymbol{u}, t)\, dt + h\left(\boldsymbol{x}(t_f), t_f\right) \tag{4.3.9}$$

and the usual augmented performance functional that includes the state variable equality constraints,

$$J_A = \int_{t_o}^{t_f} L(\boldsymbol{x}, \boldsymbol{u}, \boldsymbol{\lambda}, t)\, dt + h\left(\boldsymbol{x}(t_f), t_f\right) \tag{4.3.10}$$

with

$$L = F + \boldsymbol{\lambda}^T(\boldsymbol{f} - \dot{\boldsymbol{x}}) \tag{4.3.11}$$

Now we directly adjoin the inequality relations to the performance functional to obtain

$$J_p = \int_{t_o}^{t_f} L_p\, dt + h\left(\boldsymbol{x}(t_f), t_f\right)] \tag{4.3.12}$$

with the penalty Lagrangian defined as

$$L_p = L + \sum_{i=1}^{s} g_i^2 U[g_i] \tag{4.3.13}$$

This new penalty performance functional adds a large penalty to the minimization objective functional whenever the state variable inequality constraints are violated. In doing so, all minimum solutions of the problem will require the inequality constraints to be satisfied. The necessary conditions are as follows:

Euler–Lagrange Equation

$$\dot{\boldsymbol{\lambda}} = -\frac{\partial H_p}{\partial \boldsymbol{x}} \tag{4.3.14}$$

where

$$H_p = F + \boldsymbol{\lambda}^T \boldsymbol{f} + \sum_{i=1}^{s} g_i^2 U[g_i] \tag{4.3.15}$$

Constraint Relations

$$\dot{\boldsymbol{x}} = \boldsymbol{f} \tag{4.3.16}$$

Given Boundary Conditions

$$\boldsymbol{x}(t_o) = \boldsymbol{x}_o \tag{4.3.17}$$

Transversality Boundary Conditions

$$\left[\frac{\partial h}{\partial \boldsymbol{x}}(t_f) - \boldsymbol{\lambda}^T(t_f)\delta \boldsymbol{x}_f\right] + \left(H_p(t_f) + \frac{\partial h}{\partial t}(t_f)\right)\delta t_f = 0 \tag{4.3.18}$$

Optimal Control Conditions

$$H_p(\boldsymbol{x}, \boldsymbol{\lambda}, \boldsymbol{u}) \geq H_p(\boldsymbol{x}, \boldsymbol{\lambda}, \boldsymbol{w}) \qquad \text{for maximum} \tag{4.3.19}$$

$$H_p(\boldsymbol{x}, \boldsymbol{\lambda}, \boldsymbol{u}) \leq H_p(\boldsymbol{x}, \boldsymbol{\lambda}, \boldsymbol{w}) \qquad \text{for minimum} \tag{4.3.20}$$

$$\frac{\partial H_p}{\partial \boldsymbol{u}} = \mathbf{0} \qquad \text{when } \boldsymbol{u} \text{ not extremal} \tag{4.3.21}$$

The batch fermentation problem of Section 8.5 discusses the use of the penalty function technique when considering state variable inequality constraints.

4.4 Optimal Temperature Profile for an Ammonia Reactor

One of the most important industrial chemical reactions is the synthesis of ammonia,

$$N_2 + 3H_2 \rightleftarrows 2NH_3 + \text{heat} \tag{4.4.1}$$

Increasing the conversion is of considerable significance. The process has therefore been studied extensively (Vancini, 1971; Strelzoff, 1981). Here we consider the use of optimal control theory to calculate the temperature profile along the tubular reactor used in order to maximize the yield of ammonia (Mansson and Andresen, 1986). The model reactor is tubular with length L and constant cross-sectional area B. These are effective measures in the sense that the reactor is filled with iron oxide catalyst pellets. Nielsen *et al.* (1964) have shown that the catalytic rate of reaction can be expressed as

$$r = \frac{K_b}{K_c^{1.5}} \left[K_a^2 a_N \left(a_H^3/a_Z^2\right)^{.75} - \left(a_Z^2/a_H^3\right)^{.25} \right] \tag{4.4.2}$$

where K_a = equilibrium constant, which is a function of temperature and K_b and K_c are temperature dependent constants.

$$K_b = K_{bo} \exp(-E_b/RT) \tag{4.4.3}$$

$$K_c = K_{co} \exp(-E_c/RT) \tag{4.4.4}$$

and a_i are species activities where N is nitrogen, H is hydrogen, and Z is ammonia. Mansson and Andresen have related the activities, a_i, to mole fractions as, x_i,

$$a_i = x_i \, P \exp \left[(\beta_i + D_i)P/(RT)^2 \right] \tag{4.4.5}$$

where P is pressure and β_i and D_i are complex functions related to Beattie–Bridgeman equation of state.

For optimization, we want to maximize the conversion at the exit of the tubular reactor. It is therefore necessary to express the mole fractions, x_i, in terms of the conversion, ϕ. Here the compositions are related to the conversion by

$$x_Z = \frac{\phi}{2-\phi} \tag{4.4.6}$$

$$x_H = \frac{3\left[(1+\epsilon)(2-\sigma)-\phi(2+3\epsilon/2)-\epsilon\phi^0/2\right]}{(4+3\epsilon)(2-\phi)} \tag{4.4.7}$$

$$x_N = \frac{2(1-\phi)-\sigma-(3\epsilon/2)(\phi-\phi^0)}{(4+3\epsilon)(2-\phi)} \tag{4.4.8}$$

where

$$\sigma = (2-\phi^0)x_I^0 \tag{4.4.9}$$

and ϵ is defined as the deviation from stoichiometry

$$x_H^0 = 3(1+\epsilon)x_N^0 \tag{4.4.10}$$

The superscript zero indicates initial concentration and x_I^0 is the initial amount of inerts present.

Mansson and Andresen show that a differential material balance on the reactor yields

$$\frac{d\phi}{d\ell} = \frac{2r}{G(1+x_z)}\sum_{i=1}^{3} x_i M_i \tag{4.4.11}$$

where ℓ = the reaction length
G = constant mass flow rate
M_i = molecular weight of species i

The objective functional to maximize is

$$J = \phi(L) \tag{4.4.12}$$

subject to the equality constraint of the system response of Equation (4.4.11), which can be written

$$\frac{d\phi}{d\ell} = \phi' \tag{4.4.13}$$

The Hamiltonian is

$$H = \lambda\phi' \tag{4.4.14}$$

The necessary conditions for optimality are as follows:

Costate Equation

$$\frac{d\lambda}{d\ell} = -\frac{\partial H}{\partial\phi} \tag{4.4.15}$$

Transversality

$$\left(\frac{\partial h(L)}{\partial \phi} - \lambda(L)\right) \delta\phi_L + \left(H(L) + \frac{\partial h(L)}{\partial \ell}\right) \delta\ell(L) = 0 \tag{4.4.16}$$

Since L is fixed and ϕ is free at $\ell = L$, we have

$$\delta\ell(L) = 0 \tag{4.4.17}$$

and

$$\lambda(L) = 1 \tag{4.4.18}$$

Optimal Control

$$\frac{\partial H}{\partial T} = 0 \tag{4.4.19}$$

or

$$\frac{\partial r}{\partial T} = 0 \tag{4.4.20}$$

which is

$$\begin{aligned} &\left(\frac{E_b - 2(.75)E_c}{RT^2}\right) \left[K_a^2 a_N \left(a_H^3/a_Z^2\right)^{.75)} - \left(a_Z^2/a_H^3\right)^{.25}\right] \\ &+\frac{\partial}{\partial T} \left[K_a^2 a_N \left(a_H^3/a_Z^2\right)^{.75} - \left(a_Z^2/a_H^3\right)^{.25}\right] = 0 \end{aligned} \tag{4.4.21}$$

Since the optimal control law does not involve the costate variable λ for this problem, we do not have to compute its profile defined by Equation (4.4.15). All that is necessary is to solve the state differential equation of Equation (4.4.11) along with the optimal control law of Equation (4.4.21) simultaneously. Mansson and Andresen have performed these calculations for an inlet mixture of $x_z = .036$ and a mass flow rate of 3 kg/s, which corresponds to an industrial reactor that produces around 1000 tons/day of ammonia. Figure 4.5 presents the optimal temperature profile as well as the standard industrial temperature profile. These are quite different. The optimal profile starts at extremely high temperatures of over 1000 K. A suboptimal policy that limits the maximum temperature at 800 K is also considered, and it is noted as the T-limit curve. Figure 4.6 shows the ammonia mole fraction profiles for all three cases. There is little difference between the temperature limited and optimal ammonia profiles. These are, however, significantly better than the industrial profile. At the reactor exit, the optimal and suboptimal profiles show a 15% improvement over the industrial case.

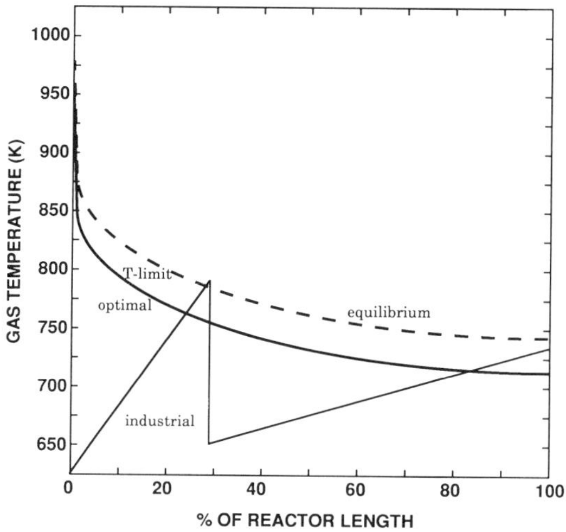

Figure 4.5 Optimal temperature profile and standard temperature profile. Reprinted with permission from Mansson and Andresen, *Ind. Eng. Chem. Process Des. Dev.* **25**, 59–65, copyright ©1986 American Chemical Society.

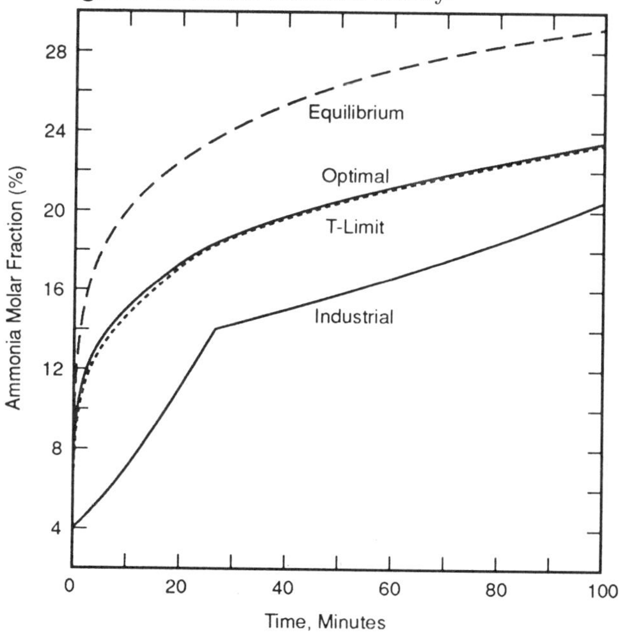

Figure 4.6. Ammonia mole fraction profiles for all three cases. Reprinted with permission from Mansson and Andresen, *Ind. Eng. Chem. Process Des. Dev.* **25**, 59–65, copyright ©1986 American Chemical Society.

4.5 Necessary Conditions for Distributed Parameter Systems

We are now ready to consider the theory of optimal control for systems described by partial differential equations. These systems are known as distributed parameter systems. Supplementary material on the optimal control for distributed parameter systems can be found in the books by Lions (1971) and Ray (1981). We will develop the necessary conditions for optimality for the one-dimensional class of problems described by

$$\frac{\partial \boldsymbol{x}}{\partial t} = \boldsymbol{f}(\boldsymbol{x}, \boldsymbol{x}_z, \boldsymbol{x}_{zz}, \boldsymbol{u}) \qquad \text{in } Q \tag{4.5.1}$$

for

$$0 < t < t_f \qquad \text{or } t \in [0, t_f] \tag{4.5.2}$$

$$0 < z < L \qquad \text{or } z \in D \text{ which is a subset of } R \tag{4.5.3}$$

where

$\boldsymbol{x}$	=	the state vector, which is $\in R^n$
$\boldsymbol{u}$	=	the controls that enter the problem over the spatial domain and are $\in R^m$
z	=	a single spatial independent variable
t	=	the time independent variable
$\boldsymbol{x}_z$	=	the partial of $\boldsymbol{x}$ with respect to z or $\dfrac{\partial \boldsymbol{x}}{\partial z}$
$\boldsymbol{x}_{zz}$	=	the second partial of $\boldsymbol{x}$ with respect to z or $\dfrac{\partial^2 \boldsymbol{x}}{\partial z^2}$
Q	=	the volumetric-time domain of the problem defined as $Q = D\, x\, [0, t_f]$

We have the initial conditions of

$$\boldsymbol{x}(t=0, z) = \boldsymbol{x}_o(z) \tag{4.5.4}$$

and the boundary conditions of

$$\boldsymbol{g}_o(\boldsymbol{x}, \boldsymbol{x}_z, \boldsymbol{v}_o) = \boldsymbol{0} \qquad \text{on } \Sigma_1 \tag{4.5.5}$$

$$\boldsymbol{g}_L(\boldsymbol{x}, \boldsymbol{x}_z, \boldsymbol{v}_L) = \boldsymbol{0} \qquad \text{on } \Sigma_2 \tag{4.5.6}$$

where

$\boldsymbol{g}_o$	and	$\boldsymbol{g}_L$ are nonlinear functions defined on the boundaries, $z = 0$ and $z = L$, respectively
$\boldsymbol{v}_o$	$=$	the controls that enter the problem on the spatial boundary $z = 0$, and are $\epsilon\ R^p$
$\boldsymbol{v}_L$	$=$	the controls that enter the problem on the spatial boundary $z = L$, and are $\epsilon\ R^q$
Σ_1	$=$	the boundary-time domain, which is defined as $\Sigma_1 = 0 \times [0, t_f]$
Σ_2	$=$	the boundary-time domain, which is defined as $\Sigma_2 = L \times [0, t_f]$

The performance functional is in the class of functionals described by

$$J = \int_0^L F_1\left(\boldsymbol{x}(t_f, z), \boldsymbol{x}(t_o, z)\right)\, dz + \int_0^{t_f} F_2\left(\boldsymbol{x}(t, L), \boldsymbol{x}(t, 0), \boldsymbol{v}_o, \boldsymbol{v}_L\right)\, dt + \int_0^{t_f}\int_0^L F(\boldsymbol{x}, \boldsymbol{u}, \boldsymbol{x}_z, \boldsymbol{x}_{zz})\, dz\, dt \tag{4.5.7}$$

The function F_1 represents an initial and terminal contribution to the performance index, the function F_2 represents the boundary contributions, and F represents contributions over the volume domain of the problem.

We are interested in finding the extremum of the functional, J, by choosing optimal values for the boundary controls, $\boldsymbol{v}_o(t)$ and $\boldsymbol{v}_L(t)$, and the volume controls, $\boldsymbol{u}(t, z)$. In order to determine the necessary conditions for the extremum of J, we first formulate an augmented performance functional that includes the equality state dynamic constraints of Equation (4.5.1):

$$J_A = J + \int_0^{t_f}\int_0^L \boldsymbol{\lambda}^T\left(\boldsymbol{f} - \frac{\partial \boldsymbol{x}}{\partial t}\right) dz\, dt \tag{4.5.8}$$

Here $\boldsymbol{\lambda}$ are the costate variables. We next obtain the first variation of the augmented performance functional by introducing variations in $\boldsymbol{\delta x}$, $\boldsymbol{\delta \dot{x}}$, $\boldsymbol{\delta x}_z$, $\boldsymbol{\delta x}_{zz}$, $\boldsymbol{\delta u}$, $\boldsymbol{\delta \lambda}$, $\boldsymbol{\delta v}_o$, $\boldsymbol{\delta v}_L$, and δt_f. As was the case earlier, it is convenient to express the results in terms of the Hamiltonian, H, which is defined as

$$H = F + \boldsymbol{\lambda}^T \boldsymbol{f} \tag{4.5.9}$$

The increment of the augmented functional is

$$\begin{aligned}\Delta J_A = \; & J_A(\boldsymbol{x} + \boldsymbol{\delta x}, \dot{\boldsymbol{x}} + \delta\dot{\boldsymbol{x}}, \boldsymbol{x}_z + \boldsymbol{\delta x}_z, \boldsymbol{x}_{zz} \\ & + \boldsymbol{\delta x}_{zz}, \boldsymbol{u} + \boldsymbol{\delta u}, \boldsymbol{\lambda} + \boldsymbol{\delta\lambda}, \boldsymbol{v}_o \\ & + \boldsymbol{\delta v}_o, \boldsymbol{v}_L + \boldsymbol{\delta v}_L, t_f + \delta t_f) \\ & - J_A(\boldsymbol{x}, \dot{\boldsymbol{x}}, \boldsymbol{x}_z, \boldsymbol{x}_{zz}, \boldsymbol{u}, \boldsymbol{\lambda}, \boldsymbol{v}_o, \boldsymbol{v}_L, t_f)\end{aligned} \tag{4.5.10}$$

The first variation of the augmented functional is

$$\delta J_A = \int_0^L \left[\left(\frac{\partial F_1(f_f, z)}{\partial \boldsymbol{x}} \right)^T \boldsymbol{\delta x}(t_f, z) + \left(\frac{\partial F_1(0, z)}{\partial \boldsymbol{x}_o} \right)^T \boldsymbol{\delta x}_o(z) \right] dz$$

$$+ \int_0^{t_f} \left[\left(\frac{\partial F_2(t, L)}{\partial \boldsymbol{x}} \right)^T \boldsymbol{\delta x}(t, L) + \left(\frac{\partial F_2(t, 0)}{\partial \boldsymbol{x}} \right)^T \boldsymbol{\delta x}(t, 0) \right.$$

$$\left. + \left(\frac{\partial F_2(t, 0)}{\partial \boldsymbol{v}_o} \right)^T \boldsymbol{\delta v}_o(t) + \left(\frac{\partial F_2(t, L)}{\partial \boldsymbol{v}_L} \right)^T \boldsymbol{\delta v}_L(t) \right] dt$$

$$+ \int_0^{t_f} \int_0^L \left[\left(\frac{\partial H}{\partial \boldsymbol{x}} \right)^T \boldsymbol{\delta x} + \left(\frac{\partial H}{\partial \boldsymbol{u}} \right)^T \boldsymbol{\delta u} + \left(\frac{\partial H}{\partial \boldsymbol{x}_z} \right)^T \boldsymbol{\delta x}_z \right.$$

$$\left. + \left(\frac{\partial H}{\partial \boldsymbol{x}_{zz}} \right)^T \boldsymbol{\delta x}_{zz} + \left(\boldsymbol{f} - \frac{\partial \boldsymbol{x}}{\partial t} \right)^T \delta \boldsymbol{\lambda} - \boldsymbol{\lambda}^T \boldsymbol{\delta \dot{x}} \right] dz\, dt$$

$$+ \left\{ \int_0^L \left[\left(\frac{\partial F_1(t_f, z)}{\partial \boldsymbol{x}} \right)^T \dot{\boldsymbol{x}}(t_f) + \left(\frac{\partial F_1(t_f, z)}{\partial t} \right)^T + F(t_f, z) \right. \right.$$

$$\left. \left. + \boldsymbol{\lambda}^T(t_f, z) \left(\boldsymbol{f}(t_f, z) - \dot{\boldsymbol{x}}(t_f, z) \right) \right] dz + F_2(t_f, L) \right\} \delta t_f \qquad (4.5.11)$$

The variations, $\boldsymbol{\delta x}_z$, $\boldsymbol{\delta x}_{zz}$, and $\boldsymbol{\delta \dot{x}}$, are not independent and can be expressed in terms of the variations, $\boldsymbol{\delta x}$, by integrating the following three terms by parts

$$\int_0^L \int_0^{t_f} \boldsymbol{\lambda}^T \quad \boldsymbol{\delta \dot{x}}\, dt\, dz = \textstyle\int_0^L \boldsymbol{\lambda}^T \boldsymbol{\delta x} \Big|_0^{t_f} dz$$

$$- \int_0^L \int_0^{t_f} \frac{\partial \boldsymbol{\lambda}^T}{\partial t} \boldsymbol{\delta x}\, dt\, dz \qquad (4.5.12)$$

$$\int_0^{t_f} \int_0^L \frac{\partial H}{\partial \boldsymbol{x}_z} \boldsymbol{\delta x}_z\, dz\, dt = \int_0^{t_f} \left(\frac{\partial H}{\partial \boldsymbol{x}_z} \right)^T \boldsymbol{\delta x} \Big|_0^L dt$$

$$- \int_0^{t_f} \int_0^L \frac{\partial}{\partial z} \left(\frac{\partial H}{\partial \boldsymbol{x}_z} \right)^T \boldsymbol{\delta x}\, dz\, dt \qquad (4.5.13)$$

$$\int_0^{t_f} \int_0^L \left(\frac{\partial H^T}{\partial \boldsymbol{x}_{zz}} \right) \boldsymbol{\delta x}_{zz}\, dz\, dt$$

$$= \int_0^{t_f} \left[\left(\left(\frac{\partial H}{\partial \boldsymbol{x}_{zz}} \right)^T \frac{\partial \boldsymbol{\delta x}}{\partial z} \right) \Big|_0^L - \left(\frac{\partial}{\partial z} \left(\frac{\partial H}{\partial \boldsymbol{x}_{zz}} \right)^T \boldsymbol{\delta x} \right) \Big|_0^L \right] dt$$

$$+ \int_0^{t_f} \int_0^L \frac{\partial^2}{\partial z_2} \left(\frac{\partial H}{\partial \boldsymbol{x}_{zz}} \right)^T \boldsymbol{\delta x}\, dz\, dt \qquad (4.5.14)$$

Using these three expressions, the first variation of Equation (4.5.11) becomes

$$\delta J_A = \int_0^{t_f}\int_0^L \left\{ \left[\frac{\partial H}{\partial \boldsymbol{x}} - \frac{\partial}{\partial z}\left(\frac{\partial H}{\partial \boldsymbol{x}_z}\right) + \frac{\partial^2}{\partial z^2}\left(\frac{\partial H}{\partial \boldsymbol{x}_{zz}}\right) + \frac{\partial \boldsymbol{\lambda}}{\partial t}\right]^T \boldsymbol{\delta x} + \left(\frac{\partial H}{\partial \boldsymbol{u}}\right)^T \boldsymbol{\delta u} \right.$$
$$\left. + \left(\boldsymbol{f} - \frac{\partial \boldsymbol{x}}{\partial t}\right)^T \delta\boldsymbol{\lambda} \right\} dz\, dt$$
$$+ \int_0^{t_f} \left\{ \left[\frac{\partial F_2(t,L)}{\partial \boldsymbol{x}} + \frac{\partial H(t,L)}{\partial \boldsymbol{x}_z} - \frac{\partial}{\partial z}\frac{\partial H(t,L)}{\partial \boldsymbol{x}_{zz}}\right]^T \boldsymbol{\delta x}(t,L) \right.$$
$$+ \left(\frac{\partial F_2(t,L)}{\partial \boldsymbol{v}_L}\right)^T \boldsymbol{\delta v}_L(t) + \left(\frac{\partial F_2(t,0)}{\partial \boldsymbol{v}_o}\right)^T \boldsymbol{\delta v}_o(t) + \left(\frac{\partial H}{\partial \boldsymbol{x}_{zz}}\right)^T \boldsymbol{\delta x}_z \Bigg|_0^L$$
$$\left. + \left[\frac{\partial F_2(t,0)}{\partial \boldsymbol{x}} - \frac{\partial H(t,0)}{\partial \boldsymbol{x}_z} + \frac{\partial H(t,0)}{\partial \boldsymbol{x}_{zz}}\right]^T \boldsymbol{\delta x}(t,0) \right\} dt$$
$$+ \int_0^L \left\{ \left[\frac{\partial F_1(t_f,z)}{\partial \boldsymbol{x}} - \boldsymbol{\lambda}(t_f,z)\right]^T \boldsymbol{\delta x}(t_f,z) \right.$$
$$\left. + \left[\frac{\partial F_1(0,z)}{\partial \boldsymbol{x}_o} + \boldsymbol{\lambda}(0,z)\right]^T \boldsymbol{\delta x}_o(z) \right\} dz + \left\{ \int_0^L \left[\left(\frac{\partial F_1(t_f,z)}{\partial \boldsymbol{x}}\right)^T \dot{\boldsymbol{x}}(t_f,z) \right.\right.$$
$$\left.\left. + \frac{\partial F_1(t_f,z)}{\partial t} + F(t_f,z) + \boldsymbol{\lambda}^T(t_f,z)\left(\boldsymbol{f}(t_f,z) - \dot{\boldsymbol{x}}(t_f,z)\right)\right] dz + F_2(t_f,L) \right\} \delta t_f \tag{4.5.15}$$

Applying Pontryagin's maximum principle, the necessary conditions for an extremum of J_A are as follows:

Definition of the Costate Equations

$$\frac{\partial \boldsymbol{\lambda}}{\partial t} = -\left[\frac{\partial H}{\partial \boldsymbol{x}} - \frac{\partial}{\partial z}\left(\frac{\partial H}{\partial \boldsymbol{x}_z}\right) + \frac{\partial^2}{\partial z^2}\left(\frac{\partial H}{\partial \boldsymbol{x}_{zz}}\right)\right] \tag{4.5.16}$$

Transversality Boundary Conditions

$$\left[\frac{\partial F_2}{\partial \boldsymbol{x}} + \frac{\partial H}{\partial \boldsymbol{x}_z} - \frac{\partial}{\partial z}\left(\frac{\partial H}{\partial \boldsymbol{x}_{zz}}\right)\right]^T \boldsymbol{\delta x}$$
$$+ \left(\frac{\partial F_2}{\partial \boldsymbol{v}_L}\right)^T \boldsymbol{\delta v}_L + \left(\frac{\partial H}{\partial \boldsymbol{x}_{zz}}\right)^T \boldsymbol{\delta x}_z = 0 \qquad \text{at } z = L \tag{4.5.17}$$

$$\left[\frac{\partial F_2}{\partial \boldsymbol{x}} - \frac{\partial H}{\partial \boldsymbol{x}_z} + \frac{\partial}{\partial z}\left(\frac{\partial H}{\partial \boldsymbol{x}_{zz}}\right)\right]^T \boldsymbol{\delta x}$$
$$+ \left(\frac{\partial F_2}{\partial \boldsymbol{v}_o}\right)^T \boldsymbol{\delta v}_o - \left(\frac{\partial H}{\partial \boldsymbol{x}_{zz}}\right)^T \boldsymbol{\delta x}_z = 0 \qquad \text{at } z = 0 \tag{4.5.18}$$

Equations (4.5.17) and (4.5.18) are used along with the variational form of the boundary conditions on the state system, namely Equations (4.5.5) and (4.5.6), to give the final costate boundary conditions. The variational form of the state boundary conditions specifies equality constraint relations between $\boldsymbol{\delta x}$, $\boldsymbol{\delta v}_o$, $\boldsymbol{\delta v}_L$, and $\boldsymbol{\delta x}_z$ at both $z = 0$ and $z = L$.

Transversality Final Conditions

$$\int_0^L \left(\frac{\partial F_1}{\partial \boldsymbol{x}} - \boldsymbol{\lambda}\right)^T \boldsymbol{\delta x}\, dz + \left\{\int_0^L \left[\left(\frac{\partial F_1}{\partial \boldsymbol{x}}\right)^T \dot{\boldsymbol{x}} + \frac{\partial F_1}{\partial t} + F + \boldsymbol{\lambda}^T(\boldsymbol{f} - \dot{\boldsymbol{x}})\right] dz + F_2\right\} \delta t_f = 0 \quad (4.5.19)$$

$$\text{at } t = t_f$$

Optimal Control

With the first three necessary conditions being satisfied, the first variation becomes

$$\delta J_A = \int_0^{t_f} \int_0^L \left(\frac{\partial H}{\partial \boldsymbol{u}}\right)^T \boldsymbol{\delta u}\, dz\, dt \qquad (4.5.20)$$

If the variation, $\boldsymbol{\delta u}$, is not constrained, then the necessary condition for an extremum is

$$\frac{\partial H}{\partial \boldsymbol{u}} = \boldsymbol{0} \qquad (4.5.21)$$

If the variation, $\boldsymbol{\delta u}$, is constrained, which means that the control is at a constraint boundary, then the necessary condition for maximizing the performance functional is

$$\text{Max } H \qquad \text{with respect to } \boldsymbol{u} \qquad (4.5.22)$$

If we seek a minimum in the performance functional, then

$$\text{Min } H \qquad \text{with respect to } \boldsymbol{u} \qquad (4.5.23)$$

State Equations

$$\frac{\partial \boldsymbol{x}}{\partial t} = \boldsymbol{f}(\boldsymbol{x}, x_z, \boldsymbol{x}_{zz}, \boldsymbol{u}) \qquad (4.5.24)$$

4.5.1 Temperature Control of a Slab

As an example of using the necessary conditions to find the optimal control policy for a distributed parameter system, let us consider the problem of adjusting the temperature distribution of a homogeneous slab to a desired temperature distribution by the control of the heat flux to the slab through the adjustment of the dimensionless environmental temperature, $v(t)$. The dimensionless temperature, $x(t,z)$, of the slab satisfies the linear heat conduction equation

$$\frac{\partial x}{\partial t} = \frac{\partial^2 x}{\partial z^2} \tag{4.5.25}$$

with the initial condition

$$x(z) = 0 \qquad \text{at } t = 0 \tag{4.5.26}$$

and the boundary conditions

$$\frac{\partial x}{\partial z} = \rho(x - v) \qquad \text{at } z = 0 \tag{4.5.27}$$

$$\frac{\partial x}{\partial z} = 0 \qquad \text{at } z = 1 \tag{4.5.28}$$

The objective functional to be minimized is

$$J = \int_0^1 \Big(x^*(z) - x(t_f, z)\Big)^2 \, dz + \frac{1}{2}\int_0^{t_f} c^2 v^2 \, dt \tag{4.5.29}$$

The final dimensionless time is $t_f = 0.4$, and the final desired temperature of the slab is $x^*(z) = 0.2$. The objective functional keeps the final temperature profile close to a desired profile and minimizes the amount of cooling control used at the boundary, $z = 0$.

The Hamiltonian for this problem is

$$H = \lambda x_{zz} \tag{4.5.30}$$

The terminal contribution to the performance functional is

$$F_1 = \Big(x^* - x(t_f)\Big)^2 \tag{4.5.31}$$

The boundary contribution to the performance functional is

$$F_2 = \frac{1}{2}\, c^2 v^2 \tag{4.5.32}$$

The costate equation is

$$\frac{\partial \lambda}{\partial t} = -\frac{\partial^2 \lambda}{\partial z^2} \tag{4.5.33}$$

The transversality boundary conditions are

$$-\frac{\partial \lambda}{\partial z}\,\delta x + \lambda\,\delta x_z = 0 \qquad \text{at } z = 1 \tag{4.5.34}$$

and

$$\frac{\partial \lambda}{\partial z}\,\delta x + c^2 v\,\delta v - \lambda\,\delta x_z = 0 \qquad \text{at } z = 0 \tag{4.5.35}$$

The boundary condition at $z = 1$ is $x_z = 0$. The variational form of this equation is

$$\delta x_z = 0 \qquad \text{at } z = 1 \tag{4.5.36}$$

Combining Equation (4.5.36) with the transversality boundary condition of Equation (4.5.34) gives

$$\frac{\partial \lambda}{\partial z}\,\delta x = 0 \qquad \text{at } z = 1 \tag{4.5.37}$$

Since δx is free and unconstrained at $z = 1$, then

$$\frac{\partial \lambda}{\partial z} = 0 \qquad \text{at } z = 1 \tag{4.5.38}$$

The variational form of the boundary condition, $z = 0$, given by Equation (4.5.27) is

$$\delta x_z = \rho\,\delta x - \rho\,\delta v \qquad \text{at } z = 0 \tag{4.5.39}$$

Combining this equation with the transversality boundary condition of Equation (4.5.35) gives

$$\left(\frac{\partial \lambda}{\partial z} - \rho\lambda\right)\delta x + (c^2 v + \rho\lambda)\,\delta v = 0 \qquad \text{at } z = 0 \tag{4.5.40}$$

Since the variations, δx and δv, are independent, then

$$\frac{\partial \lambda}{\partial z} - \rho\lambda = 0 \qquad \text{at } z = 0 \tag{4.5.41}$$

and

$$(c^2 v + \rho\lambda)\,\delta v = 0 \qquad \text{at } z = 0 \tag{4.5.42}$$

must hold. Equation (4.5.41) serves as a boundary condition for the costate equation, while Equation (4.5.42) is the condition that specifies the optimal boundary control policy.

The final time transversality condition is

$$\int_0^L \left(-2\left(x^* - x(t_f)\right) - \lambda\right)\delta x(t_f)\,dz = 0 \qquad \text{at } t = t_f \tag{4.5.43}$$

Since the final variation, $\delta x(t_f)$, is free, then the necessary condition becomes

$$\lambda = -2\left(x^* - x(t_f)\right) \qquad \text{at } t = t_f \tag{4.5.44}$$

This problem does not have an analytical solution for arbitrary functions, $x^*(z)$, so an iterative numerical solution strategy must be used to solve this split boundary value problem where the state is specified at the initial time and the costate specified at the final time. A control vector iterative method has been shown to be effective (Denn, 1969).

The control vector iterative method is as follows:

1. Select a trial or nominal control policy of $v^o(t)$.

2. Integrate the state equation forward in time with the initial condition and the appropriate boundary conditions. An implicit finite difference solution based upon the Crank–Nicolson second-order correct analogs (Ramirez, 1989) is convenient.

3. Integrate the costate equation backward in time with the specified final condition of Equation (4.5.44) and the boundary conditions of Equations (4.5.37) and (4.5.41). The final condition of Equation (4.5.44) is dependent upon the solution of the state equation. Again, an implicit finite difference method can be used.

4. Compute a new boundary control,

$$v = v^o(t) + \delta v \tag{4.5.45}$$

   which minimizes the first variation

$$\delta J = \int_{t_o}^{t_f} (\rho\lambda + c^2 v)\Big|_{z=0} \delta v \, dt \tag{4.5.46}$$

To insure that the first variation is always minimized, we choose the boundary control variation, δv, in the gradient direction,

$$\delta v = -w(\rho\lambda + c^2 v)\Big|_{z=0} \tag{4.5.47}$$

This forces δJ to a minimum when w is chosen positive.

Figure 4.7 shows the optimal control, $v(t)$, for this problem and the resulting optimal state temperature profile at the final time. It is very close (within 5%) of the desired profile of $x^*(z) = 0.2$.

4.6 Surfactant Flooding Enhanced Oil Recovery Optimization

In this section we consider the optimization of a surfactant flooding process being conducted in a linear experimental core. It is illustrated in Figure 4.8.

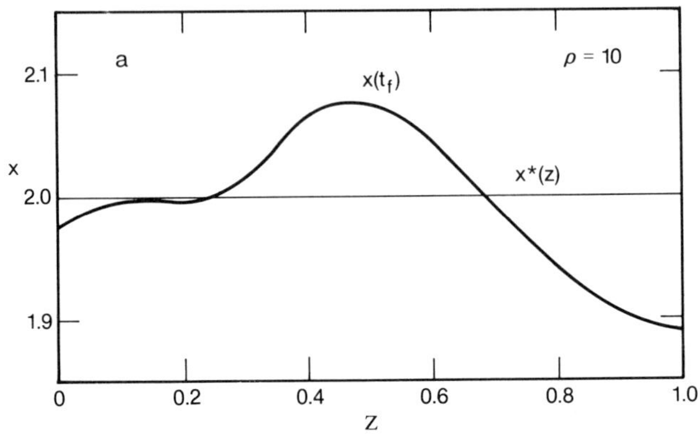

(a)

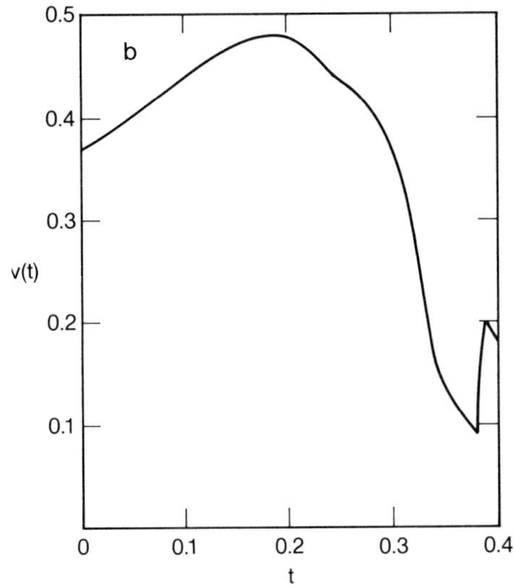

(b)

Figure 4.7(a) Optimal state profile at final time.

(b) Optimal control.

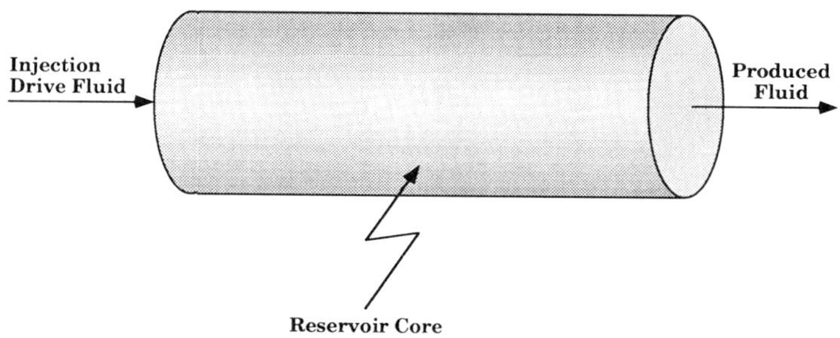

Figure 4.8 Linear core flood experiment.

This enhanced oil recovery process works on the following principles. A low interfacial tension, surfactant-water flooding process is primarily based upon the formation of very low interfacial tensions between the water phase and the reservoir crude by the addition of surfactants. The creation of very low interfacial tensions in the porous media will allow the viscous forces associated with the flow of an injected drive fluid to overcome the capillary forces holding the oil in place. Consequently, the residual oil left from a normal water flood (up to 50 to 60%) can be mobilized. The movement of the oil and water phases is characterized by the simultaneous flow of two immiscible fluids in a porous media. The mass transport of the surfactant solute is controlled by convection, dispersion, adsorption to the rock, and partitioning between the fluid phases.

Before any process is considered for field testing, extensive laboratory core flooding experiments are performed. Once a feasible fluid system has been established, that system should be optimized. Therefore, the first application of an optimization methodology would be the optimization of linear core experiments. The optimal injection policies determined for linear core applications can be easily verified experimentally. This problem has been discussed in the literature by Fathi (1981), Ramirez *et al.* (1984), Fathi and Ramirez (1984), Fathi and Ramirez (1985), and Ramirez (1987).

A two-phase (water and oil) mathematical model for surfactant flooding is presented in order to simulate one-dimensional core flooding laboratory results. This model describes the flow of both aqueous and oil phases through a one-dimensional porous media core. The model allows for the adsorption of surfactant onto the solid rock matrix, surfactant partitioning between the fluid phases, and the convective and dispersive mechanisms of mass transport.

The equation of continuity for the flow of two incompressible fluids in one spatial variable can be expressed as

$$\frac{\partial S_w}{\partial t} = -\frac{\partial f_w}{\partial z} \tag{4.6.1}$$

where S_w is the water saturation
f_w is the fractional flow of water
t is the number of pore volumes of fluid injected $t = t^* Q_T / Al\phi$
t^* is the actual time
Q_T is the constant total volumetric flow rate
A is the cross-sectional area of the core
L is the core length
ϕ is the rock porosity
z is the nondimensional length coordinate $z = z^*/L$
z^* is the actual core length

The equation of motion results in the following algebraic equation for the fractional flow of water:

$$f_w = \frac{1}{1 + \dfrac{k_{ro}\mu_w}{k_{rw}\mu_o}} \tag{4.6.2}$$

where k_{ro} and k_{rw} are oil and water relative permeabilities and μ_o and μ_w are oil and water phase viscosities.

The solute mass balance for the surfactant species is given by

$$\left[S_w + \frac{(1 - S_w)}{K_s}\right] \frac{\partial C}{\partial t} + \left[f_w + \frac{(1 - f_w)}{K_s}\right] \frac{\partial C}{\partial z} = \frac{1}{Pe} \frac{\partial^2 C}{\partial z^2} - \frac{A_{wr}}{\phi C_o} \frac{\partial \Gamma}{\partial t} \tag{4.6.3}$$

where C is the nondimensional surfactant concentration $C = C_w / C_o$
C_w is the concentration of surfactant in the water phase
C_o is a characteristic surfactant concentration
K_s is the surfactant partition coefficient
A_{wr} is the rock specific volume times the ratio of the rock density to the fluid density $A = (1 - \phi)\rho_{\text{rock}} / \rho_{\text{soln}}$
Γ is the amount of surfactant adsorbed per unit weight of the core
Pe is the Peclet number which is defined as $Pe = Lv/D$
v is the phase velocity
D is the dispersion coefficient

A Langmuir equilibrium isotherm is used to model the adsorption of the surfactant onto the solid surface,

$$\Gamma = \frac{C_o C}{M + N C_o C} \tag{4.6.4}$$

The initial conditions for the problem are

$$C = 0 \qquad \text{at } t = 0 \tag{4.6.5}$$

$$S_w = 1 - S_{OR0} \qquad \text{at } t = 0 \tag{4.6.6}$$

where S_{OR0} is a known initial residual oil saturation.

The boundary condition for the water saturation states that, at the inlet of the core, only a water phase is injected,

$$f_w = 1.0 \qquad \text{at } z = 0 \tag{4.6.7}$$

This means that

$$S_w = 1 - S_{OR} \qquad \text{at } z = 0 \tag{4.6.8}$$

The boundary conditions for the surfactant concentration are given by the Danckwerts conditions which are applicable for a solute species undergoing dispersion:

$$\frac{\partial C}{\partial z} = Pe(C - C_{\text{slug}}) \quad \text{at } z = 0 \tag{4.6.9}$$

$$\frac{\partial C}{\partial z} = 0 \qquad \text{at } z = 1 \tag{4.6.10}$$

where C_{slug} is the nondimensional surfactant concentration of the injected slug. The inlet condition is a flux balance assuming that there is no dispersion before entry into the core. The exit condition assumes equal dispersion just before and after exiting.

When there is no surfactant in the injected slug, then the inlet boundary condition is

$$C = 0 \qquad \text{at } z = 0 \tag{4.6.11}$$

It is also necessary to establish realistic relations that express the dependence of the fluid viscosities, interfacial tension, residual saturations, and relative permeabilities on the solute concentration.

The oil phase viscosity is usually assumed constant. The water phase viscosity is a function of surfactant concentration and can be modeled as

$$\mu_w = \exp(C_1 C) \qquad \mu_w \leq \mu_{\text{wmax}} \tag{4.6.12}$$

The interfacial tension between the oil and aqueous phases is assumed to be only a function of total surfactant concentration. Data can be adequately fit by straight line segments on a semilog plot. Therefore, the functional behavior is

$$\ln\, \sigma_{wo} = a_i + b_i \left(C S_w + C(1 - S_w)/K_s\right) \tag{4.6.13}$$

Typical data and a three segment fit are shown in Figure 4.9.

Based upon the work of Taber (1969), the irreducible water and residual oil saturations can be expressed as local functions of the capillary number, N_{cap}:

$$N_{\text{cap}} = \frac{v_w \mu_w}{\sigma_{wo}} \tag{4.6.14}$$

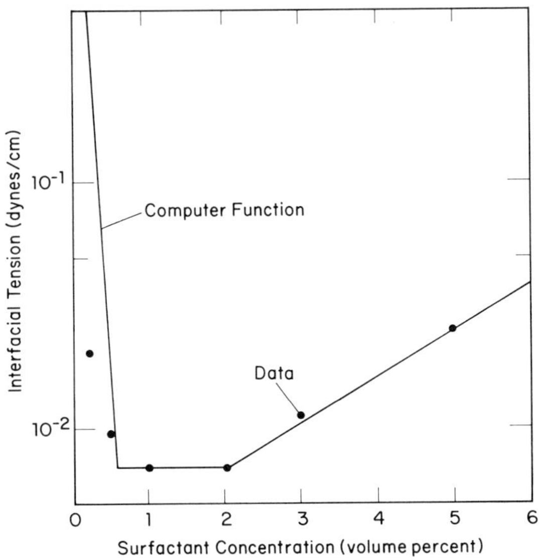

Figure 4.9 Interfacial tension behavior. From Ramirez (1987).

A specific correlation is shown in Figure 4.10. Below the lower critical capillary number of 1×10^{-5}, no enhanced oil recovery occurs. Above the upper critical capillary number of 1×10^{-2}, complete oil recovery can be obtained. The intermediate range is given by a straight line relation between S_{or} and ln N_{cap}.

Functions relating values of the relative permeabilities to the irreducible water and residual oil saturations are

$$k_{ro} = \left[\exp(4.5\ S_{OR}\right] (1 - S_{OR} - S_w)^{1.5+S_{OR}/S_{OR0}} \tag{4.6.15}$$

$$k_{rw} = (S_w - S_{OR})^{1.5+S_{OR}/S_{OR0}} \tag{4.6.16}$$

Figure 4.11 shows these relative permeability functions for $S_{or} = 0.3$ and $S_{or} = 0.0$ when S_{OR0} is 0.3.

If the oil saturation is less than that allowed by the current capillary number, then the oil left behind is trapped in place. This constraint is expressed mathematically as

$$\text{if } S_w \geq 1 - S_{OR}\ , \qquad \text{then} \quad k_{ro} = 0 \tag{4.6.17}$$

There is also a restriction on how low S_w can go in the region where the oil

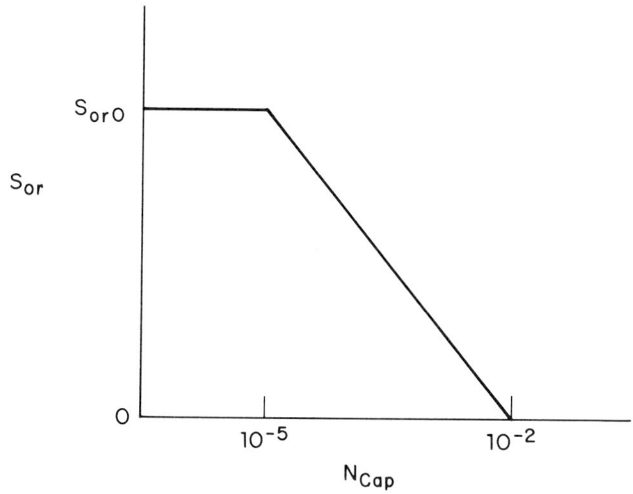

Figure 4.10 Residual oil saturation vs. capillary number. From Ramirez (1987).

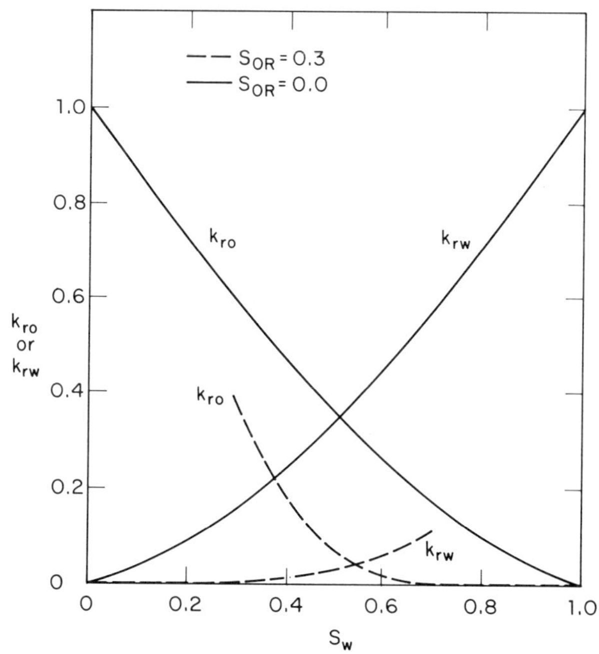

Figure 4.11 Relative permeability curves. From Ramirez (1987).

bank is formed,

$$\text{if } S_w \le S_{OR} \ , \qquad \text{then} \quad k_{rw} = 0 \tag{4.6.18}$$

The two phase surfactant flooding enhanced oil recovery model of Equations (4.6.1) and (4.6.2) can be expressed in the following matrix form:

$$\begin{pmatrix} \dfrac{\partial x_1}{\partial t} \\ \dfrac{\partial x_2}{\partial t} \end{pmatrix} = \begin{bmatrix} -\dfrac{C_2}{C_1} & 0 \\ -\dfrac{\partial f_w}{\partial x_1} & -\dfrac{\partial f_w}{\partial x_2} \end{bmatrix} \begin{pmatrix} \dfrac{\partial x_1}{\partial z} \\ \dfrac{\partial x_2}{\partial z} \end{pmatrix} + \begin{bmatrix} \dfrac{1}{PeC_1} & 0 \\ 0 & 0 \end{bmatrix} \begin{pmatrix} \dfrac{\partial^2 x_1}{\partial z^2} \\ \dfrac{\partial^2 x_2}{\partial z^2} \end{pmatrix} \tag{4.6.19}$$

or

$$\dot{\boldsymbol{x}} = \boldsymbol{A}(\boldsymbol{x})\, \boldsymbol{x}_z + \boldsymbol{B}(\boldsymbol{x})\, \boldsymbol{x}_{zz} \qquad \text{in } \boldsymbol{Q} \tag{4.6.20}$$

where $0 < t < t_f$ or $t\,[0, t_f]$
$0 < z < 1$ or $z \in D \in R$
$\boldsymbol{Q} = D \times [0, t_f]$

The boundary conditions for the problem are

$$x_{1z} = Pe\,(x_1 - v) \quad \text{on } \textstyle\sum_1 \tag{4.6.21}$$

$$x_2 = 1 - S_{OR}(x_1) \quad \text{on } \textstyle\sum_1 \tag{4.6.22}$$

$$x_{1z} = 0 \quad \text{on } \textstyle\sum_2 \tag{4.6.23}$$

where

$$v = C_{\text{slug}} \ = \ \text{the control for the problem} \tag{4.6.24}$$

$$\sum_1 \ = \ 0 \times [0, t_f] \tag{4.6.25}$$

$$\sum_2 \ = \ 1 \times [0, t_f] \tag{4.6.26}$$

For this optimal control problem we want to find the single control history, $v(t)$, contained in the set of admissible controls, V_{ad}, such that we maximize a profit performance functional. We will define the performance functional as the net profit of the surfactant flood which is assumed to be the gross revenue realized due to the oil produced minus the cost of the injected chemical. This performance functional is the maximum profit attainable. Mathematically, the performance functional is expressed as

$$J^* = \alpha_1 \int_0^{\theta_f} Q_0(\theta, L)\, d\theta - \alpha_2 \int_0^{\theta_f} C_s(\theta, 0) \rho_{\text{slug}}\, Q_w(\theta, 0)\, d\theta \tag{4.6.27}$$

where
α_1 = value of oil per unit volume
Q_0 = volumetric flow rate of oil
α_2 = cost of surfactant slug per unit weight
Q_w = volumetric flow rate of water
ρ_{slug} = slug density
C_s = surfactant concentration in weight fraction
θ = time

The first term evaluates the oil value at the outlet of the core, while the second term evaluates the cost of injecting the surfactant slug. This performance functional can be expressed in terms of dimensionless time, t (pore volumes), and dimensionless distance, z, as

$$J = \int_0^{t_f} \Big[(1 - f_w)_{z=1} - R\,v\Big]\,dt \tag{4.6.28}$$

where

$$R = \frac{\alpha_2 C_o \rho_{\text{slug}}}{\alpha_1} \tag{4.6.29}$$

In terms of the formulation of optimal control problems for distributed parameter systems, as given in Section 4.5, we have the Hamiltonian,

$$H = \boldsymbol{\lambda}^T \left[\boldsymbol{A}(\boldsymbol{x})\,\boldsymbol{x}_z + \boldsymbol{B}(\boldsymbol{x})\,\boldsymbol{x}_{zz}\right] \tag{4.6.30}$$

The terminal contribution to the performance functional is

$$F_1 = 0 \tag{4.6.31}$$

The boundary contributions to the performance functional are

$$F_2 = (1 - f_w)_{z=1} - R\,v \tag{4.6.32}$$

The contribution over the domain $\boldsymbol{Q}$ is

$$F = 0 \tag{4.6.33}$$

The necessary conditions for optimality follow:

Definition of the Costate Equations

$$\begin{aligned}\dot{\boldsymbol{\lambda}} &= \left[\frac{\partial(\boldsymbol{A}(\boldsymbol{x})\,\boldsymbol{x}_z)}{\partial \boldsymbol{x}} + \frac{\partial(\boldsymbol{B}(\boldsymbol{x})\,\boldsymbol{x}_{zz})}{\partial \boldsymbol{x}}\right]^T \boldsymbol{\lambda} \\ &\quad + \frac{\partial(\boldsymbol{A}^T(\boldsymbol{x})\,\boldsymbol{\lambda})}{\partial z} - \frac{\partial^2(\boldsymbol{B}^T(\boldsymbol{x})\,\boldsymbol{\lambda})}{\partial z^2} \qquad \text{in } Q\end{aligned} \tag{4.6.34}$$

Transversality Boundary Conditions

$$\begin{aligned}&\left[\frac{\partial f_w}{\partial \boldsymbol{x}} + \boldsymbol{A}^T(\boldsymbol{x})\,\boldsymbol{\lambda} - \frac{\partial}{\partial z}\left(\boldsymbol{B}^T(\boldsymbol{x})\,\boldsymbol{\lambda}\right)\right]^T \boldsymbol{\delta x} \\ &\quad + \Big[\boldsymbol{B}^T(\boldsymbol{x})\,\boldsymbol{\lambda}\Big]^T \boldsymbol{\delta x}_z = 0 \qquad \text{at } z = 1\end{aligned} \tag{4.6.35}$$

or

$$\begin{aligned}&\left[\frac{\partial f_w}{\partial x_1} - \frac{C_2}{C_1}\,\lambda_1 - \frac{\partial f_w}{\partial x_1}\,\lambda_2 - \frac{\partial}{\partial t}\left(\frac{\lambda_1}{PeC_1}\right)\right] \delta x_1 \\ &\quad + \left[\frac{\partial f_2}{\partial x_2} - \frac{\partial f_w}{\partial x_2}\,\lambda_2\right] \delta x_2 \\ &\quad + \left(\frac{\lambda_1}{PeC_1}\right) \delta x_{1z} = 0 \qquad \text{at } z = 1\end{aligned} \tag{4.6.36}$$

This must be coupled with the variational form of the state boundary condition at $z = 1$ (Equation (4.6.23)). The variational form is

$$\delta x_{1z} = 0 \qquad \text{at } z = 1 \tag{4.6.37}$$

Using this equation, the fact that $\partial f_w / \partial x_1 = 0$, and the fact that δx_1 and δx_2 are independent, yields the following conditions at the boundary $\Sigma_2 (z = 1)$:

$$-\frac{C_2}{C_1}\lambda_1 - \frac{\partial}{\partial z}\left(\frac{\lambda_1}{PeC_1}\right) = 0 \quad \text{on } \Sigma_2 \tag{4.6.38}$$

$$\frac{\partial f_w}{\partial x_2}(1 - \lambda_2) = 0 \qquad \text{on } \Sigma_2 \tag{4.6.39}$$

Now, on the boundary $\Sigma_1 (z = 0)$, we have the general expression

$$\left[\frac{\partial f_w}{\partial \boldsymbol{x}} - \boldsymbol{A}^T(\boldsymbol{x})\,\boldsymbol{\lambda} + \frac{\partial}{\partial z}\left(\boldsymbol{B}^T(\boldsymbol{x})\,\boldsymbol{\lambda}\right)\right]^T \boldsymbol{\delta x} + R\,\delta v$$
$$- \left[\boldsymbol{B}^T(\boldsymbol{x})\,\boldsymbol{\lambda}\right]^T \boldsymbol{\delta x}_z = 0 \qquad \text{at } z = 0 \tag{4.6.40}$$

or

$$\left[\frac{C_2}{C_1}\lambda_1 + \frac{\partial}{\partial z}\left(\frac{\lambda_1}{PeC_1}\right)\right]\delta x_1 + \left[\frac{\partial f_w}{\partial x_2} + \frac{\partial f_w}{\partial x_2}\lambda_2\right]\delta x_2 + R\,\delta v$$
$$- \left(\frac{\lambda_1}{PeC_1}\right)\delta x_{1z} = 0 \qquad \text{at } z = 0 \tag{4.6.41}$$

The state boundary condition at $z = 0$ is Equation (4.6.21). Its variational form is

$$\delta x_{1z} = Pe\,\delta x_1 - Pe\,\delta v \tag{4.6.42}$$

Using Equation (4.6.42) along with the relation that at the boundary $z = 0$

$$x_2 = 1 - S_{OR}(x_1) \tag{4.6.43}$$

and also the fact that f_w is not a function of x_z (i.e., $\partial f_w / \partial x_z = 0$), gives the following final independent relations for the transversality condition of Equation (4.6.41)

$$\frac{C_2}{C_1}\lambda_1 + \frac{\partial}{\partial z}\left(\frac{\lambda_1}{PeC_1}\right) - \frac{\lambda_1}{C_1} = 0 \qquad \text{on } \Sigma_1 \tag{4.6.44}$$

and

$$\left[R + \frac{\lambda_1}{C_1}\right]\delta v = 0 \qquad \text{on } \Sigma_1 \tag{4.6.45}$$

Transversality Final Conditions

Since the final time t_f is known, then $\delta t_f = 0$. With $F_1 = 0$, we have

$$\boldsymbol{\lambda}^T \boldsymbol{\delta x} = 0 \qquad \text{at } t = t_f \tag{4.6.46}$$

The variation $\delta \boldsymbol{x}$ at t_f is free and unconstrained; therefore, we have

$$\boldsymbol{\lambda} = \mathbf{0} \qquad \textbf{at } t = t_f \tag{4.6.47}$$

Optimal Control

Since there are no controls that enter the volume domain of this problem, there are no u control variables. Thus, we do not obtain any information from the optimal control conditions.

The control for this problem is on the boundary $\sum_1$ which is given by Equation (4.6.45). Applying the costate definition relations of Equation (4.6.34), the appropriate boundary information of Equations (4.6.38), (4.6.39), and (4.6.40), and the final conditions of Equation (4.6.47), the first variation of the functional has been reduced to

$$\delta J = \int_0^{t_f} \left[R + \frac{\lambda_1}{C_1} \right]_{\Sigma_1} \delta v \, dt \qquad v \in V_{\text{ad}} \tag{4.6.48}$$

We will use an iterative numerical technique for finding the optimal control or injected surfactant concentration history. The computational procedure is based on adjusting the estimate of the control function in order to improve the value of the objective functional. Suppose we choose any function, v, and integrate the state and costate equations with the associated initial, final, and boundary conditions. Then, for v to be optimal, the condition of Equation (4.6.45) must be satisfied. If v is not optimum, a correction δv is determined so that the functional is made larger, that is, $\delta J > 0$. The effect of a small change, δv, in the objective functional is given by Equation (4.6.48). If δv is selected as

$$\delta v = w \left[R + \frac{\lambda_1(0,t)}{C_1(0,t)} \right] \tag{4.6.49}$$

where w is an arbitrary positive weighting factor, then the first variation becomes

$$\delta J = w \int_0^t \left[R + \frac{\lambda_1(0,t)}{C_1(0,t)} \right]^2 dt \tag{4.6.50}$$

Thus, choosing δv in the gradient direction insures the local improvement in the functional J. Note that at the boundary of the admissible control set, V_{ad}, we must take δv equal to zero in order to avoid leaving the admissible set.

The computational algorithm of control vector iteration based upon gradient directions is as follows:

1) Initialization.
Make an initial guess of the control function $v(t)$.

2) Resolution of the state equations.
Using the function $v(t)$, integrate the state equations forward in time. The profit functional is evaluated, and the coefficients involved in the costate equations which are function of the state solution are computed and stored.

3) Resolution of the costate equations.
Using the stored coefficients, the costate equations are integrated numerically backwards in time. A backwards in time integration is required due to the specification of the final costate condition. We also compute and store the variation, δv, defined by Equation (4.6.49).

4) Computation of the new control law.
Using the evaluated δv, an improved control function is computed as

$$v(t)^{\text{new}} = v(t)^{\text{old}} + \delta v \qquad v(t)^{\text{new}} \in V_{\text{ad}} \tag{4.6.51}$$

or

$$v(t)^{\text{new}} = v(t)^{\text{old}} + w\left[R + \frac{\lambda_1(0,t)}{C_1(0,t)}\right] \qquad v(t)^{\text{new}} \in V_{\text{ad}} \tag{4.6.52}$$

A single variable search is used to find the value of the positive weighting factor which maximizes the improvement in the performance functional using Equation (4.6.52).

5) Termination.
The optimization algorithm is stopped when the variation δv is too small to effectively change the performance measure, that is, when

$$\mid J_{i+1} - J_i \mid < \epsilon \tag{4.6.53}$$

where ϵ is a small positive number.

The costate partial differential equations defined in Equation (4.6.34) are a system of two linear partial differential equations with nonconstant coefficients determined from the solution of the state equations. Fathi (1981) investigated many numerical schemes for the solution of this set of equations. None, however, resulted in a stable solution. The difficulty in tackling this problem is due to the existence of discontinuities in the coefficients.

There is very limited computational experience for a system of the general form given by the costate equations with discontinuous coefficients. Because of the computational difficulties encountered in solving this general system of costate equations, Fathi and Ramirez (1985) evoked a quasilinear assumption.

The quasilinear problem assumes that the operators $\boldsymbol{A}$ and $\boldsymbol{B}$ are independent of the state $\boldsymbol{x}$. Under the quasilinear assumption, the costate equations become

$$\dot{\boldsymbol{\lambda}} = \boldsymbol{A}^T(\boldsymbol{x})\,\boldsymbol{\lambda}_z - \boldsymbol{B}^T(\boldsymbol{x})\,\boldsymbol{\lambda}_{zz} \qquad \text{in } Q \tag{4.6.54}$$

or

$$\dot{\lambda}_1 = -\frac{C_2}{C_1}\,\lambda_{1z} - \frac{\partial f_w}{\partial x_1}\,\lambda_{2z} - \frac{1}{PeC_1}\,\lambda_{1zz} \tag{4.6.55}$$

$$\dot{\lambda}_1 = -\frac{\partial f_w}{\partial x_2}\,\lambda_{2z} \tag{4.6.56}$$

Therefore, under the quasilinear assumption, the costate equations have the same form as the state equations. It is the analogy between this form of the costate equations and the state equations which motivates the use of the quasilinear assumption and results in a stable numerical scheme.

In order to compute the costate variables, the coefficients C_1 and C_2 must be stored from the state solution. Also the functions $\partial f_w/\partial x_1$ and $\partial f_w/\partial x_2$ must be computed. Since analytical expressions have been used for model parameters, analytic derivatives can be obtained. For example, let's consider the term

$$f_w = (x_2 - S_{OR})^{1.5+S_{OR}/S_{OR0}} \tag{4.6.57}$$

To compute $\partial f_w/\partial x_1$, we need the following relations:

$$\begin{aligned}\frac{\partial f_w}{\partial x_1} &= \left(1.5 + \frac{S_{OR}}{S_{OR0}}\right)(x_2 - S_{OR})^{0.5+S_{OR}/S_{OR0}} \\ &\quad + \frac{\ln(x_2 - S_{OR})}{S_{OR0}}\,\frac{\partial S_{OR}}{\partial x_1}\,(x_2 - S_{OR})^{1.5+S_{OR}/S_{OR0}}\end{aligned} \tag{4.6.58}$$

$$\frac{\partial S_{OR}}{\partial x_1} = \frac{B}{N_{\text{cap}}}\,\frac{\partial N_{\text{cap}}}{\partial x_1} \tag{4.6.59}$$

$$\frac{\partial N_{\text{cap}}}{\partial x_1} = \frac{v_w\left(\dfrac{\partial \mu_w}{\partial x_1}\,\sigma_{wo} - \mu_o\,\dfrac{\partial \sigma_{wo}}{\partial x_1}\right)}{(\sigma_{wo})^2} \tag{4.6.60}$$

$$\frac{\partial \mu_w}{\partial x_1} = C_1\,\exp(C_1 x_1) \tag{4.6.61}$$

$$\frac{\partial \sigma_{wo}}{\partial x_1} = b_i\left(x_2 + \frac{(1 - x_1)}{K_s}\right)\sigma_{wo} \tag{4.6.62}$$

Therefore, these functions must be evaluated and the result of Equation (4.6.58) stored during the state solution so that it is available for use in the determination of the costates.

A slug process is chosen for optimization. Table 4.1 gives the physical parameters used during this study. The economics used to calculate the net profitability of the flood are

$$\alpha_1 = \text{value of oil/unit volume} = 1.89 \times 10^{-4}\ \$/\text{cm}^3 \quad (4.6.63)$$

$$\alpha_2 = \text{cost of surfactant slug/unit weight} = 3.59 \times 10^{-9}\ \$/\text{ppm} \quad (4.6.64)$$

Table 4.1

Model Parameters Used for Two-Phase Optimization Study

Porosity = 0.277
Adsorption Constants

M = 18 g core/g soln
N = 2900 g core/g surfactant

μ_{oil} = 12.6 cp
μ_{slug} = 22.7 cp
Residual oil waterflood saturation = 0.338
Peclet Number = 170
Density rock/Density of slug = 2.0
Partition coefficient = 1.0
Cross sectional area of core = 20.91 cm^2
Core length = 30.48 cm
Injection velocity = 4.13 ft/day
Characteristic slug surfactant concentration = 2.5% or 25000 ppm

The value of the oil has been assumed to be $30 per barrel and the surfactant slug cost of $1.63 per pound. This means that the value of R in Equation (4.6.29) is

$$R = 1.9 \times 10^{-5}\, C_o \quad (4.6.65)$$

Optimization results are obtained for an initial starting surfactant slug injection of 0.2 pore volume with concentration of 25,000 ppm or $v(t) = 1.0$ over the injection time of 0.2 pore volumes. Table 4.2 gives the results of implementing the control vector iteration scheme. Convergence was essentially reached in six iterations. The optimal injection strategy, along with the initial injection strategy is shown in Figure 4.12. The optimal injection strategy is essentially a constant slug injection of dimensionless concentration 0.55 (13,750 ppm) for a duration of 0.2 pore volumes. The maximum performance functional

Table 4.2

Convergence of Optimal Injection Policy (.2 PV slug of concentration 2.5% surfactant)

Iteration Number	Weighting Factor (W)	Performance Index	% Improvement from Original Guess	% Improvement from Last Value
0		0.2105	0	
1	.2	0.2209	4.97	4.97
2	.2	0.2306	9.54	4.35
3	.2	0.2388	13.45	3.57
4	.2	0.2450	16.42	2.61
5	.2	0.2484	17.99	1.35
6	.2	0.2490	18.29	0.25

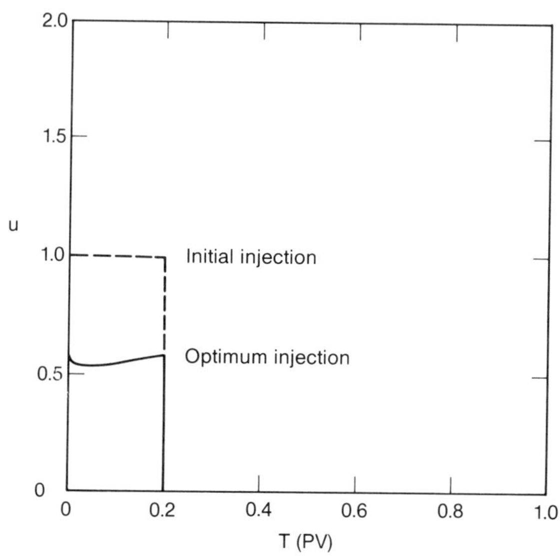

Figure 4.12 Surfactant optimization results. From Ramirez (1987).

is 0.249, which is 18.3% better than the initial strategy. The dimensionless performance functional of Equation (4.6.28) was defined so that the maximum value attainable is 1.0 when the core is initially saturated with oil and the chemical injection cost is zero. For this problem the initial saturation is 0.338 so that the maximum value attainable is 0.338. The optimum value of 0.249 is 73.7% of the value with complete recovery at no cost.

4.7 Integrating Factors and Green's Functions

In order to derive the necessary conditions for the extremization of a performance functional, we have used classical variational methods. The method presented can be summarized by the following steps:

1. Form an augmented performance functional to include all equality constraints,
2. Introduce variations, $\delta \boldsymbol{x}$, $\delta \dot{\boldsymbol{x}}$, $\delta \boldsymbol{\lambda}$, $\delta \boldsymbol{u}$, δt_f,
3. Integrate by parts to eliminate $\delta \dot{\boldsymbol{x}}$, and
4. Collect terms to get the necessary conditions

An alternative approach is to use integrating factors and Green's functions. Good references on this approach are Denn (1969) and Lions (1971). This approach is characterized by the following steps:

1. Get a variational form of the performance functional,
2. Obtain the variational Green's formula, and
3. Combine the two variational relations.

In order to discuss this alternative approach, we must first find out what is Green's formula. Let us consider the scalar equation

$$\dot{x} = a(t)x + b(t) \tag{4.7.1}$$

We introduce an integrating factor, $\Gamma(t)$, which gives

$$\Gamma(t)\dot{x} = a\Gamma(t)x + \Gamma(t)b \tag{4.7.2}$$

The integral of Equation (4.5.2) is

$$\int_{t_o}^{t} \Gamma(\tau)\dot{x}(\tau)\, d\tau = \int_{t_o}^{t} a(\tau)\Gamma(\tau)\boldsymbol{x}(\tau)\, d\tau + \int_{t_o}^{t} \Gamma(\tau)b(\tau)\, d\tau \tag{4.7.3}$$

We integrate the left-hand side of Equation (4.7.3) by parts to obtain

$$\int_{t_o}^{t} \Gamma(\tau)\dot{(x)}(\tau)\, d\tau = \Gamma(t)x(t) - \Gamma(t_o)x(t_o) - \int_{t_o}^{t} \dot{\Gamma}(\tau)x(\tau)\, d\tau \qquad (4.7.4)$$

If we define the integrating factor, $\Gamma(t)$, as

$$\dot{\Gamma} = -a\Gamma \qquad (4.7.5)$$

then the solution of this homogeneous differential equation is

$$\Gamma(t) = \Gamma(t_o)\, \exp\left[-\int_{t_o}^{t} a(\xi)\, d\xi\right] \qquad (4.7.6)$$

The integration factor defined by the differential equation of (4.7.5) is called the adjoint system. Using the solution to the adjoint system, we have for Equation (4.7.4)

$$\begin{aligned}\Gamma(t_o)x(t)\, \exp\left[-\int_{t_o}^{t} a(\xi)d\xi\right] - \Gamma(t_o)x(t_o) + \int_{t_o}^{t} a(\tau)\Gamma(\tau)x(\tau)d\tau \\ = \int_{t_o}^{t} a(\tau)\Gamma(\tau)x(\tau)d\tau + \Gamma(t_o)\int_{t_o}^{t} \exp\left[-\int_{t_o}^{\tau} a(\xi)d\xi\right] b(\tau)d\tau \end{aligned} \qquad (4.7.7)$$

or

$$x(t) = x(t_o)\, \exp\Big[\Big[\int_{t_o}^{t} a(\xi)d\xi\Big] + \int_{t_o}^{t} \exp\Big[\Big[\int_{\tau}^{t} a(\xi)d\xi\Big] b(\tau)d\tau \qquad (4.7.8)$$

The variable, $\Gamma(t)$, besides being called the integrating factor, is also known as a Green's function. Let us generalize this development for a set of differential equations

$$\dot{\boldsymbol{x}} = \boldsymbol{A}(t)\boldsymbol{x}(t) + \boldsymbol{b}(t) \qquad (4.7.9)$$

We introduce the Green's matrix, $\boldsymbol{\Gamma}$, to get

$$\boldsymbol{\Gamma}\dot{\boldsymbol{x}} = \boldsymbol{\Gamma A x} + \boldsymbol{\Gamma b} \qquad (4.7.10)$$

The integral form of Equation (4.7.10) is

$$\int_{t_o}^{t} \boldsymbol{\Gamma}\dot{\boldsymbol{x}}\, d\tau = \int_{t_o}^{t} \boldsymbol{\Gamma A x}\, d\tau + \int_{t_o}^{t} \boldsymbol{\Gamma b}\, d\tau \qquad (4.7.11)$$

We can integrate the left-hand side by parts to obtain

$$\int_{t_o}^{t} \boldsymbol{\Gamma}\dot{\boldsymbol{x}}\, d\tau = \boldsymbol{\Gamma}(t)\boldsymbol{x}(t) - \boldsymbol{\Gamma}(t_o)\boldsymbol{x}(t_o) - \int_{t_o}^{t} \dot{\boldsymbol{\Gamma}}(\tau)\boldsymbol{x}(\tau)\, d\tau \qquad (4.7.12)$$

We define the adjoint system as

$$\dot{\boldsymbol{\Gamma}}(t) = -\boldsymbol{\Gamma}(t)\boldsymbol{A}(t) \qquad (4.7.13)$$

Using the adjoint system and the integration by parts gives, for Equation (4.7.11),

$$\boldsymbol{\Gamma}(t)\boldsymbol{x}(t) - \boldsymbol{\Gamma}(t_o)\boldsymbol{x}(t_o) = \int_{t_o}^{t} \boldsymbol{\Gamma}(\tau)\boldsymbol{b}(\tau)\, d\tau \tag{4.7.14}$$

Equation (4.7.14) is referred to as a Green's identity and is a one-dimensional equivalent between the familiar relation between volume and surface integrals. Equation (4.7.14) can be solved for $\boldsymbol{x}(t)$ as

$$\boldsymbol{x}(t) = \boldsymbol{\Gamma}(t, t_o)\boldsymbol{x}(t_o) + \int_{t_o}^{t} \boldsymbol{\Gamma}(t, \tau)\boldsymbol{b}(\tau)\, d\tau \tag{4.7.15}$$

We now introduce the Green's vector, which is defined as

$$\boldsymbol{\lambda}(t) = \boldsymbol{\Gamma}^T(t, t_f)\boldsymbol{\lambda}(t_f) \tag{4.7.16}$$

Notice that this expresses $\boldsymbol{\lambda}(t)$ in terms of $\boldsymbol{\lambda}(t_f)$. Differentiating Equation (4.7.16) with respect to time gives

$$\dot{\boldsymbol{\lambda}}(t) = \dot{\boldsymbol{\Gamma}}^T(t, t_f)\boldsymbol{\lambda}(t_f) \tag{4.7.17}$$

Using the definition of the adjoint system of Equation (4.7.13), we have

$$\dot{\boldsymbol{\lambda}}(t) = -\boldsymbol{A}^T\boldsymbol{\Gamma}^T(t, t_f)\boldsymbol{\lambda}(t_f) \tag{4.7.18}$$

and then introducing the Green's vector of Equation (4.7.16) gives

$$\dot{\boldsymbol{\lambda}}(t) = -\boldsymbol{A}^T\boldsymbol{\lambda}(t) \tag{4.7.19}$$

Equation (4.7.19) is the adjoint system in terms of the Green's vector. We can also get the solution of the state differential equations of Equation (4.7.15) in terms of the Green's vector rather than the Green's matrix. This is done by multiplying Equation (4.7.19) by the Green's vector at the final time, $\boldsymbol{\lambda}^T(t_f)$. The final result is

$$\boldsymbol{\lambda}^T(t_f)\boldsymbol{x}(t_f) = \boldsymbol{\lambda}^T(t_o)\boldsymbol{x}(t_o) + \int_{t_o}^{t_f} \boldsymbol{\lambda}^T(\tau)\boldsymbol{b}(\tau)\, d\tau \tag{4.7.20}$$

Now we will develop the first-order variational Green's identity for the performance functional

$$J = h\left(\boldsymbol{x}(t_f), t_f\right) + \int_{t_o}^{t_f} F(\boldsymbol{x}, \boldsymbol{u}, t)\, dt \tag{4.7.21}$$

subject to differential equality constraints

$$\dot{\boldsymbol{x}} = \boldsymbol{f}(\boldsymbol{x}, \boldsymbol{u}, t) \tag{4.7.22}$$

and the control variable constraints that

$$\boldsymbol{u}(t) \in \Omega \qquad \text{where } \Omega \text{ is a subset of } R^m \tag{4.7.23}$$

To handle the differential equality constraints of Equation (4.7.22), we introduce the Green's vector

$$0 = \boldsymbol{\lambda}^T(\boldsymbol{f} - \dot{\boldsymbol{x}}) \tag{4.7.24}$$

Taking the integral of Equation (4.7.24) over $[t_o, t_f]$ gives

$$0 = \int_{t_o}^{t_f} \left[\boldsymbol{\lambda}^t \boldsymbol{f} - \boldsymbol{\lambda}^T \dot{\boldsymbol{x}}\right] dt \tag{4.7.25}$$

Integrating $\boldsymbol{\lambda}^T \dot{\boldsymbol{x}}$ by parts gives

$$0 = \int_{t_o}^{t_f} \boldsymbol{\lambda}^T \boldsymbol{f}\, dt + \int_{t_o}^{t_f} \dot{\boldsymbol{\lambda}}^T \boldsymbol{x}\, dt - \boldsymbol{\lambda}^T \boldsymbol{x}\Big|_{t_o}^{t_f} \tag{4.7.26}$$

This is a Green's identity. We can get a variational form of Equation (4.7.26) by introducing the variations $\delta \boldsymbol{x}$ and $\delta \boldsymbol{u}$:

$$0 = \int_{t_o}^{t_f} \boldsymbol{\lambda}^T \frac{\partial \boldsymbol{f}}{\partial \boldsymbol{x}}\, \delta \boldsymbol{x}\, dt + \int_{t_o}^{t_f} \boldsymbol{\lambda}^T \frac{\partial \boldsymbol{f}}{\partial \boldsymbol{u}}\, \delta \boldsymbol{u}\, dt + \int_{t_o}^{t_f} \dot{\boldsymbol{\lambda}}^T \delta \boldsymbol{x}\, dt - \boldsymbol{\lambda}^T \delta \boldsymbol{x}\Big|_{t_o}^{t_f} \tag{4.7.27}$$

Equation (4.7.27) is a variational Green's identity. We now develop a variational form of the performance functional of Equation (4.7.21) by introducing the variations, $\delta \boldsymbol{x}$, $\delta \boldsymbol{u}$, δt_f. This yields

$$\delta J = \frac{\partial h^T(t_f)}{\partial \boldsymbol{x}}\, \delta \boldsymbol{x}_f + \frac{\partial h(t_f)}{\partial t}\, \delta t_f + \int_{t_o}^{t_f} \left[\frac{\partial F^T}{\partial \boldsymbol{x}}\, \delta \boldsymbol{x} + \frac{\partial F^T}{\partial \boldsymbol{u}}\, \delta \boldsymbol{u}\right] dt + F(t_f)\, \delta t_f \tag{4.7.28}$$

We now combine Equations (4.7.28) and (4.7.26) to give the final variation

$$\delta J = \int_{t_o}^{t_f} \left[\dot{\boldsymbol{\lambda}} + \left(\frac{\partial \boldsymbol{f}}{\partial \boldsymbol{x}}\right)^T \boldsymbol{\lambda} + \frac{\partial F}{\partial \boldsymbol{x}}\right]^T \delta \boldsymbol{x}\, dt + \int_{t_o}^{t_f} \left[\frac{\partial F}{\partial \boldsymbol{u}} + \left(\frac{\partial \boldsymbol{f}}{\partial \boldsymbol{u}}\right)^T \boldsymbol{\lambda}\right]^T \delta \boldsymbol{u}\, dt$$
$$+ \left[\frac{\partial h}{\partial \boldsymbol{x}}(t_f)\right]^T \delta \boldsymbol{x}_f + \left[\frac{\partial h}{\partial t}(t_f) + F(t_f)\right] \delta t_f - \boldsymbol{\lambda}^T(t_f)\, \delta \boldsymbol{x}(t_f) \tag{4.7.29}$$

Since this is a free end point problem,

$$\delta \boldsymbol{x}(t_f) = \delta \boldsymbol{x}_f - \dot{\boldsymbol{x}}(t_f)\, \delta t_f \tag{4.7.30}$$

and the first variation of the performance functional is

$$\delta J = \int_{t_o}^{t_f} \left[\dot{\boldsymbol{\lambda}} + \left(\frac{\partial \boldsymbol{f}}{\partial \boldsymbol{x}}\right)^T \boldsymbol{\lambda} + \frac{\partial F}{\partial \boldsymbol{x}}\right]^T \delta \boldsymbol{x}\, dt$$

$$+ \int_{t_o}^{t_f} \left[\frac{\partial F}{\partial \boldsymbol{u}} + \left(\frac{\partial \boldsymbol{f}}{\partial \boldsymbol{u}} \right)^T \boldsymbol{\lambda} \right]^T \delta \boldsymbol{u}\, dt$$

$$+ \left[\frac{\partial h}{\partial \boldsymbol{x}} (t_f) - \boldsymbol{\lambda}(t_f) \right]^T \delta \boldsymbol{x}_f$$

$$+ \left[\frac{\partial h}{\partial t} (t_f) + F(t_f) + \boldsymbol{\lambda}^T(t_f) \boldsymbol{f}(t_f) \right] \delta t_f \qquad (4.7.31)$$

This first variation is exactly the same as that developed in Section 4.2.

References

Bilous, O. and Amundson, N. R., *Chem. Eng. Sci.* **5**, 115 (1956).

Bryson, A. E. and Ho, Y., *Applied Optimal Control: Optimization, Estimation, and Control*, Halsted Press, New York (1975).

Delancy, G. B., "An Optimal Catalyst Activation Policy for Poisoning Problems," *Chem. E. Sci.* **28**, 105–118 (1973).

Denn, M. M., *Optimization by Variational Methods*, McGraw–Hill, New York (1969).

Fathi, Z., "Optimal Injection Policies for Surfactant Flooding Enhanced Oil Recovery," M.S. Thesis, University of Colorado (1981).

Fathi, Z. and Ramirez, W.F., "Optimal Injection Policies for Enhanced Oil Recovery: Part 2—Surfactant Flooding," *Soc. of Pet. Engr. J.* **24**(3), 333 (1984).

Fathi, Z. and Ramirez, W. F., "Use of Optimal Control Theory for Computing Optimal Injection Policies for Enhanced Oil Recovery," *Automatica* **22**(1), 33–42 (1985).

Kirk, D. E., *Optimal Control Theory*, Prentice–Hall, Englewood Cliffs, New Jersey (1970).

Lewis, F. L., *Optimal Control*, Wiley and Sons, New York (1986).

Lions, J. L., *Optimal Control of Systems Governed by Partial Differential Equations*, Springer–Verlag, New York (1971).

Mansson, B. and Andresen, B., "Optimal Temperature Profile for an Ammonia Reactor," *Ind. Eng. Chem. Process Des. & Dev.* **25**, 59–65 (1986).

Nielsen, A., Kjaer, J., and Hansen, B., *J. Catal.* **3**, 68 (1964).

Pontryagin, L. S., Boltyanskii, R. V., Gamkrelidze, R. V., and Mischenko, E. F., *The Mathematical Theory of Optimal Processes*, Wiley and Sons, New York (1962).

Ramirez, W. F., *Application of Optimal Control Theory to Enhanced Oil Recovery*, Elsevier, The Netherlands (1987).

Ramirez, W. F., *Computational Methods for Process Simulation*, Butterworths, Boston, MA (1989).

Ramirez, W. F., Fathi, Z., and Cagnol, J.M., "Optimal Injection Policies for Enhanced Oil Recovery: Part 1—Theory and Computational Strategies," *Soc. of Pet. Engr. J.* **24**(3), 328 (1984).

Ray, W. H., *Advanced Process Control*, McGraw–Hill, New York (1981).

Sage, A. P. and White, C. C., III, *Optimal Systems Control*, Second Edition, Prentice–Hall, Englewood Cliffs, New Jersey (1977).

Strelzoff, S., *Technology and Manufacture of Ammonia*, Wiley and Sons, New York (1981).

Taber, J. J., "Dynamic and Static Forces Required to Remove a Discontinuous Oil Phase from Porous Media Containing Both Oil and Water," *Soc. of Pet. Engr. J.* **9**, 3 (1969).

Vancini, C. A., *Synthesis of Ammonia*, MacMillan Press, Riverside, New Jersey (1971).

Exercises

1. Find the necessary conditions for the optimal control to minimize

$$J = \frac{1}{2}\int_0^1 \left(x^2 + u^2\right)\, dt$$

for the system

$$\dot{x} = -x^3 + u \qquad x(0) = 1$$

2. Find the optimal control and trajectory that transfers the system

$$\begin{aligned} \dot{x}_1 &= x_2 & x_1(0) &= 0 \\ \dot{x}_2 &= u & x_2(0) &= 0 \end{aligned}$$

to the line

$$x_1(1) + x_2(1) = 1$$

such that

$$J = \frac{1}{2}\int_0^1 u^2(t)\, dt$$

is minimized.

3. Find the extremal to the following problem:

$$\begin{aligned} &\dot{x}_1 = x_2 \\ &\dot{x}_2 = -x_2 + u && t_f = \text{known} \\ &J = \int_{t_o}^{t_f} \frac{1}{2}\left(x_1^2 + u^2\right)\, dt && \boldsymbol{x}(t_f) = \text{free} \\ &-1 \le u \le 1 \end{aligned}$$

4. Find the control vector that minimizes

$$J = \frac{1}{2}\int_0^1 \left(x^2 + u_1^2 + u_2^2\right)\, dt$$

for the system

$$\dot{x} = u_1 + u_2 \qquad x(0) = 1$$

5. **Plug Flow Heat Exchanger**

Here we consider the system of a tubular plug flow heat exchanger with constant properties, constant velocity, no axial mixing of energy or mass, perfect radial mixing, and no wall capacitance. The system model is

$$\frac{\partial x}{\partial t} + \frac{\partial x}{\partial z} = u(z,t)$$

with initial conditions

$$\begin{array}{ll} \text{at } t = 0 & x(z) = x_0\,(z) \\ \text{at } z = 0 & x(t) = 0 \end{array}$$

Here t represents normalized time and z represents normalized axial distance from the exchange entrance. Thus, $t = 1$ is one residence time and $z = 1$ is the exchanger exit. The state variable x represents deviation of the fluid temperature from the desired final temperature profile and the control $u(z,t)$ represents deviations of the wall heat flux from the final heat flux profile.

The performance index chosen for minimization is

$$J(u) = 1/2 \int_o^{t_f} \int_o^1 \left(\mu^2 x^2 + u^2\right) \, dz \, dt$$

where μ is constant.

Derive the necessary conditions for optimal control.

Solve these equations (Linear Partial Differential Equations) using the method of lines or characteristics.

6. Verify Equations (4.1.68) and (4.1.69).

7. Determine the necessary conditions for the optimal control for the system

$$\begin{aligned} \dot{x}_1(t) &= x_2(t) \\ \dot{x}_2(t) &= -x_1(t) + \left(1 - x_1^2(t)\right) x_3(t) + u(t) \end{aligned}$$

that minimizes the performance functional

$$J = \frac{1}{2} \int_0^{t_f} \left(2x_1^2(t) + x_2^2(t) + u^2(t)\right) dt$$

   (a) The initial states are specified and $t_f = 1$.

   (b) The initial state is specified, but the final state is to be transferred to the plane

$$15\ x_1 + 20\ x_2 + 12\ t = 60$$

   The final time is free.

8. Batch binary distillation can be described by the equations

$$\begin{aligned} \dot{x}_1 &= -u \\ \dot{x}_2 &= \frac{u}{x_1} \left(x_2 - F(x_2, u)\right) \end{aligned}$$

with initial condition x_{10} and x_{20}.

Physically, the state variable x_1 denotes the total moles remaining in the still and the state variable x_2 is the mole fraction of the more volatile component in the still. The control u is the product withdrawal rate, and F is the overhead mole fraction of the more volatile component which is a function of x_2 and u.

The withdrawal rate (the control, u) is to be such that the total output amount is maximized,

$$\max J = \int_0^{t_f} u(t)dt$$

while maintaining a specified average purity

$$\bar{F} = \frac{\int_0^{t_f} F\, u\, dt}{\int_0^{t_f} u\, dt}$$

The final time, t_f, is specified.

Formulate the necessary conditions for this optimal control problem.

Note: Put the performance functional in a form that includes the average purity $\bar{F}$.

9. Develop the optimal control policy for the system

$$\dot{\boldsymbol{x}} = \begin{bmatrix} 0 & 1 \\ 0 & 0 \end{bmatrix} \boldsymbol{x} + \begin{pmatrix} 0 \\ 1 \end{pmatrix} u$$

where

$$|u| \leq 1.0$$

and we wish to reach the origin $\boldsymbol{x} = \mathbf{0}$ from a known initial condition $\boldsymbol{x}(t_o) = x_o$ in minimum time. The performance functional to be minimized is therefore

$$J = \int_0^{t_f} 1\, dt$$

10. **An Optimal Catalyst Activation Policy**

A one-dimensional model for reaction in a catalyst particle is

$$\frac{d^2\psi}{ds^2} - \Lambda^2 \frac{\left(1 - e^{-\theta/\tau_p}\right)}{\theta/\tau_p} u\psi = 0$$

where ψ = species concentration
s = dimensionless distance
Λ = Thiele reaction rate parameter
u = dimensionless catalyst concentration
θ = total operating time
τ_p = poisoning time constant

The boundary conditions for the problem are

$$\psi = 1 \qquad \text{at } s = 1$$
$$\frac{d\psi}{ds} = 0 \qquad \text{at } s = 0$$

The control variable is u and constrained by

$$0 \leq u \leq 1$$

The performance functional to be maximized is

$$J = \frac{1}{\Lambda^2} \left. \frac{d\psi}{ds} \right|_{s=1} - \alpha \int_0^1 u ds$$

and is the net return expected from the catalyst pellet operating under an average level of deactivation time θ. The first term represents the conversion and the second the cost of replacing or regenerating the catalyst.

Put the system in state variable form. Develop the necessary conditions for optimality. Solve the set of state and costate equations analytically to obtain

$$\psi = \frac{\cosh \Lambda(s - s^*)}{\cosh \Lambda(1 - s^*)}$$
$$\lambda_2 = -\frac{1}{\alpha\Lambda^2} \frac{\cosh \Lambda(s - s^*)}{\cosh \Lambda(1 - s^*)}$$

where s^* is the value of s where the control switches from u_{max} to u_{min}. Show that the Hamiltonian vanishes at

$$\Lambda^2 \lambda_2 \psi + 1 = 0$$

and that

$$\delta^* = 1 - s^* = \frac{1}{\Lambda} \cosh^{-1} \sqrt{\frac{1}{\alpha}}$$

This problem is further discussed by Delancy (1973).

Chapter 5

THE LINEAR QUADRATIC REGULATOR (LQR) PROBLEM

5.1 The Kalman Problem

This chapter develops the Kalman linear quadratic regulator (LQR) problem and shows how the proportional feedback gain matrix is easily computable using the Riccati matrix differential equation. We also show how unmeasured disturbances lead to a modified Kalman LQR problem that results in a proportional-integral feedback control structure. Techniques for the inclusion of specific response characteristics in the LQR design procedure are also presented.

Kalman was responsible for one of the first important and practical solutions to an optimal control problem, which was first presented at the first International Federation of Automatic Control (IFAC) World Congress held in Moscow in 1960. It later appeared in *SIAM J. Control, Ser. A* (1963) and *Theory of Optimal Control and the Calculus of Variation*, (1963). He considered the optimal control of a linear system with a quadratic performance functional and was able to show the optimality of the feedback control structure well established in the control literature. Kalman considered a linear system of the following form:

$$\dot{\boldsymbol{x}} = \boldsymbol{A}\boldsymbol{x}(t) + \boldsymbol{B}\boldsymbol{u}(t) \tag{5.1.1}$$

with

$$\begin{aligned} \boldsymbol{x}(t_o) &= \boldsymbol{x}_o \\ \boldsymbol{x}(t_f) &= \text{free} \\ t_f &= \text{known} \end{aligned} \tag{5.1.2}$$

We will seek the optimal control that minimizes the following quadratic regulator performance functional:

$$J = \frac{1}{2}\,\boldsymbol{x}^T(t_f)\boldsymbol{H}\boldsymbol{x}(t_f) + \frac{1}{2}\int_{t_o}^{t_f} (\boldsymbol{x}^T\boldsymbol{Q}\boldsymbol{x} + \boldsymbol{u}^T\boldsymbol{R}\boldsymbol{u})\,dt \tag{5.1.3}$$

where $\boldsymbol{H}$ and $\boldsymbol{Q}$ are real symmetric positive semidefinite matrices and $\boldsymbol{R}$ is a real symmetric positive definite matrix. This performance index tends to keep the system at the equilibrium state of $\boldsymbol{x} = \boldsymbol{0}$ during the entire time domain while minimizing the use of the control system. We are essentially trying to regulate about the equilibrium or steady state when we have an initial offset of $\boldsymbol{x}_o$. We apply the necessary conditions of Chapter 4 to determine the optimal control policy. The Hamiltonian is

$$H = F + \boldsymbol{\lambda}^T\boldsymbol{f} \tag{5.1.4}$$

or for this problem

$$H = \frac{1}{2}\,\boldsymbol{x}^T\boldsymbol{Q}\boldsymbol{x} + \frac{1}{2}\,\boldsymbol{u}^T\boldsymbol{R}\boldsymbol{u} + \boldsymbol{\lambda}^T(\boldsymbol{A}\boldsymbol{x} + \boldsymbol{B}\boldsymbol{u}) \tag{5.1.5}$$

It is significant to note that the Hamiltonian is quadratic in both the states $\boldsymbol{x}$ and the controls $\boldsymbol{u}$.

The necessary conditions for minimizing the performance functional of Equation (5.1.3) are as follows:

Euler–Lagrange Equations

$$\dot{\boldsymbol{\lambda}} = -\frac{\partial H}{\partial \boldsymbol{x}} \tag{5.1.6}$$

or

$$\dot{\boldsymbol{\lambda}} = -\boldsymbol{Q}\boldsymbol{x} - \boldsymbol{A}^T\boldsymbol{\lambda} \tag{5.1.7}$$

State Dynamics

$$\dot{\boldsymbol{x}} = \frac{\partial H}{\partial \boldsymbol{\lambda}} \tag{5.1.8}$$

or

$$\dot{\boldsymbol{x}} = \boldsymbol{A}\boldsymbol{x} + \boldsymbol{B}\boldsymbol{u} \tag{5.1.9}$$

Given Boundary Conditions

$$\boldsymbol{x}(t_o) = \boldsymbol{x}_o \tag{5.1.10}$$

Transversality Boundary Conditions

$$\left[\frac{\partial \boldsymbol{h}(t_f)}{\partial \boldsymbol{x}} - \boldsymbol{\lambda}(t_f)\right]^T \delta\boldsymbol{x}_f + \left(H(t_f) + \frac{\partial \boldsymbol{h}(t_f)}{\partial t}\right)\delta t_f = 0 \tag{5.1.11}$$

Since this is a problem with a fixed final time and a free final state, the variation $\delta t_f = 0$, while the variation $\delta \boldsymbol{x}_f \neq \mathbf{0}$. Transversality therefore becomes

$$\boldsymbol{\lambda}(t_f) = \boldsymbol{H}\boldsymbol{x}(t_f) \tag{5.1.12}$$

Optimal Control

We will assume that the variation, $\delta \boldsymbol{u}$, is unconstrained so that

$$\frac{\partial H}{\partial \boldsymbol{u}} = \mathbf{0} \tag{5.1.13}$$

or

$$\mathbf{0} = \boldsymbol{R}\boldsymbol{u} + \boldsymbol{B}^T\boldsymbol{\lambda} \tag{5.1.14}$$

which can be solved for the optimal control as

$$\boldsymbol{u} = -\boldsymbol{R}^{-1}\boldsymbol{B}^T\boldsymbol{\lambda} \tag{5.1.15}$$

This equation shows that the optimal control for this problem is proportional to the costate vector, $\boldsymbol{\lambda}$. Note that $\boldsymbol{R}^{-1}$ is assured to exist since $\boldsymbol{R}$ is square and was assumed to be a real symmetric positive definite matrix.

To proceed with the development, Kalman assumed that the costate vector is proportional to the state vector as

$$\boldsymbol{\lambda}(t) = \boldsymbol{P}(t)\boldsymbol{x}(t) \tag{5.1.16}$$

This equation is called the Riccati transformation. Using the Riccati transformation, the optimal control becomes

$$\boldsymbol{u}(t) = -\boldsymbol{R}^{-1}\boldsymbol{B}^T\boldsymbol{P}(t)\boldsymbol{x}(t) \tag{5.1.17}$$

or

$$\boldsymbol{u}(t) = -\boldsymbol{K}(t)\boldsymbol{x}(t) \tag{5.1.18}$$

Equations (5.1.17) and (5.1.18) demonstrate that the optimal controller is a multivariable time varying proportional state feedback controller. Kalman was able to show the validity of the Riccati transformation and develop a method for computing the matrix, $\boldsymbol{P}(t)$. Taking the time derivative of the Riccati transformation gives

$$\dot{\boldsymbol{\lambda}}(t) = \boldsymbol{P}(t)\dot{\boldsymbol{x}}(t) + \dot{\boldsymbol{P}}(t)\boldsymbol{x}(t) \tag{5.1.19}$$

Also using the Riccati transformation and the Euler–Lagrange equations gives

$$\dot{\boldsymbol{\lambda}}(t) = -\boldsymbol{A}^T\boldsymbol{P}(t)\boldsymbol{x}(t) - \boldsymbol{Q}\boldsymbol{x}(t) \tag{5.1.20}$$

Equating these two expressions for $\dot{\boldsymbol{\lambda}}$ yields

$$-\boldsymbol{A}^T\boldsymbol{P}(t)\boldsymbol{x} - \boldsymbol{Q}\boldsymbol{x} = \boldsymbol{P}(t)[\boldsymbol{A}\boldsymbol{x} + \boldsymbol{B}\boldsymbol{u}] + \dot{\boldsymbol{P}}(t)\boldsymbol{x} \tag{5.1.21}$$

Using the optimal control law of (5.1.17), we have

$$-\boldsymbol{A}^T\boldsymbol{P}(t)\boldsymbol{x}-\boldsymbol{Q}\boldsymbol{x}=\boldsymbol{P}(t)[\boldsymbol{A}\boldsymbol{x}-\boldsymbol{R}^{-1}\boldsymbol{B}^T\boldsymbol{P}\boldsymbol{x}]+\dot{\boldsymbol{P}}(t)\boldsymbol{x} \tag{5.1.22}$$

Since each term contains the $\boldsymbol{x}$ vector, the following matrix equality must hold:

$$\dot{\boldsymbol{P}}(t)=-\boldsymbol{P}(t)\boldsymbol{A}-\boldsymbol{A}^T\boldsymbol{P}(t)-\boldsymbol{Q}+\boldsymbol{P}(t)\boldsymbol{B}\boldsymbol{R}^{-1}\boldsymbol{B}^T\boldsymbol{P}(t) \tag{5.1.23}$$

Equation (5.1.23) is referred to as the Riccati matrix differential equation and is a quadratic matrix set of differential equations. Riccati was an Italian mathematician who studied quadratic algebraic equations. His "opere," which summarizes his main work, was published by his son after his death in 1765 (Editrice, 1989). The boundary conditions for the Riccati matrix differential equation of (5.1.23) can be derived from the transversality conditions

$$\boldsymbol{\lambda}(t_f)=\boldsymbol{H}\boldsymbol{x}(t_f) \tag{5.1.24}$$

Since the Riccati transformation gives

$$\boldsymbol{\lambda}(t_f)=\boldsymbol{P}(t_f)\boldsymbol{x}(t_f) \tag{5.1.25}$$

we have the condition that

$$\boldsymbol{P}(t_f)=\boldsymbol{H} \tag{5.1.26}$$

Notice that we know $\boldsymbol{P}$ at the final time, t_f. Therefore, we must integrate the Riccati differential equations backwards in time. However, Equation (5.1.23) is only a function of system parameters, $\boldsymbol{A}$ and $\boldsymbol{B}$, and performance parameters, $\boldsymbol{Q}$ and $\boldsymbol{R}$. It is not a function of the state, $\boldsymbol{x}$, or the control, $\boldsymbol{u}$. It can therefore be computed independently from the state differential equations. This means that the optimal control gain function, $\boldsymbol{K}(t)$, can be computed separately from the state dynamic response. In fact, it can be precomputed and stored for later use.

A significant simplification occurs when we let the final time, t_f, become infinity. In this case, the feedback gain function, $\boldsymbol{K}(t)$, becomes a constant, $\boldsymbol{K}_{ss}$. This fact is illustrated in Figure 5.1. If we start the calculation of $\boldsymbol{P}(t)$ at $t=\infty$, then eventually the set of Riccati differential equations will approach a steady state. This steady-state value, $\boldsymbol{K}_{ss}$, will be applicable for all finite times when starting from $t=0$. For the infinite final time problem, we have

$$\boldsymbol{u}(t)=-\boldsymbol{K}_{ss}\boldsymbol{x}(t) \tag{5.1.27}$$

where

$$\boldsymbol{K}_{ss}=\boldsymbol{R}^{-1}\boldsymbol{B}^T\boldsymbol{P}_{ss} \tag{5.1.28}$$

This steady-state form of the state feedback regulator is the one that is often implemented since we run regulator problems for many system time constants.

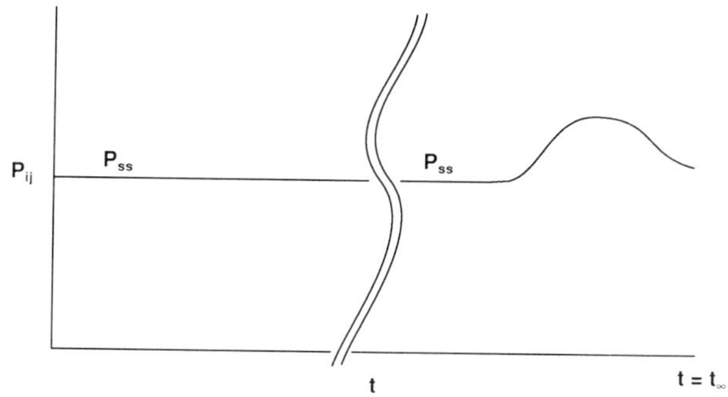

Figure 5.1 Steady-state Riccati matrix calculation.

It should be noted that Equation (5.1.27) is extremely easy to implement. It is also important to stress that we have found a control law that has feedback properties. The optimal regulator is a proportional state feedback controller. The block diagram for the LQR optimal controller is given in Figure 5.2. It shows that the controller is state feedback and assumes regulation about a steady-state set point of $\boldsymbol{x}_{sp} = \mathbf{0}$.

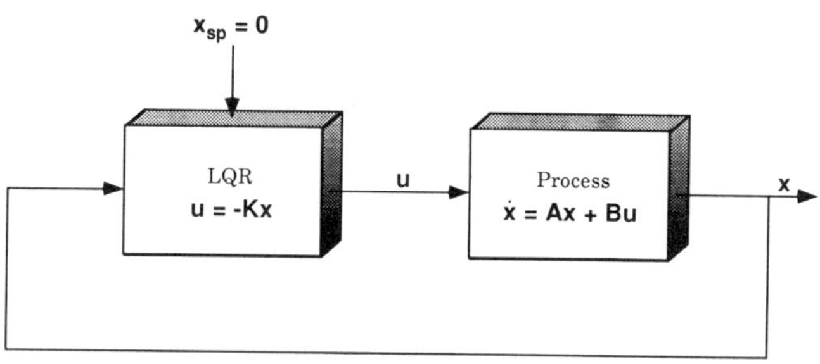

Figure 5.2 LQR feedback controller.

5.1.1 Output Regulation

There are times when we want to regulate about the system outputs rather than the system states. The system outputs are defined by

$$\boldsymbol{y} = \boldsymbol{M}\boldsymbol{x} \tag{5.1.29}$$

and the quadratic performance functional in terms of the system outputs is

$$J = \frac{1}{2}\,\boldsymbol{y}^T(t_f)\boldsymbol{H}'\boldsymbol{y}(t_f) + \frac{1}{2}\int_{t_o}^{t_f} \left(\boldsymbol{y}^T\boldsymbol{Q}'\boldsymbol{y} + \boldsymbol{u}^T\boldsymbol{R}\boldsymbol{u}\right)\,dt \tag{5.1.30}$$

Substituting Equation (5.1.29) into (5.1.30) gives

$$J = \frac{1}{2}\,\boldsymbol{x}^T(t_f)\boldsymbol{M}^T\boldsymbol{H}'\boldsymbol{M}\boldsymbol{x}(t_f) + \frac{1}{2}\int_{t_o}^{t_f} \left(\boldsymbol{x}^T\boldsymbol{M}^T\boldsymbol{Q}'\boldsymbol{M}\boldsymbol{x} + \boldsymbol{u}^T\boldsymbol{R}\boldsymbol{u}\right)\,dt \tag{5.1.31}$$

Therefore, with

$$\boldsymbol{Q} = \boldsymbol{M}^T\boldsymbol{Q}'\boldsymbol{M} \tag{5.1.32}$$

and

$$\boldsymbol{H} = \boldsymbol{M}^T\boldsymbol{H}'\boldsymbol{M} \tag{5.1.33}$$

we have transformed this output regulation problem into the original Kalman problem. The optimal control law is therefore proportional state variable feedback control with $\boldsymbol{Q}$ and $\boldsymbol{H}$ defined by Equations (5.1.32) and (5.1.33). Figure 5.3 gives the block diagram of output regulation. It shows that regulation is based upon the state, $\boldsymbol{x}$, even though the quadratic performance index is based upon the measurements, $\boldsymbol{y}$.

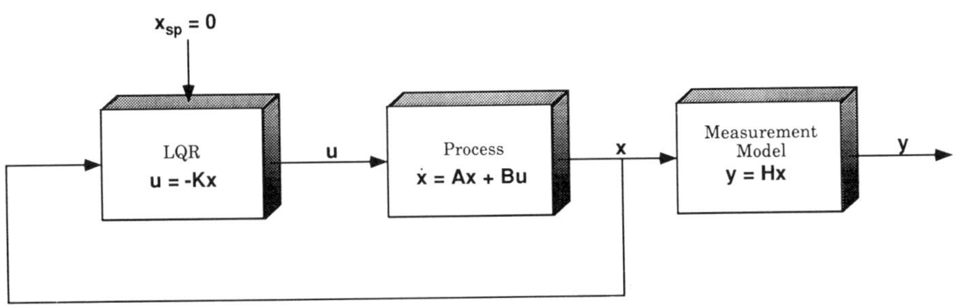

Figure 5.3 Optimal LQR output regulation.

5.2 Analytical Solution of the Riccati Equation

Section 5.1 has developed the Riccati differential equations and has shown that backwards integration of the matrix set of differential equations is feasible and can be an off-line operation. In this section we develop an analytical solution to the Riccati matrix, $\boldsymbol{P}(t)$. The necessary conditions for extremizing the quadratic performance functional for the linear regulator problem can be written

$$\dot{\boldsymbol{x}} = \boldsymbol{A}\boldsymbol{x} - \boldsymbol{B}\boldsymbol{R}^{-1}\boldsymbol{B}^T\boldsymbol{\lambda} \tag{5.2.1}$$

$$\dot{\boldsymbol{\lambda}} = -\boldsymbol{Q}\boldsymbol{x} - \boldsymbol{A}^T\boldsymbol{\lambda} \tag{5.2.2}$$

These homogeneous differential equations can be expressed as

$$\begin{pmatrix} \dot{\boldsymbol{x}} \\ \dot{\boldsymbol{\lambda}} \end{pmatrix} = \begin{bmatrix} \boldsymbol{A} & -\boldsymbol{B}\boldsymbol{R}^{-1}\boldsymbol{B}^T \\ -\boldsymbol{Q} & -\boldsymbol{A}^T \end{bmatrix} \begin{pmatrix} \boldsymbol{x} \\ \boldsymbol{\lambda} \end{pmatrix} \tag{5.2.3}$$

or

$$\begin{pmatrix} \dot{\boldsymbol{x}} \\ \dot{\boldsymbol{\lambda}} \end{pmatrix} = \boldsymbol{Z} \begin{pmatrix} \boldsymbol{x} \\ \boldsymbol{\lambda} \end{pmatrix} \tag{5.2.4}$$

Since Equation (5.2.4) is linear and homogeneous, its solution can be expressed as

$$\begin{pmatrix} \boldsymbol{x}(T) \\ \boldsymbol{\lambda}(T) \end{pmatrix} = e^{\boldsymbol{Z}\,\Delta t} \begin{pmatrix} \boldsymbol{x}(t) \\ \boldsymbol{\lambda}(t) \end{pmatrix} \tag{5.2.5}$$

where Δt is the time interval $(T - t)$ or

$$\begin{pmatrix} \boldsymbol{x}(T) \\ \boldsymbol{\lambda}(T) \end{pmatrix} = \boldsymbol{\theta}(\Delta t) \begin{pmatrix} \boldsymbol{x}(t) \\ \boldsymbol{\lambda}(t) \end{pmatrix} \tag{5.2.6}$$

Note that the transition matrix $\boldsymbol{\theta}(\Delta t)$ is only a function of the time interval Δt or $(T - t)$ and not the actual value of time t or T.

The transition matrix, $\boldsymbol{\theta}(\Delta t)$, can be partitioned as

$$\boldsymbol{\theta}(\Delta t) = \begin{bmatrix} \boldsymbol{\theta}_{11}(\Delta t) & \boldsymbol{\theta}_{12}(\Delta t) \\ \boldsymbol{\theta}_{21}(\Delta t) & \boldsymbol{\theta}_{22}(\Delta t) \end{bmatrix} \tag{5.2.7}$$

Therefore, the state and costate solutions are expressed as

$$\boldsymbol{x}(T) = \boldsymbol{\theta}_{11}(\Delta t)\boldsymbol{x}(t) + \boldsymbol{\theta}_{12}(\Delta t)\boldsymbol{\lambda}(t) \tag{5.2.8}$$

$$\boldsymbol{\lambda}(T) = \boldsymbol{\theta}_{21}(\Delta t)\boldsymbol{x}(t) + \boldsymbol{\theta}_{22}(\Delta t)\boldsymbol{\lambda}(t) \tag{5.2.9}$$

Also from the Riccati Equation, we have

$$\boldsymbol{P}(T)\boldsymbol{x}(T) = \boldsymbol{\theta}_{21}(\Delta t)\boldsymbol{x}(t) + \boldsymbol{\theta}_{22}(\Delta t)\boldsymbol{\lambda}(t) \tag{5.2.10}$$

Multiplying Equation (5.2.8) by $\boldsymbol{P}(T)$ and equating it with Equation (5.2.10) gives

$$\boldsymbol{P}(T)\boldsymbol{\theta}_{11}(\Delta t)\boldsymbol{x}(t) + \boldsymbol{P}(T)\boldsymbol{\theta}_{12}(\Delta t)\boldsymbol{\lambda}(t) = \boldsymbol{\theta}_{21}(\Delta t)\boldsymbol{x}(t) + \boldsymbol{\theta}_{22}(\Delta t)\boldsymbol{\lambda}(t) \tag{5.2.11}$$

or

$$\boldsymbol{\lambda}(t) = \left[\boldsymbol{\theta}_{22}(\Delta t) - \boldsymbol{P}(T)\boldsymbol{\theta}_{12}(\Delta t)\right]^{-1} \left[\boldsymbol{P}(T)\boldsymbol{\theta}_{11}(\Delta t) - \boldsymbol{\theta}_{21}(\Delta t)\right] \boldsymbol{x}(t) \quad (5.2.12)$$

Using the Riccati transformation

$$\boldsymbol{\lambda}(t) = \boldsymbol{P}(t)\boldsymbol{x}(t) \quad (5.2.13)$$

gives

$$\boldsymbol{P}(t) = \left[\boldsymbol{\theta}_{22}(\Delta t) - \boldsymbol{P}(T)\boldsymbol{\theta}_{12}(\Delta t)\right]^{-1} \left[\boldsymbol{P}(T)\boldsymbol{\theta}_{11}(\Delta t) - \boldsymbol{\theta}_{21}(\Delta t)\right] \quad (5.2.14)$$

and from the transversality condition at $T = T_f$

$$\boldsymbol{P}(T_f) = \boldsymbol{H} \quad (5.2.15)$$

Equation (5.2.14) can be used as a recursive method for calculating the Riccati matrix from its known final condition to its initial condition in time increments of $\Delta t = (T - t)$. The solution is in terms of a transition matrix, $\boldsymbol{\theta}(\Delta t)$. The time increment Δt should be chosen such that accurate numerical results are obtained for the transition matrix $\boldsymbol{\theta}(\Delta t)$.

5.3 MATLAB Program for the Linear Regulator Problem

The MATLAB computer package has several subroutines for the solution of the linear regulator problem. These include LQR. The purpose of LQR is to solve the steady-state linear regulator problem. The description of LQR is given in Appendix A. We will illustrate the on-line use of LQR for the solution of the stirred tank reactor problem of the next section.

5.4 Regulator Control of a Stirred Tank Reactor

As a focus for the development of practical concepts needed for the formulation and implementation of optimal control, let us consider the continuous stirred tank reactor studied by Nyquist and Ramirez (1971). The reactor was a 16-liter vessel with an internal cooling coil used for cooling the exothermic reaction. A schematic of the laboratory reactor is shown in Figure 5.4.

The reaction studied was the decomposition of hydrogen peroxide to oxygen and water:

$$H_2O_2 \rightarrow H_2O(\ell) + \frac{1}{2}\,O_2(g) \quad (5.4.1)$$

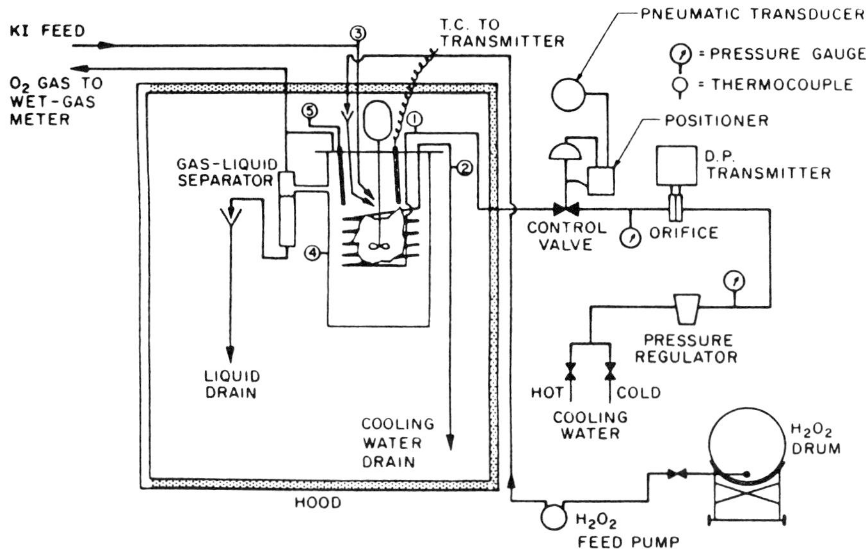

Figure 5.4 Laboratory reactor system.

The temperature and hydrogen peroxide concentration in the reactor are the state variables of interest. The temperature is measured by an iron constantan thermocouple. The m.V. signal is converted to an m.A. signal with a transmitter. A voltage signal (0.5 – 2.5 V, dc) is also available. The hydrogen peroxide concentration is not directly measurable, but rather the reaction rate is measured continuously by monitoring the oxygen gas evolution rate.

The cooling water flow was controlled by a 1/2 inch air-to-close pneumatic valve equipped with a valve positioner to overcome inherent valve hysteresis. The signal to operate the valve varied between 20 to 100 V. Temperature control of the reactor is achieved by varying the cooling valve control signal to control the cooling water flow through the cooling coil.

Nyquist and Ramirez have developed and tested a mathematical model of the process and have shown that it describes the process to within 2%. Their model includes the peroxide material balance

$$\frac{dC}{dt} = \frac{F_p C_o}{V_x} - \frac{FC}{V_x} - R \tag{5.4.2}$$

The reaction rate is linear in both the peroxide concentration and the homogeneous catalyst concentration of potassium iodide

$$R = k' C\, C_I \tag{5.4.3}$$

The rate constant follows the Arrhenius temperature dependency

$$k' = k_o \exp\left(-E/(T + 2.73.16)\right) \tag{5.4.4}$$

The material balance for the iodide ion catalyst concentration is

$$C_I = \frac{(F - F_p)}{F} C_{I_0} \tag{5.4.5}$$

The energy balance for the reactor is

$$V_H \rho c_P \frac{dT}{dt} = F\rho c_p(T_i - T) + Q_s + (-H)RV_x - Q_c \tag{5.4.6}$$

Nyquist and Ramirez have shown that the cooling rate can be modeled as

$$Q_c = \beta(T - T_{c_i}) \tag{5.4.7}$$

where the effective cooling rate coefficient, β, is a unique function of the cooling valve signal, V_s:

$$\beta = 1.51 \times 10^3 - 27.26\, V_s + 0.1499\, V_s^2 - 2.38 \times 10^{-4} V_s^3 \tag{5.4.8}$$

The reactor was equipped with a relatively powerful stirrer, so the conversion of mechanical energy to thermal energy resulted in a stirrer transfer rate, Q_s. This term was shown to be given adequately by a first-order model

$$Q_s = 1288 - 27.06\, T \tag{5.4.9}$$

Specific model parameters that were used to verify the model experimentally are given in Table 5.1. Figure 5.5 shows typical model predictions versus experimental data in running the reactor system. The model was considered verified by these types of results.

If we linearize the model and express it in terms of normalized deviation variables, we have

$$\dot{\boldsymbol{x}} = \boldsymbol{A}\boldsymbol{x} + \boldsymbol{b}u \tag{5.4.10}$$

where

$$\boldsymbol{A} = \begin{bmatrix} -0.05156 & -0.05877 \\ 0.01503 & -0.005887 \end{bmatrix}$$

$$\boldsymbol{b} = \begin{pmatrix} 0 \\ -0.03384 \end{pmatrix}$$

$$\boldsymbol{x} = \begin{pmatrix} \hat{C} \\ \hat{T} \end{pmatrix}$$

Table 5.1 Parameters Used in Simulations

Operating parameters	
$V_H = 16.74$ l.	$-H = -22,600$ cal/g mole H_2O_2
$V_X = 15.78$ l.	$C_0 = 8.53$ g mole/l.
$F = 0.1312$ l.	$T_i = 24.5°C$
$F_p = 0.032$ l./min	$T_s = 25.0°C$
$V_{ss} = 60\ V$	
$K_p = 15\ V/°C$	
$T(\theta = 0) = 34.33°C$	
$C(\theta = 0) = \dfrac{0.310 \text{ g mole}}{l.}$	

Inlet water temperature

θ (min)	T_c (°C)
0	10.0
15	10.9
30	12.0
45	11.8
60	12.2

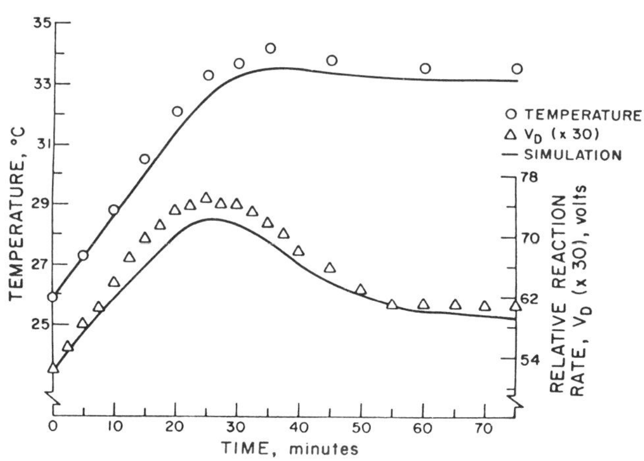

Figure 5.5 Model and data for an experimental run.

$$
\begin{aligned}
u &= \hat{\beta} \\
\hat{C} &= \frac{C - C_{ss}}{\Delta C} \\
C_{ss} &= \text{steady-state concentration, 0.361 gmoles}/\ell \\
\Delta C &= \text{characteristic concentration range, 0.492 gmoles}/\ell \\
&\quad \text{(the amount the steady-state value varies under max} \\
&\quad \text{and min cooling)} \\
\hat{T} &= \frac{T - T_{ss}}{\Delta T} \\
T_{ss} &= \text{steady-state temperature, 30 °C} \\
\Delta T &= \text{characteristic temperature range, 28.9 °C} \\
\hat{\beta} &= \frac{\beta - \beta_{ss}}{\Delta \beta} \\
\beta_{ss} &= \text{steady-state effective heat transfer coefficient,} \\
&\quad \text{256 cal/min °C} \\
\Delta\beta &= \text{characteristic } \beta \text{ range, 978 cal/min °C}
\end{aligned}
$$

Figure 5.6 gives the MATLAB interactive program that computes the steady-state Riccati matrix and steady-state feedback gain matrix for this problem. The program also uses the MATLAB function IMPULSE to determine the transient response of the system. Appropriate plots are also prepared. The LQR results for $\boldsymbol{Q} = \boldsymbol{R} = \boldsymbol{I}$ are

$$
\boldsymbol{P}_{ss} = \begin{bmatrix} 9.237 & -1.493 \\ -1.493 & 27.31 \end{bmatrix} \tag{5.4.11}
$$

$$
\boldsymbol{K}_{ss} = \begin{bmatrix} 5.05 \times 10^{-2} & -9.24 \times 10^{-1} \end{bmatrix} \tag{5.4.12}
$$

The response of the linearized system using the optimal state LQR feedback control law is shown in Figure 5.7. These results show that the linear system using the LQR control law is very satisfactory. Figure 5.7 also shows the response of the control variable as a function of time. In implementing this control law, it is important to realize that measurements must be available for both the system temperature and the hydrogen peroxide concentration, since the feedback law is a function of both state variables. Knowledge of the entire state is certainly a limitation to the direct usage of the Kalman optimal regulator. However, in Chapter 10 we show how state estimation theory can give optimal estimates of the entire state using the available measurements and the mathematical model of the system.

```
aa =

   -0.0516   -0.0588
    0.0167   -0.0372

>> aa=a-b*k

aa =

   -0.0516   -0.0588
    0.0389   -0.1195

>> y=impulse (aa,bb,c,d,1,t);
>> u=-k*y';
>> plot (t,y,t,u')
>> meta plot2
>> title('Plot of optimal state and contro vs time')
>> ylabel('State vector components and control variable')
>> xlabel('Time, min')
>> meta plot2
>>% redifining the Q and R weighting matrices
>> q(1,1)=10;
>> r=1;
>> [k,p]=lqr (a,b,q,r)

k =

    1.0172   -1.9623


p =

   78.1773  -30.0600
  -30.0600   57.9868

>> aa=a-b*k

aa =

   -0.0516   -0.0588
    0.0495   -0.0723

>> y=impulse (aa,bb,c,d,1,t);
>> u = -k*y';
>> plot (t,y,t,u')
>> title('Plot of optimal state and control vs time')
>> ylabel('State vector components and control variable')
>> xlabel('Time,min')
>> meta plot3
>> quit

 17423 flop(s).
```

Figure 5.6 MATLAB program for LQR of Nyquist–Ramirez reactor.

```
>> help lqr

 LQR    Linear quadratic regulator design for continuous systems.
        [K,S,E] = LQR(A,B,Q,R)  calculates the optimal feedback gain
        matrix K such that the feedback law  u = -Kx  minimizes the cost
        function:
                J = Integral {x'Qx + u'Ru} dt

        subject to the constraint equation:
                .
                x = Ax + Bu

        Also returned is S, the steady-state solution to the associated
        algebraic Riccati equation and the closed loop eigenvalues E:
                                      -1
                0 = SA + A'S - SBR  B'S + Q      E = EIG(A-B*K)

        [K,S,E] = LQR(A,B,Q,R,N) includes the cross-term N that relates
        u to x in the cost function.

                J = Integral {x'Qx + u'Ru + 2*x'Nu}

        The controller can be formed with REG.

        See also: LQRY, LQR2, and REG.

>> %input the a and b matrices
>> a =  [-0.05156  -0.05877;
          0.01503  -0.005887];
>> b = [0  -0.03384]';
>> help eye

EYE     Identity matrix. EYE(N) is the N-by-N identity matrix.
        EYE(M,N) is an M-by-N matrix with 1's on the diagonal and
        zeros elsewhere. EYE(A) is the same size as A.

>> % imput the q and r maatrices
>>
>> q=eye(2)

q =
     1     0
     0     1
>> r=1;
>> % compute the Kalman gain and Riccati  maatrix
>>
>>  [k,p] = lqr(a,b,q,r)

k =

    0.0505   -0.9241


p =

    9.2375   -1.4929
   -1.4929   27.3078

>> % use IMPLUSE to get the response of the linear system
>>
>> help impulse

 IMPULSE Impulse response of continuous-time linear systems.
```

Figure 5.6 continued.

```
        IMPULSE(A,B,C,D,IU)  plots the time response of the linear system
                .
                x = Ax + Bu
                y = Cx + Du
        to an impulse applied to the single input IU.  The time vector is
        automatically determined.

        IMPULSE(NUM,DEN) plots the impulse response of the polynomial
        transfer function  G(s) = NUM(s)/DEN(s)  where NUM and DEN contain
        the polynomial coefficients in descending powers of s.

        IMPULSE(A,B,C,D,IU,T) or IMPULSE(NUM,DEN,T) uses the user-supplied
        time vector T which must be regularly spaced.  When invoked with
        left hand arguments,
                [Y,X,T] = IMPULSE(A,B,C,D,...)
                [Y,X,T] = IMPULSE(NUM,DEN,...)
        returns the output and state time history in the matrices Y and X.
        No plot is drawn on the screen.  Y has as many columns as there
        are outputs and length(T) rows.  X has as many columns as there
        are states.

        See also: STEP,INITIAL,LSIM and DIMPULSE.

>> % redefine some matrices to put in proper form so that the immpulse is
>> % applied to both state variables with magnitude unity
>> aa = a-b*k

aa =

   -0.0516   -0.0588
    0.0167   -0.0372

>>
>> bb = [1 1]';
>> c = eye(2)

c =

     1     0
     0     1

>> d =  [0  0]'
>> t = [0:1:100];
>> y = impulse (aa,bb,c,d,1,t);
>> u = -k*y';
>> uu=u';
>> plot (t,y,t,uu)
>> title ('Plot of optimal state and control vs time')
>> xlabel ('Time, min')
>> ylabel ('State vector components and control variable')
>> meta plot1
>>% readjusting the R weighting
>> r=.1;
>> [k,p]=lqr(a,b,q,r)

k =

    0.7054   -3.3588


p =

    8.6073   -2.0845
   -2.0845    9.9255

>> aa
```

Figure 5.6 continued.

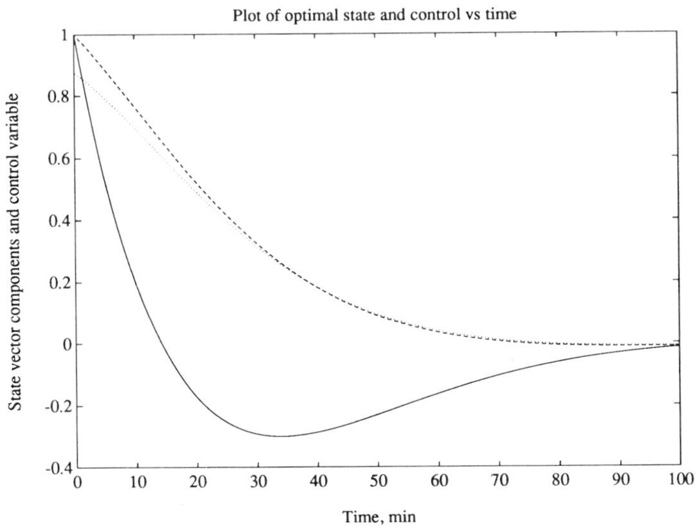

Figure 5.7 Optimal LQR response; $\boldsymbol{Q} = \boldsymbol{I}$, $\boldsymbol{R} = \boldsymbol{I}$ $\boldsymbol{K} = [.0505 - .9241]$. Concentration (—), temperature (- - -), cooling flow (···).

The control engineer has to choose appropriate values for the weighting matrices $\boldsymbol{Q}$ and $\boldsymbol{R}$. They represent the relative importance of quadratic terms in the various states and quadratic terms in the controls to the overall quadratic objective functional of Equation (5.1.3). As a designer, a control engineer will usually use diagonal $\boldsymbol{Q}$ and $\boldsymbol{R}$ matrices as a matter of convenience. For this problem the performance functional becomes, assuming diagonal weighting matrices,

$$J = \int_{t_0}^{t_f} \left(q_{11} x_1^2 + q_{22} x_2^2 + r_{11} u_1 \right) dt \tag{5.4.13}$$

For the base case we chose $q_{11} = q_{22} = r_{11} = 1.0$, which implies that there is equal importance in quadratic errors in each of these terms. It could be that it is more important to minimize the quadratic terms in the state and not penalize the use of the controls as heavily. Figure 5.8 shows the response of the states and the control variables when $\boldsymbol{Q} = \boldsymbol{I}$ and $\boldsymbol{R} = .1\boldsymbol{I}$. The system responds faster, but larger values for the control variable are used. The designer must be cognizant of potential limits on the controls in choosing relatively small $\boldsymbol{R}$ weightings.

It is also possible that it is more important to more tightly control one of the state variables. For this problem it is probably more important that the concentration deviations, x_1, be minimized in preference to temperature

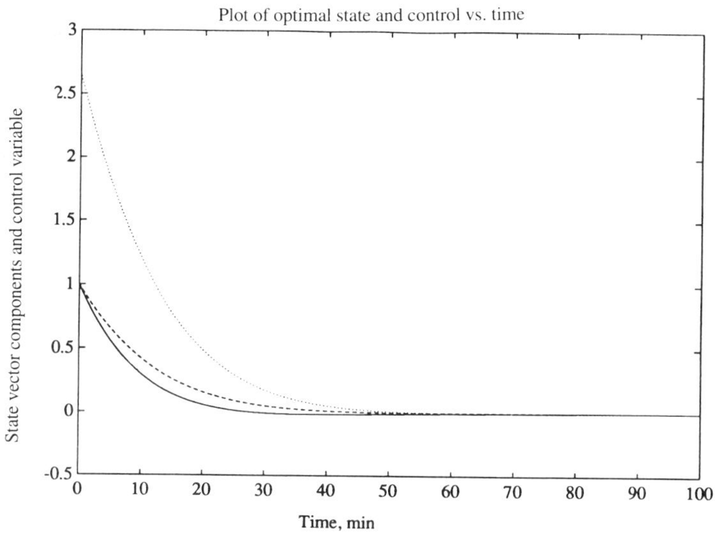

Figure 5.8. Optimal LQR response; $\boldsymbol{Q} = \boldsymbol{I}$, $\boldsymbol{R} = .1\boldsymbol{I}$ $\boldsymbol{K} = [.7054 \quad 3.359]$. Concentration (—), temperature (- - -), cooling flow (···).

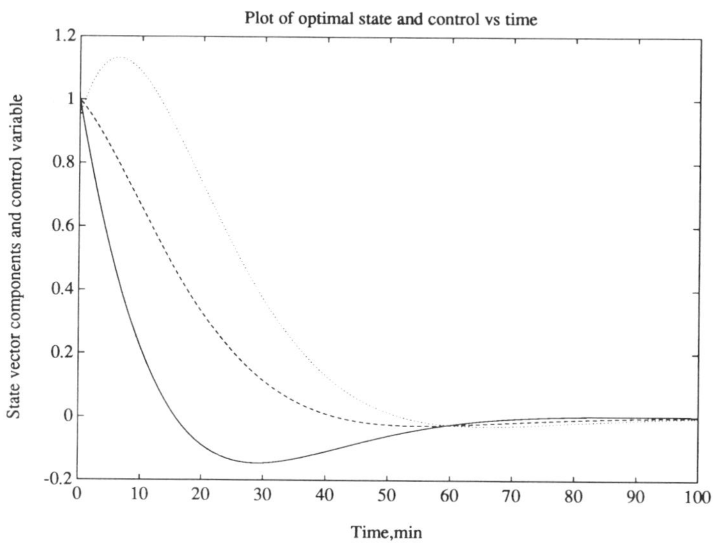

Figure 5.9. Optimal LQR response; $q_{11} = 10$, $q_{22} = 1$, $r_{11} = 1$ $\boldsymbol{K} = [1.0172 \quad -1.9623]$. Concentration (—), temperature (- - -), cooling flow (···).

deviations, since the outlet concentration is the process product of importance. Figure 5.9 shows the response of the states and the control variables when $q_{11} = 10$, $q_{22} = 1$, and $r_{11} = 1$ as well as the optimal feedback gain matrix for this case. Indeed there is an improved regulatory response of the concentration state since more relative weight was put on the concentration deviation in the performance functional.

5.5 The Tracking Problem with Disturbances

One of the major drawbacks of Kalman's original work was the fact that load disturbances, that is, disturbances that affect the system over the operational period, are not included in the development. Most process control problems need to respond to disturbances that enter the system. A linear model of process that includes disturbances is

$$\dot{\boldsymbol{x}} = \boldsymbol{A}\boldsymbol{x}(t) + \boldsymbol{B}\boldsymbol{u}(t) + \boldsymbol{d}(t) \tag{5.5.1}$$

where $\boldsymbol{d}(t)$ are the time-varying disturbances that enter the problem. The initial conditions are

$$\boldsymbol{x}(t_o) = \boldsymbol{x}_o \tag{5.5.2}$$

and the output variables that we desire to track are

$$\boldsymbol{y}(t) = \boldsymbol{M}\boldsymbol{x}(t) \tag{5.5.3}$$

If the disturbances are constant, then Equation (5.5.1) can be solved for the steady state by setting $\dot{\boldsymbol{x}} = \boldsymbol{0}$.

If we use the Kalman proportional feedback controller

$$u_{ss} = -\boldsymbol{K}_{ss}\boldsymbol{x}_{ss} \tag{5.5.4}$$

then the steady state becomes

$$\boldsymbol{x}_{ss} = -\left[\boldsymbol{A} - \boldsymbol{B}\boldsymbol{K}_{ss}\right]^{-1}\boldsymbol{d} \tag{5.5.5}$$

which shows that there is a steady-state error in all the states using the standard Kalman proportional feedback control law. It is well known that proportional controllers yield a steady-state error when constant load disturbances are experienced by a system. A new control law therefore must be developed to handle load disturbances.

We will assume that there is a desired path that the output variables should follow, which we denote as $\boldsymbol{\eta}$. As a performance measure for the optimal control problem, we use a form quadratic in the final error, $\boldsymbol{\eta}(t_f) - \boldsymbol{y}(t_f)$, quadratic in the integral of the instantaneous error, $\boldsymbol{\eta}(t) - \boldsymbol{y}(t)$, and quadratic in the control. The performance functional is therefore

$$J = \frac{1}{2}\left\|\boldsymbol{\eta}(t_f) - \boldsymbol{y}(t_f)\right\|_{\boldsymbol{H}}^2 + \frac{1}{2}\int_{t_o}^{t_f}\left[\left\|\boldsymbol{\eta}(t) - \boldsymbol{y}(t)\right\|_{\boldsymbol{Q}}^2 + \left\|\boldsymbol{u}(t)\right\|_{\boldsymbol{R}}^2\right] dt \quad (5.5.6)$$

We now want to develop necessary conditions for the optimal control policy for this tracking problem with disturbances. The Hamiltonian is

$$H = \frac{1}{2}\left\|\boldsymbol{\eta} - \boldsymbol{M}\boldsymbol{x}\right\|_{\boldsymbol{Q}}^2 + \frac{1}{2}\left\|\boldsymbol{u}\right\|_{\boldsymbol{R}}^2 + \boldsymbol{\lambda}^T\left[\boldsymbol{A}\boldsymbol{x}(t) + \boldsymbol{B}\boldsymbol{u}(t) + \boldsymbol{d}(t)\right] \quad (5.5.7)$$

The costate equation dynamics are given by

$$\frac{\partial H}{\partial \boldsymbol{x}} = -\dot{\boldsymbol{\lambda}} \quad (5.5.8)$$

or

$$-\dot{\boldsymbol{\lambda}} = \boldsymbol{M}^T\boldsymbol{Q}\left[\boldsymbol{M}\boldsymbol{x} - \boldsymbol{\eta}\right] + \boldsymbol{A}^T\boldsymbol{\lambda} \quad (5.5.9)$$

The transversality conditions are

$$\boldsymbol{\lambda}(t_f) = \boldsymbol{M}^T\boldsymbol{H}\left[\boldsymbol{M}\boldsymbol{x}(t_f) - \boldsymbol{\eta}(t_f)\right] \quad (5.5.10)$$

The optimal control policy is given by

$$\frac{\partial H}{\partial \boldsymbol{u}} = \boldsymbol{0} \quad (5.5.11)$$

or

$$\boldsymbol{u} = -\boldsymbol{R}^{-1}\boldsymbol{B}^T\boldsymbol{\lambda} \quad (5.5.12)$$

If we assume a Riccati transformation of the form

$$\boldsymbol{\lambda}(t) = \boldsymbol{P}(t)\boldsymbol{x}(t) - \boldsymbol{\xi}(t) \quad (5.5.13)$$

then this leads to the following Riccati equations:

$$\dot{\boldsymbol{P}} = -\boldsymbol{P}\boldsymbol{A} - \boldsymbol{A}^T\boldsymbol{P} + \boldsymbol{P}\boldsymbol{B}\boldsymbol{R}^{-1}\boldsymbol{B}^T\boldsymbol{P} - \boldsymbol{M}^T\boldsymbol{Q}\boldsymbol{M} \quad (5.5.14)$$

with

$$\boldsymbol{P}(t_f) = \boldsymbol{M}^T\boldsymbol{H}\boldsymbol{M} \quad (5.5.15)$$

and

$$\dot{\boldsymbol{\xi}} = \left[\boldsymbol{A} + \boldsymbol{B}\boldsymbol{R}^{-1}\boldsymbol{B}^T\boldsymbol{P}\right]^T\boldsymbol{\xi} + \boldsymbol{P}\boldsymbol{d}(t) - \boldsymbol{M}^T\boldsymbol{Q}\boldsymbol{\eta}(t) \quad (5.5.16)$$

with

$$\boldsymbol{\xi}(t_f) = \boldsymbol{M}^T\boldsymbol{H}\boldsymbol{\eta}(t_f) \quad (5.5.17)$$

The optimal control law is

$$\boldsymbol{u} = -\boldsymbol{R}^{-1}\boldsymbol{B}^T\left[\boldsymbol{P}(t)\boldsymbol{x}(t) - \boldsymbol{\xi}(t)\right] \tag{5.5.18}$$

This Riccati transformation has also led to a decoupling of the state response from the computation of the Riccati matrix differential equations of Equation (5.5.14). Also the Riccati vector, $\boldsymbol{\xi}(t)$, is not a function of the state response; however, it is a function of the disturbances, $\boldsymbol{d}(t)$. Unfortunately, to compute $\boldsymbol{\xi}(t)$ from Equation (5.5.16), we must integrate the equations backward in time since the transversality condition has specified $\boldsymbol{\xi}(t_f)$ in Equation (5.5.17). This means that we need to have knowledge of the disturbances, $\boldsymbol{d}(t)$, over the time domain, $[t_o, t_f]$. It is interesting to look at the form of the controller of Equation (5.5.18). It is a state feedback controller with a predictor or feed-forward term, $\boldsymbol{\xi}(t)$. This feed-forward term is a function of known or measurable disturbances, $\boldsymbol{d}(t)$; however, these disturbances must be known over the entire time domain, $[t_o, t_f]$. Therefore, for most problems, this formulation of the problem is not practical. However, in Section 6.9.3 we illustrate that this approach is feasible for trajectory-following tracking problems.

5.6 Unmeasurable Disturbances

In the late 1960s and early 1970s, Johnson (1968, 1972) proposed a method for considering unmeasurable disturbances. The system he considered is the linear plant

$$\dot{\boldsymbol{x}} = \boldsymbol{A}\boldsymbol{x} + \boldsymbol{B}\boldsymbol{u} + \boldsymbol{W}\boldsymbol{d} \tag{5.6.1}$$

where $\boldsymbol{d}$ are unmeasurable step disturbances. If we define $\boldsymbol{y}$ as

$$\boldsymbol{y} = \boldsymbol{B}\boldsymbol{u} + \boldsymbol{W}\boldsymbol{d} \tag{5.6.2}$$

then for piecewise constant disturbances, we have

$$\dot{\boldsymbol{y}} = \boldsymbol{B}\dot{\boldsymbol{u}} \tag{5.6.3}$$

With this definition of the vector, $\boldsymbol{y}$ (Equation (5.6.2)), the original process can be expressed as

$$\dot{\boldsymbol{x}} = \boldsymbol{A}\boldsymbol{x} + \boldsymbol{y} \tag{5.6.4}$$

Combining Equations (5.6.3) and (5.6.4) yields

$$\begin{pmatrix} \dot{\boldsymbol{x}} \\ \dot{\boldsymbol{y}} \end{pmatrix} = \begin{bmatrix} \boldsymbol{A} & \boldsymbol{I} \\ 0 & 0 \end{bmatrix} \begin{pmatrix} \boldsymbol{x} \\ \boldsymbol{y} \end{pmatrix} + \begin{bmatrix} 0 \\ \boldsymbol{B} \end{bmatrix} \dot{\boldsymbol{u}} \tag{5.6.5}$$

If we let

$$\tilde{\boldsymbol{x}} = \begin{pmatrix} \boldsymbol{x} \\ \boldsymbol{y} \end{pmatrix} \tag{5.6.6}$$

$$\tilde{\boldsymbol{u}} = \dot{\boldsymbol{u}} \tag{5.6.7}$$

$$\tilde{\boldsymbol{A}} = \begin{bmatrix} \boldsymbol{A} & \boldsymbol{I} \\ 0 & 0 \end{bmatrix} \tag{5.6.8}$$

$$\tilde{\boldsymbol{B}} = \begin{bmatrix} 0 \\ \boldsymbol{B} \end{bmatrix} \tag{5.6.9}$$

then the original problem has been transformed to the equation set

$$\dot{\tilde{\boldsymbol{x}}} = \tilde{\boldsymbol{A}}\tilde{\boldsymbol{x}} + \tilde{\boldsymbol{B}}\tilde{\boldsymbol{u}} \tag{5.6.10}$$

We want to develop the optimal control for the quadratic performance functional

$$J = \frac{1}{2}\,\tilde{\boldsymbol{x}}^T(t_f)\tilde{\boldsymbol{H}}\tilde{\boldsymbol{x}}(t_f) + \frac{1}{2}\int_{t_o}^{t_f} \left(\tilde{\boldsymbol{x}}^T\tilde{\boldsymbol{Q}}\tilde{\boldsymbol{x}} + \tilde{\boldsymbol{u}}^T\tilde{\boldsymbol{R}}\tilde{\boldsymbol{u}}\right)\,dt \tag{5.6.11}$$

where

$$\tilde{\boldsymbol{Q}} = \begin{bmatrix} \boldsymbol{Q} & 0 \\ 0 & 0 \end{bmatrix} \tag{5.6.12}$$

$$\tilde{\boldsymbol{R}} = \boldsymbol{R} \tag{5.6.13}$$

$$\tilde{\boldsymbol{H}} = \begin{bmatrix} \boldsymbol{H} & 0 \\ 0 & 0 \end{bmatrix} \tag{5.6.14}$$

Equations (5.6.10) and (5.6.11) show that we have transformed the problem into an equivalent Kalman problem so that the optimal control is

$$\tilde{\boldsymbol{u}} = -\boldsymbol{K}(t)\tilde{\boldsymbol{x}} \tag{5.6.15}$$

where $\boldsymbol{K}(t)$ can be computed from the appropriate Riccati matrix differential equations. The infinite-time or steady-state solution to the Riccati equations leads to

$$\tilde{\boldsymbol{u}} = -\boldsymbol{K}_{ss}\tilde{\boldsymbol{x}} \tag{5.6.16}$$

Since the system described by Equation (5.6.10) is often not controllable (although the original problem described by Equation (5.6.1) is controllable), the MATLAB algorithm LQR cannot be used to compute the steady-state Kalman feedback gains. Instead, the recursive algorithm given by Equation (5.2.14) should be used.

In terms of the original problem variables, we have

$$\dot{\boldsymbol{u}} = -\left[\begin{array}{c|c} \boldsymbol{K}_1 & \boldsymbol{K}_2 \end{array}\right]\begin{pmatrix} \boldsymbol{x} \\ \boldsymbol{y} \end{pmatrix} \tag{5.6.17}$$

If we rearrange Equation (5.6.4), then

$$\boldsymbol{y} = \dot{\boldsymbol{x}} - \boldsymbol{A}\boldsymbol{x} \tag{5.6.18}$$

Using Equation (5.6.18) in (5.6.17) gives

$$\dot{\boldsymbol{u}} = -\boldsymbol{K}_1\boldsymbol{x} - \boldsymbol{K}_2\,(\dot{\boldsymbol{x}} - \boldsymbol{A}\boldsymbol{x}) \tag{5.6.19}$$

This is an important relationship since it expresses the control as a unique function of the state. Integrating both sides of Equation (5.6.19) yields

$$\boldsymbol{u} = -\boldsymbol{K}_2\boldsymbol{x} - [\boldsymbol{K}_1 - \boldsymbol{K}_2\boldsymbol{A}] \int \boldsymbol{x}\, dt \tag{5.6.20}$$

This shows that the optimal regulator for a linear system with unmeasurable constant disturbances is a state feedback proportional-integral controller. The block diagram for the Johnson controller is given in Figure 5.10. It shows that the Johnson LQR algorithm for handling unmeasured disturbances has the same feedback structure as the Kalman algorithm (Figure 5.2). The only difference is the inclusion of the integral term in the optimal controller.

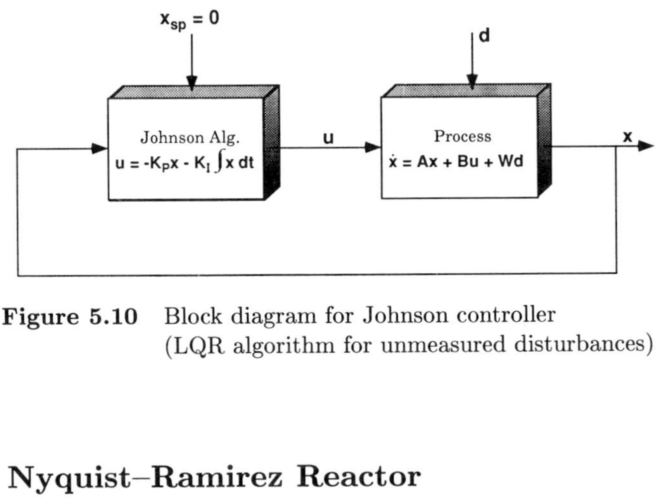

Figure 5.10 Block diagram for Johnson controller (LQR algorithm for unmeasured disturbances).

5.6.1 Nyquist–Ramirez Reactor

Let us apply this technique to the Nyquist–Ramirez reactor. The actual disturbances for the system are the inlet concentration, C_i, and the inlet temperature, T_i. In terms of normalized deviation variables

$$\hat{C}_i = \frac{C_i - C_{iss}}{\Delta C_I} \tag{5.6.21}$$

$$\hat{T}_i = \frac{T_i - T_{is}}{\Delta T_i} \tag{5.6.22}$$

where $\Delta C_i = .114$ g mole/liter and $\Delta T_i = 6.7°C$, the mathematical model of the system is

$$\dot{\boldsymbol{x}} = \boldsymbol{A}\boldsymbol{x} + \boldsymbol{B}u + \boldsymbol{W}\boldsymbol{d} \tag{5.6.23}$$

with

$$\boldsymbol{d} = \begin{pmatrix} \hat{C}_i \\ \hat{T}_i \end{pmatrix} \tag{5.6.24}$$

and

$$\boldsymbol{W} = \begin{bmatrix} .0125 & 0 \\ 0 & .0125 \end{bmatrix} \tag{5.6.25}$$

For this problem,

$$\tilde{\boldsymbol{A}} = \begin{bmatrix} a_{11} & a_{12} & 1 & 0 \\ a_{21} & a_{22} & 0 & 1 \\ 0 & 0 & 0 & 0 \\ 0 & 0 & 0 & 0 \end{bmatrix} \tag{5.6.26}$$

$$\tilde{\boldsymbol{B}} = \begin{bmatrix} 0 \\ 0 \\ 0 \\ b_2 \end{bmatrix} \tag{5.6.27}$$

$$\tilde{\boldsymbol{R}} = r_{11} \tag{5.6.28}$$

$$\tilde{\boldsymbol{Q}} = \begin{bmatrix} q_{11} & 0 & 0 & 0 \\ 0 & q_{22} & 0 & 0 \\ 0 & 0 & 0 & 0 \\ 0 & 0 & 0 & 0 \end{bmatrix} \tag{5.6.29}$$

When we desire to regulate for constant concentration, then the appropriate weighting matrix $\boldsymbol{Q}$ is $q_{11} = 1$ and $q_{22} = 0$. If we desire to regulate for constant temperature, then $q_{22} = 1$ and $q_{11} = 0$. For this system, because we have two state variables and only one control variable, only one of them can be regulated about the original steady-state value. With a load disturbance, the other state will go to a new steady-state value.

The MATLAB program for solving the PI optimal control of the Nyquist–Ramirez reactor is given in Figure 5.11. This program is an M file which can be run in MATLAB. The program calculates the steady-state feedback gain matrix using the recursive algorithm given in Section 5.2. It calculates the transition matrix for a specified time increment, $\Delta t = 1$ min, and contains a logical loop for implementing the backwards recursive formula of Equation (5.2.14). The results of the algorithm are given in Figure 5.12 when $q_{11} = 1$ and $q_{22} = 0$ (composition regulation). Figure 5.13 shows the results for the case when $q_{11} = 0$ and $q_{22} = 1$ (temperature regulation). The steady-state

```
% input the a and b matrices
a = [-0.05156  -0.05877;
      0.01503  -0.005887];
b = [0  -0.03384]';
% input the w matrix
w = [0.0125  0;
     0        0.0125];

% define the Johnson transformation

aa= [a    eye(2);
     zeros(2) zeros(2)];
bb= [0;
     0;
     b];
% define the weighting matrices

q = [1  0;
     0  0];

r = 1;

% Johnson transformation weighting matrices

qq = [q       zeros(2);
      zeros(2) zeros(2)];

% calculate the feedback gain using recursive method of section 5.2

z = [aa      -bb*inv(r)*bb';
     -qq     -aa'          ];

% chose the integration time interval dt and calculate transition matrix

dt = 1;

theta = expm(z);

% partition the transition matrix

theta11 = theta(1:4,1:4);
theta12 = theta(1:4,5:8);
theta21 = theta(5:8,1:4);
theta22 = theta(5:8,5:8);

% define the final Raccati matrix p(T)

pt = zeros(4);

% define t max

tmax = 70;
%loop for calcuating feedback gain matrix k

for t = tmax:-1:1,
        pt1 = inv(theta22 - pt*theta12)*(pt*theta11 - theta21);
        kt1 = inv(r)*bb'*pt1;
%       this loop is to transform the  matrix into a vector
        for col = 1:4
                k(t,col) = kt1(1,col);
        end
        pt=pt1;
```

Figure 5.11 MATLAB program for the Johnson algorithm control of the Nyquist–Ramirez reactor.

```
% plot k vs time

t = 0:(tmax-1);
plot (t,k);
title('Gain Matrix vs Time'),ylabel('Gain Components'),xlabel('Time (min)');
meta pigainmat;
% compute proportional and integral feedback matrices
%
k1 = kt1(1,1:2);
k2 = kt1(1,3:4);
kp = k2
ki = k1 - k2*a
%
% preparing system for step change response
%
api = [a-b*kp    -b*ki;
        eye(2)    zeros(2)];
t = [0:1:60];
wi= .2*[w          zeros(2);
          zeros(2) zeros(2)];
c = eye(4);
d = zeros(4);
xpi = step(api,wi,c,d,1,t);
x = xpi(1:61,1:2);
plot (t,x);
title('Johnson Algorithm State Response');
ylabel('Normalized composition and temperature');
xlabel('Time(min)');
meta xpi;
```

Figure 5.11 continued

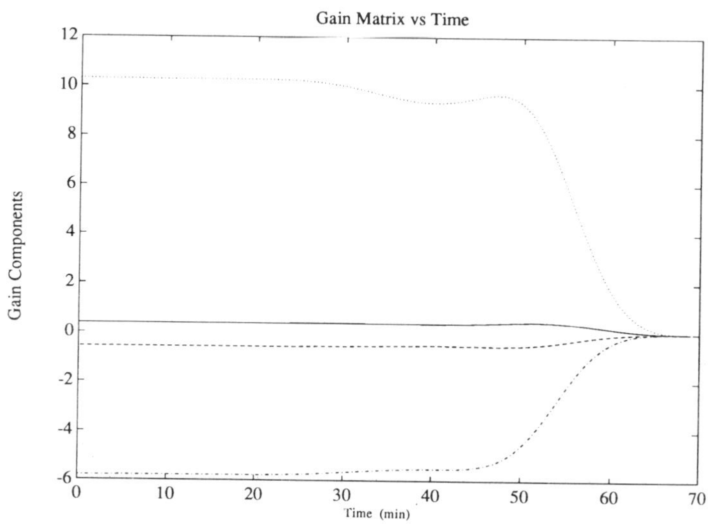

Figure 5.12 Feedback gain matrix convergence; $q_{11} = 1$, $q_{22} = 0$. K_{13} $(\cdots)$, K_{11} (—), K_{12} (- - - -), K_{14} $(- \cdot -\cdot)$.

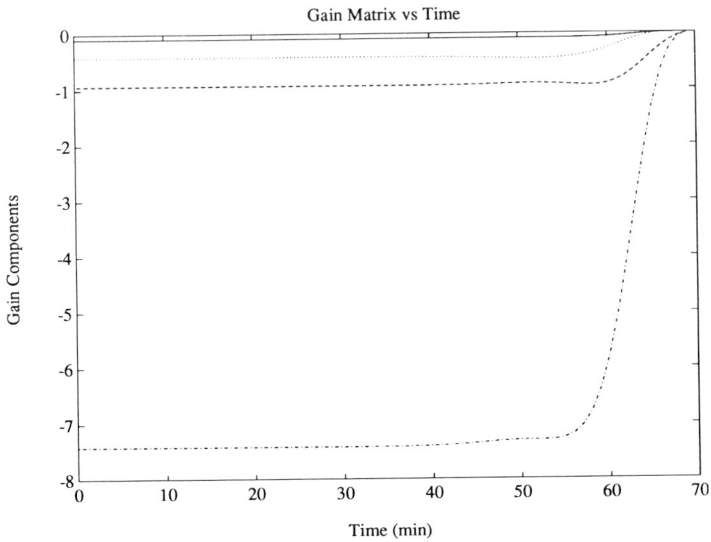

Figure 5.13 Feedback gain matrix convergence; $q_{11} = 0$, $q_{22} = 1$. K_{11} (—), K_{13} (···), K_{12} (- - - -), K_{14} (— · —·).

values are those at $t = 0$, since the recursive algorithm works backwards in time starting at $t = t_f = 70$ min.

Once the feedback gain for the Johnson problem is computed (Equation (5.6.16)), then the proportional gain and integral gain are computed. For composition regulation, the results are

$$K_p = [10.3 \quad -5.81] \tag{5.6.30}$$
$$K_I = [1.0 \quad 0] \tag{5.6.31}$$

For temperature regulation, the results are

$$K_p = [-.414 \quad -7.42] \tag{5.6.32}$$
$$K_I = [0 \quad -1.0] \tag{5.6.33}$$

It is interesting to note that the integral term associated with temperature for composition regulation is zero and the one associated with composition for temperature regulation is zero. This makes good physical sense.

Finally, the program uses the function STEP to compute the step response of the system to normalized disturbances of 0.2 in magnitude. In order to do this, auxilliary state variables are defined to handle the integral term in the control law of Equation (5.6.20). The system of equations that is solved is

$$\dot{\boldsymbol{x}} = (\boldsymbol{A} - \boldsymbol{B}\boldsymbol{K}_p)\boldsymbol{x} - \boldsymbol{B}\boldsymbol{K}_I\boldsymbol{x}_I \tag{5.6.34}$$

$$\dot{\boldsymbol{x}}_I = \boldsymbol{x} \tag{5.6.35}$$

The response under optimal composition regulation is shown in Figure 5.14 and that for optimal temperature regulation in Figure 5.15.

Figure 5.14 shows that the composition response has no steady-state off-set under optimal composition regulation. Also, the order of magnitude for the transient error is quite small. The figure also shows that the nonregulated state of temperature reaches a new steady-state value which is not that of its initial value. Figure 5.15 shows that the temperature regulation is almost perfect. Figure 5.16 expands the scale in order to observe the actual temperature response under temperature regulation.

5.7 Regulation for Naturally Unstable Systems

When a linear system is naturally unstable, the $\boldsymbol{A}$ matrix has some positive real-part eigenvalues for the system

$$\dot{\boldsymbol{x}} = \boldsymbol{A}\boldsymbol{x} + \boldsymbol{B}\boldsymbol{u} + \boldsymbol{W}\boldsymbol{d} \tag{5.7.1}$$

In the design of a feedback regulator, it is important that the closed loop response of the system be stable. That is, we desire to guarantee that the response of the system under control will approach a fixed operating region. The condition of asymptotic stability assures that the system will approach a desired operational strategy. Mathematically, we have

$$\boldsymbol{x}(t) - \boldsymbol{r}(t) \rightarrow \boldsymbol{0} \tag{5.7.2}$$

as t approaches infinity. Here $\boldsymbol{x}(t)$ is the state response and $\boldsymbol{r}(t)$ the desired state response.

It is interesting to note that the finite horizon regulator with the quadratic performance index of

$$J = \frac{1}{2}\int_0^{t_f} \left(\boldsymbol{x}^T\boldsymbol{Q}\boldsymbol{x} + \boldsymbol{u}^T\boldsymbol{R}\boldsymbol{u}\right) dt \tag{5.7.3}$$

does not consider the question of asymptotic stability since the time of interest is finite. The infinite time regulator, however, does necessitate the consideration of the closed loop asymptotic stability of the system since the final time in the quadratic performance functional is infinity,

$$J = \frac{1}{2}\int_0^{\infty} \left(\boldsymbol{x}^T\boldsymbol{Q}\boldsymbol{x} + \boldsymbol{u}^T\boldsymbol{R}\boldsymbol{u}\right) dt \tag{5.7.4}$$

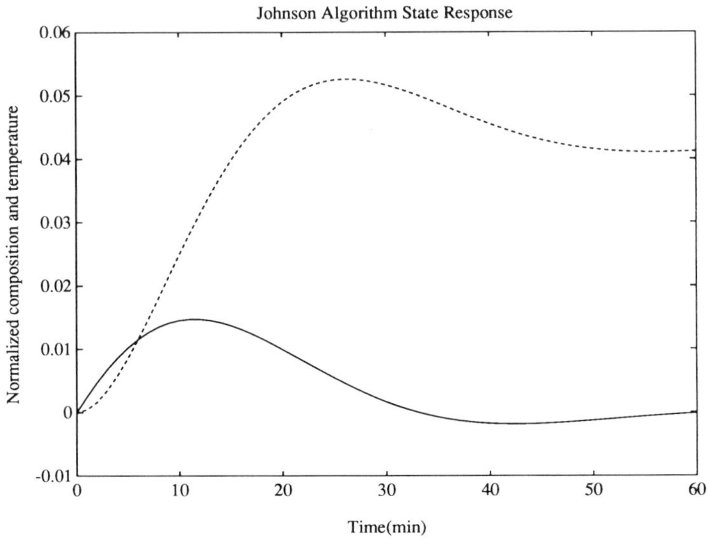

Figure 5.14 Optimal composition regulation state response. Temperature (- - - -), composition (—).

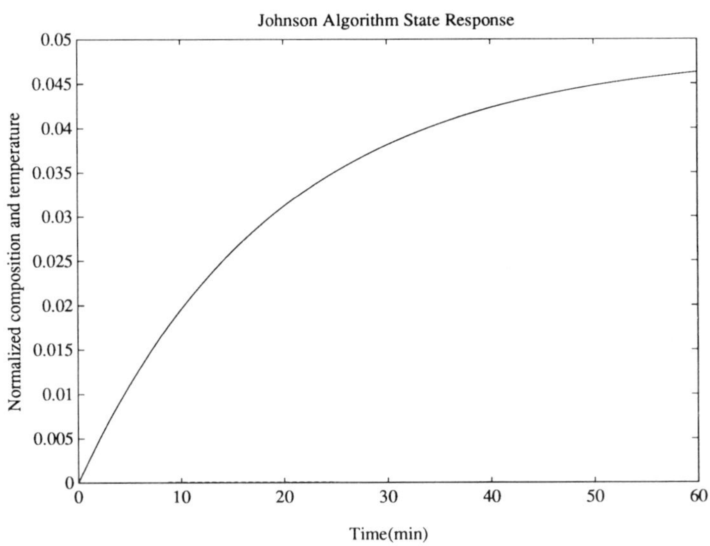

Figure 5.15 Optimal temperature regulation state response. Composition (—), temperature (- - - -); temperature remains constant, dotted line on x–axis.

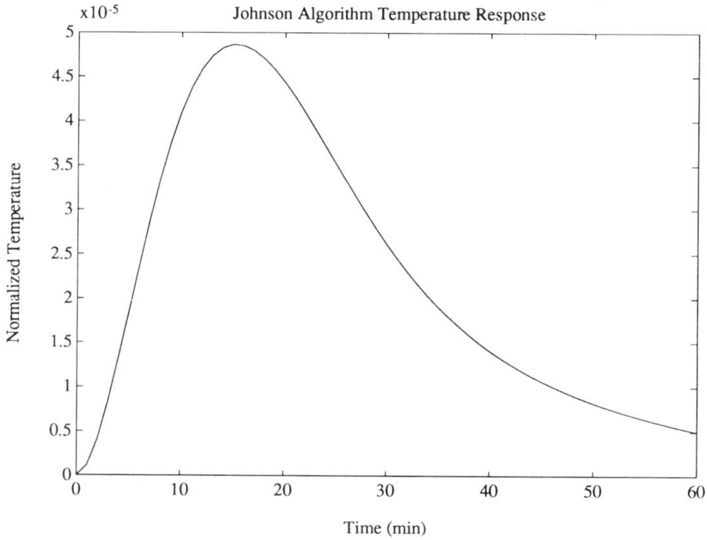

Figure 5.16 Temperature response under optimal temperature regulation.

Here we consider the conditions for stability of the closed loop system formed when the infinite time LQR problem is considered. The state dynamics are given by the linear system

$$\dot{\boldsymbol{x}} = \boldsymbol{A}\boldsymbol{x} + \boldsymbol{B}\boldsymbol{u} \tag{5.7.5}$$

The optimal infinite time LQR controller from Section 5.1 is

$$\boldsymbol{u} = -\boldsymbol{K}_{ss}\boldsymbol{x} \tag{5.7.6}$$

Accordingly, the closed loop system response becomes

$$\dot{\boldsymbol{x}} = (\boldsymbol{A} - \boldsymbol{B}\boldsymbol{K}_{ss})\,\boldsymbol{x} \tag{5.7.7}$$

For the closed loop system response to be stable, the eigenvalues of $(\boldsymbol{A} - \boldsymbol{B}\boldsymbol{K}_{ss})$ must not have positive real parts.

In designing a stabilizing, optimal feedback regulator, we can apply the LQR Kalman approach of Section 5.1 for systems without load disturbances, $\boldsymbol{d}$. The inclusion of unmeasurable disturbances, $\boldsymbol{d}$, leads to the Johnson PI regulator of Section 5.6. To compute either the steady-state or dynamic feedback gain matrices, the appropriate Riccati Equation (Equation (5.1.23)) must be solved. The steady-state Kalman gain matrix for controllable systems can be calculated using the LQR MATLAB routine. Since the Johnson algorithm performs a linear transformation that usually results in the transformed system not being controllable (although the original physical system is controllable), the LQR routine cannot be used. In Section 5.6, we suggested the recursive algorithm of Equation (5.2.14) as an alternative computational procedure. This algorithm works fine when the process is stable, but due to positive eigenvalues in the exponential matrix, it often will fail for unstable systems. The best solution is to directly compute the Riccati matrix differential equations of Equation (5.1.23) by backwards integration with the known final condition of Equation (5.1.26). This method is computationally more intensive but does yield accurate dynamic and steady-state values for the Riccati matrix.

5.7.1 Regulation for an Unstable Stirred Tank Reactor

Ramirez and Turner (1969) considered carrying out the decomposition of hydrogen peroxide,

$$H_2O_2 \rightarrow H_2O(\ell) + 1/2O_2(g) \tag{5.7.8}$$

in a stirred tank reactor. The basic material and energy balances are the same as those given in Section 5.4 for the Nyquist–Ramirez reactor. Again, the system consists of two state variables,

$$\begin{aligned} x_1 &= \text{normalized composition} && (5.7.9)\\ x_2 &= \text{normalized temperature} && (5.7.10) \end{aligned}$$

and one control variable,

$$u = \text{normalized coolant flow rate} \tag{5.7.11}$$

Table 5.2 defines the state variables and desired operating conditions. The desired steady-state condition is unstable, which means that the reactor cannot operate about that condition without some type of stabilizing control. After linearization, the system model becomes

$$\begin{pmatrix} \dot{x}_1 \\ \dot{x}_2 \end{pmatrix} = \boldsymbol{A}\begin{pmatrix} x_1 \\ x_2 \end{pmatrix} + \boldsymbol{b}u + \boldsymbol{W}\boldsymbol{d} \tag{5.7.12}$$

where

$$\boldsymbol{A} = \begin{bmatrix} -0.03318 & -0.002511 \\ 0.05237 & 0.010216 \end{bmatrix} \tag{5.7.13}$$

$$\boldsymbol{b} = \begin{pmatrix} 0 \\ -0.0988 \end{pmatrix} \tag{5.7.14}$$

$$\boldsymbol{W} = \begin{bmatrix} .030864 & 0 \\ 0 & 0.030864 \end{bmatrix} \tag{5.7.15}$$

Table 5.2 Process Conditions for the Turner–Ramirez Unstable Reactor

Steady-State Conditions	Characteristic Values
$C_{ss} = 10.697$ gmole/liter	$\Delta C = 5$ gmole/liter
$T_{ss} = 30°\text{C}$	$\Delta T = 5°\text{C}$
$C_{iss} = 11.5$ gmole/liter	$\Delta C_i = 5$ gmole/liter
$T_{iss} = 22°\text{C}$	$\Delta T_i = 5°\text{C}$
$\beta_{ss} = 253.66$ cal/min °C	$\Delta\beta = 400$ cal/min °C

Definition of Dimensionless, Normalized Variables

$$
\begin{aligned}
x_1 &= \frac{C - C_{ss}}{\Delta C} \\
x_2 &= \frac{T - T_{ss}}{\Delta T} \\
u &= \frac{\beta - \beta_{ss}}{\Delta \beta} \\
d_1 &= \frac{C_i - C_{iss}}{\Delta C_i} \\
d_2 &= \frac{T_i - T_{iss}}{\Delta T_i}
\end{aligned}
$$

Two load disturbances are considered to be potentially important. These are

$$d_1 = \text{normalized inlet feed composition of hydrogen peroxide} \quad (5.7.16)$$
$$d_2 = \text{normalized cooling water temperature} \quad (5.7.17)$$

The MATLAB program for studying the regulation of this system is given in Figure 5.17. We start off by entering the system model parameters and investigating the stability of the system. The eigenvalues of the $\boldsymbol{A}$ matrix are

$$\begin{aligned} \lambda_1 &= -.0299 \\ \lambda_2 &= .0069 \end{aligned} \quad (5.7.18)$$

The positive λ_2 value shows that the uncontrolled system is unstable. The response of the uncontrolled system given in Figure 5.18 to a unit impulse in both state variables (an initial unit off-set) does demonstrate that the system is indeed unstable.

Next we investigate the LQR Kalman controller for this system. Here we are assuming that there are essentially no load disturbances that affect the system, so that an optimal proportional controller is sufficient. We calculate the optimal feedback gain matrix for the weighting matrices $\boldsymbol{Q} = \boldsymbol{I}$, $\boldsymbol{R} = \boldsymbol{I}$ using the routine LQR. The results are

$$K = [.4015 \quad 1.0985] \quad (5.7.19)$$

The closed loop eigenvalues for the matrix

$$A - BK \quad (5.7.20)$$

are

$$\begin{aligned} \lambda_1 &= -0.0978 \\ \lambda_2 &= -0.0337 \end{aligned} \quad (5.7.21)$$

which shows that the closed loop, feedback controlled system is stable. The LQR design has stabilized the originally unstable reactor system. The response of the LQR feedback controlled system is shown in Figure 5.19 to a unit impulse in both state variables. The results show that the system is well controlled with reasonable control effort.

Next, we want to investigate the inclusion of unmeasurable load disturbances. This leads to the implementation of the LQR Johnson algorithm. In order to do this, we need to augment the original system matrices. Also, since there is only one control variable for this system, only one state can be regulated about its original steady-state value. The other state must be free to reach a new operating condition when load disturbances are present. We

```
>>
>> aa = a - b*k;
>>
>> y = impulse(aa,bb,c,d,1,t);
>> u = -k*y';
>> plot (t,y,t,u');
>> title('Optimal State and Control Variables');
>> xlabel ('Time');
>> ylabel ('State Vector Components and Control Variables');
>> meta optimal2
>>
>> % implement the Johnson algorithm
>>
>> aa = [a         eye(2);
        zeros(2)   zeros(2)];
>> bb = [0;
        0;
        b];
>> q = [1   0;
       0   0];
>> qq = [q         zeros(2);
        zeros(2)  zeros(2)];
>>
>> % integrate the Riccati Equation backwards in time
>>
>> t0 = 0; tf = 200;
>> p0 =[0  0  0  0  0  0  0  0  0  0]';
>> [t,pv] = ode45('xprime',t0,tf,p0);
>> plot (t,pv)
>> title ('Riccati Matrix vs tf - t');
>> xlabel('tf - t'); ylabel('Riccati Gain Components');
>> meta riccati1
>>
>> % preparing for the response of the system to unit step disturbances
>>
>> % steady state values of the Riccati matrix elements
>>
>> pv(65,1:10)

ans =

   1.0e+03 *

  Columns 1 through 7

    0.0128   -0.0005    0.1786   -0.0032    0.0000   -0.0174    0.0006

  Columns 8 through 10

    6.1447   -0.1913    0.0110
```

Figure 5.17 MATLAB solution to regulation of unstable reactor.

```
>> p(1,1) = pv(65,1);
>> p(1,2) = pv(65,2);
>> p(1,3) = pv(65,3);
>> p(1,4) = pv(65,4);
>> p(2,1) = p(1,2);
>> p(2,2) = pv(65,5);
>> p(2,3) = pv(65,6);
>> p(2,4) = pv(65,7);
>> p(3,1) = p(1,3);
>> p(3,2) = p(2,3);
>> p(3,3) = pv(65,8);
>> p(3,4) = pv(65,9);
>> p(4,1) = p(1,4);
>> p(4,2) = p(2,4);
>> p(4,3) = p(3,4);


>> % input the a, b, and w matrices
>> a = [-0.033181   -0.002511;
         0.05237     0.010216];
>> b = [0    0.0988]';
>> w = [0.030864   0;
        0           0.030864];
>>
>> % check the eigenvalues of a
>> eig(a)

ans =

   -0.0299
    0.0069

>> % the positive eigenvalue shows that the system is unstable
>>
>> % let's look at the response of the uncontrolled system
>>
>> % use a unit impulse on each state
>> % define appropriate matrices
>>
>> bb = [1   1]';
>> c = eye(2);
>> t = [0:1:100];
>> d = [0  0]';
>> y = impulse(a,bb,c,d,1,t);
>> x1 = y(1:101,1);
>> x2 = y(1:101,2);
>> plot (t,y)
>> title('Uncontrolled Response');
>> xlabel('Time');
>> ylabel('State Vector Components');
>> meta uncontroll
>>
>> % compute the Kalman feedback gain matrix for the system
>> % input the weighting matrices
>>
>> q = [1  0;
        0  1];
>> r = 1;
>> [k,p,e] = lqr(a,b,q,r)
```

Figure 5.17 continued.

```
k =

    0.4015    1.0985


p =

   19.0538    4.0642
    4.0642   11.1187


e =

   -0.0978
   -0.0337

>> % these results show that a stable optimal proportional feedback system
>> % has been designed
>>
>> % now let's display the results of the controlled system to an unit
>> % impulse disturbance to each state variable
>>
>> % define the closed loop response matrix
```

Figure 5.17 continued

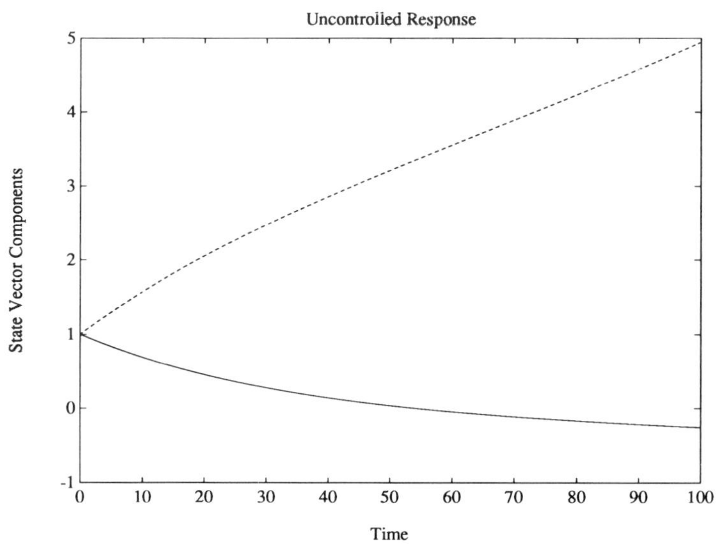

Figure 5.18 Uncontrolled system response.
Temperature (- - - -), composition (—).

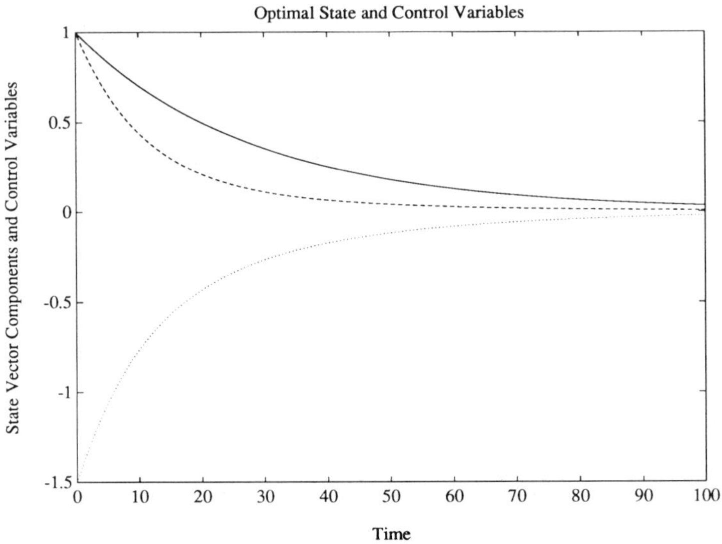

Figure 5.19 LQR controlled response. Composition (—), temperature (- - - -), cooling flow (···).

choose to regulate the system composition (x_1) so that the state weighting matrix is

$$Q = \begin{bmatrix} 1 & 0 \\ 0 & 0 \end{bmatrix} \tag{5.7.22}$$

Since the augmented system is uncontrollable, we cannot use the LQR MATLAB routine. Due to the instability of the original system, we will directly backwards integrate the Riccati Equations (5.1.23). In order to do this, the MATLAB ordinary differential equation routine ODE 45 is used. The Riccati differential equations are defined in an external M file, xprime.m given in Figure 5.20. A plot of the elements of the Riccati Matrix elements using backwards integration is shown in Figure 5.21. This figure proves that a steady-state value is reached, so that a constant steady-state gain matrix can be used for the feedback control of the system assuming that $T_f \to \infty$. Finally, the program investigates the response of the system to unit step disturbances. The proportional and integral gain constants are computed as

$$\begin{aligned} K_p &= [-18.8991 \quad 1.0890] \\ K_I &= \quad [-1.0 \qquad 0] \end{aligned} \tag{5.7.23}$$

The feedback response of the system is given in Figure 5.22. It shows that the composition is well regulated and that the temperature is free to reach a

```
function pdot = xprime(t,pv)
%
% define the system and weighting matrices
%
a = [-0.033181  -0.002511;
     0.05237     0.010216];
b = [0  0.0988]';
q = [1  0;
     0  0];
%
% define the Johnson transformation
%
r = 1;
aa = [a       eye(2);
      zeros(2) zeros(2)];
bb = [0;
      0;
      b];
qq = [q        zeros(2);
      zeros(2) zeros(2)];
%
%define the Riccati matrix in terms of a Riccati vector of variables
%
p(1,1)= pv(1);
p(1,2)= pv(2);
p(1,3)= pv(3);
p(1,4)= pv(4);
p(2,1)= pv(2);
p(2,2)= pv(5);
p(2,3)= pv(6);
p(2,4)= pv(7);
p(3,1)= pv(3);
p(3,2)= pv(6);
p(3,3)= pv(8);
p(3,4)= pv(9);
p(4,1)= pv(4);
p(4,2)= pv(7);
p(4,3)= pv(9);
p(4,4)= pv(10);
%
% the Riccati Matrix differential equation prepared for backwards integration
%
pd = p*aa + aa'*p + qq -p*bb*inv(r)*bb'*p;
%
% defining the appropriate derivative vector
%
pdot(1)= pd(1,1);
pdot(2)  = pd(1,2);
pdot(3)  = pd(1,3);
pdot(4)  = pd(1,4);
pdot(5)  = pd(2,2);
pdot(6)  = pd(2,3);
pdot(7)  = pd(2,4);
pdot(8)  = pd(3,3);
pdot(9)  = pd(4,3);
pdot(10)  = pd(4,4);
```

Figure 5.20 M file for defining differential equations.

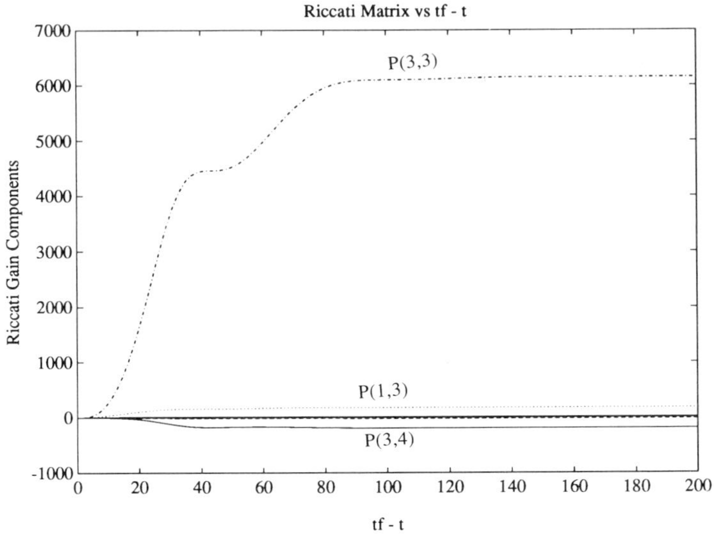

Figure 5.21 Convergence of Riccati matrix.

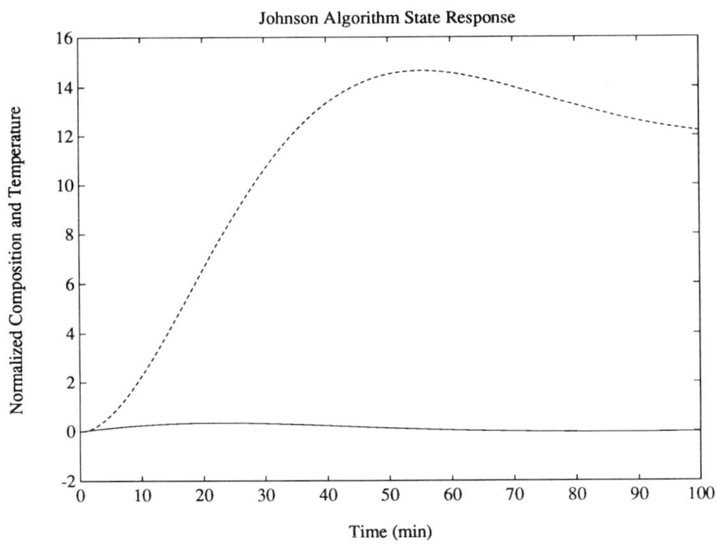

Figure 5.22 State response using Johnson algorithm. Temperature (- - - -), composition (—).

new steady-state value which is approximately 60°C above the normal operating temperature of 30°C. There is some practical concern with this result since the system boiling point is around 100°C. Other $\boldsymbol{Q}$ weighting matrices should be studied in order to improve upon this design before actual implementation. Such a study is left for the interested reader.

5.8 Performance Design of Cost Weighting Matrices

Up until now we have been able to design the optimal LQR problem using weighting matrices $\boldsymbol{Q}$ and $\boldsymbol{R}$ based upon the relative importance of the various state and control variables. In order to make that process feasible, we usually nondimensionalize and normalize the state, control, and disturbance variables before we assign weights to them. Using this relative importance concept, we normally are restricted to diagonal weighting matrices. Sometimes there are certain performance characteristics that should be included into the design procedure. This section investigates two techniques for achieving system performance characteristics, namely eigenstructure assignment and modal analysis (Grimble and Johnson, 1988).

5.8.1 Optimal Eigenstructure Assignment

In this section, we wish to devise a design procedure that will assure that the closed loop response of the system is stable and will decay at least as fast as $e^{-\alpha t}$. The procedure is due to Anderson and Moore (1969). We redefine the state and control variables through the following transformation,

$$\tilde{\boldsymbol{x}} = e^{-\alpha t}\boldsymbol{x} \tag{5.8.1}$$

$$\tilde{\boldsymbol{u}} = e^{-\alpha t}\boldsymbol{u} \tag{5.8.2}$$

Using these definitions, the linear system

$$\dot{\boldsymbol{x}} = \boldsymbol{A}\boldsymbol{x} + \boldsymbol{B}\boldsymbol{u} \tag{5.8.3}$$

is transformed into the system

$$\dot{\tilde{\boldsymbol{x}}} = \tilde{\boldsymbol{A}}\tilde{\boldsymbol{x}} + \tilde{\boldsymbol{B}}\tilde{\boldsymbol{u}} \tag{5.8.4}$$

with

$$\tilde{\boldsymbol{A}} = (\boldsymbol{A} + \alpha\boldsymbol{I}) \tag{5.8.5}$$

and

$$\tilde{\boldsymbol{B}} = \boldsymbol{B} \tag{5.8.6}$$

For a quadratic performance index of

$$J = \frac{1}{2}\int_0^\infty \left(\tilde{\boldsymbol{x}}^T \boldsymbol{Q}\tilde{\boldsymbol{x}} + \tilde{\boldsymbol{u}}^T \boldsymbol{R}\tilde{\boldsymbol{u}}\right) dt \tag{5.8.7}$$

we can compute the optimal LQR controller since we have transformed the original problem into an equivalent Kalman problem,

$$\tilde{\boldsymbol{u}} = -\tilde{\boldsymbol{K}}_{ss}\tilde{\boldsymbol{x}} \tag{5.8.8}$$

or in terms of the original state and control variables, we have

$$\boldsymbol{u} = -\tilde{\boldsymbol{K}}_{ss}\boldsymbol{x} \tag{5.8.9}$$

The important property of this controller is that it achieves a closed loop response with the real parts of the eigenvalues less than $-\alpha$. Of course, the larger α the more stable the system, but that stability and speed of response is achieved only by the use of stronger control action. The designer must be aware of the cost of control action capacity or the possibility of maximum or minimum control constraints that could affect implementation.

5.8.2 Optimal Modal Control

The objective of modal control is to determine a state feedback law,

$$\boldsymbol{u} = -\boldsymbol{K}_{ss}\boldsymbol{x} \tag{5.8.10}$$

so that the closed loop system response matrix

$$(\boldsymbol{A} - \boldsymbol{B}\boldsymbol{K}) \tag{5.8.11}$$

has a prescribed set of eigenvalues. The parameters of the quadratic weighting matrices $\boldsymbol{Q}$ and $\boldsymbol{R}$ are chosen so as to achieve this objective. The approach is due to Solheim (1972). The system matrix $\boldsymbol{A}$ is assumed to have real distinct eigenvalues, so that a modal transformation can be performed. The modal transformation is to introduce the matrix of normalized right eigenvectors, $\boldsymbol{T}$, such that

$$\boldsymbol{x} = \boldsymbol{T}\boldsymbol{\xi} \tag{5.8.12}$$

where

$$\boldsymbol{T} = (\boldsymbol{t}_1,\ \boldsymbol{t}_2, \cdots \boldsymbol{t}_n) \tag{5.8.13}$$

The state system therefore becomes

$$\dot{\boldsymbol{\xi}} = \boldsymbol{\Lambda}\boldsymbol{\xi} + \tilde{\boldsymbol{B}}\boldsymbol{u} \tag{5.8.14}$$

with

$$\boldsymbol{\Lambda} = \text{diag}\ (\lambda_1,\ \lambda_2, \cdots \lambda_n) \tag{5.8.15}$$

$$\tilde{\boldsymbol{B}} = \boldsymbol{T}^{-1}\boldsymbol{B} \tag{5.8.16}$$

The MATLAB function EIG computes the eigenvalues of a matrix as well as the $\boldsymbol{T}$ matrix of normalized right eigenvectors.

If we use a quadratic performance functional

$$J = \frac{1}{2}\int_0^\infty \left(\boldsymbol{\xi}^T\tilde{\boldsymbol{Q}}\boldsymbol{\xi} + \boldsymbol{u}^T\boldsymbol{R}\boldsymbol{u}\right) dt \tag{5.8.17}$$

with

$$\tilde{\boldsymbol{Q}} = \left(\boldsymbol{T}^T\boldsymbol{Q}\boldsymbol{T}\right) \tag{5.8.18}$$

then we have transformed the problem into the Kalman LQR form. We assume that $\tilde{\boldsymbol{Q}}$ has a diagonal structure of

$$\tilde{\boldsymbol{Q}} = \text{diag}\ (0, 0, \cdots, \tilde{q}_i, \cdots 0, 0) \tag{5.8.19}$$

This fixes the eigenvalues of the closed loop system except for the ith element. For a desired response time for the ith state of α_i, we can use the appropriate value of $\tilde{q}_i$ given by Solheim (1972),

$$\tilde{q}_i = \frac{(\alpha_i^2 - \lambda_i^2)}{\tilde{h}_{ij}} \tag{5.8.20}$$

where

$$\tilde{\boldsymbol{H}} = \tilde{\boldsymbol{B}}\boldsymbol{R}^{-1}\tilde{\boldsymbol{B}}^T \tag{5.8.21}$$

The design algorithm is therefore as follows:

1. Select a response time α_1 and compute the appropriate $\tilde{q}_1$ from Equation (5.8.20). Solve the steady-state Riccati Equation

$$\boldsymbol{P}_1\boldsymbol{A} + \boldsymbol{A}^T\boldsymbol{P}_1 - \boldsymbol{P}_1\boldsymbol{B}\boldsymbol{R}^{-1}\boldsymbol{B}^T\boldsymbol{P}_1 + \boldsymbol{Q}_1 = \boldsymbol{0} \tag{5.8.22}$$

Note that this equation is in terms of the original state, control, and weighting matrices, not the transformed ones. Compute a new state matrix $\boldsymbol{A}_1$ as

$$\boldsymbol{A}_1 = \boldsymbol{A} - \boldsymbol{B}\boldsymbol{K}_1 \tag{5.8.23}$$

with

$$\boldsymbol{K}_1 = \boldsymbol{R}^{-1}\boldsymbol{B}^T\boldsymbol{P}_1 \tag{5.8.24}$$

2. Now select the next response time α_2 and compute the appropriate $\tilde{q}_2$ from Equation (5.8.20). Solve the steady-state Riccati Equation

$$\boldsymbol{P}_2\boldsymbol{A}_1 + \boldsymbol{A}_1^T\boldsymbol{P}_2 - \boldsymbol{P}_2\boldsymbol{B}\boldsymbol{R}^{-1}\boldsymbol{B}\boldsymbol{P}_2 + \boldsymbol{Q}_2 = \mathbf{0} \tag{5.8.25}$$

   Compute a new state matrix $\boldsymbol{A}_2$ as

$$\boldsymbol{A}_2 = \boldsymbol{A}_1 - \boldsymbol{B}\boldsymbol{K}_2 \tag{5.8.26}$$

   with

$$\boldsymbol{K}_2 = \boldsymbol{R}^{-1}\boldsymbol{B}^T\boldsymbol{P}_2 \tag{5.8.27}$$

   The mechanisms of the selection of $\tilde{q}_1$ and $\tilde{q}_2$ have insured that the desired eigenvalue lactations α_1 and α_2 have been achieved without being changed by subsequent steps.

3. Continue the process until all response characteristics are set.

References

Anderson, B. D. O. and Moore, J. B., "Linear System Optimization with Prescribed Degree of Stability," *Proc. IEE* **116**, No. 12, 2083–2087 (1969).

Douglas, J. M., "The Use of Optimization Theory to Design Simple Multivariable Control Systems," *Chem. Eng. Sci.* **21**, 519–532 (1966).

Editrice, P., "Count Riccati and the Early Days of the Riccati Equation," Lecture Notes from Workshop on The Riccati Equation in Control, Systems, and Signals, S. Bittanti, ed., Pitagora Editrice, Bologna (1989).

Grimble, M. J. and Johnson, M. A., *Optimal Control and Stochastic Estimation Theory and Applications, Volume 1*, Wiley and Sons, New York (1988).

Johnson, C. D., *IEEE Trans. Auto. Control* **AC–13**, 416 (1968).

Johnson, C. D., *Int. J. Control*, **15**, No. 2, 209 (1972).

Kalman, R., *SIAM J. Control, Ser. A*, 152 (1963).

Kalman, R., "Theory of Optimal Control and the Calculus of Variation," R. Bellman, ed., *Math. Opt. Tech.*, Univ. of California Press, Berkeley, CA (1963).

Newell, R. B., Fisher, D. G., and Seborg, D. E., "Computer Control Using Optimal Multivariable Feedforward-Feedback Algorithms," *AIChE J.* **18**, 976–983 (1972).

Nyquist, J. K. and Ramirez, W. F., "Time Optimal Control of an Experimental Continuous Stirred Tank Reactor," *Chem. E. Sci.* **26**, 1673–1688 (1971).

Ramirez, W. F., Lewis, D. W., and Mickley, M. C., "Optimal Control of the Artificial Kidney-Patient System," *Med. & Bio. Engr.* **11**, 743–746 (1973).

Ramirez, W. F. and Turner, B. A., "The Dynamic Modeling, Stability, and Control of a Continuous Stirred Tank Chemical Reactor," *AIChE Journal* **15**, No. 6, 853 (1969).

Ramirez, W. F., Beyer, K. P., and Abedi, H., "Optimal Regulator Control for Systems with Time–Varying Measurable and Partially Measurable Disturbances," *Optimal Control Appl. & Meth.* **8**, 159–174 (1987).

Solheim, O. A., "Design of Optimal Control Systems with Prescribed Eigenvalues," *Int. J. Control* **15**, No. 1, 143–160 (1972).

Exercises

1. Douglas (1966) has also studied the control of a continuous stirred tank reactor. His linearized model is

$$\dot{\boldsymbol{x}} = \begin{bmatrix} .4245 & .1137 \\ -.4145 & -.0987 \end{bmatrix} \boldsymbol{x} + \begin{bmatrix} 9.764 & 0 \\ -6.32 & -3.455 \end{bmatrix} \times 10^{-4} \boldsymbol{u}$$

where

$$\boldsymbol{x} = \begin{pmatrix} \text{composition} \\ \text{temperature} \end{pmatrix} \quad \boldsymbol{u} = \begin{pmatrix} \text{feed rate} \\ \text{coolant feed rate} \end{pmatrix}$$

   a) Compute the LQR Kalman optimal feedback controller when

$$\boldsymbol{Q} = \begin{bmatrix} 84.5 & 0 \\ 0 & 6.16 \end{bmatrix} \quad \boldsymbol{R} = \begin{bmatrix} 10^{-3} & 0 \\ 0 & 10^{-3} \end{bmatrix}$$

   b) Determine the unit impulse response for this system under LQR control.

2. Newell *et al.* (1972) study the control of a double effect evaporator. Their linearized mathematical model is

$$\begin{aligned} \boldsymbol{x} &= \begin{bmatrix} 0 & -.00156 & -.1711 & 0 & 0 \\ 0 & -.1419 & .1711 & 0 & 0 \\ 0 & -.00875 & -1.102 & 0 & 0 \\ 0 & -.00128 & -.1489 & 0 & .00013 \\ 0 & .0605 & .1489 & 0 & -.0591 \end{bmatrix} \boldsymbol{x} \\ &+ \begin{bmatrix} 0 & -.143 & 0 \\ 0 & 0 & 0 \\ .392 & 0 & 0 \\ 0 & .108 & -.0592 \\ 0 & -.0486 & 0 \end{bmatrix} \boldsymbol{u} \\ &+ \begin{bmatrix} .2174 & 0 & 0 \\ -.074 & .1434 & 0 \\ -.036 & 0 & .1814 \\ 0 & 0 & 0 \\ 0 & 0 & 0 \end{bmatrix} \boldsymbol{d} \end{aligned}$$

where

$$\boldsymbol{x} = \begin{pmatrix} \text{hold up in first effect} \\ \text{first effect product concentration} \\ \text{enthalpy of first effect} \\ \text{hold up in second effect} \\ \text{second effect product concentration} \end{pmatrix}$$

$$\boldsymbol{u} = \begin{pmatrix} \text{steam flow rate} \\ \text{first effect bottoms flow rate} \\ \text{second effect bottoms flow rate} \end{pmatrix}$$

$$\boldsymbol{d} = \begin{pmatrix} \text{feed flow rate} \\ \text{feed composition} \\ \text{feed enthalpy} \end{pmatrix}$$

a) Check to see if the system is controllable.

b) Compute the LQR Kalman optimal feedback controller when $\boldsymbol{Q} = \boldsymbol{R} = \boldsymbol{I}$.

c) Determine the unit impulse response of the system under LQR control.

d) Compute the Johnson optimal PI controller.

e) Determine the unit step disturbance response of the system using the Johnson controller.

3. Ramirez *et al.* (1973) discuss the optimal control of an artificial kidney-patient system. Their linearized model is

$$\dot{\boldsymbol{x}} = \begin{bmatrix} -4.266 & 3.78 & .0058 \\ 2.268 & -2.268 & 0 \\ .6296 & 0 & -.6442 \end{bmatrix} \boldsymbol{x} + \begin{bmatrix} -.0346 \\ 0 \\ 0 \end{bmatrix} u$$

where

$$\boldsymbol{x} = \begin{pmatrix} \text{extracellular urea concentration} \\ \text{intracellular urea concentration} \\ \text{cerebrospinal fluid concentration} \end{pmatrix}$$

$$u = \text{artificial kidney clearance effectiveness}$$

a) Check the controllability of the system.

b) Compute the LQR Kalman optimal feedback controller when $\boldsymbol{Q} = \boldsymbol{I}$ and $R = 50$.

c) Determine the unit impulse response for this system under LQR control.

4. Use the recursive algorithm of Equation (5.2.14) to compute the steady-state gain matrices of Problems 5.1 and 5.3.

Chapter 6

MODEL PREDICTIVE CONTROL

6.1 Introduction

One of the advances in control theory in recent years has been the inclusion of model predictions in the design of regulator control systems. Predictive methods have many names including Dynamic Matrix Control (DMC), Internal Model Control (IMC), Model Algorithmic Control (MAC), and Model Predictive Control (MPC). A good review is that of Garcia *et al.* (1986). All of these approaches can be cast as an LQR type of problem. The classical LQR proportional feedback control design illustrated in the block diagram of Figure 5.2, or the Johnson PI algorithm given in Figure 5.10, shows that the controller design is only a function of the actual state of the system, $\boldsymbol{x}$. There is no predictive action based upon the process model. In contrast to the standard feedback design philosophy, the model predictive controller design is based upon feedback information that is the difference between the actual state of the system, $\boldsymbol{x}$, and a predictive state $\boldsymbol{x}_p$. A general block diagram for the model predictive control (MPC) design is given in Figure 6.1.

From Figure 6.1 we see that the feedback signal is the error, $\boldsymbol{e}$, between the actual state, $\boldsymbol{x}$, and the predicted state, $\boldsymbol{x}_p$. This feedback signal will go to zero if there are no unmeasured process disturbances, $\boldsymbol{d}_u$, to the system and if the process model is completely accurate. Otherwise, there is a disturbance feedback signal that goes to the model predictive controller (MPC). We therefore observe that the structure of an MPC system is a feed-forward element given by the predictive state $\boldsymbol{x}_p$ and a feedback element that corrects for model uncertainty and unmeasured disturbances. There are a number of design philosophies that are possible in order to synthesize a specific MPC controller (Morari and Zafiriou, 1989). In this chapter, we will develop design strategies for completely measurable static disturbances, completely measurable dynamic disturbances, and systems with both measurable and unmeasur-

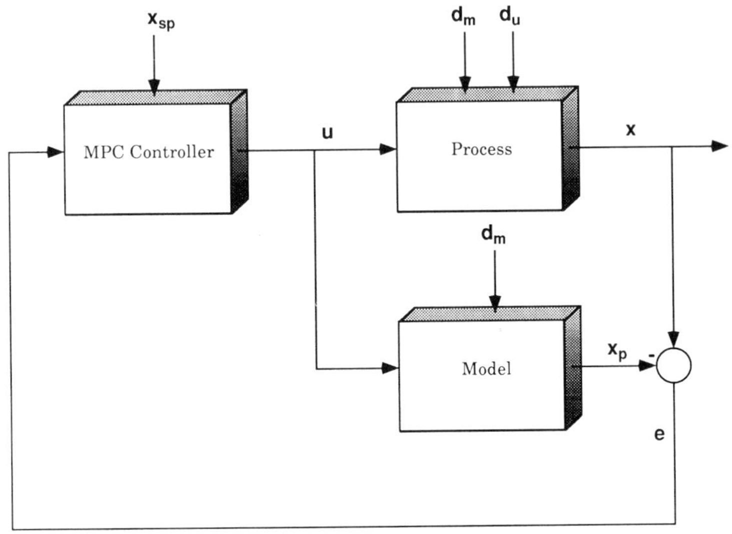

$$\begin{aligned} \boldsymbol{x} &= \text{process state} \\ \boldsymbol{x}_p &= \text{predicted state} \\ \boldsymbol{x}_{sp} &= \text{set point state} \\ \boldsymbol{d}_m &= \text{measured process disturbances} \\ \boldsymbol{d}_u &= \text{unmeasured process disturbances} \\ \boldsymbol{u} &= \text{control variables} \\ \boldsymbol{e} &= \text{error between process state and predicted state} \end{aligned}$$

Figure 6.1 Model predictive controller block diagram.

able (partially measurable) disturbances. We also consider the effect of time delays and trajectory following on the design of optimal regulatory feedback controllers.

6.2 Measurable Step Disturbances

In 1969, Anderson introduced a practical means of handling measurable step disturbances to the optimal regulator problems (Anderson, 1969). The system that we will consider is a linear plant with constant measurable disturbances, $\boldsymbol{d}_m$,

$$\dot{\boldsymbol{x}} = \boldsymbol{A}\boldsymbol{x} + \boldsymbol{B}\boldsymbol{u} + \boldsymbol{W}\boldsymbol{d}_m \tag{6.2.1}$$

We now introduce deviation variables, $\bar{\boldsymbol{x}}$ and $\bar{\boldsymbol{u}}$, given by

$$\bar{\boldsymbol{x}} = \boldsymbol{x} - \boldsymbol{x}_e \tag{6.2.2}$$

$$\bar{\boldsymbol{u}} = \boldsymbol{u} - \boldsymbol{u}_e \tag{6.2.3}$$

where $\boldsymbol{x}_e$ and $\boldsymbol{u}_e$ are new equilibrium conditions that the system will reach with the constant disturbances, $\boldsymbol{d}_m$. These new equilibrium conditions are computed by the equilibrium or steady-state equations

$$\boldsymbol{0} = \boldsymbol{A}\boldsymbol{x}_e + \boldsymbol{B}\boldsymbol{u}_e + \boldsymbol{W}\boldsymbol{d}_m \tag{6.2.4}$$

Now we can introduce the deviation variables into the original problem of Equation (6.2.1) using the fact that $\dot{\boldsymbol{x}} = \dot{\bar{\boldsymbol{x}}}$. This gives

$$\dot{\bar{\boldsymbol{x}}} = \boldsymbol{A}(\bar{\boldsymbol{x}} + \boldsymbol{x}_e) + \boldsymbol{B}(\bar{\boldsymbol{u}} + \boldsymbol{u}_e) + \boldsymbol{W}\boldsymbol{d}_m \tag{6.2.5}$$

or, using Equation (6.2.4), we have

$$\dot{\bar{\boldsymbol{x}}} = \boldsymbol{A}\bar{\boldsymbol{x}} + \boldsymbol{B}\bar{\boldsymbol{u}} \tag{6.2.6}$$

If we use the performance functional of

$$J = \frac{1}{2}\,\bar{\boldsymbol{x}}^T(t_f)\boldsymbol{H}\bar{\boldsymbol{x}}(t_f) + \frac{1}{2}\int_{t_o}^{t_f} \left(\bar{\boldsymbol{x}}^T\boldsymbol{Q}\bar{\boldsymbol{x}} + \bar{\boldsymbol{u}}^T\boldsymbol{R}\bar{\boldsymbol{u}}\right)\,dt \tag{6.2.7}$$

then Equations (6.2.6) and (6.2.7) are simply a restatement of the Kalman problem of Section 5.1. We have therefore transformed the original problem into an equivalent Kalman problem, and the optimal control is given by

$$\bar{\boldsymbol{u}} = -\boldsymbol{K}(t)\bar{\boldsymbol{x}} \tag{6.2.8}$$

where $\boldsymbol{K}(t)$ is computable from the Riccati matrix differential equations of Sections 5.1 and 5.2. The infinite time or steady-state version is

$$\bar{\boldsymbol{u}} = -\boldsymbol{K}_{ss}\bar{\boldsymbol{x}} \tag{6.2.9}$$

which, expressed in terms of the original problem variables, is

$$\boldsymbol{u} = \boldsymbol{u}_e - \boldsymbol{K}_{ss}(\boldsymbol{x} - \boldsymbol{x}_e) \tag{6.2.10}$$

The structure of the Anderson algorithm is that of a model predicted controller (MPC). This is shown in Figure 6.2. The feed-forward model is that of the steady-state or equilibrium predicted state, $\boldsymbol{x}_e$, and the steady-state or equilibrium control, $\boldsymbol{u}_e$. The feedback error signal, $\boldsymbol{e}$, is the difference between the actual state, $\boldsymbol{x}$, and the predicted equilibrium state $\boldsymbol{x}_e$.

Now let us consider the computation of the equilibrium state, $\boldsymbol{x}_e$, and the equilibrium control, $\boldsymbol{u}_e$. We can partition the equilibrium state into

$$\boldsymbol{x}_e = \begin{bmatrix} \boldsymbol{x}_{se} \\ \boldsymbol{x}_{fe} \end{bmatrix} \tag{6.2.11}$$

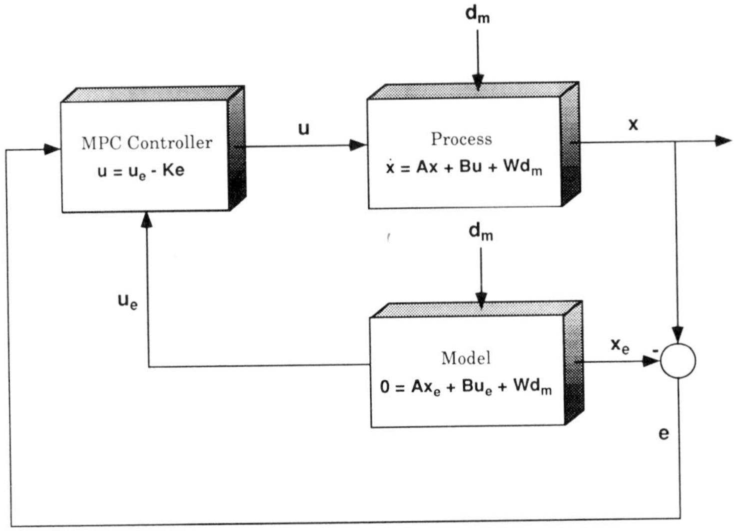

Figure 6.2 Anderson MPC algorithm.

where $\boldsymbol{x}_{se}$ = the state variables that are specified or to be brought to a desired value

$\boldsymbol{x}_{fe}$ = the state variables free to attain new equilibrium values

The specified variables, $\boldsymbol{x}_{se}$, can also be considered as the system outputs to be regulated. The steady-state condition is therefore

$$\begin{bmatrix} \boldsymbol{A}_1 & \vdots & \boldsymbol{A}_2 \end{bmatrix} \begin{bmatrix} \boldsymbol{x}_{se} \\ \boldsymbol{x}_{fe} \end{bmatrix} + \boldsymbol{B}\boldsymbol{u}_e + \boldsymbol{W}\boldsymbol{d}_m = \boldsymbol{0} \tag{6.2.12}$$

Rearranging so that the unknowns are on the left-hand side and the knowns are on the right-hand side gives,

$$-\begin{bmatrix} \boldsymbol{A}_2 & \vdots & \boldsymbol{B} \end{bmatrix} \begin{bmatrix} \boldsymbol{x}_{fe} \\ \boldsymbol{u}_e \end{bmatrix} = \boldsymbol{A}_1\boldsymbol{x}_{se} + \boldsymbol{W}\boldsymbol{d}_m \tag{6.2.13}$$

Assuming that the number of specified state variables to be brought to desired conditions, $\boldsymbol{x}_{se}$, equals the number of control variables, m, the matrix, $[\boldsymbol{A}_2 \vdots \boldsymbol{B}]$, is square. If $[\boldsymbol{A}_2 \vdots \boldsymbol{B}]^{-1}$ exists, then

$$\begin{bmatrix} \boldsymbol{x}_{fe} \\ \boldsymbol{u}_e \end{bmatrix} = -\begin{bmatrix} \boldsymbol{A}_2 & \vdots & \boldsymbol{B} \end{bmatrix}^{-1} (\boldsymbol{A}_1\boldsymbol{x}_{se} + \boldsymbol{W}\boldsymbol{d}_m) \tag{6.2.14}$$

and Equation (6.2.10) becomes

$$\boldsymbol{u} = -\boldsymbol{K}_{ss}\boldsymbol{x} + \left[\boldsymbol{u}_e + \boldsymbol{K}_{ss}\begin{pmatrix} \boldsymbol{x}_{se} \\ \boldsymbol{x}_{fe} \end{pmatrix}\right] \tag{6.2.15}$$

This equation clearly shows that the optimal controller that we have computed for constant measurable step disturbances is a state proportional feedback controller with a multivariable static predictor or feed-forward term.

6.2.1 Nyquist–Ramirez Reactor

Let us apply the Anderson algorithm to the Nyquist–Ramirez reactor of Section 5.4. We will assume that there are two measurable disturbances for this system. These are the inlet concentration, C_i, and the inlet temperature, T_i. In terms of normalized deviation variables, the linearized system is given by

$$\dot{\boldsymbol{x}} = \boldsymbol{A}\boldsymbol{x} + \boldsymbol{b}u + \boldsymbol{W}\boldsymbol{d}_m \tag{6.2.16}$$

with

$$\boldsymbol{A} = \begin{bmatrix} -0.05156 & -0.05877 \\ 0.01503 & -0.005887 \end{bmatrix} \tag{6.2.17}$$

$$\boldsymbol{b} = \begin{bmatrix} 0 \\ -0.03384 \end{bmatrix} \tag{6.2.18}$$

$$\boldsymbol{W} = \begin{bmatrix} 0.0125 & 0 \\ 0 & 0.0125 \end{bmatrix} \tag{6.2.19}$$

$$\boldsymbol{x} = \begin{bmatrix} \hat{C} \\ \hat{T} \end{bmatrix} \tag{6.2.20}$$

$$u = \hat{\beta} \tag{6.2.21}$$

$$\boldsymbol{d}_m = \begin{bmatrix} \hat{C}_i \\ \hat{T}_i \end{bmatrix} \tag{6.2.22}$$

For this system there are two state variables and only one control variable so that only one state variable can be specified as the specified or output vector, $\boldsymbol{x}_{se}$, and the other state variable is free to attain new conditions. Let us assume that the temperature is the output state or state to be specified so that $x_2 = 0$. This yields for the equilibrium equations,

$$0 = a_{11}x_{1e} + w_{11}d_1 \tag{6.2.23}$$

and

$$0 = a_{21}x_{1e} + b_2u_e + w_{22}d_2 \tag{6.2.24}$$

Equation (6.2.23) allows for the computation of the free equilibrium state, x_{1e}, and Equation (6.2.24) gives the equilibrium value of the control.

Since we are trying to regulate about a specified temperature, the weighting matrix in the quadratic performance functional is chosen as

$$\boldsymbol{Q} = \begin{bmatrix} 0 & 0 \\ 0 & 1 \end{bmatrix} \tag{6.2.25}$$

The MATLAB program for implementing the Anderson algorithm is given in Figure 6.3.

The optimal infinite time or steady-state controller gain matrix using MATLAB is computed assuming $\boldsymbol{R} = \boldsymbol{I}$ as

$$\boldsymbol{K}_{ss} = [-0.1028 \qquad -0.6464] \tag{6.2.26}$$

and the optimal control law is

$$u = -\boldsymbol{K}_{ss}\boldsymbol{x} + \left[u_e + \boldsymbol{K}_{ss}\begin{pmatrix} x_{1e} \\ 0 \end{pmatrix}\right] \tag{6.2.27}$$

The response of the system for a 20 percent step-normalized disturbance vector is shown in Figure 6.4. In general, the temperature response of the system is excellent using this static predictor with state feedback. There is a small transient error in temperature. This is due to the fact that the predicted state coming from the model is not exactly that of the process. This means that the feedback error, $\boldsymbol{e}$, of Figure 6.2 does exist and is caused by modeling errors. The feedback element trims and eventually eliminates these modeling errors. The composition response of the system is also shown in Figure 6.4. As expected, it does not stay constant but attains new values due to the load disturbances imposed upon the system.

6.3 Time Varying Measurable Disturbances

Ramirez *et al.* (1987) considered the development of the optimal regulator control problem for a linear system with time varying measurable disturbances. The system is given by

$$\dot{\boldsymbol{x}} = \boldsymbol{A}\boldsymbol{x} + \boldsymbol{B}\boldsymbol{u} + \boldsymbol{W}\boldsymbol{d}_m(t) \tag{6.3.1}$$

where $\boldsymbol{d}_m(t)$ is a vector of measurable time varying disturbances. The objective is to define a stable, dynamic feed-forward or predictive state, $\boldsymbol{x}_p$, and predictive control, $\boldsymbol{u}_p$, on the basis of available information, $\boldsymbol{d}_m(t)$. The predictive state can be partitioned into

$$\boldsymbol{x}_p = \begin{bmatrix} \boldsymbol{x}_{sp} \\ \boldsymbol{x}_{fp} \end{bmatrix} \tag{6.3.2}$$

where $\boldsymbol{x}_{sp}$ are the m state variables that will be maintained at specified values. The specified states can be considered as desired process outputs that need to remain at constant values or follow a prescribed trajectory. $\boldsymbol{x}_{fp}$ are the free or unspecified states. For regulation, we desire that the specified states,

```
>> % input the system matrices
>> a = [-0.05156   -0.05877;
         0.01503   -0.005887];
>> b = [0  -0.03384]';
>> w = [0.0125  0;
        0       0.0125];
>>
>> %  input the weighting matrices
>> q = [0  0;
        0  1];
>> r = 1;
>>
>> % compute the LQR feedback gain matrix
>> [k,p] = lqr(a,b,q,r)

k =

   -0.1028   -0.6464


p =

    0.7834    3.0391
    3.0391   19.1024

>>
>> % compute the equilibrium control and unspecified state
>> d = [.2  .2]';
>> x1e = -w(1,1) * d(1) / a(1,1)

x1e =

    0.0485

>> ue = -( a(2,1) * x1e + w(2,2) * d(2)) / b(2)

ue =

    0.0954

>> % compute the predicted feedforward control
>> xe = [x1e  0]';
>> up = ue  + k * xe

up =

    0.0904

>> % prepare for plotting step disturbance response
>> aa = a - b*k;
>> bb = w*d +b*up

bb =

    0.0025
   -0.0006

>> t = [0:1:100];
>> c = eye(2);
>> d = [0 0]';
>> xc = step(aa,bb,c,d,1,t);
>> plot (t,xc);
>> title ('Anderson Algorithm MPC State Response');
>> ylabel ('Normalized Composition and Temperature');
>> xlabel ('Time (min)');
>> meta anderson;
```

Figure 6.3 MATLAB implementation of Anderson MPC algorithm.

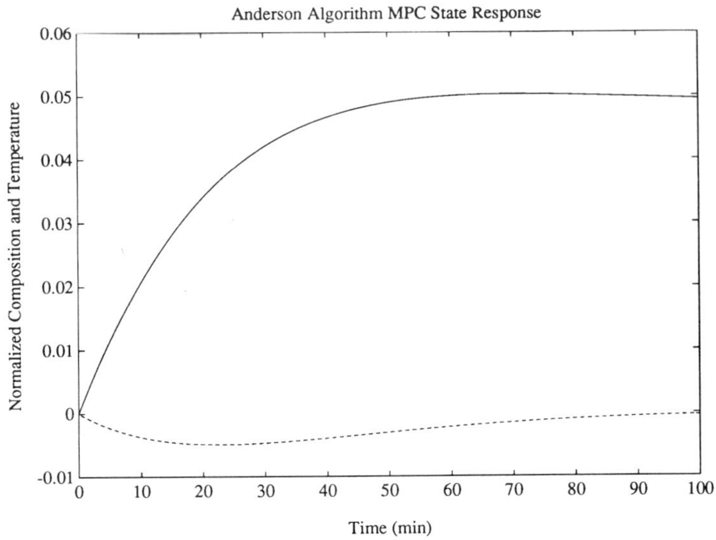

Figure 6.4 Anderson algorithm MPC state response. Composition (—), temperature (- - - -).

$\boldsymbol{x}_{sp}$, remain at the original steady-state condition of zero. The design of the predictive states and controls is therefore given by

$$\begin{pmatrix} \mathbf{0} \\ \dot{\boldsymbol{x}}_{fp} \end{pmatrix} = \begin{bmatrix} \boldsymbol{A}_{11} & \boldsymbol{A}_{12} \\ \boldsymbol{A}_{21} & \boldsymbol{A}_{22} \end{bmatrix} \begin{pmatrix} \boldsymbol{x}_{sp} \\ \boldsymbol{x}_{fp} \end{pmatrix} + \begin{bmatrix} \boldsymbol{B}_1 \\ \boldsymbol{B}_2 \end{bmatrix} \boldsymbol{u}_p + \begin{bmatrix} \boldsymbol{W}_1 \\ \boldsymbol{W}_2 \end{bmatrix} \boldsymbol{d}_m(t) \quad (6.3.3)$$

Equation (6.3.3) is a set of m linear algebraic and $(n - m)$ linear differential equations that must be solved for the $(n-m)$ unknown variables of the free or unspecified predictive state trajectories, $\boldsymbol{x}_{fp}(t)$, and the m predictive control trajectories, $\boldsymbol{u}_p(t)$.

In order to formulate the optimal regulator problem, we define deviation variables

$$\hat{\boldsymbol{x}} = \boldsymbol{x} - \boldsymbol{x}_p \quad (6.3.4)$$

$$\hat{\boldsymbol{u}} = \boldsymbol{u} - \boldsymbol{u}_p \quad (6.3.5)$$

The original state differential equation of Equation (6.3.1) is therefore transformed into

$$\dot{\hat{\boldsymbol{x}}} = \boldsymbol{A}\hat{\boldsymbol{x}} + \boldsymbol{B}\hat{\boldsymbol{u}} \quad (6.3.6)$$

which is identical to the model of the original Kalman problem. If we desire to minimize the quadratic performance functional

$$J = \frac{1}{2}\,\hat{\boldsymbol{x}}^T(t_f)\boldsymbol{H}\hat{\boldsymbol{x}}(t_f) + \frac{1}{2}\int_{t_o}^{t_f} \left(\hat{\boldsymbol{x}}^T\boldsymbol{Q}\hat{\boldsymbol{x}} + \hat{\boldsymbol{u}}^T\boldsymbol{R}\hat{\boldsymbol{u}}\right)\, dt \quad (6.3.7)$$

then we have transformed the original problem statement into an equivalent Kalman LQR control problem and the optimal control is given by

$$\hat{\boldsymbol{u}} = -\boldsymbol{K}_{ss}\hat{\boldsymbol{x}} \tag{6.3.8}$$

In terms of the original variables, the optimal regulator control law with measurable disturbances is

$$\boldsymbol{u} = -\boldsymbol{K}_{ss}\boldsymbol{x} + \left[\begin{array}{c:c} \boldsymbol{I} & \boldsymbol{K}_{ss} \end{array}\right] \begin{pmatrix} \boldsymbol{u}_p(t) \\ \boldsymbol{x}_p(t) \end{pmatrix} \tag{6.3.9}$$

This indicates that this optimal regulator is a multivariable proportional feedback controller with a dynamic feed-forward or predictive term defined by Equation (6.3.3). As with the Anderson regulator, which has an equilibrium and not a dynamic predictive term, the gain matrix, $\boldsymbol{K}_{ss}$, is computed using the Riccati equations. The block diagram for this control system is given in Figure 6.5. Again the structure is that of model predictive control (MPC). The main difference between this algorithm and the Anderson algorithm is that here we are using a dynamic predictive model that is a better representation of the actual process response.

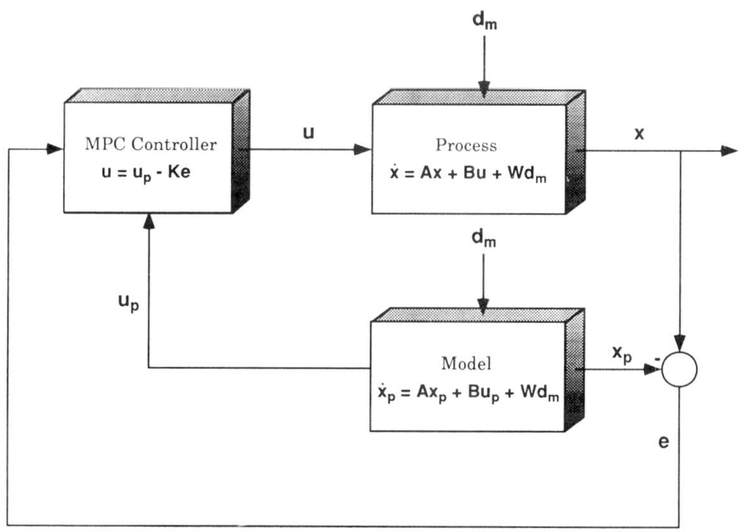

Figure 6.5 Dynamic predictor MPC algorithm.

We now need to consider the methods for solving Equation (6.3.3), which defines the dynamic predictive states and predictive controls. One possible solution strategy is to solve for the m $\boldsymbol{u}_p$ variables using the m algebraic equations so that

$$\boldsymbol{u}_p = \boldsymbol{B}_1^{-1}\left[-\boldsymbol{A}_{11}\boldsymbol{x}_{sp} - \boldsymbol{A}_{12}\boldsymbol{x}_{fp} - \boldsymbol{W}_1\boldsymbol{d}_m(t)\right] \tag{6.3.10}$$

This solution depends upon the existence of $\boldsymbol{B}_1^{-1}$. Equation (6.3.10) shows that in order to solve for the predictive control vector, $\boldsymbol{u}_p$, instantaneous values of the measured disturbances, $\boldsymbol{d}_m(t)$, are needed as well as the predictive state, $\boldsymbol{x}_{fp}$. Using Equations (6.3.3) and (6.3.10), the $(n-m)$ free predictive states are given by the $(n-m)$ differential equations

$$\dot{\boldsymbol{x}}_{fp} = \left[\boldsymbol{A}_{22} - \boldsymbol{B}_2\boldsymbol{B}_1^{-1}\boldsymbol{A}_{12}\right]\boldsymbol{x}_{fp} + \left[\boldsymbol{A}_{21} - \boldsymbol{B}_2\boldsymbol{B}_1^{-1}\boldsymbol{A}_{11}\right]\boldsymbol{x}_{sp} + \left[-\boldsymbol{B}_2\boldsymbol{B}_1^{-1}\boldsymbol{W}_1 + \boldsymbol{W}_2\right]\boldsymbol{d}_m(t) \tag{6.3.11}$$

The following conditions must exist for the free predictive state, $\boldsymbol{x}_{fp}$, to be stable. First, the matrix, $\hat{\boldsymbol{A}}_{22}$, must be negative semi–definite, where

$$\hat{\boldsymbol{A}}_{22} = \left[\boldsymbol{A}_{22} - \boldsymbol{B}_2\boldsymbol{B}_1^{-1}\boldsymbol{A}_{12}\right] \tag{6.3.12}$$

Also, $\boldsymbol{B}_1^{-1}$ must exist.

For the case when $\boldsymbol{B}_1^{-1}$ does not exist, the predictive states and controls depend not only on measured disturbances but also on the time derivatives of the disturbances. The development of the necessary conditions for a solution requires extensive partitioning of Equation (6.3.3). The partitioning is motivated by isolating the singular and nonsingular parts of $\boldsymbol{B}_1$. In doing this, upper and lower sections of the matrices must also be defined. The m specified state variables, $\boldsymbol{x}_{sp}$, remain intact, but the $(n-m)$ free predicted state, $\boldsymbol{x}_{fp}$, and m control variables, $\boldsymbol{u}_p$, are both subdivided into singular and nonsingular elements. Equation (6.3.3) becomes, under partitioning

$$\begin{pmatrix} \boldsymbol{0} \\ \boldsymbol{0} \\ \dot{\boldsymbol{x}}_{fp}^{s} \\ \dot{\boldsymbol{x}}_{fp}^{n} \end{pmatrix} = \begin{bmatrix} \boldsymbol{A}_{11}^{u} & \boldsymbol{A}_{12}^{s} & \boldsymbol{A}_{12}^{u} \\ \boldsymbol{A}_{11}^{\ell} & \boldsymbol{A}_{12}^{\ell} & \boldsymbol{A}_{12}^{n} \\ \boldsymbol{A}_{21}^{u} & \boldsymbol{A}_{22}^{s} & \boldsymbol{A}_{22}^{u} \\ \boldsymbol{A}_{21}^{\ell} & \boldsymbol{A}_{22}^{\ell} & \boldsymbol{A}_{22}^{n} \end{bmatrix} \begin{pmatrix} \boldsymbol{x}_{sp} \\ \boldsymbol{x}_{fp}^{s} \\ \boldsymbol{x}_{fp}^{n} \end{pmatrix} + \begin{bmatrix} \boldsymbol{B}_1^{s} & \boldsymbol{B}_1^{u} \\ \boldsymbol{B}_1^{\ell} & \boldsymbol{B}_1^{n} \\ \boldsymbol{B}_2^{s} & \boldsymbol{B}_2^{u} \\ \boldsymbol{B}_2^{\ell} & \boldsymbol{B}_2^{n} \end{bmatrix} \begin{pmatrix} \boldsymbol{u}_p^{s} \\ \boldsymbol{u}_p^{n} \end{pmatrix} + \begin{bmatrix} \boldsymbol{W}_1^{s} \\ \boldsymbol{W}_1^{n} \\ \boldsymbol{W}_2^{s} \\ \boldsymbol{W}_2^{n} \end{bmatrix} \boldsymbol{d}_m(t) \tag{6.3.13}$$

where $\boldsymbol{B}_1^{s}$ is the singular part of $\boldsymbol{B}_1$, $\boldsymbol{B}_1^{n}$ is the nonsingular part of $\boldsymbol{B}_1$, $\boldsymbol{B}_1^{u}$ is the upper partition of $\boldsymbol{B}_1$, and $\boldsymbol{B}_1^{\ell}$ is the lower partition of $\boldsymbol{B}_1$.

We want to solve the set of linear algebraic equations for $\boldsymbol{x}_{fp}^{s}$, which is the singular part of the differential equations of (6.3.13), and for the controls not associated with the singular part of (6.3.13), that is, $\boldsymbol{u}_p^{n}$. The m algebraic equations are therefore

$$\begin{pmatrix} \boldsymbol{x}_{fp}^{s} \\ \boldsymbol{u}_p^{n} \end{pmatrix} = -\begin{bmatrix} \boldsymbol{A}_{12}^{s} & \boldsymbol{B}_1^{u} \\ \boldsymbol{A}_{12}^{\ell} & \boldsymbol{B}_1^{n} \end{bmatrix}^{-1} \left\{ \begin{bmatrix} \boldsymbol{A}_{11}^{u} \\ \boldsymbol{A}_{11}^{\ell} \end{bmatrix} \boldsymbol{x}_{sp} \right.$$

$$+ \left[\begin{array}{cc} \boldsymbol{A}_{12}^{u} & \boldsymbol{B}_{1}^{s} \\ \boldsymbol{A}_{12}^{n} & \boldsymbol{B}_{1}^{\ell} \end{array} \right] \left(\begin{array}{c} \boldsymbol{x}_{fp}^{n} \\ \boldsymbol{u}_{p}^{s} \end{array} \right) + \left[\begin{array}{c} \boldsymbol{W}_{1}^{s} \\ \boldsymbol{W}_{1}^{n} \end{array} \right] \boldsymbol{d}_{m}(t) \Bigg\} \quad (6.3.14)$$

or, by appropriate redefinition,

$$\left(\begin{array}{c} \boldsymbol{x}_{fp}^{s} \\ \boldsymbol{u}_{p}^{n} \end{array} \right) = \left[\begin{array}{c} \bar{\boldsymbol{A}}_{1} \\ \bar{\boldsymbol{A}}_{2} \end{array} \right] \boldsymbol{x}_{sp} + \left[\begin{array}{cc} \bar{\boldsymbol{B}}_{11} & \bar{\boldsymbol{B}}_{12} \\ \bar{\boldsymbol{B}}_{21} & \bar{\boldsymbol{B}}_{22} \end{array} \right] \left(\begin{array}{c} \boldsymbol{x}_{fp}^{n} \\ \boldsymbol{u}_{p}^{s} \end{array} \right) + \left[\begin{array}{c} \bar{\boldsymbol{W}}_{1} \\ \bar{\boldsymbol{W}}_{2} \end{array} \right] \boldsymbol{d}_{m}(t) \quad (6.3.15)$$

For Equation (6.3.14) to be realizable,

$$\left[\begin{array}{cc} \boldsymbol{A}_{12}^{s} & \boldsymbol{B}_{1}^{u} \\ \boldsymbol{A}_{12}^{\ell} & \boldsymbol{B}_{1}^{n} \end{array} \right]^{-1} \quad (6.3.16)$$

must exist.

The $(n-m)$ differential equations of Equation (6.3.13) are given by

$$\left(\begin{array}{c} \dot{\boldsymbol{x}}_{fp}^{s} \\ \dot{\boldsymbol{x}}_{fp}^{n} \end{array} \right) = \left[\begin{array}{c} \boldsymbol{A}_{21}^{u} \\ \boldsymbol{A}_{21}^{\ell} \end{array} \right] \boldsymbol{x}_{sp} + \left[\begin{array}{cc} \boldsymbol{A}_{22}^{u} & \boldsymbol{B}_{2}^{s} \\ \boldsymbol{A}_{22}^{n} & \boldsymbol{B}_{2}^{\ell} \end{array} \right] \left(\begin{array}{c} \boldsymbol{x}_{fp}^{n} \\ \boldsymbol{u}_{p}^{s} \end{array} \right)$$
$$+ \left[\begin{array}{cc} \boldsymbol{A}_{22}^{s} & \boldsymbol{B}_{2}^{u} \\ \boldsymbol{A}_{22}^{\ell} & \boldsymbol{B}_{2}^{n} \end{array} \right] \left(\begin{array}{c} \boldsymbol{x}_{fp}^{s} \\ \boldsymbol{u}_{p}^{s} \end{array} \right) + \left[\begin{array}{c} \boldsymbol{W}_{2}^{s} \\ \boldsymbol{W}_{2}^{n} \end{array} \right] \boldsymbol{d}_{m}(t) \quad (6.3.17)$$

Using Equation (6.3.15) to eliminate $\boldsymbol{x}_{fp}^{s}$ gives the following $(n-m)$ differential equations which need to be solved for the variables $\boldsymbol{x}_{fp}^{n}$ and $\boldsymbol{u}_{p}^{s}$:

$$\left[\begin{array}{cc} \bar{\boldsymbol{B}}_{11} & \bar{\boldsymbol{B}}_{12} \\ \boldsymbol{I} & \boldsymbol{0} \end{array} \right] \left(\begin{array}{c} \dot{\boldsymbol{x}}_{fp}^{n} \\ \dot{\boldsymbol{u}}_{p}^{s} \end{array} \right) = \left\{ \left[\begin{array}{cc} \boldsymbol{A}_{22}^{u} & \boldsymbol{B}_{2}^{s} \\ \boldsymbol{A}_{22}^{n} & \boldsymbol{B}_{2}^{\ell} \end{array} \right] + \left[\begin{array}{cc} \boldsymbol{A}_{22}^{s} & \boldsymbol{B}_{2}^{u} \\ \boldsymbol{A}_{22}^{\ell} & \boldsymbol{B}_{2}^{n} \end{array} \right] \left[\begin{array}{cc} \bar{\boldsymbol{B}}_{11} & \bar{\boldsymbol{B}}_{12} \\ \bar{\boldsymbol{B}}_{21} & \bar{\boldsymbol{B}}_{22} \end{array} \right] \right\}$$
$$\left(\begin{array}{c} \boldsymbol{x}_{fp}^{n} \\ \boldsymbol{u}_{p}^{s} \end{array} \right) + \left[\begin{array}{c} \boldsymbol{A}_{21}^{u} \\ \boldsymbol{A}_{21}^{\ell} \end{array} \right] \boldsymbol{x}_{sp} + \left[\begin{array}{c} \boldsymbol{W}_{1}^{s} \\ \boldsymbol{W}_{2}^{n} \end{array} \right] \boldsymbol{d}_{m}(t) + \left[\begin{array}{c} \bar{\boldsymbol{W}}_{1} \\ \boldsymbol{0} \end{array} \right] \dot{\boldsymbol{d}}_{m}(t) \quad (6.3.18)$$

In order to ensure the stability of the solution to Equation (6.3.18), we need to look at the solution of the homogeneous part. It is convenient to redefine matrices and vectors such that

$$\tilde{\boldsymbol{A}}\dot{\tilde{\boldsymbol{x}}} = \tilde{\boldsymbol{B}}\tilde{\boldsymbol{x}} + \tilde{\boldsymbol{C}}\boldsymbol{x}_{sp} + \tilde{\boldsymbol{D}}\boldsymbol{d}_{m}(t) + \tilde{\boldsymbol{E}}\dot{\boldsymbol{d}}_{m}(t) \quad (6.3.19)$$

where

$$\tilde{\boldsymbol{x}} = \left(\begin{array}{c} \boldsymbol{x}_{fp}^{n} \\ \boldsymbol{u}_{p}^{s} \end{array} \right) \quad (6.3.20)$$

The matrix, $\tilde{\boldsymbol{A}}$, can be decomposed so that the homogeneous part of the differential equations of Equation (6.3.19) can be written as

$$\boldsymbol{U}\boldsymbol{S}\boldsymbol{V}^{T}\dot{\tilde{\boldsymbol{x}}} = \tilde{\boldsymbol{B}}\boldsymbol{x} \quad (6.3.21)$$

where $\boldsymbol{U}$ and $\boldsymbol{V}$ are orthogonal matrices, $\boldsymbol{V}^{-1} = \boldsymbol{V}^T$, $\boldsymbol{U}^{-1} = \boldsymbol{U}^T$, and $\boldsymbol{S}$ is a diagonal matrix. The MATLAB function SVD can perform this singular value decomposition. Letting

$$\begin{aligned} \boldsymbol{y} &= \boldsymbol{V}^T \tilde{\boldsymbol{x}} \qquad \text{and} \\ \boldsymbol{D} &= \boldsymbol{U}^T \tilde{\boldsymbol{B}} \boldsymbol{V} \end{aligned} \tag{6.3.22}$$

yields for the homogeneous problem of Equation (6.3.21),

$$\begin{bmatrix} \boldsymbol{S}_d & \boldsymbol{0} \\ \boldsymbol{0} & \boldsymbol{0} \end{bmatrix} \begin{pmatrix} \dot{\boldsymbol{y}}_1 \\ \dot{\boldsymbol{y}}_2 \end{pmatrix} = \begin{bmatrix} \boldsymbol{D}_{11} & \boldsymbol{D}_{12} \\ \boldsymbol{D}_{21} & \boldsymbol{D}_{22} \end{bmatrix} \begin{pmatrix} \boldsymbol{y}_1 \\ \boldsymbol{y}_2 \end{pmatrix} \tag{6.3.23}$$

where $\boldsymbol{S}_d$ is the nonzero part of the diagonalized $\tilde{\boldsymbol{A}}$ matrix $\boldsymbol{S}$. Equation (6.3.23) can be decomposed into a set of algebraic equations

$$\boldsymbol{y}_2 = -\boldsymbol{D}_{22}^{-1} \boldsymbol{D}_{21} \boldsymbol{y}_1 \tag{6.3.24}$$

and a set of differential equations

$$\dot{\boldsymbol{y}}_1 = \boldsymbol{S}_d^{-1} \left[\boldsymbol{D}_{11} + \boldsymbol{D}_{12} \boldsymbol{D}_{22}^{-1} \boldsymbol{D}_{21} \right] \boldsymbol{y}_1 \tag{6.3.25}$$

Therefore, the necessary conditions to ensure the stability of the predictive states, $\boldsymbol{x}_{fp}^n$, and controls, $\boldsymbol{u}_p^s$, are that $\boldsymbol{D}_{22}^{-1}$ exists and that $\boldsymbol{S}_d^{-1}[\boldsymbol{D}_{11} + \boldsymbol{D}_{12}\boldsymbol{D}_{22}^{-1} \boldsymbol{D}_{21}]$ must be a negative semi–definite matrix.

6.3.1 Nyquist–Ramirez Reactor

Let us apply this method to the Nyquist–Ramirez reactor defined in Section 6.2.1. Since there is only one control variable, only one state variable can be specified. The remaining state variable is left unspecified. When temperature is the specified or output variable, then $\hat{T}_p = 0$ and $\hat{C}_p$ is unspecified. Equation (6.3.3) for the predictor equations becomes

$$0 = a_{21}\hat{C}_p + b_2 u_p + w_{22}\hat{T}_i(t) \tag{6.3.26}$$

and

$$\dot{\hat{C}}_p = a_{11}\hat{C}_p + w_{11}\hat{C}_i(t) \tag{6.3.27}$$

In this case, $\boldsymbol{B}_1^{-1} = b_2^{-1}$, which exists and satisfies the necessary condition for an algebraic solution of $\boldsymbol{u}_p$ from Equation (6.3.26). The $\hat{\boldsymbol{A}}_{22}$ matrix, defined from Equation (6.3.12), becomes

$$\hat{\boldsymbol{A}}_{22} = a_{11} = -0.05156 \tag{6.3.28}$$

Therefore, $\hat{\boldsymbol{A}}_{22}$ is negative semi–definite. This ensures the stability of the differential equation (6.3.27) for the predicted concentration state, $\hat{C}_p$.

For a state weighting matrix

$$Q = \begin{bmatrix} 0 & 0 \\ 0 & 1 \end{bmatrix} \tag{6.3.29}$$

we can compute the optimal feedback gain matrix to be

$$K_{ss} = [-0.1028 \qquad -0.6464] \tag{6.3.30}$$

It is identical to the gain matrix of Example 6.2.1. If we apply this linear control law to the nonlinear problem with 20 percent step changes in the disturbance variables, we obtain the response of Figure 6.6. This figure shows the response of both the Anderson algorithm and the dynamic predictive algorithm. Figure 6.6 shows that the response of the system with the dynamic predictive algorithm is superior and is essentially perfect. This is because the system is linear in concentration and only nonlinear in temperature. By keeping the temperature close to the condition about which the system was linearized, the nonlinear response is essentially that of a linear system.

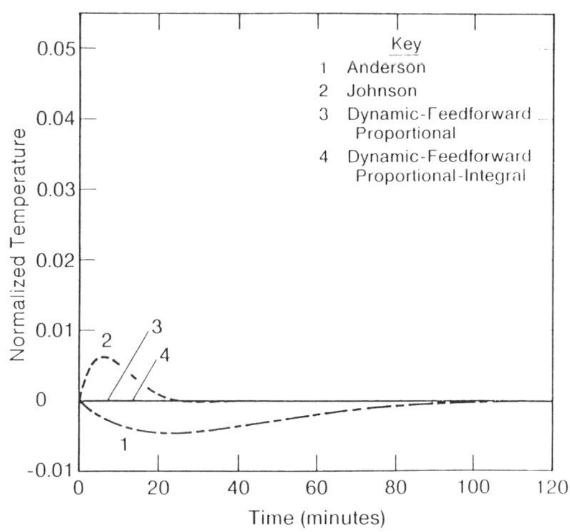

Figure 6.6. Outlet temperature regulated with completely measurable disturbances. From Ramirez *et al.* (1987).

When the concentration is chosen as the output or specified variable, $\hat{C}_p = 0$ and $\hat{T}_p$ is unspecified. The set of predictor equations is

$$0 = a_{12}\hat{T}_p + w_{11}\hat{C}_i(t) \tag{6.3.31}$$

and

$$\dot{\hat{T}}_p = a_{22}\hat{T}_p + b_2 u_p + w_{22}\hat{T}_i(t) \tag{6.3.32}$$

In this case, $\boldsymbol{B}_1^{-1}$ does not exist since the control does not appear in the algebraic equation (6.3.31). The strategy is to solve for the predicted temperature state using the algebraic equation of (6.3.31) and the predicted control from the differential equation of (6.3.32). We therefore partition the system consistent with Equation (6.3.13). To solve the algebraic equations, the matrix inverse of Equation (6.3.16) must exist. For this case, we have

$$\begin{bmatrix} \boldsymbol{A}_{12}^s & \boldsymbol{B}_1^u \\ \boldsymbol{A}_{12}^\ell & \boldsymbol{B}_1^n \end{bmatrix} = a_{12}^{-1} = -0.05877^{-1} \tag{6.3.33}$$

The first necessary condition is satisfied because a_{12}^{-1} exists. The algebraic solution to Equation (6.3.31) is

$$\hat{T}_p = -\frac{w_{11}}{a_{12}}\hat{C}_i(t) \tag{6.3.34}$$

From the differential equation of (6.3.32), the predicted control must be computed and is

$$u_p = \frac{1}{b_2}\left[-\frac{w_{11}}{a_{12}}\dot{\hat{C}}_i(t) + \frac{a_{22}w_{11}}{a_{12}}\hat{C}_i(t) - w_{22}\hat{T}_i(t)\right] \tag{6.3.35}$$

The solution exists because $\boldsymbol{D}_{22}^{-1} = b_2^{-1}$. In this case, the $\boldsymbol{y}$ vector does not exist so there are no stability concerns from Equation (6.3.25). The predictive control is a function of the disturbances, $\hat{C}_i(t)$ and $\hat{T}_i(t)$, as well as the time derivative of the concentration disturbance. The implementation of the algorithm for this case requires that the disturbances be piecewise continuous functions. If not, impulses in the controls will occur. Step disturbances can be approximated as ramps over sampling intervals.

We test the algorithm using ramp disturbances lasting 10 minutes, reaching 20 percent of normalized unity values. The response of the nonlinear system for both the Anderson algorithm and the dynamic predictor algorithm are shown in Figure 6.7. The dynamic predictor algorithm is able to perform much better than the static predictor of Anderson.

In general, the dynamic predictor will perform better than the Anderson algorithm, but it is harder to compute and implement.

6.4 Partially Measurable Disturbances

Ramirez *et al.* (1987) have discussed the development of an optimal regulator for the case of partially measurable disturbances. The dynamics of a linear system are given by

$$\dot{\boldsymbol{x}} = \boldsymbol{A}\boldsymbol{x} + \boldsymbol{B}\boldsymbol{u} + \boldsymbol{W}_m\boldsymbol{d}_m(t) + \boldsymbol{W}_u\boldsymbol{d}_u \tag{6.4.1}$$

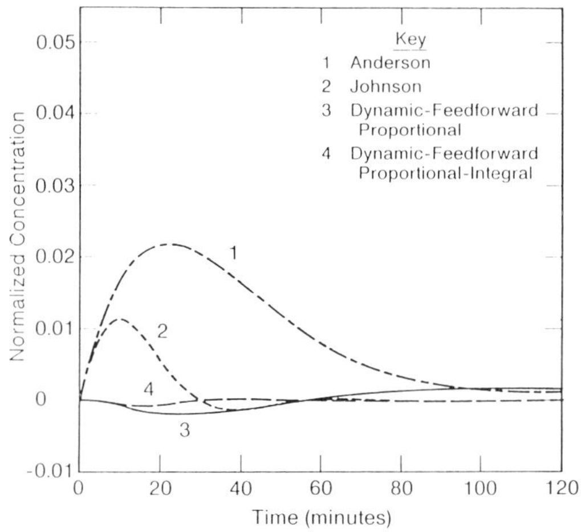

Figure 6.7. Outlet concentration regulated with completely measurable disturbances. From Ramirez *et al.* (1987).

Here we have two classes of disturbances, those that are measurable, $\boldsymbol{d}_m(t)$, and those that are unmeasurable, $\boldsymbol{d}_u(t)$. We first transform the problem in terms of specified or output states, $\boldsymbol{x}_{sp}$, and predictive states, $\boldsymbol{x}_{fp}$, which use the measurable disturbances, $\boldsymbol{d}_m(t)$. The predictive states and controls are defined by the solution of

$$\begin{pmatrix} \mathbf{0} \\ \dot{\boldsymbol{x}}_{fp} \end{pmatrix} = \begin{bmatrix} \boldsymbol{A}_{11} & \boldsymbol{A}_{12} \\ \boldsymbol{A}_{21} & \boldsymbol{A}_{22} \end{bmatrix} \begin{pmatrix} \boldsymbol{x}_{sp} \\ \boldsymbol{x}_{fp} \end{pmatrix} + \begin{bmatrix} \boldsymbol{B}_1 \\ \boldsymbol{B}_2 \end{bmatrix} \boldsymbol{u}_p + \begin{bmatrix} \boldsymbol{W}_{m1} \\ \boldsymbol{W}_{m2} \end{bmatrix} \boldsymbol{d}_m(t) \quad (6.4.2)$$

The solution of the algebraic and differential equations of (6.4.2) yield values for the free unspecified states, $\boldsymbol{x}_{fp}$, and the dynamic predictive controls, $\boldsymbol{u}_p(t)$. Defining the deviation variables

$$\hat{\boldsymbol{x}} = \boldsymbol{x} - \boldsymbol{x}_p \quad (6.4.3)$$

and

$$\hat{\boldsymbol{u}} = \boldsymbol{u} - \boldsymbol{u}_p \quad (6.4.4)$$

we have, for the system description,

$$\dot{\hat{\boldsymbol{x}}} = \boldsymbol{A}\hat{\boldsymbol{x}} + \boldsymbol{B}\hat{\boldsymbol{u}} + \boldsymbol{W}_u\boldsymbol{d}_u \quad (6.4.5)$$

We have now transformed the problem into a form that is equivalent to the problem considered by Johnson in Section 5.6. Therefore, for the variable vector

$$\tilde{\boldsymbol{x}} = \begin{pmatrix} \hat{\boldsymbol{x}} \\ \boldsymbol{y} \end{pmatrix} \tag{6.4.6}$$

we know from the development of Section 5.6 that the optimal regulator control is

$$\dot{\hat{\boldsymbol{u}}} = -\left[\, \boldsymbol{K}_1 \;\vdots\; \boldsymbol{K}_2 \,\right] \begin{pmatrix} \hat{\boldsymbol{x}} \\ \boldsymbol{y} \end{pmatrix} \tag{6.4.7}$$

Expressing this result in terms of the original variables, we have

$$\boldsymbol{u} = \boldsymbol{u}_p - \boldsymbol{K}_2(\boldsymbol{x} - \boldsymbol{x}_p) - [\boldsymbol{K}_1 - \boldsymbol{K}_2\boldsymbol{A}] \int (\boldsymbol{x} - \boldsymbol{x}_p)\, dt \tag{6.4.8}$$

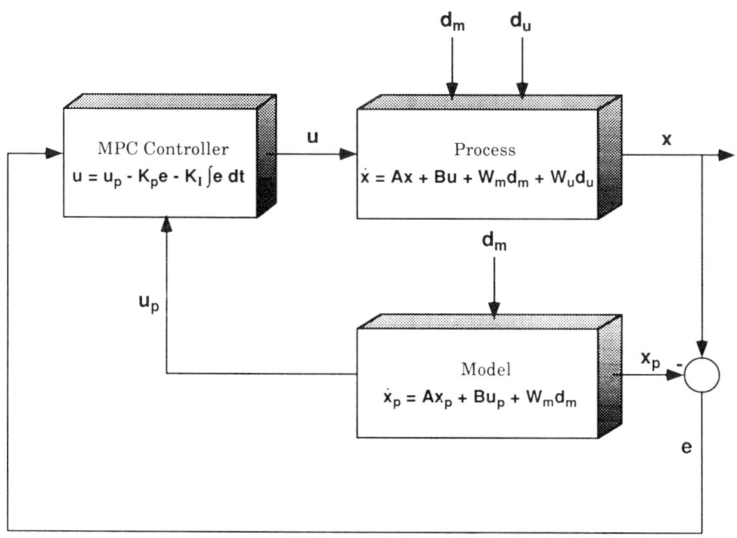

Figure 6.8 Partially measurable MPC algorithm.

This is a state proportional-integral feedback controller with a dynamic predictive feed-forward element. The structure of the controller is given in Figure 6.8. The figure shows the MPC structure. It also clearly indicates that the model even if physically correct is operating only on the measured disturbances while the process is affected by both measured and unmeasured disturbances. The integral term in the feedback loop compensates for unmeasurable disturbances. When all disturbances are measurable, the controller is equivalent to the dynamic predictive controller of Section 6.3 for linear systems. For nonlinear systems, the integral term of the partially measurable algorithm can be an advantage even if all the measurements are measured since it can help compensate for modeling errors. Figure 6.6 shows the temperature regulation of the nonlinear Nyquist–Ramirez reactor. The problem is essentially linear in concentration so when the temperature is held constant, there is no difference in including the integral term. However, as shown in Figure 6.7, when the nonlinearities show up, inclusion of the integral terms tends to improve control. When all disturbances are unmeasurable, the partially measurable controller is equivalent to that developed by Johnson and presented in Section 5.6. The controller no longer has MPC structure since the predicted state and control will stay at the zero equilibrium condition since there are no measurable disturbances to drive it from that condition. The error vector, $\boldsymbol{e}$, becomes equal to the output state vector $\boldsymbol{x}$.

6.4.1 Nyquist–Ramirez Reactor

Let us apply this algorithm to the Nyquist–Ramirez reactor when all the disturbances are not measurable. Figure 6.9 shows the case when outlet temperature is regulated with only inlet temperature being measurable. Figure 6.10 is for temperature regulation when only inlet concentration is measurable. For the first case, the MPC partially measurable algorithm which includes the dynamic feed-forward term yields the best performance when compared to the LQR Johnson PI algorithm and the MPC dynamic predictor without the integral term to account for the unmeasurable disturbance. The case of Figure 6.10 shows that the Johnson and partially measurable algorithms are equivalent and superior to the dynamic proportional feed-forward algorithm when unmeasurable disturbances are present. In Figures 6.11 and 6.12 we present the results for concentration regulation with measurable inlet temperature and inlet concentration disturbances, respectively. When the inlet concentration is measurable, the partially measurable algorithm is superior, but when the inlet temperature is measurable, the Johnson algorithm is slightly superior to the partially measurable algorithm. The feed-forward action of the partially measurable algorithm enhances the performance only if the measured disturbance affects the desired state strongly. Using incomplete measurement information can sometimes lead to a deterioration of system performance as illustrated in Figure 6.11.

6.5 Optimal Distillation Column Regulation

A laboratory distillation column, consisting of five trays, a total condenser and a reboiler, has been controlled by optimal regulators by Hu and Ramirez (1972). A linear model of that column will be used to evaluate the simulated performance of the dynamic feed-forward algorithms along with the Johnson

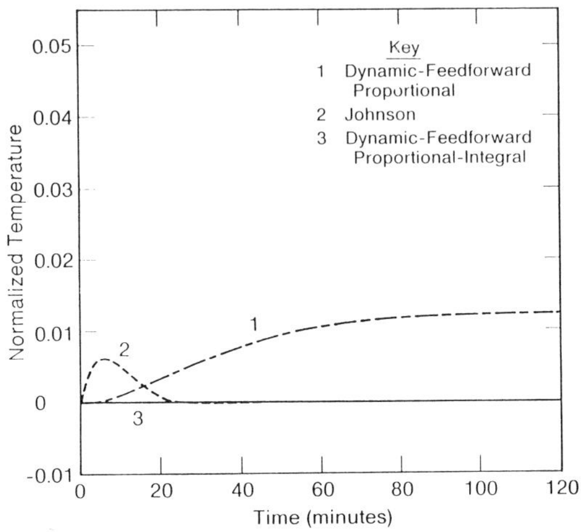

Figure 6.9 Outlet temperature regulated with only inlet temperature measurable. From Ramirez *al.* (1987).

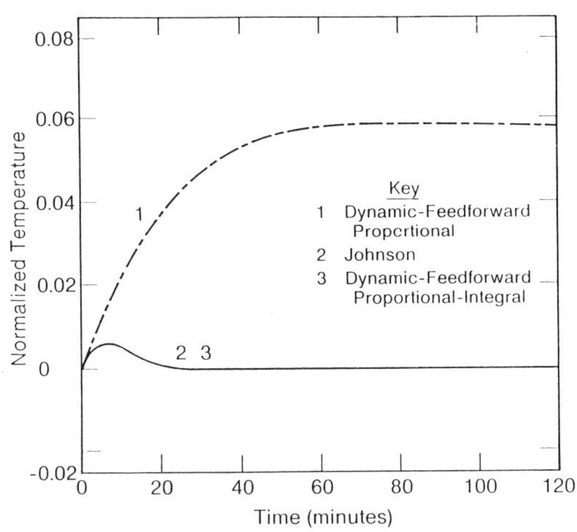

Figure 6.10 Outlet temperature regulated with only inlet concentration measurable. From Ramirez *et al.* (1987).

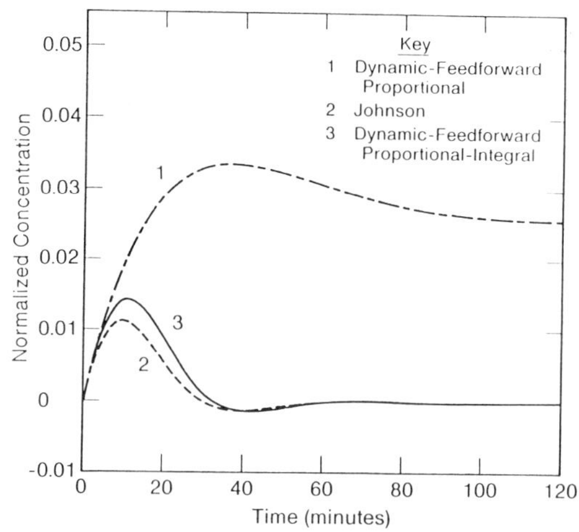

Figure 6.11 Outlet concentration regulated with only inlet temperature measurable. From Ramirez *et al.* (1987).

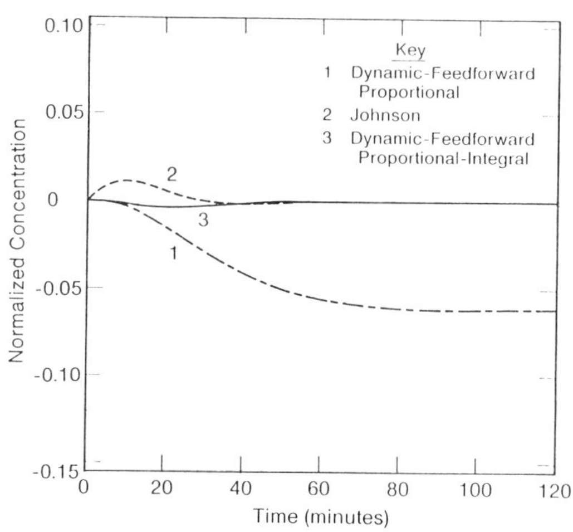

Figure 6.12 Outlet concentration regulated with only inlet concentration measurable. From Ramirez *et al.* (1987).

and Anderson control laws. An input feed stream is fed to the system at the third tray from the top. On the basis of the non-linear distillation tower model of Franke *et al.* (1963), a linear model for the separation of methanol/water was developed by Abedi (1976) and is used in this study. The system is controlled by the reflux ratio after the condenser and by the heat duty of the reboiler. Since there are two control variables, two states can be considered the specified or output variables for regulation. The compositions of the overhead and the bottoms are chosen to be the states to be maintained at desired values since they are the actual outputs of the column that affect other operating units.

The linearized model has been shown to reasonably represent the response of the non-linear system and, therefore, will be used to test the various optimal regulator algorithms we have developed for distillation control. The linear model is

$$\dot{\boldsymbol{x}} = \boldsymbol{A}\boldsymbol{x} + \boldsymbol{B}\boldsymbol{u} + \boldsymbol{W}\boldsymbol{d} \tag{6.5.1}$$

where $\boldsymbol{x}$ represents the state variables of the seven methanol liquid phase deviation mole fractions (gmol/gmol); $\boldsymbol{u}$ represents the control variables of the normalized deviations in the reflux ratio, $\widehat{RR}$ ((gmol/min)/gmol/min); and heat duty, $\hat{Q}$ ((cal/min)/cal/min); and $\boldsymbol{d}$ represents the disturbance variables, which are assumed to be normalized deviations in feed flow, $\hat{F}$ (gmol min^{-1}/gmol min^{-1}), feed composition, $\hat{x}_f$ (gmol/gmol), feed temperature, $\hat{T}_f$ (°C/°C), and reflux temperature, $\hat{T}_r$ (°C/°C).

Table 6.1 lists the elements for $\boldsymbol{A}$, $\boldsymbol{B}$, and $\boldsymbol{W}$. The system as defined is naturally stable because the $\boldsymbol{A}$ matrix is negative semi–definite. The eigenvalues of $\boldsymbol{A}$ are

$$\begin{bmatrix} -7.71 \\ -2.46 \\ -1.76 \\ -3.88 \\ -0.586 \\ -5.04 \\ -0.0208 \end{bmatrix} \tag{6.5.2}$$

For this application, the state weighting matrix, $\boldsymbol{Q}$, is Diag(1, 0, 0, 0, 0, 0, 1), and the control weighting matrix, $\boldsymbol{R}$, is Diag(0.1, 0.1). The states that we desire to maintain at specified values are the overhead composition, x_1, and the bottoms composition, x_7. Therefore

$$\boldsymbol{x}_{sp} = \begin{bmatrix} x_1 \\ x_7 \end{bmatrix} \tag{6.5.3}$$

The existence and stability for the dynamic predictor must be satisfied. For this problem, an additional state, x_2, is also specified because of the collinearity of x_1 and x_2. This collinearity is shown from the first row of the system

Table 6.1
Elements for A, B and W

$$
A = \begin{bmatrix}
-2.271 & 0.9535 & 0 & 0 & 0 & 0 & 0 \\
1.449 & -2.428 & 0.9472 & -1.68 \times 10^{-4} & 2.36 \times 10^{-5} & 2.0 \times 10^{-4} & -1.99 \times 10^{-4} \\
-1.37 \times 10^{-3} & 1.387 & -2.217 & 0.9259 & 3.53 \times 10^{-5} & 2.99 \times 10^{-4} & -2.98 \times 10^{-4} \\
-1.85 \times 10^{-3} & -1.71 \times 10^{-2} & 1.1825 & -4.836 & 0.9891 & 1.17 \times 10^{-3} & -6.61 \times 10^{-4} \\
-1.27 \times 10^{-4} & -1.17 \times 10^{-3} & 0 & 3.98 & -4.944 & 1.007 & 9.9 \times 10^{-6} \\
-4.88 \times 10^{-4} & -4.52 \times 10^{-3} & 0 & -1.42 \times 10^{-4} & 3.822 & -4.739 & 1.203 \\
-1.89 \times 10^{-5} & -1.75 \times 10^{-4} & 0 & -5.48 \times 10^{-6} & 0 & 4.09 \times 10^{-2} & -3.39 \times 10^{-2}
\end{bmatrix}
$$

$$
B = \begin{bmatrix}
0 & 0 \\
3.412 \times 10^{-2} & 3.11 \times 10^{-5} \\
5.105 \times 10^{-2} & 4.654 \times 10^{-5} \\
6.895 \times 10^{-2} & 2.176 \times 10^{-1} \\
4.715 \times 10^{-3} & 3.142 \times 10^{-2} \\
1.817 \times 10^{-2} & 1.21 \times 10^{-1} \\
7.04 \times 10^{-4} & 4.687 \times 10^{-3}
\end{bmatrix}
$$

$$
W = \begin{bmatrix}
0 & 0 & 0 & 0 \\
3.12 \times 10^{-5} & -3.85 \times 10^{-7} & -7.99 \times 10^{-7} & -4.187 \times 10^{-3} \\
4.665 \times 10^{-5} & -5.75 \times 10^{-7} & -1.20 \times 10^{-6} & -6.265 \times 10^{-3} \\
-0.7362 & 2.593 & -2.434 \times 10^{-2} & -8.4653 \times 10^{-3} \\
3.149 \times 10^{-2} & -3.88 \times 10^{-4} & -8.07 \times 10^{-4} & -5.79 \times 10^{-4} \\
0.1213 & -1.497 \times 10^{-3} & -3.11 \times 10^{-3} & -2.23 \times 10^{-3} \\
4.699 \times 10^{-3} & -5.80 \times 10^{-5} & -1.205 \times 10^{-4} & -8.63 \times 10^{-5}
\end{bmatrix}
$$

dynamics:

$$0 = a_{11}x_{1p} + a_{12}x_{2p} \tag{6.5.4}$$

If

$$x_{1p} = 0 \tag{6.5.5}$$

then

$$x_{2p} = 0 \tag{6.5.6}$$

and from the second row of the system dynamics the model predictions become

$$\dot{x}_{2p} = 0 \tag{6.5.7}$$

For this distillation problem, the matrix, $\boldsymbol{B}_1^{-1}$, exists:

$$\boldsymbol{B}_1^{-1} = \begin{bmatrix} 29.312337 & -0.19449833 \\ -4.402918 & 213.38531 \end{bmatrix} \tag{6.5.8}$$

Also the matrix, $\hat{\boldsymbol{A}}_2 = [\boldsymbol{A}_{22} - \boldsymbol{B}_2\boldsymbol{B}_1^{-1}\boldsymbol{A}_{12}]$, is a negative semi–definite. The eigenvalues of $\hat{\boldsymbol{A}}_2$ are

$$\begin{bmatrix} -1.11 \\ -7.23 + 1.99i \\ -7.23 - 1.99i \\ -3.64 \end{bmatrix} \tag{6.5.9}$$

Therefore, $\hat{\boldsymbol{A}}_2$ is negative semi–definite and the dynamic predictor exists and is stable.

The four control algorithms were tested by introducing a positive 10 percent step change in feed composition and negative 10 percent step changes in the feed flow, feed temperature, and reflux temperature. This yields a maximum state deviation for 10 percent disturbance changes. The system response was simulated for 40 minutes by using the linear equations. In all cases we desire to maintain the bottoms and overhead compositions. Either the disturbance vector was assumed completely measurable or the feed composition unmeasurable.

When the disturbances are completely measurable, all four algorithms are able to maintain both the bottoms and overhead composition to within 0.1 percent of the desired operating point, as shown in Figures 6.13 and 6.14. Yet, the dynamic feed-forward algorithms are able to control the composition immediately and without any deviation.

When the feed composition is unmeasurable, the dynamic feed-forward proportional-integral algorithm controls the bottoms and overhead compositions with less deviation than the Johnson algorithm (Figures 6.15 and 6.16). The dynamic feed-forward proportional regulator is unable to maintain the desired operating point without the feed composition measurement and without integral action. This proportional algorithm causes a one percent offset

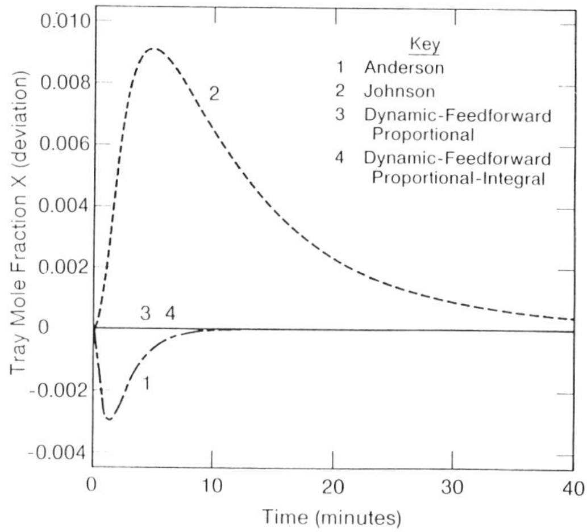

Figure 6.13 Bottoms product response with completely measurable disturbances. From Ramirez *et al.* (1987).

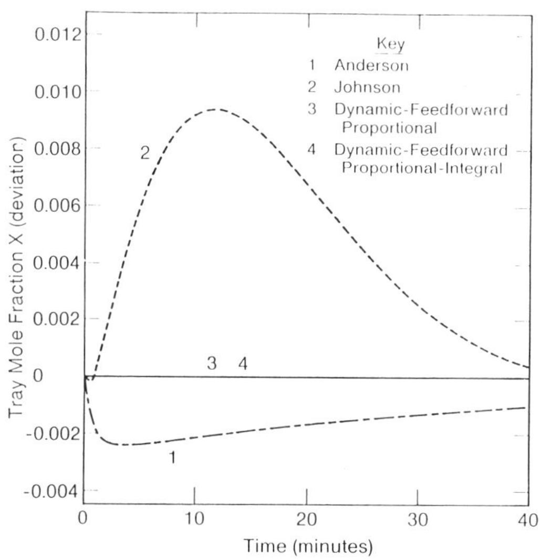

Figure 6.14 Overhead product response with completely measurable disturbances. From Ramirez *et al.* (1987).

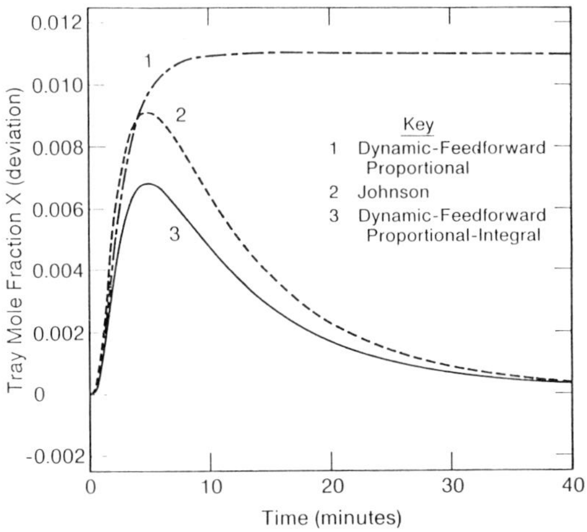

Figure 6.15 Bottoms product response with feed composition unmeasurable. From Ramirez *et al.* (1987).

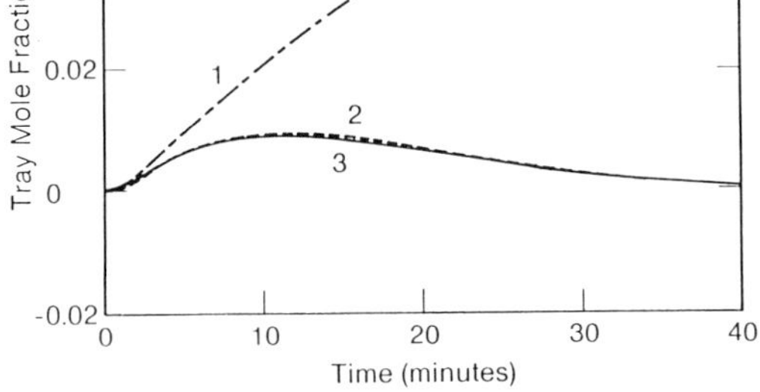

Figure 6.16 Overhead product response with feed composition unmeasurable. From Ramirez *et al.* (1998).

in the bottoms composition and a six percent offset in the overhead composition. Excellent composition control (maintaining both overhead and bottoms purity) is achievable when the major disturbances affecting binary distillation columns are measurable and used properly in controller design. The dynamic feed-forward predictive ability yields improvements over the static predictor of Anderson. The partially measurable formulation yields the advantages of dynamic feed-forward control for measured disturbance variables and integral error compensation characteristic of the Johnson approach for unmeasured disturbances.

6.6 Optimal Regulation with Control Constraints

Let us now consider the effect that constraints on control variables have on optimal regulatory control. Since we have been able to transform practical optimal regulatory problems back to the Kalman problem, we will therefore assume the equivalent form of the Kalman problem:

$$\dot{\boldsymbol{x}} = \boldsymbol{A}\boldsymbol{x} + \boldsymbol{B}\boldsymbol{u} \tag{6.6.1}$$

with

$$\begin{aligned} \boldsymbol{x}(0) &= \boldsymbol{x}_0 \\ \boldsymbol{x}(t_f) &= \text{free} \\ t_f &= \text{known} \end{aligned} \tag{6.6.2}$$

and the constraints

$$\boldsymbol{u}_{\min} \leq \boldsymbol{u} \leq \boldsymbol{u}_{\max} \tag{6.6.3}$$

The quadratic performance index is

$$J = \frac{1}{2}\boldsymbol{x}^T(t_f)\boldsymbol{H}\boldsymbol{x}(t_f) + \frac{1}{2}\int_0^{t_f} \left(\boldsymbol{x}^T\boldsymbol{Q}\boldsymbol{x} + \boldsymbol{u}^T\boldsymbol{R}\boldsymbol{u}\right) dt \tag{6.6.4}$$

The Hamiltonian for this problem is

$$H = \frac{1}{2}\boldsymbol{x}^T\boldsymbol{Q}\boldsymbol{x} + \frac{1}{2}\boldsymbol{u}^T\boldsymbol{R}\boldsymbol{u} + \boldsymbol{\lambda}^T(\boldsymbol{A}\boldsymbol{x} + \boldsymbol{B}\boldsymbol{u}) \tag{6.6.5}$$

The necessary conditions for a minimum in the performance index are as follows:

Costate Equations

$$\dot{\boldsymbol{\lambda}} = -\boldsymbol{Q}\boldsymbol{x} - \boldsymbol{A}^T\boldsymbol{\lambda} \tag{6.6.6}$$

Transversality Conditions

$$\boldsymbol{\lambda}(t_f) = \boldsymbol{H}\boldsymbol{x}(t_f) \tag{6.6.7}$$

Optimal Control

Minimize H with respect to $\boldsymbol{u}$. If $\boldsymbol{u}$ is unconstrained and not at a constraint value, then

$$\frac{\partial H}{\partial \boldsymbol{u}} = \boldsymbol{0} \tag{6.6.8}$$

This leads to the solution of the Kalman problem with

$$\boldsymbol{u} = \boldsymbol{R}^{-1}\boldsymbol{B}^T\boldsymbol{\lambda} \tag{6.6.9}$$

Or, from Section 5.1 using the Riccati transformation, we have the state feedback control law,

$$\boldsymbol{u} = -\boldsymbol{K}(t)\boldsymbol{x}(t) \tag{6.6.10}$$

If $\boldsymbol{u}$ is on the constraint boundary, then from the minimum principle of Section 4.2 we have the optimal control policy as the minimization of H with respect to $\boldsymbol{u}$. Since

$$\boldsymbol{u}^T\boldsymbol{R}\boldsymbol{u} > 0 \tag{6.6.11}$$

the minimization of the Hamiltonian (Equation (6.6.5)) is equivalent to

$$\text{if} \quad \boldsymbol{B}\boldsymbol{\lambda}^T > \boldsymbol{0} \quad \text{then} \quad \boldsymbol{u} = \boldsymbol{u}_{\min} \tag{6.6.12}$$
$$\text{if} \quad \boldsymbol{B}\boldsymbol{\lambda}^T < \boldsymbol{0} \quad \text{then} \quad \boldsymbol{u} = u_{\max} \tag{6.6.13}$$

The optimal control is therefore state feedback control when the control does not exceed the constraint values of $\boldsymbol{u}_{\min}$ or $\boldsymbol{u}_{\max}$ but is constrained to those minimum and maximum values when the inequalities of (6.6.12) and (6.6.13) are in effect. An efficient algorithm for finding optimal regulatory control with control constrains is as follows:

1. Assume an initial control policy of

$$u = -\boldsymbol{K}_{ss}\boldsymbol{x} \tag{6.6.14}$$

2. Solve the state equations forward in time assuming the current control law, but constrain each control according to

$$\boldsymbol{u}_{\min} \leq \boldsymbol{u} \leq \boldsymbol{u}_{\max} \tag{6.6.15}$$

   Calculate the objective functional J.

3. Solve the costate equations backwards in time from Equation (6.6.6) with the initial conditions of Equation (6.6.7) using stored $\boldsymbol{x}(t)$ values.

4. Using this $\boldsymbol{\lambda}(t)$ solution, develop a new unconstrained law, $\boldsymbol{u}(t)$, from Equation (6.6.9). Constrain that control law to

$$\boldsymbol{u}_{\min} \leq \boldsymbol{u} \leq \boldsymbol{u}_{\max} \tag{6.6.16}$$

5. Check for convergence by comparing the new control law versus the old control law. Stop if convergence is obtained, but go to step 2 if convergence is not obtained.

A suboptimal control algorithm that is usually adequate is as follows:

1. Assume an initial control policy of

$$\boldsymbol{u} = -\boldsymbol{K}_{ss}\boldsymbol{x} \tag{6.6.17}$$

applying the constraints

$$\boldsymbol{u}_{\min} \leq \boldsymbol{u} \leq \boldsymbol{u}_{\max} \tag{6.6.18}$$

This noniterativc algorithm is usually a close approximation to the result of the complete iterative algorithm. It is the algorithm recommended for most process applications.

6.7 Optimal Fermentor Temperature Control with Cooling Jacket Dynamics

In the fermentation industry and in industry as a whole, it is often necessary to maintain process vessels at constant temperatures. In order to do this, the vessel must be equipped with some type of heating/cooling mechanism by which the temperature can be controlled. For many applications, a heating/cooling jacket is used around all or part of the vessel to provide the required temperature control. Gee and Ramirez (1991) considered a fully jacketed fermentor is used for isothermal batch beer fermentation (Figure 6.17). Coolant is circulated through this jacket to provide the cooling control for this reactor.

The results of experiments utilizing PI temperature control are typified by the behavior shown in Figure 6.18. As can be seen in the figure, the PI controller was unable to maintain close control about the set point, but oscillated by as much 2°C above and below the desired fermentation temperature. It was first thought that this behavior may have resulted from the use of wrong

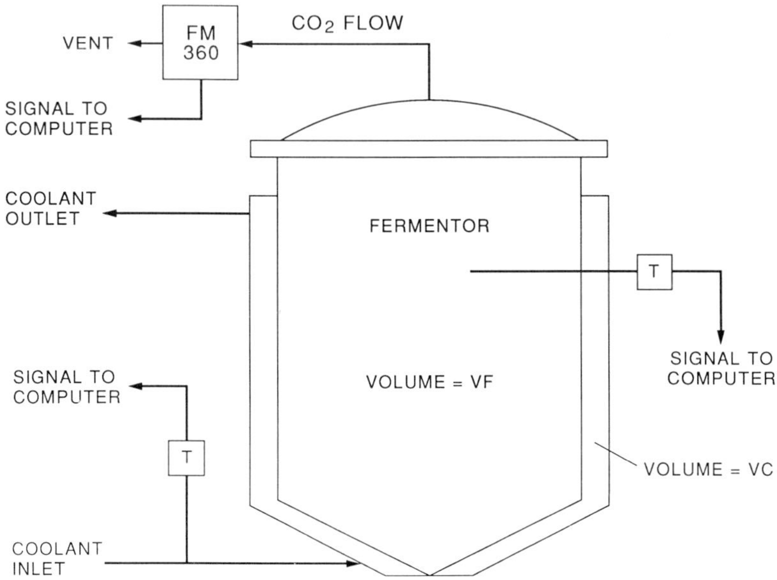

Figure 6.17 Fermentor configuration. From Gee and Ramirez (1991).

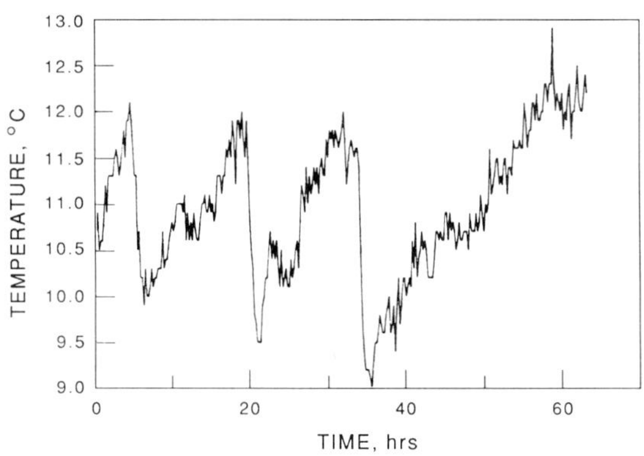

Figure 6.18 Observed temperature fluctuations under PI control set point temperature was 10°C. From Gee and Ramirez (1991).

gains in the controller, but experimentation over a wide range of gain values proved fruitless in improving the performance of the controller.

From a thorough examination of the data and apparatus, Gee and Ramirez determined that these temperature oscillations were the result of a combination of factors including process dead time, cooling dynamics, and disturbances. The dead time and cooling dynamics problems were the direct result of the design used for the fermentation vessel, and the process disturbances were caused by both apparatus and inherent process upsets.

The design configuration of the fermentation vessel, shown in Figure 6.17, was cylindrico-conical in shape and was fully encompassed along the bottom and sides by a cooling jacket. The volume of the cooling jacket was significant in comparison with that of the vessel itself. When the temperature inside the vessel rose above the set point, cooling was activated. However, the coolant flow rate was low enough so that a considerable amount of time was required to lower the temperature in the jacket to a level where significant cooling began (the dead time problem). During this time, the fermentation reaction generated enough heat to raise the vessel temperature 2°C. Once cooling began to take effect, the vessel temperature was brought back to the set point fairly quickly, at which time cooling was stopped. At this point, the cooling jacket was filled with coolant at a low temperature ($\sim$ 0°C), which equilibrated with the fluid inside the vessel, resulting in roughly a 2°C temperature drop below the set point inside the vessel (the design problem). This cycle repeated itself periodically during the fermentation resulting in the undesirable temperature behavior observed.

Along with the phenomena described above, there were also disturbances to the process which affected the control. These disturbances were a changing heat generation rate in the fermentation and an oscillating inlet coolant temperature to the fermentor. The heat generation rate variation shown in Figure 6.19 resulted from the process dynamics, while the inlet coolant temperature oscillations arose due to the nature of the refrigeration system itself. This refrigeration system was designed as a chiller with a ten gallon coolant tank maintained at a set temperature which served as a supply manifold for fermentation cooling. Whenever the fluid temperature in this tank rose above the set point, the refrigeration system activated and lowered the fluid temperature to a point well below the set point. This caused the oscillations in the coolant temperature, which are shown in Figure 6.20.

An ideal controller for this particular problem should have the attributes of simplicity of implementation, effective disturbance rejection, tight regulatory control, and dead time compensation. Optimal dynamic regulation for use with time-varying measurable and unmeasurable disturbances was developed in Section 6.4. This MPC controller, which utilizes a feed-forward dynamic predictor to compensate for dead time and cooling dynamics, and performs disturbance rejection for measured disturbances, should be able to handle this

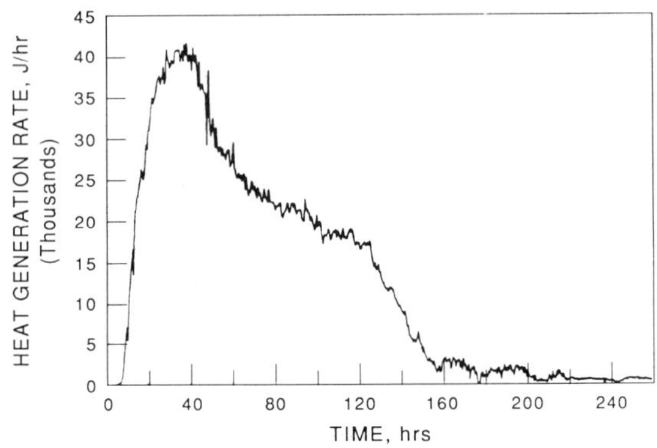

Figure 6.19 Heat generation profile. From Gee and Ramirez (1991).

cooling problem well. The controller also contains a PI feedback term which contributes robustness in the form of compensation for modeling errors in the feed-forward control prediction and elimination of regulator offset. Although the theoretical development of this strategy, like that of many optimal MPC controllers, is somewhat complicated, the actual application is quite simple once a process model has been obtained.

6.7.1 Process Model

As seen in Figure 6.17, the actual coolant temperature was not known in the cooling jacket. However, the inlet coolant temperature could be measured and used in estimating the jacket temperature. For modeling purposes, it was assumed that

$$T_c = \frac{T_{\text{in}} + T_{\text{out}}}{2} \tag{6.7.1}$$

where T_c = effective bulk coolant temperature.

Note that for simplicity, the jacket was modeled as a lumped–parameter system despite its obviously distributed nature. Based on this effective coolant temperature, an energy balance on the cooling jacket yields

$$\frac{d}{dt}\left(\rho_c c_{p_c} V_c T_c\right) = F_c c_{p_c}\left(T_{\text{in}} - T_{\text{out}}\right) + hA_c(T - T_c) \tag{6.7.2}$$

which, with Equation (6.7.1) substituted in for T_{out} and by assuming constant ρ_c, c_{p_c}, V_c, becomes

$$\frac{dT_c}{dt} = \frac{2F_c}{\rho_c V_c}\left(T_{\text{in}} - T_c\right) + \frac{hA_c}{\rho_c V_c c_{p_c}}(T - T_c) \tag{6.7.3}$$

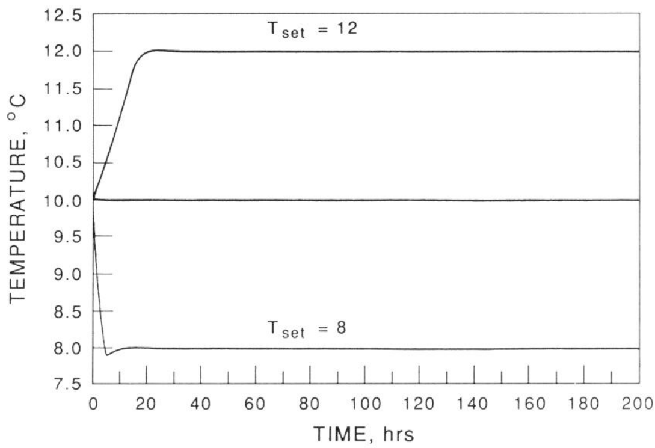

Figure 6.20 Typical dynamic profile of the inlet coolant temperature, T_{in}, during fermentation. From Gee and Ramirez (1991).

A similar energy balance on the reactor vessel yields

$$\frac{dT}{dt} = \frac{\Delta H_F}{\rho_F V_F c_{p_F}} - \frac{hA_c}{\rho_F V_F c_{p_F}}(T - T_c) \tag{6.7.4}$$

where ΔH_f refers to the heat generation rate during the fermentation. This rate is not directly measurable but is related directly to the rate of CO_2 evolution:

$$\Delta H_F = R\frac{dCO_2}{dt} \tag{6.7.5}$$

where R is an effective yield coefficient.

Using a mass flow meter attached to the exhaust gas line for the fermentor, the CO_2 evolution rate was easily measurable. As described by Gee and Ramirez (1988), beer fermentation involves three sugar substrates, glucose, maltose, and maltotriose. Each of these sugars contributes differently to the yield coefficient, R, in Equation (6.7.5) due to stoichiometric differences. An average value was chosen. The value for maltose, which is the most abundant sugar in wort and which strikes an average between the values for glucose and maltotriose, was used. Therefore Equation (6.7.4) becomes

$$\frac{dT}{dt} = \frac{R}{\rho_F V_F c_{p_F}}\frac{dCO_2}{dt} - \frac{hA_c}{\rho_F V_F c_{p_F}}(T - T_c) \tag{6.7.6}$$

This equation and Equation (6.7.3) comprise the dynamic model for this process.

To verify the dynamic interaction of the two equations in this model, the eigenvalues of the linearized model were calculated for representative values of the parameters (given in Table 6.2). The eigenvalues for the tank and jacket dynamics were −2.9 and −15.6, respectively. The ratio of these numbers, which is 5.4 to 1, shows that the cooling jacket dynamics are faster but not more than an order of magnitude faster than the vessel dynamics for this system, thus causing difficulty for the conventional PI control algorithm.

Table 6.2

Model Parameters

$V_F = 0.235$ m^3	$V_c = 0.032$ m^3
$\rho_F = 1040$ Kg/m^3	$\rho_c = 1062$ kg/m^3
$c_{p_F} = 4016$ J/kg·°C	$c_{p_c} = 3768$ J/kg·°C
$h = 879$ KJ/hr·m^2·°C	$A_c = 1.746$ m^2
$R = 58$ KG·m^3/mol	$0 \leq F_c \leq 168$ kg/hr

6.7.2 The Controller

Since in Equations (6.7.3) and (6.7.6), the values of h, $d\ CO_2/dt$, and T_{in} are subject to change during the process, these parameters are considered local disturbances to the system. As such, both T_{in} and $d\ CO_2/dt$ are measurable while h is unmeasurable. This suggested the need to use the optimal regulator control algorithm that considers time varying, partially measurable disturbances. For the linear system,

$$\dot{\boldsymbol{x}} = \boldsymbol{A}\boldsymbol{x} + \boldsymbol{B}\boldsymbol{u} + \boldsymbol{W}_m\boldsymbol{d}_m + \boldsymbol{W}_u\boldsymbol{d}_u \tag{6.7.7}$$

where $\boldsymbol{x}$ = process state vector
$\boldsymbol{u}$ = control vector
$\boldsymbol{d}_m$ = measured disturbances
$\boldsymbol{d}_u$ = unmeasured disturbances

The optimal regulator was shown in Section 6.4 to be

$$\boldsymbol{u} = \boldsymbol{u}_p - \boldsymbol{K}_2(\boldsymbol{x} - \boldsymbol{x}_p) - (\boldsymbol{K}_1 - \boldsymbol{K}_2\boldsymbol{A}) \int (\boldsymbol{x} - \boldsymbol{x}_p)\, dt \tag{6.7.8}$$

where $\boldsymbol{u}_p$ = a dynamic predicted control vector based upon measured disturbances
$\boldsymbol{x}_p$ = a dynamic predicted state vector based upon measured disturbances

Equation (6.7.8) is an MPC multivariable, dynamic feed-forward controller with multivariable, proportional-integral state feedback. The feed-forward element compensates for measurable time-varying disturbances, and the integral element compensates for unmeasurable piecewise constant disturbances.

To compute the predicted states, $\boldsymbol{x}_p$, and the predicted controls, $\boldsymbol{u}_p$, the predicted state vector is partitioned into

$$\boldsymbol{x}_p = \begin{bmatrix} \boldsymbol{x}_{s_p} \\ \boldsymbol{x}_{f_p} \end{bmatrix} \tag{6.7.9}$$

where $\boldsymbol{x}_{s_p}$ are the state variables that will be maintained at a specified value, and $\boldsymbol{x}_{f_p}$ are unspecified free predicted states. The maximum number of state variables that can usually be specified, $\boldsymbol{x}_{s_p}$, is equal to the number of control variables.

For regulation we desire that the specified states be constant; therefore the unspecified free predicted states, $\boldsymbol{x}_{f_p}$, and the predicted controls, $\boldsymbol{u}_p$, are computed from the following set of differential and algebraic equations:

$$\begin{bmatrix} \mathbf{0} \\ \dot{\boldsymbol{x}}_{f_p} \end{bmatrix} = \begin{bmatrix} \boldsymbol{A}_{11} & \boldsymbol{A}_{12} \\ \boldsymbol{A}_{21} & \boldsymbol{A}_{22} \end{bmatrix} \begin{bmatrix} \boldsymbol{x}_{s_p} \\ \boldsymbol{x}_{f_p} \end{bmatrix} + \begin{bmatrix} \boldsymbol{B}_1 \\ \boldsymbol{B}_2 \end{bmatrix} \boldsymbol{u}_p + \begin{bmatrix} \boldsymbol{W}_{m_1} \\ \boldsymbol{W}_{m_2} \end{bmatrix} \boldsymbol{d}_m \tag{6.7.10}$$

For the model presented for batch fermentation, we can define

$$\begin{aligned} x_{s_p} &= T_p \\ x_{f_p} &= T_{c_p} \\ u_p &= F_{cp} \end{aligned} \tag{6.7.11}$$

Using the nonlinear model (Equations (6.7.3) and (6.7.6)), we can calculate the predicted coolant temperature and predicted control to be

$$T_{c_p} = x_{s_p} - \frac{d_{m_1}}{hA_c} \tag{6.7.12}$$

$$u_p = \frac{\rho_c V_c c_{p_c} \dot{d}_{m_2} + hA_c d_{m_2}}{2c_{p_c}\left(hA_c x_{s_p} - hA_c d_{m_1} - d_{m_2}\right)} \tag{6.7.13}$$

where the measured disturbances are

$$\begin{aligned} d_{m_1} &= \Delta H_F \\ d_{m_2} &= T_{in} \end{aligned} \tag{6.7.14}$$

The optimal control is

$$u = u_p - \boldsymbol{K}_2 (\boldsymbol{x} - \boldsymbol{x}_p) - (\boldsymbol{K}_1 - \boldsymbol{K}_2 \boldsymbol{A}) \int (\boldsymbol{x} - \boldsymbol{x}_p)\, dt \tag{6.7.15}$$

where $\boldsymbol{K}_1$ and $\boldsymbol{K}_2$ are obtained by solution of:

$$\boldsymbol{K} = \left[\boldsymbol{K}_1 \;\vdots\; \boldsymbol{K}_2 \right] = \tilde{\boldsymbol{R}}^{-1} \tilde{\boldsymbol{B}}^T \boldsymbol{P} \tag{6.7.16}$$

where

$$\dot{\boldsymbol{P}} = -\boldsymbol{P}\tilde{\boldsymbol{A}} - \tilde{\boldsymbol{A}}^T \boldsymbol{P} - \tilde{\boldsymbol{Q}} + \boldsymbol{P}\tilde{\boldsymbol{B}}\tilde{\boldsymbol{R}}^{-1}\tilde{\boldsymbol{B}}^T \boldsymbol{P} \tag{6.7.17}$$

The values for $\boldsymbol{A}$, $\boldsymbol{B}$, $\boldsymbol{W}_m$, $\boldsymbol{W}_u$, $\boldsymbol{d}_m$, $\boldsymbol{d}_u$, $\tilde{\boldsymbol{A}}$, $\tilde{\boldsymbol{B}}$, $\tilde{\boldsymbol{R}}$, and $\tilde{\boldsymbol{W}}$ along with the necessary initial conditions are given in Table 6.3.

6.7.3 Control Simulations

A set of simulations have been performed by Gee and Ramirez (1991) to investigate the usefulness of the optimal controller. First, the controller gains were determined by solution of the Riccati matrix differential equations. For this problem, the dynamic Riccati matrices (Equation (6.7.17)) quickly converged to a steady-state solution which was used to calculate steady-state controller gains using Equation (6.7.15). Results of these calculations are given in Table 6.4.

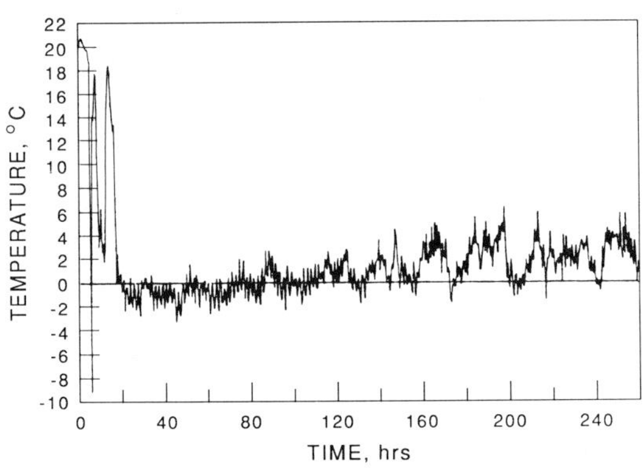

Figure 6.21 Optimal control simulations—curves are (a) $T_{\text{set}} = 12°\text{C}$ (b) $T_{\text{set}} = 10°\text{C}$ (c) $T_{\text{set}} = 8°\text{C}$. From Gee and Ramirez (1991).

These steady-state gains were then used in simulations of temperature control using various starting conditions. Typical results of these simulations are given in Figure 6.21. As shown in Figure 6.21, this MPC optimal controller was capable of controlling the fermentation temperature very well (within $\pm 0.001°\text{C}$) once the system had been brought to the set-point temperature. This is a marked improvement as compared to the PI control of Figure 6.18.

Table 6.3

Model and Controller Matrices*

$$\boldsymbol{A} = \begin{bmatrix} h_0 D_0 & -h_0 D_0 \\ h_0 B_0 & -A_0 F_{c0} - h_0 B_0 \end{bmatrix} \qquad \boldsymbol{B} = \begin{bmatrix} 0 \\ A_0(T_{\text{in}0} - T_{c0}) \end{bmatrix}$$

$$\boldsymbol{W}_m = \begin{bmatrix} 0 & C_0 \\ A_0 F_{c0} & 0 \end{bmatrix} \qquad \boldsymbol{W}_u = \begin{bmatrix} D_0(T_0 - T_{c0}) \\ B_0(T_0 - T_{c0} \end{bmatrix}$$

$$\boldsymbol{D}_m^T = [(\Delta H_F - \Delta H_{F0}) \ , \ (T_{\text{in}} - T_{\text{in}0})] \qquad \boldsymbol{d}_u = [h - h_0]$$

$$\tilde{\boldsymbol{A}} = \begin{bmatrix} \boldsymbol{A} & \boldsymbol{I} \\ \boldsymbol{0} & \boldsymbol{0} \end{bmatrix} \qquad \tilde{\boldsymbol{b}} = \begin{bmatrix} \boldsymbol{0} \\ \boldsymbol{B} \end{bmatrix}$$

$$\tilde{\boldsymbol{R}} = [\boldsymbol{R}] \qquad \tilde{\boldsymbol{Q}} = \begin{bmatrix} \boldsymbol{Q} & \boldsymbol{0} \\ \boldsymbol{0} & \boldsymbol{0} \end{bmatrix}$$

*where subscript zero indicates evaluation at the linearization point and

$$A = \frac{2}{\rho_c V_c} \ , \quad B = \frac{A_c}{\rho_c V_c c_{p_c}} \ , \quad C = \frac{1}{\rho_F V_F c_{p_F}} \ , \quad , D = -\frac{A_c}{\rho_F V_F c_{p_F}}$$

Table 6.4

Weighting Matrices and Computed Controller Gains

$$\boldsymbol{Q} = \begin{bmatrix} 1 & 0 \\ 0 & 1 \end{bmatrix} \qquad \boldsymbol{R} = [1]$$

$$\boldsymbol{P}_{ss} = \begin{bmatrix} 6.698 & 0.8981 & 14.560 & 2.118 \\ 0.8981 & 0.1578 & 1.903 & 0.2794 \\ 14.560 & 1.903 & 86.250 & 9.139 \\ 2.118 & 0.2794 & 9.139 & 1.270 \end{bmatrix}$$

$$\boldsymbol{K}_{ss} = \left[\boldsymbol{K}_{1ss} \vdots \boldsymbol{K}_{2ss}\right] = [-1.2462, \ -0.1644 \ \vdots \ -5.3781, \ -0.7476]$$

One thing which should be noted is that this is an unconstrained controller. This is to say that the controller assumed the system is capable of both cooling the process and also heating it. The introduction of the control constraint (a minimum value for the control variable) is not considered in the optimal regulator theory of Section 6.4. Section 6.6 discusses the introduction of control constraints. The optimal constrained control policy at sub–set-point temperatures is for the control to remain off. An offshoot of this fact is that if started at a sub–set-point temperature, the integral terms in Equation (6.7.8) tend to wind up, thus causing overshoot and oscillations in the controller. The approximate suboptimal solution for the control constraint is dealt with by setting these integral terms to zero whenever the process temperature is below the set point.

6.7.4 On-Line Control

One of the only problems encountered in implementing the controller on-line was the fact that no thermocouples could easily be placed in the cooling jacket to provide the necessary feedback on the effective coolant temperature. This problem was alleviated by measuring T and T_{in} and then integrating Equation (6.7.3) for T_c. Admittedly, this scheme relies heavily on the reliability of the heat-transfer model. However, it was the simplest solution to the problem and worked quite well.

Figure 6.22 shows the temperature control results for a typical run. The $\pm 0.5°\text{C}$ fluctuations found using the optimal controller are very acceptable considering the A/D accuracy, tank size, and the inaccuracies inherent in the mathematical model.

The optimal MPC controller requires two extra measurements not required with the PI controller, those of T_{in} and the CO_2 flow rate. With these measurements and the off-line computation of the gain, the optimal controller can be implemented with the same ease as the PI algorithm. It allows for much better regulatory control when cooling system dynamics are significant and there are significant measurable input disturbances. For highly temperature-sensitive processes such as fermentation, good temperature control is critical so that the microorganisms used will not undergo thermal shocks resulting from large temperature fluctuations.

In the solution of the predicted Equations (6.7.12)–(6.7.13), the nonlinear mathematical model was used. It was found through simulation that using a linearized model yielded a final temperature off-set which depended upon the linearization operating condition. It was possible to eliminate this off-set by choosing the correct linearization point. However, the gain calculations depend on the linearization point and so a change in set point will again yield a control off-set unless the gains are recomputed using a new flow linearization value based on the new set point. Using the nonlinear model to compute the

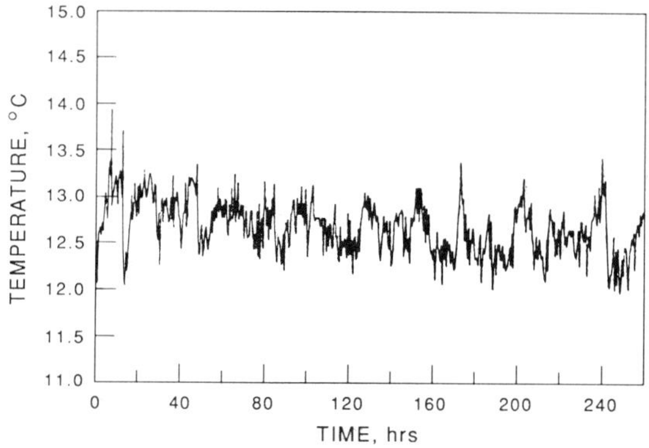

Figure 6.22. On–line beer fermentation control run temperature profile. From Gee and Ramirez (1991).

predicted states and controls rather than a linearized model eliminated the off-set problem regardless of the linearization point used in the gain calculations.

6.8 Time Delayed Systems

Often processes are modeled as low-order linear systems with dead-time. This simplified model structure has been shown to approximate complex physical behavior quite well and this approach to modeling is often taken in practice. It lacks physical insight that models based upon first principles have, but such models are often easy to develop and can effectively form the basis of predictive models for regulatory control. For example, Wood and Berry (1973) modeled a binary distillation column by the following multivariable time delayed transfer function

$$\bar{\boldsymbol{y}}(s) = \boldsymbol{G}(s)\bar{\boldsymbol{u}}(s) + \boldsymbol{G}_d(s)\bar{\boldsymbol{d}}(s) \tag{6.8.1}$$

where y_1 = overhead mole fraction of methanol
y_2 = bottoms mole fraction of methanol
u_1 = overhead reflux flow rate
u_2 = bottoms steam flow rate
d = column feed flow rate

Based upon pulse testing, the transfer functions they report are

$$\boldsymbol{G}(s) = \begin{bmatrix} \dfrac{12.8e^{-s}}{16.7s+1} & -\dfrac{18.9e^{-3s}}{21.0s+1} \\ \\ \dfrac{6.6e^{-7s}}{10.9s+1} & -\dfrac{19.4e^{-3s}}{14.4s+1} \end{bmatrix} \tag{6.8.2}$$

$$\boldsymbol{G}_d(s) = \begin{bmatrix} \dfrac{3.8e^{-8.1s}}{14.9s+1} \\ \\ \dfrac{4.9e^{-3.4s}}{12.3s+1} \end{bmatrix} \tag{6.8.3}$$

Transfer functions of the form

$$G_{ij} = \frac{k_{ij}e^{-\theta_{ij}s}}{\tau_{ij}s+1} \tag{6.8.4}$$

can be represented in state variable form as (Kailath, 1980)

$$\dot{x}_{ij}(t) = A_{ij}x_{ij}(t) + B_{ij}u_j(t-\theta_{ij}) \tag{6.8.5}$$

$$y_{ij}(t) = C_{ij}x_{ij}(t) \tag{6.8.6}$$

$$y_i(t) = \sum_j y_{ij}(t) \tag{6.8.7}$$

where

$$A_{ij} = -\frac{1}{\tau_{ij}} \qquad B_{ij} = \frac{k_{ij}}{\tau_{ij}} \tag{6.8.8}$$

This means that the Wood and Berry model results in the following state space formulation:

$$\begin{aligned} \dot{x}_1(t) &= -.06\ x_1(t) + .768\ u_1(t-1) \\ \dot{x}_2(t) &= -.04762\ x_2(t) + .9\ u_2(t-3) \\ \dot{x}_3(t) &= -.09174\ x_3(t) + .6055\ u_1(t-7) \\ \dot{x}_4(t) &= -.06944\ x_4(t) + 1.3472\ u_2(t-3) \end{aligned} \tag{6.8.9}$$

$$\begin{aligned} \dot{x}_{d1}(t) &= -.0671\ x_{d1}(t) + .255\ d(t-8.1) \\ \dot{x}_{d_2}(t) &= -.07576\ x_{d2}(t) + .3712\ d(t-3.4) \end{aligned} \tag{6.8.10}$$

$$\begin{aligned} y_1(t) &= x_1(t) - x_2(t) + x_{d1}(t) \\ y_2(t) &= x_3(t) - x_4(t) + x_{d2}(t) \end{aligned} \tag{6.8.11}$$

The model can be generalized as

$$\dot{\boldsymbol{x}}(t) = \boldsymbol{A}\boldsymbol{x}(t) + \sum_i \boldsymbol{B}_i\boldsymbol{u}(t-\alpha_i) + \sum_j \boldsymbol{W}_j\boldsymbol{d}(t-\beta_j) \tag{6.8.12}$$

Research on time delayed systems has demonstrated that an effective approach to regulation is to base the feedback control on a predicted state that is compensated for the time delays. Smith (1957, 1959) considered the time delay compensation for a single delay in a single control loop. Moore *et al.* (1970) presented a state space realization of the Smith controller. Ogunnaike and Ray (1979) and Ray (1981) considered the multivariable, multiple time delay problem. They showed that feedback control should be based upon a predicted output that predicts states an appropriate distance into the future. For the Wood and Berry distillation column problem, we need a predictive output of

$$\begin{aligned} y_1^*(t) &= x_1(t+1) - x_2(t+3) \\ y_2^*(t) &= x_3(t+7) - x_4(t+3) \end{aligned} \tag{6.8.13}$$

A predicted state $\boldsymbol{x}^*$ can therefore be defined as

$$\boldsymbol{x}^*(t) = \begin{pmatrix} x_1(t+1) \\ x_2(t+3) \\ x_3(t+7) \\ x_4(t+3) \end{pmatrix} \tag{6.8.14}$$

This predicted state $\boldsymbol{x}^*(t)$ corresponds to a state vector that has control action $\boldsymbol{u}(t)$ at time t rather than the original state vector $\boldsymbol{x}(t)$ formulation that has the controls $u_j(t-\alpha_j)$ acting at times delayed by $(t-\alpha_j)$. By introducing the predicted state $\boldsymbol{x}^*(t)$, we have placed the state and the controls in the same time frame.

If we reformulate the optimal control problem in terms of the variables $\boldsymbol{y}^*$, $\boldsymbol{x}^*$, and $\boldsymbol{u}$, the quadratic performance index for regulation in terms of the predicted system outputs is

$$J = \frac{1}{2}\int\left(\boldsymbol{y}^{*T}\boldsymbol{Q}\boldsymbol{y}^* + \boldsymbol{u}^T\boldsymbol{R}\boldsymbol{u}\right)\,dt \tag{6.8.15}$$

with

$$\boldsymbol{y}^* = \boldsymbol{H}\boldsymbol{x}^* \tag{6.8.16}$$

The performance index can be put in terms of the predicted states as

$$J = \frac{1}{2}\int\left(\boldsymbol{x}^{*T}\boldsymbol{H}^T\boldsymbol{Q}\boldsymbol{H}\boldsymbol{x}^* + \boldsymbol{u}^T\boldsymbol{R}\boldsymbol{u}\right)\,dt \tag{6.8.17}$$

with

$$\dot{\boldsymbol{x}}^*(t) = \boldsymbol{A}^*\boldsymbol{x}^*(t) + \boldsymbol{B}^*\boldsymbol{u}(t) \tag{6.8.18}$$

for the Wood and Berry model

$$\boldsymbol{A}^* = \boldsymbol{A} \qquad \text{and} \qquad \boldsymbol{B}^* = \boldsymbol{B} \tag{6.8.19}$$

but these relations are not general and the specific relationship between $\boldsymbol{A}^*$ and $\boldsymbol{A}$ and $\boldsymbol{B}^*$ and $\boldsymbol{B}$ must be developed for each specific problem.

The optimal LQR Kalman problem for Equations (6.8.17) and (6.8.18) gives the following optimal controller,

$$\boldsymbol{u} = -\boldsymbol{K}_{ss}\boldsymbol{x}^* \tag{6.8.20}$$

This shows that the optimal controller is feedback in the predicted time compensated state and not feedback in the actual state. This again is an example of model predictive control.

6.8.1 Wood and Berry Distillation Control

In applying the time delayed predictive algorithm (Equation (6.8.20)) to the Wood and Berry model of a distillation column, we need to first compute the predicted state. The first state is predicted forward into the future one minute, the second and fourth states predicted forward three minutes, and the third state seven minutes. In simulating the delayed response dynamics of Equation (6.8.9), either the controls must be stored or three additional time domains defined in the differential equation integrations. We take the latter approach in the MATLAB program that simulated this algorithm. Figure 6.23 gives the MATLAB program that first considers a set-point change of -0.01 mole fraction in the output overhead concentration. This problem results in an initial deviation of 0.01 in the first state variable. We compute the one minute forward in time prediction for the first state variable. The other state variables are not affected by this set-point change. Next we define the augmented state vector (xx) composed of the real states and the other three time domain states. The program then calculates the process response for all four time domains. The actual outputs are synthesized and plotted. The M files which define the differential equations that are integrated are given in Figure 6.24. The results of the simulation are given in Figure 6.25. This figure shows that both outputs are affected by a set-point change in the overhead composition. In fact, there is a fairly large disturbance that is introduced into the bottoms. The multivariable, time compensated controller has definitely regulated both outputs well. The gain matrix, k, that is used in the M file xprime.m was computed from the MATLAB program LQR. It is based upon an identity weighting matrix $\tilde{\boldsymbol{Q}}$ for both output states and an identity matrix $\boldsymbol{R}$ as given below:

$$J = \frac{1}{2}\int\left(\boldsymbol{y}^{*^T}\tilde{\boldsymbol{Q}}\boldsymbol{y}^* + \boldsymbol{u}^T\boldsymbol{R}\boldsymbol{u}\right)\,dt \tag{6.8.21}$$

The actual state weighting matrix is therefore

$$\boldsymbol{Q} = \boldsymbol{H}^T\tilde{\boldsymbol{Q}}\boldsymbol{H} \tag{6.8.22}$$

```
>> % compute the initial time compensated state with an initial disturbance
>> % only on the first state variable
>> [t1,x1] = ode45('x1prime',0,1,.01);
>> size (x1)

ans =

     7     1

>> x1(7)

ans =

    0.0094

>> xs = [ .0094 0 0 0];
>> x0 = [.01 0 0 0];
>> xx = [x0 xs xs xs];
>> xx0 = xx;
>> size(xx0)

ans =

     1    16

>> [t,xx] = ode45('xprime',0,50,xx0);
>> size (xx)

ans =

    56    16

>> y(1:56,1) = xx(1:56,1) - xx(1:56,2);
>> y(1:56,2) = xx(1:56,3) - xx(1:56,4);
>> plot(t,y)
>> meta y1
>> % compute the initial time compensated state with an initial disturbance
>> % only in the third state variable
>> [t2,x3] = ode45('x3prime',0,7,.2);
>> size (x3)

ans =

     7     1

>> x3(7)

ans =

    0.1052

>> xs = [0 0 .1052 0];
>> x0 = [0 0 .2 0];
>> xx = [x0 xs xs xs];
>> xx0 = xx;
>> [t3,xx3] = ode45('xprime',0,50,xx0);
Singularity likely at t = 1.002787
>> size(xx3)

ans =

     8    16

>> y3(1:8,1) = xx3(1:8,1) - xx3(1:8,2);
>> y3(1:8,2) = xx3(1:8,3) - xx3(1:8,4);
>> [t31,xx31] = ode45('xprime',1,50,xx0);
Singularity likely at t = 3.001935
>> size(xx31)

ans =
```

Figure 6.23 MATLAB program for simulation of control of the Wood–Berry distillation column.

```
    16    16

>> y3(9:24,1) = xx31(1:16,1) - xx31(1:16,2);
>> y3(9:24,2) = xx31(1:16,3) - xx31(1:16,4);
>> xx10 = xx3(8,1:16);
>> [t31,xx31] = ode45('xprime',1,50,xx10);
Singularity likely at t = 3.000060
>> size(xx31)

ans =

    15    16

>> size(t31)

ans =

    15     1

>> y3(9:23,1) = xx31(1:15,1) - xx31(1:15,2);
>> y3(9:23,2) = xx31(1:15,3) - xx31(1:15,4);
>> xx20 = xx31(15,1:16);
>> [t32,xx32] = ode45('xprime',3,50,xx20);
Singularity likely at t = 7.002121
>> size(xx32)
ans =

    20    16

>> y3(24:43,1) = xx32(1:20,1) - xx32(1:20,2);
>> y3(24:43,2) = xx32(1:20,3) - xx32(1:20,4);
>> xx30 = xx32(20,1:16);
>> [t33,xx33] = ode45('xprime',7,50,xx30);
>> size(xx33)

ans =

    35    16

>> y3(44:78,1) = xx33(1:35,1) - xx33(1:35,2);
>> y3(44:78,2) = xx33(1:35,3) - xx33(1:35,4);
>> tt = [t3;
         t31;
         t32;
         t33];
>> plot(tt,y3)
>>% now chech the effect that using feedback in the non compensated
>>% has on this case
>> xnc0 = [x0,x0,x0,x0];
>> [t3,xx4] = ode45('xprime',0,50,xnc0);
Singularity likely at t = 1.000810
>> size(xx4)

ans =
    11    16
>> y4(1:11,1) = xx4(1:11,1) - xx4(1:11,2);
>> y4(1:11,2) = xx4(1:11,3) - xx4(1:11,4);
>> xx410 = xx4(11,1:16);
>> [t41,xx41] = ode45('xprime',1,50,xx410);
Singularity likely at t = 3.000324
>> size(xx41)
ans =

    16    16
```

Figure 6.23 continued.

```
>> p(4,4) = pv(65,10);
>> k = inv(r)*bb'*p;
>> kp = k(1,3:4)

kp =

  -18.8991    1.0890

>> ki = k(1,1:2) -k(1,3:4)*a

ki =

   -1.0000   -0.0000

>> api = [a-b*kp  -b*ki;
          eye(2)   zeros(2)];
>> wi = [w        zeros(2);
         zeros(2) zeros(2)];
>> c = eye(4);
>> d = zeros(4);
>> t = [0:1:100];
>> xpi = step(api,wi,c,d,1,t);
>> x = xpi(1:101,1:2);
>> plot (t,x)
>> title('Johnson Algorithm State Response');
>> ylabel('Normalized Composition and Temperature');
>> xlabel('Time (min)');
>> meta xpi3;

>> y4(12:27,1) = xx41(1:16,1) - xx41(1:16,2);
>> y4(12:27,2) = xx41(1:16,3) - xx41(1:16,4);
>> xx420 = xx41(16,1:16);
>> [t42,xx42] = ode45('xprime',3,50,xx420);
Singularity likely at t = 7.000081
>> size(xx42)

ans =

    18    16

>> y4(28:45,1) = xx42(1:18,1) - xx42(1:18,2);
>> y4(28:45,2) = xx42(1:18,3) - xx42(1:18,4);
>> xx430 = xx42(18,1:16);
>> [t43,xx43] = ode45('xprime',7,50,xx430);
>> size(xx43)

ans =

    36    16

>> y4(46:81,1) = xx43(1:36,1) - xx43(1:36,2);
>> y4(46:81,2) = xx43(1:36,3) - xx43(1:36,4);
>> tp4 = [t3;
         t41;
         t42;
         t43];
>> plot(tt,y3,tp4,y4)
>> plot(tt,y3)
>> meta y1
>> plot(tt,y3,tp4,y4)
>> meta y2
>> diary off
```

Figure 6.23 continued.

```
function xxdot = xprime(t,xx)
x(1) = xx(1);
x(2) = xx(2);
x(3) = xx(3);
x(4) = xx(4);
xs1(1) = xx(5);
xs1(2) = xx(6);
xs1(3) = xx(7);
xs1(4) = xx(8);
xs2(1) = xx(9);
xs2(2) = xx(10);
xs2(3) = xx(11);
xs2(4) = xx(12);
xs3(1) = xx(13);
xs3(2) = xx(14);
xs3(3) = xx(15);
xs3(4) = xx(16);
a = [-0.06      0        0        0;
      0   -0.04762     0        0;
      0          0 -0.09174     0;
      0          0        0 -0.06944];
b = [.768  0;
     0    .9;
     .6055 0;
     0    1.3472];
k = [.843   -0.8657 -0.0353 0.0583;
    -0.2291 0.2214 -0.8664 -0.8883];
k1 = k(1,1:4);
k2 = k(2,1:4);
%
% compute the first state variable
%
if t<1
   xdot(1) = -.06*x(1);
   xxdot(5) = 0;
   xxdot(6) = 0;
   xxdot(7) = 0;
   xxdot(8) = 0;
else
   xs1dot = (a -b*k)*xs1';
   u(1) = -k1*xs1';
   xdot(1) = -.06*x(1) + .768*u(1);
   xxdot(5) = xs1dot(1);
   xxdot(6) = xs1dot(2);
   xxdot(7) = xs1dot(3);
   xxdot(8) = xs1dot (4);
end
xxdot(1) = xdot(1);
%
% compute the second and fourth state variables
%
if t<3
   xdot(2) = -.04762*x(2);
   xdot(4) = -.06944*x(2);
   xxdot(9) = 0;
   xxdot(10) = 0;
   xxdot(11) = 0;
   xxdot(12) = 0;
else
   xs2dot = (a -b*k)*xs2';
   u(2) = -k2*xs2';
   xdot(2) = -.04762*x(2) + .9*u(2);
   xdot(4) = -.06944*x(4) + 1.3472*u(2);
   xxdot(9) = xs2dot(1);
   xxdot(10) = xs2dot(2);
   xxdot(11) = xs2dot(3);
   xxdot(12) = xs2dot(4);
end
xxdot(2) = xdot(2);
xxdot(4) = xdot(4);
```

Figure 6.24 M files for Wood–Berry distillation column simulation.

function xxdot – simulates the states for various time domains

```
%
% compute the third state variable
%
if t<7
   xdot(3)= -0.09174*x(3);
   xxdot(13) = 0;
   xxdot(14) = 0;
   xxdot(15) = 0;
   xxdot(16) = 0;
else
   xs3dot = (a -b*k)*xs3';
   u(1) = -k1*xs3';
   xdot(3) = -0.09174*x(3) + .6055*u(1);
   xxdot(13) = xs3dot(1);
   xxdot(14) = xs3dot(2);
   xxdot(15) = xs3dot(3);
   xxdot(16) = xs3dot(4);
end
xxdot(3) = xdot(3);
```

function x1dot – projects the first state forward in time by one minute

```
function x1dot = x1prime(t,x1)
%
% defines the first differential equation without control

x1dot = -.06*x1;
```

function x3dot – projects the third state forward in time by seven minutes

```
function x3dot = x3prime(t,x3)

x3dot = -.09174*x3;
```

Figure 6.24 continued.

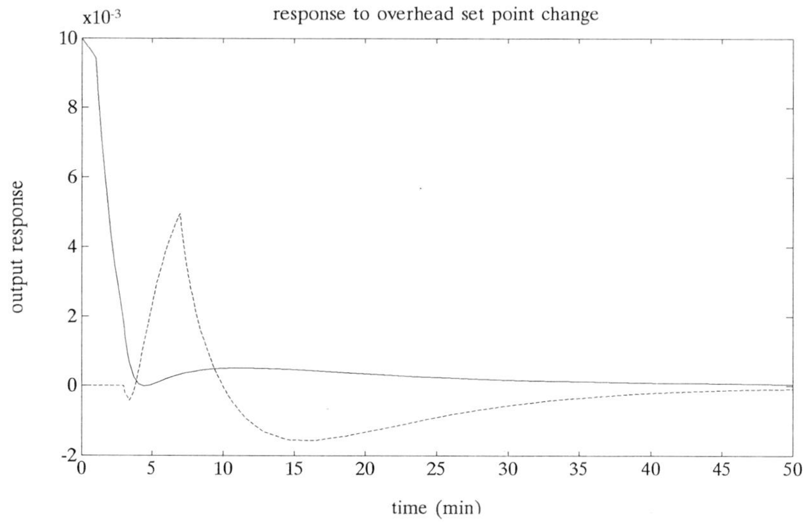

Figure 6.25 Output response to overhead set-point change of −.01 mole fraction (— overhead response, - - - bottoms response).

with

$$\boldsymbol{H} = \begin{bmatrix} 1 & -1 & 0 & 0 \\ 0 & 0 & 1 & -1 \end{bmatrix} \tag{6.8.23}$$

The MATLAB program also investigates the effect of a $-.2$ mole fraction set-point change in the bottoms composition. Here we look at feedback control based upon the time compensated state and also the non-time-compensated state. The program shows that the integration routine needs to be restarted after each time delay due to sudden changes that are introduced into the state response. The results of the simulations are shown in Figure 6.26. Again there is significant interaction between the control loops. A bottoms set-point change has introduced an effect in the overhead response, although both outputs are well regulated.

The response of the system not using the time compensated predicted state is definitely worse for both overhead and bottoms responses. As the delay time increases, the need for the time compensated predicted state is increased.

6.8.2 General Theory for Time-Delayed Systems

A more complete development of the optimal control of time-delayed systems is given by Soliman and Ray (1972) and Koivo and Lee (1972). They consider

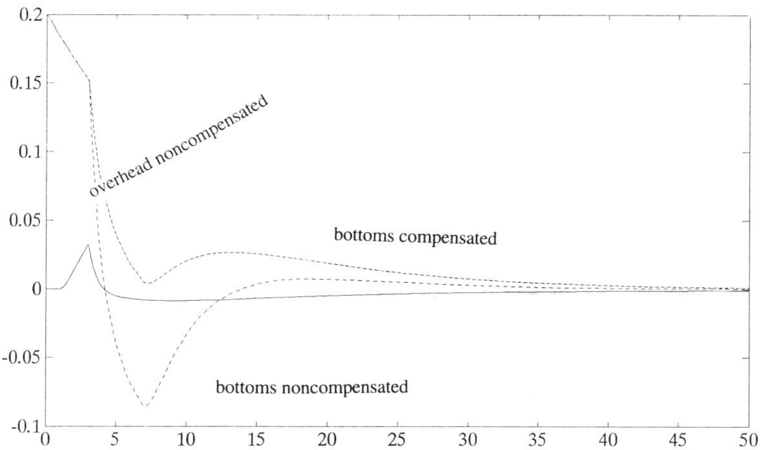

Figure 6.26 Output response to bottoms set point change of $-.2$ mole fraction.

the general time delayed system given by

$$\dot{\boldsymbol{x}}(t) = \boldsymbol{A}_0\boldsymbol{x}(t) + \sum_{i=1}^{\delta} \boldsymbol{A}_i\boldsymbol{x}(t-\alpha_i) + \boldsymbol{B}_0\boldsymbol{u}(t) + \sum_{j=1}^{\mu} \boldsymbol{B}_j\boldsymbol{u}(t-\beta_j) \tag{6.8.24}$$

For the quadratic performance functional of

$$J = \frac{1}{2}\int_0^\infty \left(\boldsymbol{x}^T(t)\boldsymbol{Q}\boldsymbol{x}(t) + \boldsymbol{u}^T(t)\boldsymbol{R}\boldsymbol{u}(t)\right) dt \tag{6.8.25}$$

the optimal controller takes the form

$$\begin{aligned}\boldsymbol{u} &= -\boldsymbol{R}^{-1}\left[\left(\boldsymbol{B}_0^T\boldsymbol{P}_0 + \boldsymbol{P}_3(0)\right)\boldsymbol{x} + \int_{-\alpha_\delta}^{0}\left(\boldsymbol{B}_0^T\boldsymbol{P}_1(s) + \boldsymbol{P}_5^T(s,0)\right)\boldsymbol{x}(t+s)ds\right.\\ &\quad\left. + \int_{-\beta_\mu}^{0}\left(\boldsymbol{B}_0^T\boldsymbol{P}_3(s) + \boldsymbol{P}_4(0,s)\right)\boldsymbol{u}(t+s)ds\right] \end{aligned}\tag{6.8.26}$$

and the Riccati matrices $\boldsymbol{P}_i$ are given by

$$\begin{aligned}\boldsymbol{0} &= -\boldsymbol{P}_0\boldsymbol{A}_0 - \boldsymbol{A}_0^T\boldsymbol{P}_0 - \boldsymbol{P}_1^T(0) - \boldsymbol{P}_1(0)\\ &\quad + \left(\boldsymbol{P}_0\boldsymbol{B}_0 + \boldsymbol{P}_3(0)\right)\boldsymbol{R}^{-1}\left(\boldsymbol{B}_0^T\boldsymbol{P}_0 + \boldsymbol{P}_3^T(0)\right) - \boldsymbol{Q}\end{aligned}\tag{6.8.27}$$

$$\begin{aligned}\boldsymbol{0} &= \frac{d\boldsymbol{P}_1}{ds} - \boldsymbol{A}_0^T\boldsymbol{P}_1(s) - \boldsymbol{P}_2(0,s)\\ &\quad + \left(\boldsymbol{P}_0\boldsymbol{B}_0 + \boldsymbol{P}_3(0)\right)\boldsymbol{R}^{-1}\left(\boldsymbol{B}_0^T\boldsymbol{P}_1(s) + \boldsymbol{P}_5^T(s,0)\right)\end{aligned}\tag{6.8.28}$$

$$
\begin{aligned}
\mathbf{0} &= \frac{\partial \boldsymbol{P}_2(r,s)}{\partial r} + \frac{\partial \boldsymbol{P}_2(r,s)}{\partial s} \\
&+ \left(\boldsymbol{P}_1^T(r)\boldsymbol{B}_0 + \boldsymbol{P}_5(r,0)\right)\boldsymbol{R}^{-1}\left(\boldsymbol{B}_0^T\boldsymbol{P}_1(s) + \boldsymbol{P}_5(s,0)\right) \qquad (6.8.29) \\
\mathbf{0} &= \frac{d\boldsymbol{P}_3}{ds} - \boldsymbol{A}_0^T\boldsymbol{P}_3(s) - \boldsymbol{P}_5(0,s) \\
&+ \left(\boldsymbol{P}_0\boldsymbol{B}_0 + \boldsymbol{P}_3(0)\right)\boldsymbol{R}^{-1}\left(\boldsymbol{B}_0^T\boldsymbol{P}_3(s) + \boldsymbol{P}_4(0,s)\right) \qquad (6.8.30) \\
\mathbf{0} &= \frac{\partial \boldsymbol{P}_4(r,s)}{\partial r} + \frac{\partial \boldsymbol{P}_4(r,s)}{\partial s} \\
&+ \left(\boldsymbol{P}_3^T(r)\boldsymbol{B}_0 + \boldsymbol{P}_4(r,0)\right)\boldsymbol{R}^{-1}\left(\boldsymbol{B}_0^T\boldsymbol{P}_3(s) + \boldsymbol{P}_4(0,s)\right) \qquad (6.8.31) \\
\mathbf{0} &= \frac{\partial \boldsymbol{P}_5(r,s)}{\partial r} + \frac{\partial \boldsymbol{P}_5(r,s)}{\partial s} \\
&+ \left(\boldsymbol{P}_1^T(r)\boldsymbol{B}_0 + \boldsymbol{P}_5(r,0)\right)\boldsymbol{R}^{-1}\left(\boldsymbol{B}_0^T\boldsymbol{P}_3(s) + \boldsymbol{P}_4(0,s)\right) \qquad (6.8.32)
\end{aligned}
$$

This problem reduces to the standard LQR problem for systems without time delays. The real difficulty in implementing these optimal control results is due to the fact that the Riccati equations become partial differential equations in the dummy time variables r and s. These partial differential equations describe the exact predictive effects of delays in the states, α_i, and the delays in the controls, β_j. Due to computational and storage problems, the complete optimal formulation is rarely used. Instead, regulation based upon the time compensated predictor of Section 6.7 is a more practical and often a nearly equivalent approach.

6.9 Trajectory Following Problems

Another class of model predictive control (MPC) problems is that of trajectory following problems. There are three types of trajectory following problems. These are servo problems, tracking problems, and model following problems. A servo problem is one where the system is to follow a class of desired trajectories. The class might be, for example, trajectories described by polynomials. A tracking problem is one where the system is to follow a prescribed function of time. Model following problems are those where the system is to follow that of another plant or model of a plant. An excellent treatment of this topic is found in the book by Anderson and Moore (1989). For all such systems we wish to control the process such that the system output

$$\boldsymbol{y} = \boldsymbol{H}\boldsymbol{x} \qquad (6.9.1)$$

follows a desired trajectory $\overline{\boldsymbol{y}}$ and minimizes a quadratic performance functional

$$J = \frac{1}{2}\int_0^T \left(\boldsymbol{u}^T\boldsymbol{R}\boldsymbol{u} + (\boldsymbol{y}-\overline{\boldsymbol{y}})^T\boldsymbol{Q}(\boldsymbol{y}-\overline{\boldsymbol{y}})\right) dt \tag{6.9.2}$$

This problem can be converted into a state following problem through the introduction of the following transformation:

$$\overline{\boldsymbol{x}} = \boldsymbol{L}\overline{\boldsymbol{y}} \tag{6.9.3}$$

with

$$\boldsymbol{L} = \boldsymbol{H}^T(\boldsymbol{H}\boldsymbol{H}^T)^{-1} \tag{6.9.4}$$

and

$$\boldsymbol{y} = \overline{\boldsymbol{H}}\boldsymbol{x} \tag{6.9.5}$$

where

$$\overline{\boldsymbol{H}} = (\boldsymbol{I} - \boldsymbol{L}\boldsymbol{H}^T) \tag{6.9.6}$$

The quadratic performance functional is generalized to

$$J = \frac{1}{2}\int_0^T \left(\boldsymbol{u}^T\boldsymbol{R}\boldsymbol{u} + \overline{\boldsymbol{y}}^T\boldsymbol{Q}_1\overline{\boldsymbol{y}} + (\boldsymbol{y}-\overline{\boldsymbol{y}})^T\boldsymbol{Q}_2(\boldsymbol{y}-\overline{\boldsymbol{y}})\right) dt \tag{6.9.7}$$

With

$$\boldsymbol{Q} = \overline{\boldsymbol{H}}^T\boldsymbol{Q}_1\overline{\boldsymbol{H}} + \boldsymbol{H}^T\boldsymbol{Q}_2\boldsymbol{H} \tag{6.9.8}$$

the original problem becomes

$$J = \frac{1}{2}\int_0^T \left(\boldsymbol{u}^T\boldsymbol{R}\boldsymbol{u} + (\boldsymbol{x}-\overline{\boldsymbol{x}})\boldsymbol{Q}(\boldsymbol{x}-\overline{\boldsymbol{x}})\right) dt \tag{6.9.9}$$

where $\overline{\boldsymbol{x}}$ is defined by Equation (6.9.3) and is the desired state trajectory.

6.9.1 The Servo Problem

The servo problem is characterized by a desired output $\overline{\boldsymbol{y}}$ generated by a model of the form

$$\overline{\boldsymbol{y}} = \boldsymbol{C}\boldsymbol{z} \tag{6.9.10}$$

with

$$\dot{\boldsymbol{z}} = \boldsymbol{F}\boldsymbol{z} \tag{6.9.11}$$

For example, if $\boldsymbol{C}$ is a vector and $\boldsymbol{F}$ a p by p matrix of the form

$$\boldsymbol{F} = \begin{bmatrix} 0 & 1 & & & \\ & 0 & 1 & & \\ & & & \ddots & \\ & & & & 1 \\ 0 & & \cdots & & 0 \end{bmatrix} \tag{6.9.12}$$

then the desired output $\overline{\boldsymbol{y}}$ leads to the class of all polynomials of degree $(p-1)$.

The state and measurement systems are described by

$$\dot{\boldsymbol{x}} = \boldsymbol{A}\boldsymbol{x} + \boldsymbol{B}\boldsymbol{u} \tag{6.9.13}$$

$$\boldsymbol{y} = \boldsymbol{H}\boldsymbol{x} \tag{6.9.14}$$

We introduce the transformations

$$\overline{\boldsymbol{x}} = \begin{pmatrix} \boldsymbol{x} \\ \boldsymbol{z} \end{pmatrix} \tag{6.9.15}$$

$$\overline{\boldsymbol{A}} = \begin{bmatrix} \boldsymbol{A} & \boldsymbol{0} \\ \boldsymbol{0} & \boldsymbol{F} \end{bmatrix} \qquad \overline{\boldsymbol{B}} = \begin{bmatrix} \boldsymbol{B} \\ \boldsymbol{0} \end{bmatrix} \tag{6.9.16}$$

$$\overline{\boldsymbol{Q}} = \begin{bmatrix} \boldsymbol{Q} & -\boldsymbol{QLC} \\ -\boldsymbol{C}^T\boldsymbol{LQ} & \boldsymbol{C}^T\boldsymbol{LQL}^T\boldsymbol{C} \end{bmatrix} \tag{6.9.17}$$

which yield the Kalman regulator problem

$$\dot{\overline{\boldsymbol{x}}} = \overline{\boldsymbol{A}}\overline{\boldsymbol{x}} + \overline{\boldsymbol{B}}\boldsymbol{u} \tag{6.9.18}$$

with

$$J = \frac{1}{2}\int_0^T \left(\boldsymbol{u}^T\boldsymbol{R}\boldsymbol{u} + \overline{\boldsymbol{x}}^T\overline{\boldsymbol{Q}}\overline{\boldsymbol{x}}\right) dt \tag{6.9.19}$$

The optimal controller is therefore

$$\boldsymbol{u} = \boldsymbol{K}\overline{\boldsymbol{x}} = \boldsymbol{K}_1\boldsymbol{x} + \boldsymbol{K}_2\boldsymbol{z} \tag{6.9.20}$$

which indicates a feedback–feed-forward structure. The predictive term follows the servo model $\boldsymbol{z}$.

6.9.2 The Model Following Problem

The model following problem is attacked through a similar development. For this class of problems, the predictive model is

$$\begin{aligned} \overline{\boldsymbol{y}} &= \boldsymbol{C}_1\boldsymbol{z}_1 \\ \dot{\boldsymbol{z}}_1 &= \boldsymbol{F}_1\boldsymbol{z} + 1 + \boldsymbol{G}_1\boldsymbol{r} \end{aligned} \tag{6.9.21}$$

where the command inputs $\boldsymbol{r}$ belong to the class described by

$$\begin{aligned} \boldsymbol{r} &= \boldsymbol{C}_2\boldsymbol{z}_2 \\ \dot{\boldsymbol{z}}_2 &= \boldsymbol{F}_2\boldsymbol{z}_2 \end{aligned} \tag{6.9.22}$$

These two systems can be combined as

$$\begin{aligned} \overline{\boldsymbol{y}} &= \boldsymbol{C}\boldsymbol{z} \\ \dot{\boldsymbol{z}} &= \boldsymbol{F}\boldsymbol{z} \end{aligned} \tag{6.9.23}$$

with

$$z = \begin{pmatrix} z_1 \\ z_2 \end{pmatrix} \tag{6.9.24}$$

$$F = \begin{bmatrix} F_1 & G_1C_2 \\ 0 & F_2 \end{bmatrix} \tag{6.9.25}$$

$$C = \begin{bmatrix} C_1 & 0 \end{bmatrix} \tag{6.9.26}$$

This now is in the form of the servo problem of Section 6.9.1, and the optimal controller is given by Equation (6.9.20).

6.9.3 The Tracking Problem

For the tracking problem, it is desirable to follow the known trajectory

$$\overline{y}(t) \qquad t_o \leq t \leq T \tag{6.9.27}$$

To solve this problem, we first set up a temporary servo problem and then show that with knowledge of $\overline{y}(t)$, it is not necessary to have available the actual servo model. We make the temporary assumption that

$$\begin{aligned} \overline{y} &= Cz \\ \dot{z} &= Fz \end{aligned} \tag{6.9.28}$$

which gives

$$u = -Kx + u_p \tag{6.9.29}$$

with

$$\begin{aligned} u_p &= -R^{-1}B^Tb \\ b &= P_{12}z \end{aligned} \tag{6.9.30}$$

The Riccati submatrix P_{12} depends upon the actual model parameters R and C, but the variable b can be determined directly if only $\overline{y}(t)$ is known. This is proved by showing that

$$-\frac{d}{dt}(P_{12}z) = -\dot{P}_{12}z - P_{12}\dot{z} \tag{6.9.31}$$

leads to

$$-\dot{b} = \left(A - BR^{-1}B^TP\right)^T b - Q\overline{x} \tag{6.9.32}$$

The optimal tracking problem is therefore

$$u = -R^{-1}B^T(Px + b) \tag{6.9.33}$$

or

$$u = -Kx - R^{-1}B^Tb \tag{6.9.34}$$

with

$$-\dot{\boldsymbol{P}} = \boldsymbol{P}\boldsymbol{A} + \boldsymbol{A}^T\boldsymbol{P} + \boldsymbol{P}\boldsymbol{B}\boldsymbol{R}^{-1}\boldsymbol{B}^T\boldsymbol{P} + \boldsymbol{Q} \qquad \boldsymbol{P}(T) = \boldsymbol{0} \tag{6.9.35}$$

$$-\dot{\boldsymbol{b}} = \left(\boldsymbol{A} - \boldsymbol{B}\boldsymbol{R}^{-1}\boldsymbol{B}^T\boldsymbol{P}\right)^T \boldsymbol{b} - \boldsymbol{Q}\overline{\boldsymbol{x}} \qquad \boldsymbol{b}(T) = \boldsymbol{0} \tag{6.9.36}$$

$$\overline{\boldsymbol{x}} = \boldsymbol{H}^T(\boldsymbol{H}\boldsymbol{H}^T)^{-1}\overline{\boldsymbol{y}}(t) \tag{6.9.37}$$

Again the optimal controller is a proportional feedback–feed-forward controller with the feed-forward term tracking the known desired state $\overline{\boldsymbol{x}}(t)$. It is important to recognize that the $\boldsymbol{b}$ prediction is computed backwards in time from $t = T$.

References

Abedi, H., "Optimal Control and Identification of Distillation Towers," Ph.D. Thesis, University of Colorado (1976).

Anderson, J. H., "Dynamic Control of a Power Boiler," *IEE* **116**, No. 7, 1257 (1969).

Anderson, B. D. O. and Moore, J. B., *Optimal Control—Linear Quadratic Methods*, Prentice–Hall, Englewood Cliffs, NJ (1989).

Franke, F. R., May, F. P., and Huckaba, C. E., *Chem. Eng. Progr. Symp. Ser.* **59**, 38 (1963).

Garcia, C. E., Prett, D. M., and Morari, M., *Model Predictive Control in Chemical Process Control—CPC3*, Morari, M. and McAvoy, T. J. (eds.), Elsevier, New York, pp. 249–294 (1986).

Gee, D. A. and Ramirez, W. F., "Optimal Temperature Control for Batch Beer Fermentation," *Biotech. Bioeng.* **31**, 224–234 (1988).

Gee, D. A. and Ramirez, W. F., "Optimal Temperature Control of Jacket Cooled Fermentation Reactors," *Optimal Control Appl. & Meth.* **12**, 49–62 (1991).

Hu, Y. C. and Ramirez, W. F., "Application of Modern Control Theory to Distillation Columns," *AIChE Journal* **18**, No. 3, 479–485 (1972).

Kailath, T., *Linear Systems*, Prentice–Hall, Englewood Cliffs, NJ (1980).

Koivo, H. N. and Lee, E. B., *Automatica* **8**, 203 (1972).

Moore, C. F., Smith, C. L., and Murrill, P. W., "Improved Algorithm for DDC," *Inst. Control Sys.* **43**, No. 1, 70 (1970).

Morari, M. and Zafiriou, E., *Robust Process Control*, Prentice Hall, Englewood Cliffs, NJ (1989).

Nyquist, J. K. and Ramirez, W. F., "Tiime Optimal Control of an Experimental Continuous Stirred Tank Reactor," *Chem. Eng. Sci.* **26**, 1673 (1971).

Ogunnaike, B. A. and Ray, W. H., "Multivariable Controller Design for Linear Systems Having Multiple Time Delays," *AIChE J.* **25**, No. 6, 1043–1057 (1979).

Ramirez, W. F., Beyer, K. P., and Abedi, H., "Optimal Regulator Control for Systems with Time-Varying Measurable and Partially Measurable Disturbances," *Optimal Control Appl. & Meth.* **8**, 159–174 (1987).

Ray, W. H., *Advanced Process Control*, McGraw–Hill, Hightstown, NJ (1981).

Smith, O. J. M., "Closer Control of Loops with Dead Time," *Chem. Eng. Prog.* **53**, 217 (1957).

Smith, O. J. M., "A Controller to Overcome Dead Time," *ISA J.* **6**, No. 2, 28 (1959).

Soliman, M. A. and Ray, W. H., "Optimal Feedback Control for Linear Quadratic Systems Having Time Delays," *Int. J. Control* **15**, 609 (1972).

Wood, R. K. and Berry, M. W., *Chem. Eng. Sci.* **28**, 1707 (1973).

Exercises

1. For the Newell, Fisher, and Seborg evaporation of Exercise 5.2,

    a) Implement the Anderson MPC algorithm, assuming that all disturbances are measurable.

    b) Implement the time–varying MPC algorithm, assuming that all disturbances are measurable.

2. In this problem we want to study regulator control of the following system:

$$\begin{aligned}\dot{x}_1 &= -x_1 + 3x_2 + u_1 + d_1 \\ \dot{x}_2 &= 2x_1 - x_2 - u_1\end{aligned}$$

    - Assuming that d_1 is unmeasurable, what is the optimal control for a generalized quadratic performance index? Be specific!
    - Assuming that d_1 is measurable and that you want to maintain x_{1e} (x_1 equilibrium) equal to zero, what is the optimal control? Again use a generalized quadratic performance index. Again be specific.
    - Assuming that d_1 is measurable and that you desire to maintain the dynamic response of x_1 equal to zero, what is the optimal control?
    - Can you physically implement the third problem? If not, why not? Do you have any suggestions on how to overcome this problem?

3. For the Wood and Berry model of binary distillation, determine the optimal time compensated response to step disturbances of 20% in each of the disturbance variables.

Chapter 7

ROBUST CONTROL

7.1 Introduction

Models of industrial processes are only approximate representations of actual systems; therefore, process control designs should accommodate parameter uncertainties and variations. The concept of robustness refers to the preservation of closed-loop stability under allowable variations in system parameters.

Feedback systems are inherently robust; that is, the relevant properties of the closed-loop system tend to be preserved under parameter variations. The aim of the theory of robust control is to establish design methods that guarantee robustness. The feedback system for robust control is modeled and designed on the basis of incomplete information about the process. This contrasts with traditional design principles which assume that a sufficiently accurate model of the process exists and forms the basis for the control system design.

Robust controller design has been divided into four steps by Lunze (1989):

1. Stability—The operating point or region of the closed-loop system needs to be asymptotically stable.
2. Asymptotic regulation—For a given class of command inputs and external disturbances, the asymptotic regulation $\boldsymbol{y}(t) - \boldsymbol{v}(t) \to \mathbf{0}$ as $t \to \infty$ needs to occur. Here $\boldsymbol{y}$ is the output vector and $\boldsymbol{v}$ the desired response vector.
3. Dynamical requirements—The performance of the closed-loop system has to satisfy a set of specific specifications.
4. Robustness—The closed-loop system properties of steps 1 to 3 need to be preserved under a given class of parameter variations.

The LQR design principles introduced in Chapters 5 and 6 address steps 1 to 3. New criteria need to be developed in order to address step 4. These criteria are developed in this chapter.

7.2 Robust Stability Results for the LQR Problem

Safonov and Athans (1977) and Safonov (1980) have developed an important stability theorem. Their work has shown that a multiloop feedback system is closed-loop stable if the function space on which the system's dynamical relations are defined can be topologically separated into two disjoint sets. The relations for the open loop lie in one part of the separated space and the relations for the feedback loop lie in the other. The Safonov stability theorem has a direct bearing on the robust stability of the feedback control problem.

We apply the Safonov stability result to the linear quadratic regulator (LQR) problem for the model or predictive system

$$\dot{\boldsymbol{x}}_p = \boldsymbol{A}_p\boldsymbol{x}_p + \boldsymbol{B}_p\boldsymbol{u}_p \tag{7.2.1}$$

with a performance index of

$$J = \frac{1}{2}\int_0^\infty \left(\boldsymbol{x}_p^T\boldsymbol{Q}\boldsymbol{x}_p + \boldsymbol{u}_p^T\boldsymbol{R}\boldsymbol{u}_p\right) dt \tag{7.2.2}$$

The optimal feedback control comes from the solution of the Riccati equation as

$$\boldsymbol{u}_p = -\boldsymbol{K}\boldsymbol{x}_p \tag{7.2.3}$$

with

$$\boldsymbol{K} = \boldsymbol{R}^{-1}\boldsymbol{B}_p^T\boldsymbol{P} \tag{7.2.4}$$

The actual uncertain system is given as

$$\dot{\boldsymbol{x}} = \boldsymbol{A}\boldsymbol{x} + \boldsymbol{B}\boldsymbol{u} \tag{7.2.5}$$

The Safonov stability theorem yields the fact that if

$$\boldsymbol{P}\left[(\boldsymbol{A}_p - \boldsymbol{A}) - (\boldsymbol{B}_p - \boldsymbol{B})\boldsymbol{K}\right] + \frac{1}{2}\left[\boldsymbol{Q} + \boldsymbol{P}\boldsymbol{B}_p\boldsymbol{R}^{-1}\boldsymbol{B}_p^T\boldsymbol{P}\right] > \boldsymbol{0} \tag{7.2.6}$$

then the actual system of Equation (7.2.5) is closed-loop finite gain stable.

To more fully appreciate the implications of this result, Safonov introduces the following special case of

$$\boldsymbol{A} = \boldsymbol{A}_p \tag{7.2.7}$$

and

$$\boldsymbol{B}\boldsymbol{u} = \boldsymbol{B}_p(\boldsymbol{N}\boldsymbol{u}) \tag{7.2.8}$$

where

$$\boldsymbol{N}\boldsymbol{u} = \begin{pmatrix} N_1u_1 \\ \vdots \\ N_mu_m \end{pmatrix} \tag{7.2.9}$$

All the differences between the open-loop dynamics of the actual system of (7.2.5) and the approximate linear model of (7.2.1) are lumped into m dynamical nonlinear perturbations N_i. This actuator or control parameter perturbation model is shown in the block diagram of Figure 7.1. This system is a state feedback regulator with noninteracting control parameter perturbations in each plant input channel.

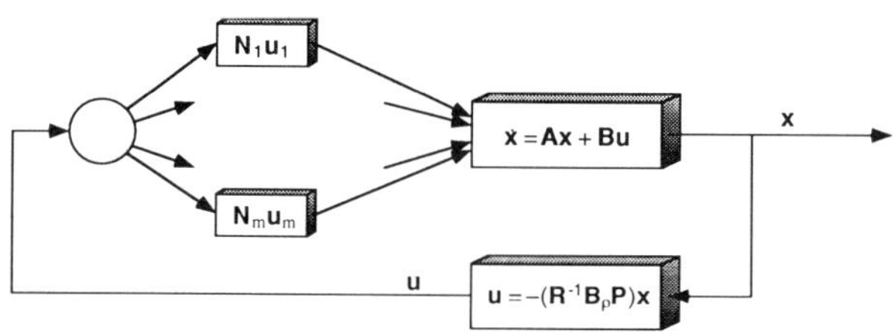

Figure 7.1 Block diagram of actuator perturbation model.

If we further assume that $\boldsymbol{R}$ is diagonal

$$\boldsymbol{R} = \text{diag}\ (r_1, \cdots r_m) \tag{7.2.10}$$

then the stability condition reduces to

$$\boldsymbol{PB}\left[\text{diag}\left((N_i - \frac{1}{2})r_1^{-1}, \cdots, (N_m - \frac{1}{2})r_m^{-1}\right)\right]\boldsymbol{B}^T\boldsymbol{P} + \frac{1}{2}\ \boldsymbol{Q} \geq \boldsymbol{0} \tag{7.2.11}$$

This relation is satisfied if

$$N_i - \frac{1}{2} \leq 0 \qquad\qquad i = 1, \cdots m \tag{7.2.12}$$

Equation (7.2.12) shows the inherent robustness of the LQR problem by establishing a lower bound on the stability margins of the design. All LQR problems can tolerate at least perturbations or uncertainties N_i satisfying Equation (7.2.12). Figure 7.2 is a graphical representation of the class of uncertainties that are allowable. The line $u_i = u_i$ is the system without uncertainty. That is, N_i is equal to unity. A typical uncertain $N_i u_i$ which is a nonlinear function of u_i is also shown on the figure. Essentially the slope functions N_i which characterize the actuator model uncertainties can vary from $+1/2$ to infinity. This is a large stability margin and therefore the LQR design is robust in the sense that it is closed-loop stable for a large class of control parameter variations which represent substantial differences between the assumed model and the actual system.

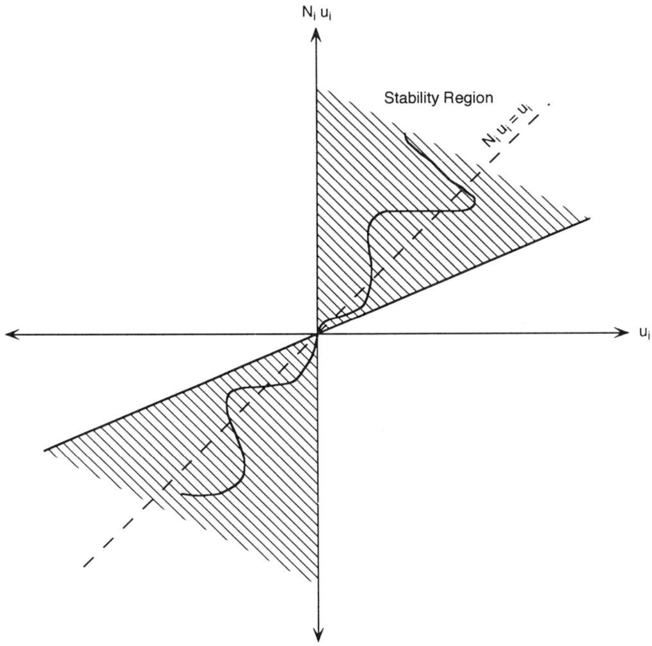

Figure 7.2 Robustness of LQR.

7.3 Robust Integrity

The following result by Wong *et al.* (1978) enables us to design for robust integrity. We assume that the control parameter perturbations comprise real gains $\boldsymbol{N} = \text{diag}\ (n_1,\ n_2, \cdots, n_m)$ and the control weighting matrix $\boldsymbol{R}$ is also diagonal (Equation (7.2.10)). The LQR controller is closed-loop stable for all $\boldsymbol{N}$ such that

$$n_i > \frac{1}{2}\ (1 - \lambda_0) \qquad i = 1, \cdots m \tag{7.3.1}$$

where λ_0 is the minimum eigenvalue of

$$\left(\boldsymbol{R}^{1/2}\boldsymbol{K}^T\boldsymbol{Q}^{-1}\boldsymbol{K}\boldsymbol{R}^{1/2}\right)^{-1} > \mathbf{0} \tag{7.3.2}$$

Clearly if a high degree of integrity and robustness is required, then λ_0 must be maximized by the choice of $\boldsymbol{Q}$ and $\boldsymbol{R}$. The design procedure is to first establish the feedback gain matrix $\boldsymbol{K}$ based upon the normal design methods of Chapters 5 and 6. Then the minimum eigenvalue of Equation (7.3.2) is computed and a decision is reached as to whether it allows for stability for perturbations n_i that are sufficient. If not, new weighting matrices $\boldsymbol{Q}$ and $\boldsymbol{R}$ are used in order to enlarge the stability margin.

As Safonov (1980) points out, by restricting the robust stability analysis to systems with only control parameter perturbations, we are not restricting our attention to only that class of systems. We are merely stipulating that the actual system's open loop dynamics have the same input-output behavior as such a system with only control parameter uncertainties. In practice, experimental input-output data would be used to deduce typical input perturbations $\boldsymbol{N}$.

If we apply the stability criteria of Equation (7.3.1) to the Nyquist–Ramirez reactor example of Section 5.4 with

$$\boldsymbol{Q} = \boldsymbol{R} = \boldsymbol{I} \tag{7.3.3}$$

and

$$\boldsymbol{K} = \begin{bmatrix} .0505 \\ -.9241 \end{bmatrix} \tag{7.3.4}$$

then the minimum and only eigenvalue is

$$\lambda_0 = 1.168 \tag{7.3.5}$$

and

$$n_1 > -.0803 \tag{7.3.6}$$

This result shows that the system is stable for disturbances n_1 of loops greater than -0.0838. A graphical interpretation is given in Figure 7.3. The region of stability is very large and the system is very robust to uncertainties in the control parameters $\boldsymbol{B}$.

If we apply the criteria to the unstable reactor of Section 5.7.1, then

$$n_1 > .1345 \tag{7.3.7}$$

The region of stability for this problem is less than that for the Nyquist–Ramirez reactor, but this system is still very robust. Both of these results, of course, fall within the Safonov criteria of Equation (7.2.12).

References

Lunze, J., *Robust Multivariable Feedback Control*, Prentice Hall, Englewood Cliffs, NJ (1989).

Safonov, M. G. and Athans, M., "Gain and Phase Margin for Multiloop LQG Regulators," *IEEE Trans. Auto Control* **AC-22**, 173–179 (1977).

Safonov, M. G., *Stability and Robustness of Multivariable Feedback Systems*, MIT Press, Cambridge, MA (1980).

Wong, P. K., Stein, G., and Athans, M., "Structural Reliability and Robustness Properties of Optimal Linear Quadratic Multivariable Regulators," *IFAC Preprints Helsinki*, 1797-1805 (1978).

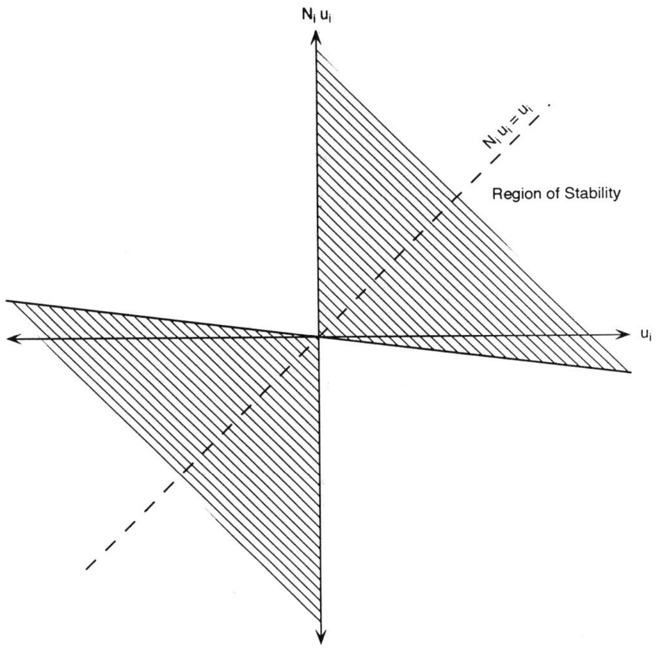

Figure 7.3 Robustness of Nyquist–Ramirez reactor.

Exercise

1. Compute the robust stability regions for the following:

    1. The PI control problem of Section 5.6.1.
    2. The Douglas reactor problem 5.1.
    3. The Newell *et al.* evaporator problem 5.2.
    4. The Ramirez *et al.* dialysis problem 5.3.
    5. The distillation column of Section 6.5.

Chapter 8

PROBLEMS LINEAR IN THE CONTROL VARIABLES

In this chapter we consider control problems that are linear in the control variables. The optimal control solution for these problems results in bang–bang control strategies. The existence of singular or intermediate control must also be investigated. We will study time optimal control and minimum integral square error problems as typical of problems that are linear in the controls.

8.1 Time Optimal Control

Time optimal control has been treated by many authors. The theoretical concepts were first considered by Pontryagin *et al.* (1962), LaSalle (1960), and Bellman *et al.* (1956).

The time optimal control problem is to take a system from an initial state, $\boldsymbol{x}$, to some specified state, $\boldsymbol{x}_f$, or target set, N, in minimum time. To study the problem, let's focus on a specific example.

8.1.1 Time Optimal Control of a Continuous Stirred Tank Reactor

As a focus for the development of practical concepts needed for the formulation and implementation of time optimal control, let us consider the continuous stirred tank reactor studied by Nyquist and Ramirez (1971). The system was introduced in Chapter 5.

The reaction studied was the decomposition of hydrogen peroxide to oxygen and water:

$$H_2O_2 \rightarrow H_2O(\ell) + \frac{1}{2}\,O_2(g) \tag{8.1.1}$$

The temperature and hydrogen peroxide concentration in the reactor are the state variables of interest.

Nyquist and Ramirez have developed and tested a mathematical model of the process and have shown that it describes the process to within 2%. Their model is a material balance

$$\frac{dC}{dt} = \frac{F_p C_o}{V} - \frac{FC}{V} - R \tag{8.1.2}$$

The reaction rate is linear in the peroxide concentration

$$R = kC_I \tag{8.1.3}$$

The rate constant follows the Arrhenius temperature dependency

$$k = k_o \exp\left(-E/(T + 2.73.16)\right) \tag{8.1.4}$$

The energy balance for the reactor is

$$V\rho c_P \frac{dT}{dt} = F\rho c_p(T_i - T) + Q_s + HRV - Q_c \tag{8.1.5}$$

Nyquist and Ramirez have shown that the cooling rate can be modeled as

$$Q_c = \beta(T - T_{c_i}) \tag{8.1.6}$$

where β is the effective cooling rate coefficient.

The reactor was equipped with a relatively powerful stirrer, so the conversion of mechanical energy to thermal energy resulted in a stirrer transfer rate, Q_s. This term was shown to be given adequately by a first-order model

$$Q_s = 1288 - 27.06\,T \tag{8.1.7}$$

Specific model parameters are given in Table 8.1.

Table 8.1
Parameters for Nyquist–Ramirez Reactor

Operating parameters		
V=16.7 l.	C_0 = 9.3 g mole/l.	β_{max}=1023 cal/min°C
F =0.2 l./min	T_i=25°C	β_{min}=45 cal/min°C
F_p=0.032 l./min	k_0=8.699×10^7 min^{-1}	ρ=1000 g/l
H=22,600 cal/g mole H_2O_2	E =6525 (°K)$^{-1}$	C_p=1.0 cal/°C g
		T_c=14°C

Let us now consider the time optimal control of this reactor. We wish to bring the reactor from any initial starting condition, $\boldsymbol{x}_o$, to a desired operating state, $\boldsymbol{x}_f$. The model of the system can be expressed as

$$\frac{dC}{dt} = f_1(C,T) \tag{8.1.8}$$

$$\frac{dT}{dt} = f_2(C,T) + b(C,T)\,\beta \tag{8.1.9}$$

We see that the model is nonlinear in the state but linear in the control, β. The objective functional is the minimum time objective

$$J = \int_o^{t_f} dt \tag{8.1.10}$$

and the necessary conditions for optimal control are given in terms of the Hamiltonian, which, for this problem, is

$$H = 1 + f_1\lambda_1 + f_2\lambda_2 + b\beta\lambda_2 \tag{8.1.11}$$

The necessary conditions are as follows:

Definition of the State Equations

$$\dot{\boldsymbol{x}} = \frac{\partial H}{\partial \boldsymbol{\lambda}} \tag{8.1.12}$$

For this problem we have the description of Equations (8.1.8) and (8.1.9).

Definition of the Costate Equations

$$\dot{\boldsymbol{\lambda}} = -\frac{\partial H}{\partial \boldsymbol{x}} \tag{8.1.13}$$

which become, for this problem

$$\dot{\lambda}_1 = -\lambda_1 \frac{\partial f_1}{\partial x_1} - \lambda_2 \frac{\partial f_2}{\partial x_1} - \lambda_2 u \frac{\partial b}{\partial x_1} \tag{8.1.14}$$

$$\dot{\lambda}_2 = -\lambda_1 \frac{\partial f_1}{\partial x_2} - \lambda_2 \frac{\partial f_2}{\partial x_2} - \lambda_2 u \frac{\partial b}{\partial x_2} \tag{8.1.15}$$

Optimal Control

The general statement is to minimize the Hamiltonian with respect to the control. Since the Hamiltonian is linear in the control, then

$$\beta = \beta_{\min} \quad \text{when} \quad \lambda_2 b > 0 \tag{8.1.16}$$

$$\beta = \beta_{\max} \quad \text{when} \quad \lambda_2 b < 0 \tag{8.1.17}$$

We have therefore developed a bang–bang control strategy as being the optimal control law. There is also the possibility that $\lambda_2 b$ remains zero for a finite period of time. In that case, the costate variable, $\lambda_2 = 0$, and $\partial H/\partial \beta = 0$ also. This is called singular control, and the solution is a singular arc. We have to investigate the existence of singular control for this problem. If $\lambda_2 = 0$ and stays zero, then $\dot{\lambda}_2 = 0$. From the Hamiltonian

$$H = 1 + f_1\lambda_1 + f_2\lambda_2 + bu\lambda_2 \tag{8.1.18}$$

so that

$$\lambda_1 = \frac{H-1}{f_1} \tag{8.1.19}$$

along the singular arc.

Transversality

The general transversality necessary condition is

$$-\boldsymbol{\lambda}^T(t_f)\delta\boldsymbol{x}_f + H(t_f)\delta t_f = 0 \tag{8.1.20}$$

This is a fixed end-point problem since $\boldsymbol{x}(t_f)$ is known. It is also a free final time problem since t_f is free. This means that the final variations are

$$\delta\boldsymbol{x}_f = 0 \qquad \text{and} \qquad \delta t_f \neq 0 \tag{8.1.21}$$

The transversality relation of (8.1.20) therefore becomes

$$H(t_f) = 0 \tag{8.1.22}$$

Now let's return to the question of whether or not a singular solution is possible for this problem. The time derivative of the Hamiltonian is

$$\frac{dH}{dt} = \left(\frac{\partial H}{\partial \boldsymbol{u}}\right)^T \dot{\boldsymbol{u}} + \left(\frac{\partial H}{\partial \boldsymbol{x}}\right)^T \dot{\boldsymbol{x}} + \left(\frac{\partial H}{\partial \boldsymbol{\lambda}}\right)^T \dot{\boldsymbol{\lambda}} \tag{8.1.23}$$

or

$$\frac{dH}{dt} = \left(\frac{\partial H}{\partial \boldsymbol{u}}\right)^T \dot{\boldsymbol{u}} + \left(-\dot{\boldsymbol{\lambda}}^T\dot{\boldsymbol{x}} + \dot{\boldsymbol{x}}^T\boldsymbol{\lambda}\right) = \left(\frac{\partial H}{\partial \boldsymbol{u}}\right)^T \dot{\boldsymbol{u}} \tag{8.1.24}$$

Along a singular arc, $\frac{\partial H}{\partial \boldsymbol{u}} = \boldsymbol{0}$, and along the bang–bang control, $\dot{\boldsymbol{u}} = \boldsymbol{0}$; therefore

$$\frac{dH}{dt} = 0 \tag{8.1.25}$$

which is

$$H = \text{const} \tag{8.1.26}$$

Since the terminal condition is $H(t_f) = 0$, we have

$$H = 0 = 1 + \lambda_1(t_f)f_1(t_f) \tag{8.1.27}$$

Can the singular arc be a solution for this problem? From Equation (8.1.15), we have

$$\dot{\lambda}_2 = -\lambda_1 \frac{\partial f_1}{\partial x_2} - \lambda_2 \frac{\partial f_2}{\partial x_2} - \lambda_2 u \frac{\partial b}{\partial x_2} \tag{8.1.28}$$

which means that along the singular arc ($\lambda_2 = 0$ and $\dot{\lambda}_2 = 0$), we require

$$0 = \lambda_1 \frac{\partial f_1}{\partial x_2} \tag{8.1.29}$$

so that either

$$\lambda_1 = 0 \tag{8.1.30}$$

or

$$\frac{\partial f_1}{\partial x_2} = 0 \tag{8.1.31}$$

The variable, λ_1, cannot be zero. With $\lambda_1 = 0$, the costate dynamic relation of Equation (8.1.14) implies that $\dot{\lambda}_1 = 0$. The costate $\lambda_1(t_f)$ would then also be zero. With both $\lambda_1(t_f)$ and $\lambda_2(t_f)$ being zero, we would violate the transversality condition which states that the Hamiltonian must be zero at the final time. The expression for $\partial f_1/\partial x_2$ is

$$\frac{\partial f_1}{\partial x_2} = k_o C_I C \frac{E}{T^2} e^{-E/T} = 0 \tag{8.1.32}$$

The only solutions to this relationship are

$$C = 0 \tag{8.1.33}$$

or

$$T = \pm\infty \tag{8.1.34}$$

These are not feasible solutions; therefore the singular arc for this problem does not exist, and we do not have an intermediate control policy. Bang–bang control is the only optimal minimum time control policy for this system.

Now we wish to develop a convenient computational procedure for determining the time to switch from one control extreme to the other. Since we know at the final time we must be either at $\beta_{\min}$ or $\beta_{\max}$, we can integrate the state differential equations backward in time using both $\beta_{\min}$ and $\beta_{\max}$. This computation develops the switching curve shown in Figure 8.1. The switching curve separates the phase plane into two regions, one where the control is $\beta_{\min}$ and the other where the control is $\beta_{\max}$. For any specific starting point, the optimal control is defined as well as the switching condition. Figure 8.1 also shows the time optimal trajectories for several initial starting conditions.

Nyquist and Ramirez also considered the time optimal control of this stirred tank reactor to a target set. The specific target set chosen was an elliptical

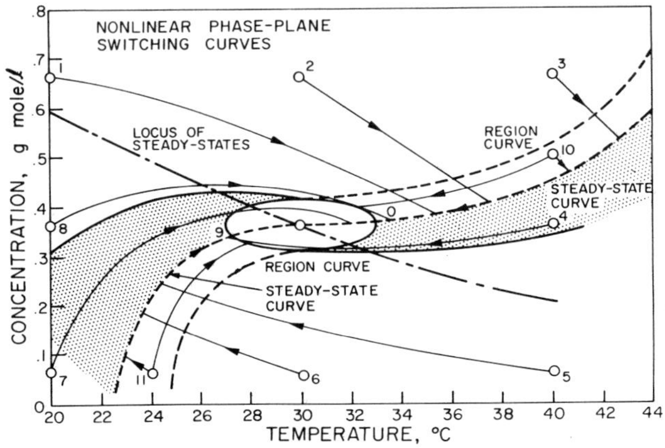

Figure 8.1 Time-optimal control phase plane. From Nyquist and Ramirez (1971).

region about the steady state with major axes of 0.05 g moles/liter in concentration and 3°C in temperature. A target set constraint relation can be written as

$$N(t_f) = 1 - \frac{x_1^2(t_f)}{.05^2} - \frac{x_2^2(t_f)}{3^2} = 0 \tag{8.1.35}$$

The transversality relation of Section 4.2.3 becomes

$$\left[\frac{\partial h}{\partial x}(t_f) + \frac{\partial \boldsymbol{N}^T(t_f)}{\partial \boldsymbol{x}}\boldsymbol{\nu} - \boldsymbol{\lambda}(t_f)\right] \delta \boldsymbol{x}_f = 0 \tag{8.1.36}$$

or

$$\left(\frac{\partial \boldsymbol{N}}{\partial \boldsymbol{x}}(t_f)\right)^T \boldsymbol{\nu} = \boldsymbol{\lambda}(t_f) \tag{8.1.37}$$

For this case we have

$$\frac{\partial N}{\partial x_1} = -\frac{2x_1(t_f)}{.05^2} \tag{8.1.38}$$

and

$$\frac{\partial N}{\partial x_2} = -\frac{2x_2(t_f)}{3^2} \tag{8.1.39}$$

which gives

$$\begin{pmatrix} -\dfrac{2x_1(t_f)}{.05^2} \\ -\dfrac{2x_2(t_f)}{3^2} \end{pmatrix} \nu = \begin{pmatrix} \lambda_1(t_f) \\ \lambda_2(t_f) \end{pmatrix} \tag{8.1.40}$$

This relation means that the costate vector at the final time is either in the direction of the normal vector to the target set (for $\nu > 0$) or in the direction opposite to the normal vector (for $\nu < 0$). For a minimization problem, we have $\nu < 0$ so that the costate vector must be in the direction opposite to the normal vector. For a maximization problem, the condition is $\nu > 0$ so that the normal and costate vectors are in the same direction. The relationship of the final variation, $\delta \boldsymbol{x}_f$, the normal vector, $\boldsymbol{n}$, and the final costate vector, $\boldsymbol{\lambda}(t_f)$, are shown in Figure 8.2.

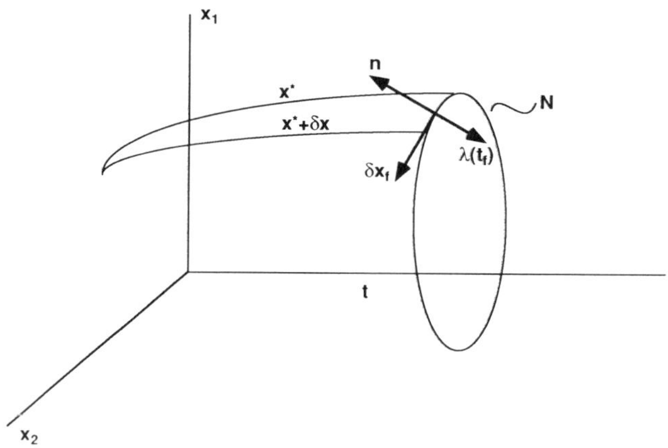

Figure 8.2. Relation between the final variation and the final costate.

For this case, when there is only one terminal constraint relation, we have

$$\boldsymbol{\lambda}(t_f) = -\frac{\partial N}{\partial \boldsymbol{x}}\, d \tag{8.1.41}$$

This states that the final costate is in the direction opposite that of the normal vector to the target set.

An easier parameterization of the target set is

$$x_1(t_f) = .05\ \sin\phi + x_{1c} \tag{8.1.42}$$

where x_{1c} is the state at the center of the ellipse and ϕ is the angle from the horizontal. Also,

$$x_2(t_f) = 3\ \cos\phi + x_{2c} \tag{8.1.43}$$

where x_{2c} is the state at the center of the ellipse. This yields

$$\lambda_1(t_f) = \frac{d\ \sin\phi}{.05} \tag{8.1.44}$$

$$\lambda_2(t_f) = \frac{d \cos\phi}{3} \tag{8.1.45}$$

The Hamiltonian at the final time is zero, so

$$H(t_f) = 0 = 1 + \boldsymbol{\lambda}^T(t_f)\dot{\boldsymbol{x}}(t_f) \tag{8.1.46}$$

This means that for a specific starting point, there will be a specific ending condition or value of d that will satisfy Equation (8.1.46).

Nyquist and Ramirez propose an efficient computational procedure that circumvents the solution of the two point boundary problem implied by the necessary conditions. At the final time we know that we will be on the target set. We can therefore integrate both the state and costate equations backwards in time assuming different starting conditions on the target set for an arbitrary value of d, say 10. We choose the control that satisfies the bang–bang optimal control law of Equations (8.1.16) and (8.1.17). The state and costate equations are integrated backwards in time until λ_2 changes sign or it is clear that the system is no longer in a region of interest. At the point that λ_2 changes sign, the state variables are saved as the coordinates of a point along the switching curve. The process is repeated for different starting conditions until the switching curve in the phase plane is established. The results are shown on Figure 8.1 along with the switching curve to the final steady-state condition.

Once a time optimal control problem reaches the final state or the final target set, then the solution is complete. However, it is important that the system remain at or close to these desired operating conditions. The control problem then becomes one of regulation, and the analysis of Chapters 5 and 6 allows for the implementation of optimal regulator control policies.

8.2 The Number of Switches for Time Optimal Control

We can show that for linear systems the number of switches for each control variable is $\leq N - 1$, where N is the order of the system and $\boldsymbol{A}$ has real eigenvalues. The system is

$$\dot{\boldsymbol{x}} = \boldsymbol{A}\boldsymbol{x} + \boldsymbol{B}\boldsymbol{u} \tag{8.2.1}$$

and the performance functional is

$$J = \int_{t_o}^{t_f} dt \tag{8.2.2}$$

The costate variables are

$$\dot{\boldsymbol{\lambda}} = -\boldsymbol{A}^T\boldsymbol{\lambda} \tag{8.2.3}$$

which have the solution

$$\boldsymbol{\lambda} = e^{-\boldsymbol{A}^T t}\boldsymbol{\lambda}(0) \tag{8.2.4}$$

The optimal control is given by minimizing the Hamiltonian with respect to the control variables. The Hamiltonian is

$$H = 1 + \boldsymbol{\lambda}^T(\boldsymbol{A}\boldsymbol{x} + \boldsymbol{B}\boldsymbol{u}) \tag{8.2.5}$$

The conditions for a switch are therefore

$$\boldsymbol{\lambda}^T \boldsymbol{B} = \boldsymbol{0} \tag{8.2.6}$$

or

$$\boldsymbol{\lambda}^T(0)e^{-\boldsymbol{A}^T t}\boldsymbol{B} = \boldsymbol{0} \tag{8.2.7}$$

Equation (8.2.7) is a sum of exponential terms in the eigenvalues of $\boldsymbol{A}$. When the eigenvalues are real, then the number of zeros for each control is a sum of N real exponential functions, and the number of zeros is at most $N - 1$.

When some of the eigenvalues are complex, then the number of switches is finite, but it depends upon the distances between the initial state and the target. Let us consider the specific case of

$$\begin{pmatrix} \dot{x}_1 \\ \dot{x}_2 \end{pmatrix} = \begin{bmatrix} 0 & 1 \\ -1 & 0 \end{bmatrix} \begin{pmatrix} x_1 \\ x_2 \end{pmatrix} + \begin{pmatrix} 0 \\ 1 \end{pmatrix} u \tag{8.2.8}$$

with

$$|u| \le 1 \tag{8.2.9}$$

The costate variables are

$$\dot{\boldsymbol{\lambda}} = -\boldsymbol{A}^T \boldsymbol{\lambda} \tag{8.2.10}$$

or

$$\boldsymbol{\lambda}(t) = e^{-\boldsymbol{A}^T(t-t_f)}\boldsymbol{\lambda}(t_f) \tag{8.2.11}$$

The transition matrix is given by

$$e^{\boldsymbol{A}\tau} = \begin{bmatrix} \cos\tau & \sin\tau \\ -\sin\tau & \cos\tau \end{bmatrix} \tag{8.2.12}$$

or

$$e^{-\boldsymbol{A}^T\tau} = \begin{bmatrix} \cos\tau & -\sin\tau \\ \sin\tau & \cos\tau \end{bmatrix} \tag{8.2.13}$$

The optimal control is

$$u = -\text{sign}\left[\boldsymbol{\lambda}^T \boldsymbol{b}\right] \tag{8.2.14}$$

or

$$u = -\text{sign}\left[\boldsymbol{\lambda}^T(t_f)e^{\boldsymbol{A}(t_f-t)}\boldsymbol{b}\right] \tag{8.2.15}$$

This means that the control will change sign each time interval of π units. Let us investigate the solution to the problem when $u = +1$. The state equations become

$$\dot{x}_1 = x_2 \tag{8.2.16}$$

$$\dot{x}_2 = -x_1 + 1 \tag{8.2.17}$$

which can be combined to give

$$\frac{dx_1}{dx_2} = \frac{x_2}{1 - x_1} \tag{8.2.18}$$

and can be integrated to yield

$$(x_1 - 1)^2 + x_2^2 = R^2 \tag{8.2.19}$$

This is a solution of circles of radius R centered about $x_1 = 1$ and $x_2 = 0$. When $u = -1$, the solution to the problem is

$$(x_1 + 1)^2 + x_2^2 = R^2 \tag{8.2.20}$$

which is a series of circles of radius R centered about $x_1 = -1$ and $x_2 = 0$. The switching curves generated by such a solution are shown in Figure 8.3.

8.3 Existence of Singular Arcs for Linear Time Optimal Problems

We want to investigate the conditions when singular arcs are possible for linear systems under time optimal control. The system is

$$\dot{\boldsymbol{x}} = \boldsymbol{A}\boldsymbol{x} + \boldsymbol{B}\boldsymbol{u} \tag{8.3.1}$$

with the objective functional

$$J = \int_{t_o}^{t_f} dt \tag{8.3.2}$$

The Hamiltonian for this problem is

$$H = 1 + \boldsymbol{\lambda}^T \boldsymbol{A}\boldsymbol{x} + \boldsymbol{\lambda}^T \boldsymbol{B}\boldsymbol{u} \tag{8.3.3}$$

The control will switch from one extreme to another when $\boldsymbol{\lambda}^T \boldsymbol{B}$ changes sign. Singular control is possible when

$$\boldsymbol{\lambda}^T \boldsymbol{B} = \boldsymbol{0} \tag{8.3.4}$$

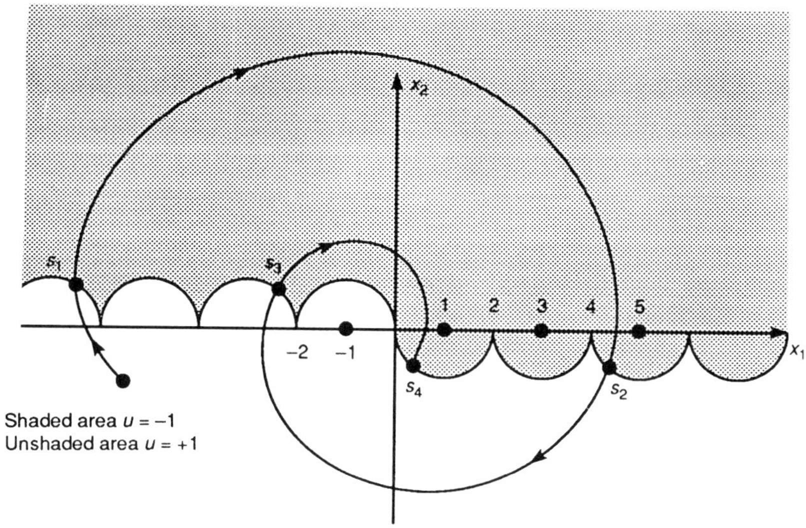

Figure 8.3 Optimal switching for complex eigenvalues of $\boldsymbol{A}$. From Sage and White, *OPTIMUM SYSTEMS CONTROL*, 2e, ©1977, p. 108. Reprinted by permission of Prentice Hall, Englewood Cliffs, New Jersey.

Since the matrix, $\boldsymbol{B}$, is not a null matrix, then we need to investigate the possibility that $\boldsymbol{\lambda}$ is zero. If $\boldsymbol{\lambda}$ is zero, then the Hamiltonian becomes $H = 1$. This is impossible since the Hamiltonian at the final time is zero from transversality, and using Equation (8.1.24), the Hamiltonian is a constant which must be zero, not unity. Therefore, the costate, $\boldsymbol{\lambda}$, is not zero. Finally, we need to investigate the possibility that the product

$$\boldsymbol{\lambda}^T \boldsymbol{B} = \mathbf{0} \tag{8.3.5}$$

This says that the two vector components are orthogonal and remain orthogonal for a finite period of time. This implies that

$$\frac{d^k}{dt^k}\left[\boldsymbol{\lambda}^T \boldsymbol{B}\right] = \mathbf{0} \tag{8.3.6}$$

Let's consider the first derivative. From the costate defining relations, we have

$$\dot{\boldsymbol{\lambda}} = -\boldsymbol{A}^T \boldsymbol{\lambda} \tag{8.3.7}$$

which has the solution

$$\boldsymbol{\lambda} = e^{-\boldsymbol{A}^T t} \boldsymbol{c} \tag{8.3.8}$$

Equation (6.3.5) therefore becomes

$$\mathbf{0} = -\left[e^{-\boldsymbol{A}^T t} \boldsymbol{C}\right]^T \boldsymbol{B} \tag{8.3.9}$$

This means that

$$\dot{\boldsymbol{\lambda}}^T \boldsymbol{B} = -\left[\boldsymbol{A}^T \boldsymbol{\lambda}\right]^T \boldsymbol{B} = \boldsymbol{0} \tag{8.3.10}$$

or

$$\boldsymbol{0} = -\left[e^{-\boldsymbol{A}^T t}\boldsymbol{c}\right]^T \boldsymbol{AB} \tag{8.3.11}$$

The second derivative term is

$$\ddot{\boldsymbol{\lambda}}\boldsymbol{B} = -\left[\boldsymbol{A}^T \dot{\boldsymbol{\lambda}}\right]^T \boldsymbol{B} = \boldsymbol{0} \tag{8.3.12}$$

which can be expressed as

$$\boldsymbol{0} = -\left[e^{-\boldsymbol{A}^T t}\boldsymbol{c}\right]^T \boldsymbol{A}^2\boldsymbol{B} \tag{8.3.13}$$

If we continue this process, we arrive at the matrix

$$\boldsymbol{E} = \left[\ \boldsymbol{B}\ \vdots\ \boldsymbol{AB}\ \vdots\ \boldsymbol{A}^2\boldsymbol{B}\ \vdots\ \cdots\ \vdots\ \boldsymbol{A}^{n-1}\boldsymbol{B}\ \right] \tag{8.3.14}$$

The matrix, $\boldsymbol{E}$, must be singular for the singular arc to exist. $\boldsymbol{E}$ is also the definition of the controllability matrix for the original problem. This means the only condition that can exist for the singular arc to exist for a linear system under time optimal control is that the system is not controllable. Therefore, for controllable systems, singular control is impossible for the time optimal control of linear systems.

8.4 Minimum Integral Square Error Problems

The linearized Nyquist–Ramirez reactor can be expressed as

$$\begin{aligned} \dot{x}_1 &= a_{11}x_1 + a_{12}x_2 \\ \dot{x}_2 &= a_{21}x_1 + a_{22}x_2 + b_2u_1 \end{aligned} \tag{8.4.1}$$

Following Denn (1969), we can simplify the form of the equations by defining new variables, y_1 and y_2, as

$$x_1 = -y_1 \tag{8.4.2}$$

$$x_2 = \frac{a_{11}}{a_{12}}y_1 - \frac{1}{a_{12}}y_2 \tag{8.4.3}$$

This gives for the defining dynamics the following equations:

$$\dot{y}_1 = y_2 \tag{8.4.4}$$

$$\dot{y}_2 = (a_{12}a_{21} - a_{11}a_{22})y_1 + (a_{11} + a_{22})y_2 + (-a_{12}b_2u_1) \tag{8.4.5}$$

With the obvious notation, we have

$$\dot{y}_1 = y_2 \tag{8.4.6}$$

$$\dot{y}_2 = -a_2 y_1 - a_1 y_2 + u \tag{8.4.7}$$

The integral square error criterion is

$$J = \frac{1}{2}\int_{t_o}^{t_f} (q_{11}y_1^2 + 2q_{12}y_1y_2 + q_{22}y_2^2)\, dt \tag{8.4.8}$$

This is similar to the general quadratic performance functional of Chapter 5 except that there is no quadratic cost of control term. We will assume that the control is bounded by

$$u_{\min} \leq u \leq u_{\max} \tag{8.4.9}$$

The Hamiltonian for this problem is

$$\begin{aligned} H = & \frac{1}{2}(q_{11}y_1^2 + 2q_{12}y_1y_2 + q_{22}y_2^2) + \lambda_1 y_2 - \lambda_2 a_2 y_1 \\ & -\lambda_2 a_1 y_2 + \lambda_2 u \end{aligned} \tag{8.4.10}$$

Using the necessary conditions for optimal control, the costate variables are defined by

$$\dot{\lambda}_1 = a_2\lambda_2 - q_{11}y_1 - q_{12}y_2 \tag{8.4.11}$$

$$\dot{\lambda}_2 = -\lambda_1 + a_1\lambda_2 - q_{12}y_1 - q_{22}y_2 \tag{8.4.12}$$

Since the Hamiltonian is linear in the control, the optimal control is given by

$$u = \begin{cases} u_{\max} & \lambda_2 < 0 \\ u_{\min} & \lambda_2 > 0 \end{cases} \tag{8.4.13}$$

Also the singular control defined by

$$\frac{\partial H}{\partial u} = 0 = \lambda_2 \tag{8.4.14}$$

is a possibility that needs to be investigated. For singular control, $\lambda_2 = 0$ and $\dot{\lambda}_2 = 0$. Under these conditions, the defining relation for the costate variable, λ_2, is

$$\dot{\lambda}_2 = -\lambda_1 - q_{12}y_1 - q_{22}y_2 = 0 \tag{8.4.15}$$

or

$$\lambda_1 = -q_{12}y_1 - q_{22}y_2 \tag{8.4.16}$$

Taking the time derivative, we have

$$\dot{\lambda}_1 = -q_{12}\dot{y}_1 - q_{22}\dot{y}_2 \tag{8.4.17}$$

The defining relation (Equation (8.4.11)) for λ_1 reduces to

$$\dot{\lambda}_1 = -q_{11}y_1 - q_{12}y_2 \tag{8.4.18}$$

Combining Equations (8.4.17) and (8.4.18) yields

$$q_{11}y_1 + q_{12}y_2 = q_{12}\dot{y}_1 + q_{22}\dot{y}_2 \tag{8.4.19}$$

Substituting the definition of the state dynamics into Equation (8.4.19) gives

$$q_{11}y_1 + q_{12}y_2 = q_{12}y_2 - q_{22}a_2y_1 - q_{22}a_1y_2 + q_{22}u \tag{8.4.20}$$

Solving for the singular control, we have

$$u = \left(a_2 + \frac{q_{11}}{q_{22}}\right) y_1 + a_1 y_2 \tag{8.4.21}$$

The singular control is a proportional feedback controller. The question that remains is how can we get on the singular arc? From transversality, we have

$$H(t_f) = 0 \tag{8.4.22}$$

and since

$$\frac{dH}{dt} = \frac{\partial H}{\partial u}\dot{u} \tag{8.4.23}$$

we know that the Hamiltonian must remain constant both on the singular arc and during bang–bang operation. Evaluating the Hamiltonian while on the singular arc, we have

$$H = 0 = \frac{1}{2}\left(q_{11}y_1^2 + q_{22}y_2^2\right) + q_{12}y_1y_2 + \left(-q_{12}y_1 - q_{22}y_2\right) y_2 \tag{8.4.24}$$

which simplifies to

$$\frac{1}{2}\left(q_{11}y_1^2 - q_{22}y_2^2\right) = 0 \tag{8.4.25}$$

This means that either

$$\sqrt{q_{11}}\, y_1 + \sqrt{q_{22}}\, y_2 = 0 \tag{8.4.26}$$

or

$$\sqrt{q_{11}}\, y_1 - \sqrt{q_{22}}\, y_2 = 0 \tag{8.4.27}$$

must hold while the system is under singular control. The relation of Equation (8.4.27) is impossible because it leads to a solution with positive exponentials. This means that Equation (8.4.26) must hold during singular control. This relation is a single line in the phase plane and defines the singular arc. Figure 8.4 is a graphical representation of the solution. It shows the singular arc and the bang–bang switching curves. When the system hits the singular arc, we compute the singular control

$$u = \left(a_2 + \frac{q_{11}}{q_{22}}\right) y_1 + a_1 y_2 \qquad (8.4.28)$$

and determine if it is admissible ($u_{\min} < u < u_{\max}$). If so, we take the singular solution since it always leads to a smaller objective functional, ($\delta u = 0$). If we can't take the singular arc, then we continue with bang–bang control.

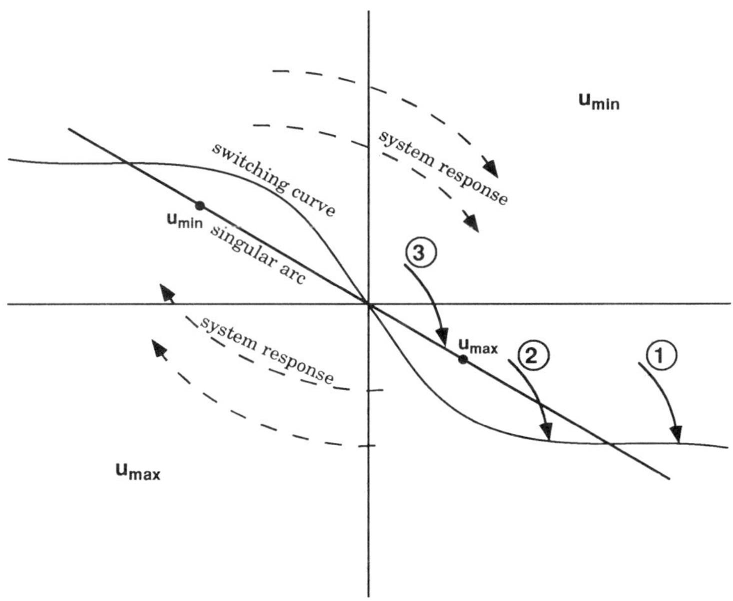

Figure 8.4 Optimal integral square error CSTR control.

Three specific solutions are illustrated. The curve labeled (1) switches from $u_{\min}$ to $u_{\max}$ at the bang–bang switching curve. The curve labeled (2) cannot switch from $u_{\min}$ to the singular arc since the singular control is larger than $u_{\max}$. It also switches at the bang–bang switching curve. The curve labeled (3) switches from $u_{\min}$ to u_{singular} since u_{singular} is allowable.

8.5 Batch Fermentation—A Problem with State Variable Inequality Constraints

To illustrate the methods needed to include state variable inequality constraints into an optimal control problem, let us consider the temperature control of a batch fermentation. In particular, we will consider the substrate model of Engasser *et al.* (1981) modified by Gee and Ramirez (1988). The fermenting medium is assumed to contain three sugars—glucose, maltose, and maltotriose—as the limiting nutrients. The rates of uptake of the sugars follow the Michaelis–Menton enzymatic rate expressions. Maltose is inhibited by glucose, and maltotriose is inhibited by both glucose and maltose. The describing equations are as follows:

Glucose

$$\frac{dG}{dt} = -\mu_1 X \tag{8.5.1}$$

Maltose

$$\frac{dM}{dt} = -\mu_2 X \tag{8.5.2}$$

Maltotriose

$$\frac{dN}{dt} = -\mu_3 X \tag{8.5.3}$$

where

$$\mu_1 = \frac{V_G G}{K_G + G} \tag{8.5.4}$$

$$\mu_2 = \frac{V_M M}{K_M + M}\,\frac{K'_G}{K'_G + G} \tag{8.5.5}$$

$$\mu_3 = \frac{V_N N}{K_N + N}\,\frac{K'_G}{K'_G + G}\,\frac{K'_M}{K'_M + M} \tag{8.5.6}$$

and
G = glucose concentration, gmol/m^3
M = maltose concentration, gmol/m^3
N = maltotriose concentration, gmol/m^3
X = biomass concentration, gmol/m^3
μ_i = specific rate of sugar uptake, h^{-1}
V_i = maximum reaction velocity, h^{-1}
K_i = Michaelis constant, gmol/m^3
K'_i = inhibition constant, gmol/m^3

The rate of temperature change in the fermenting medium is given by

$$\frac{dT}{dt} = \frac{1}{\rho C_p}\left[\Delta H_G \frac{dG}{dt} + \Delta H_M \frac{dM}{dt} + \Delta H_N \frac{dN}{dt} - u(T - T_c)\right] \tag{8.5.7}$$

where u is the fermentor cooling control rate on the system and is equal to

$$u = \frac{UA}{V} \tag{8.5.8}$$

where
C_p = mixture heat capacity, kJ/kg°C
ρ = mixture density, kg/m^3
T = temperature, °C
ΔH_i = heat of fermentation, kJ/gmol
u = fermentor cooling rate, kJ/h m^3 °C
U = overall heat transfer coefficient, kJ/h m^2 °C
A = heat transfer area, m^2
V = volume of fermenting mixture, m^3
T = coolant temperature, °C

The rates of biomass and ethanol production are proportional to the uptakes of the individual sugars and are given in terms of yield coefficients

$$\begin{aligned} X &= X(t_0) + R_{X_G}\Big(G(t_0) - G\Big) + R_{X_M}\Big(M(t_0) - M\Big) \\ &\quad + R_{X_N}\Big(N(t_0) - N\Big) \end{aligned} \tag{8.5.9}$$

$$\begin{aligned} E &= R_{E_G}\Big(G(t_0) - G\Big) + R_{E_M}\Big(M(t_0) - M\Big) \\ &\quad + R_{E_N}\Big(N(t_0) - N\Big) \end{aligned} \tag{8.5.10}$$

where
R_{X_i} = stoichiometric yield of biomass
R_{E_i} = stoichiometric yield of ethanol

The reaction rate parameters are assumed to follow the Arrhenius temperature dependency

$$V_i = V_{i0} \exp\left[-E_{V_i}/R(T + 273.16)\right] \tag{8.5.11}$$

$$K_i = k_{i0} \exp\left[-E_{K_i}/R(T + 273.16)\right] \tag{8.5.12}$$

$$K_i' = k_{i0}' \exp\left[-E_{K_i'}/R(T + 273.16)\right] \tag{8.5.13}$$

As the optimal control problem, Gee and Ramirez considered obtaining a maximum final ethanol concentration within a minimum batch time. Therefore, the objective functional to be minimized is

$$J = -E(t_f) - w \int_0^{t_f} dt \tag{8.5.14}$$

where w is a weighting factor. Expressing the model equations in state variable form gives

$$\dot{\boldsymbol{x}} = \boldsymbol{f}(\boldsymbol{x}, \dot{\boldsymbol{x}}, \boldsymbol{u}, t) \tag{8.5.15}$$

where

$$x_1 = G \qquad x_2 = M \qquad x_3 = N \qquad x_4 = T \tag{8.5.16}$$

Note the presence of $\dot{\boldsymbol{x}}$ on the RHS of Equation (8.5.15). This means that we will need to use the original Lagrangian and not the Hamiltonian approach to derive the necessary conditions for optimality. The objective functional becomes

$$J = \sum_{i=1}^{3} R_{E_i} \left(x_i(t_f) - x_i(0) \right) + w \int_0^{t_f} dt \tag{8.5.17}$$

There are both state variable and control variable inequality constraints that must be considered. There are minimum and maximum values for the temperature as well as minimum and maximum values for the control. These inequality constraints are expressed as

$$T \leq T_{\max} \tag{8.5.18}$$

and

$$u_{\min} \leq u \leq u_{\max} \tag{8.5.19}$$

Pontryagin's minimum principle handles the control variable inequality constraints. To include the state variable inequality constraint, we will use the penalty function method discussed in Section 4.3. An augmented performance functional which includes the penalty function is

$$J' = \sum_{i=1}^{3} R_{E_i} \left(x_i(t_f) - x_i(t_0) \right) - \int_0^{t_f} L \, dt \tag{8.5.20}$$

where L is the Lagrangian defined by

$$L = w + \boldsymbol{\lambda}^T \left(\boldsymbol{f} - \dot{\boldsymbol{x}} \right) + Q \left(x_4 - T_{\max} \right)^2 U \left(x_4 - T_{\max} \right) \tag{8.5.21}$$

$U(x_4 - T_{\max})$ is the unit step function defined by

$$U \left(x_4 - T_{\max} \right) = \begin{cases} 0 & \text{if} \quad x_4 \leq T_{\max} \\ 1 & \text{if} \quad x_4 > T_{\max} \end{cases} \tag{8.5.22}$$

and Q is a large positive weighting constant. The necessary conditions for minimizing the performance functional are as follows:

Costate Equations

The general Euler–Lagrange equation was given in Section 4.1 as

$$\frac{\partial L}{\partial x} - \frac{d}{dt} \frac{\partial L}{\partial \dot{x}} = 0 \tag{8.5.23}$$

For this problem, f_4 is

$$f_4 = \frac{\Delta H_1}{\rho C_p}\,\dot{x}_1 + \frac{\Delta H_2}{\rho C_p}\,\dot{x}_2 + \frac{\Delta H_3}{\rho C_p}\,\dot{x}_3 + \frac{u(T - T_c)}{\rho C_p} \tag{8.5.24}$$

so that the defining costate equations are

$$\dot{\lambda}_1 = \frac{\Delta H_1}{\rho C_p}\,\dot{\lambda}_4 - \sum_{i=1}^{3} \lambda_i \frac{\partial f_i}{\partial x_1} \tag{8.5.25}$$

$$\dot{\lambda}_2 = \frac{\Delta H_2}{\rho C_p}\,\dot{\lambda}_4 - \sum_{i=1}^{3} \lambda_i \frac{\partial f_i}{\partial x_2} \tag{8.5.26}$$

$$\dot{\lambda}_3 = \frac{\Delta H_3}{\rho C_p}\,\dot{\lambda}_4 - \sum_{i=1}^{3} \lambda_i \frac{\partial f_i}{\partial x_3} \tag{8.5.27}$$

$$\begin{aligned}\dot{\lambda}_4 = {} & -\sum_{i=1}^{4} \lambda_i \frac{\partial f_i}{\partial x_4} - 2Q\left(x_4 - T_{\max}\right) U\left(x_4 - T_{\max}\right) \\ & + Q\left(x_4 - T_{\max}\right)^2 \delta\left(x_4 - T_{\max}\right)\end{aligned} \tag{8.5.28}$$

The last term of Equation (8.5.28) comes from taking the derivative of the step function, which yields the Dirac delta function, δ.

Transversality

The general transversality condition is

$$\left[\frac{\partial h}{\partial \boldsymbol{x}}(t_f) - \boldsymbol{\lambda}^T(t_f)\right] \delta \boldsymbol{x}_f + H(t_f)\,\delta t_f = 0 \tag{8.5.29}$$

which gives, for a free final state and final time

$$\delta \boldsymbol{x}_f \neq 0 \qquad\qquad \delta t_f \neq 0 \tag{8.5.30}$$

where the Hamiltonian is

$$H = L + \boldsymbol{\lambda}^T \dot{\boldsymbol{x}} \tag{8.5.31}$$

Since the final time and final state are independent, we have

$$\lambda_1(t_f) = R_{E_1} \tag{8.5.32}$$

$$\lambda_2(t_f) = R_{E_2} \tag{8.5.33}$$

$$\lambda_3(t_f) = R_{E_3} \tag{8.5.34}$$

$$\lambda_4(t_f) = 0 \tag{8.5.35}$$

and the final time is given by the condition

$$H(t_f) = 0 \tag{8.5.36}$$

Optimal Control

The optimal control is given by minimization of the Hamiltonian with respect to the control, which yields

$$\lambda_4 > 0 \quad \text{then} \quad u = u_{\max} \tag{8.5.37}$$

$$\lambda_4 < 0 \quad \text{then} \quad u = u_{\min} \tag{8.5.38}$$

$$\lambda_4 = 0 \quad \text{and stays zero; then} \quad u = u_{\text{sing}} \tag{8.5.39}$$

We want to investigate the singular control policy further. Combining Equation (8.5.39) and (8.5.28), we have

$$\begin{aligned} 0 &= -\sum_{i=1}^{3} \lambda_i \frac{\partial f_i}{\partial x_4} - 2Q\left(x_4 - T_{\max}\right) U\left(x_4 - T_{\max}\right) \\ &\quad + Q\left(x_4 - T_{\max}\right)^2 \delta\left(x_4 - T_{\max}\right) \end{aligned} \tag{8.5.40}$$

The nontrivial solution to Equation (8.5.40) is isothermal operation at the maximum temperature, that is

$$x_4 = T_{\max} \tag{8.5.41}$$

Along an isothermal path, $\partial f_i / \partial x_4 = 0$. From the temperature state equation, Equation (8.5.7), we have

$$u_{\text{sing}} = \frac{1}{\left(T_{\max} - T_c\right) \rho C_p} \left(\Delta H_1 \dot{x}_1 + \Delta H_2 \dot{x}_2 + \Delta H_3 \dot{x}_3\right) \tag{8.5.42}$$

The following algorithm can be used to find the optimal control for this problem:

- A constant control is assumed as $u = u_{\max}$.
- The state equations are integrated forward in time until the final time is reached, as indicated by Equation (8.5.36).
- The costate equations are integrated backward in time from the final time to the initial time. The new control action is calculated based upon the sign of the fourth costate variable (Equations (8.5.37) to (8.5.39)). When the penalty function effect was detected ($x_4 > T_{\max}$), the singular control policy of Equation (8.5.42) was implemented, and λ_4 was set to zero. Along the singular arc, it is possible that the control constraint could be violated. In that case, the control variable is set to $u_{\max}$.
- The control policy of step 3 is implemented by integrating the states as in step 2.

- Steps 3 and 4 are repeated until the control policy converges to the optimal control.

Gee and Ramirez give the optimal results for the operation of a 100 liter fermentor. Table 8.2 gives the parameters used in their study. The results for the temperature state variable are shown in Figure 8.5 and for the control in Figure 8.6. As shown by the case when $u_{max} = 40$, the optimal control policy is to let $u = u_{max}$ until the maximum allowable temperature is attained. At this point, the singular control policy is initiated. If operation along the singular arc violates the maximum cooling, then, as shown by the case when $u_{max} = 30$, the system switches earlier and operates along the u_{max} extremal control for a period before operation on the singular arc.

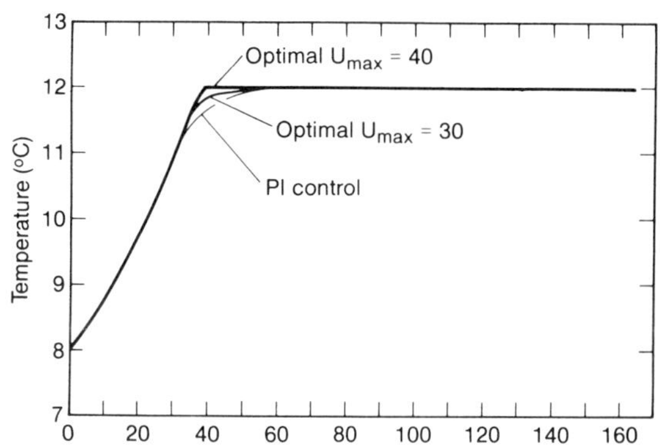

Figure 8.5 Temperature profiles for three cases. Reprinted from Gee and Ramirez, *Biotech. & Bioengr.* **31**, 224–234, ©1988, John Wiley and Sons.

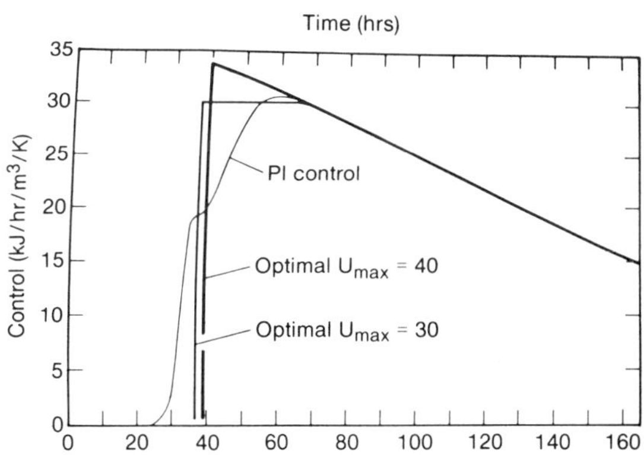

Figure 8.6 Controller profiles for three cases. Reprinted from Gee and Ramriez, *Biotech. & Bioeng.* **31**, 224–234, ©1988, John Wiley and Sons.

Table 8.2

Reaction Parameters

Sugar	$R_{E/i}$	$R_{X/i}$	ΔH_{R_i} kJ/mol
Glucose	1.92	0.134	− 91.2
Maltose	3.84	0.268	−226.3
Maltotriose	5.76	0.402	−361.3

Arrhenius Constants

Parameter	Activation Energy (kcal/mol)	Natural Logarithm of Frequency Factor
V_G	22.6	35.77
V_M	11.3	16.40
V_N	7.16	10.59
K_G	−68.6	121.3
K_M	−14.4	− 19.15
K_N	−19.9	− 26.78
K'_G	10.2	23.33
K'_M	26.3	55.61

Initial Conditions and Equipment Parameters

$G(t_0)$	=	70 mol/m^3	$X(t_0)$	=	125 mol/m^3
$M(t_0)$	=	220 mol/m^3	$T(t_0)$	=	8°C
$N(t_0)$	=	40 mol/m^3			
A	=	0.188 m^2	$u_{\min}$	=	0 J/m^{3}h °C
C_p	=	4016 J/kg °C	$u_{\max}$	=	4 × 10^4 J/m^3 h °C
ρ	=	1040 K g/m^3	V	=	0.1 m^3

Table 8.3
Effect of Weighting Factor w on Optimal Fermentation

w	$E(t_f)(\times 10^{-3}$ mol/m$^3)$	t_f (h)	J' $(\times 10^{-2})$
0.1	1.20	314	−11.7
0.5	1.18	236	−10.7
1.0	1.16	203	−9.58
2.0	1.11	171	−7.73
5.0	0.957	123	−3.40
6.0	0.895	112	−2.22
7.0	0.287	36.1	−0.346
8.0	0	0	0

The weighting factor, w, is the most important variable affecting the final product of the optimal fermentation. Typical values of $E(t_f)$ are on the order of 10^3 gmol/m, while the final time is usually on the order of 10^2 hours in magnitude. Therefore, for small values of w (i.e., $w = 0$ to $w = 1$), the effect of ethanol in the objective functional is considerably more important than that of the final time. For these weightings, the computed optimal control results in high final ethanol concentrations and long batch times. For $w = 0.1$, the fermentation proceeds essentially to completion. Table 8.3 gives final ethanol concentrations and final times for various values of w. Once w gets large enough ($w = 8$), the final time portion of the objective functional becomes so important that there is no fermentation operation. Therefore, the choice of the weighting factor must be made based upon the relative importance of the batch time and the sugar conversion to ethanol.

8.6 Optimal Production of Secreted Foreign Protein—Multiple Singular Arcs

The emerging field of biotechnology offers the opportunity to formulate and solve new and important optimal control problems. One area where optimal control theory can be successfully applied is that of batch and fed-batch reactor operation. Genetic engineering has given us the ability to modify the genetic code of microorganisms so that foreign substances can be produced by bacteria and yeast. One class of foreign substances is that of proteins. Most foreign proteins must be secreted out of the cell in order for them to be active. Secretion pathways and the major secretion macromolecular events have been identified by Schekman and his associates (Novik *et al.*, 1980). Park and Ramirez (1989) have determined and verified a mathematical model for the dynamics of production secreted proteins. Park and Ramirez (1988) have also treated the optimal production of secreted protein in fed-batch reactors. A fed-batch reactor is one where the yeast is initially placed in the reactor, but the growth medium (the sugar source) is fed continuously to the reactor. The mathematical model for this problem involves a material balance for the secreted protein concentration, the total protein concentration, the yeast concentration or biomass concentration, and the sugar concentration, as well as an overall material balance to determine the volume of the reactor. The model of Park and Ramirez (1989) is

$$\dot{P}_M = \phi(P_T - P_M) - \frac{q}{V} P_M \tag{8.6.1}$$

$$\dot{P}_T = f_p X - \frac{q}{V} P_T \tag{8.6.2}$$

$$\dot{X} = \mu_x X + \frac{q}{V} X \tag{8.6.3}$$

$$\dot{S} = -Y\mu_x X + \frac{q}{V}(m - S) \tag{8.6.4}$$

$$\dot{V} = q \quad \text{where} \quad q_{\min} \leq q \leq q_{\max} \tag{8.6.5}$$

where

$$\phi = \text{the protein secretion rate} = \frac{4.75\ \mu_x}{0.12 + \mu_x} \quad ; \quad 1/\text{hr} \tag{8.6.6}$$

$$f_p = \text{the protein synthesis rate} = \frac{Se^{-5.0\ S}}{0.1 + S} \quad ; \quad 1/\text{hr} \tag{8.6.7}$$

$$\mu_x = \text{the growth rate} = \frac{21.87\ S}{(S + 0.4)(S + 62.5)} \quad ; \quad 1/\text{hr} \tag{8.6.8}$$

and where
P_M = the secreted protein concentration
P_T = the total produced protein
X = the yeast or biomass concentration
S = the glucose concentration
V = the volume of the fermentor
q = the feed flow rate to the fermentor
m = glucose concentration of the feed
Y = the yield of glucose (g/g cell mass)

We have five state variables and a single control variable. The control for the system is the feed flow rate of the glucose concentration to the reactor, q. The objective of the optimal control problem is to determine the feed flow rate policy that will result in the maximization of the total secreted protein produced at the end of running the fed-batch reactor. The final time is when the reactor is full. This objective can be expressed by the minimization of the following functional:

$$J = -P_M(t_f)\, V(t_f) \tag{8.6.9}$$

This is a terminal cost problem. The necessary conditions for minimization of this performance functional are given by Pontryagin's minimum principle and are expressed in terms of the Hamiltonian, which is, for this problem

$$\begin{aligned} H &= \lambda_1 \left(\phi(P_T - P_M) - \frac{q}{V} P_M\right) + \lambda_2 \left(f_p X - \frac{q}{V} P_T\right) + \lambda_3 \left(\mu_x X - \frac{q}{V} X\right) \\ &\quad + \lambda_4 \left(-Y\mu_x X + \frac{q}{V} m - \frac{q}{V} S\right) + \lambda_5 q \end{aligned} \tag{8.6.10}$$

Combining terms, we have

$$\begin{aligned} H &= \lambda_1 \phi (P_T - P_M) + \lambda_2 f_p X + \lambda_3 \mu_x X - \lambda_4 Y \mu_x X \\ &\quad + \frac{q}{V} \left[-\lambda_1 P_M - \lambda_2 P_T - \lambda_3 X + \lambda_4 (m - S) + \lambda_5 V\right] \end{aligned} \tag{8.6.11}$$

The necessary conditions for optimality are as follows:

Costate Equations

$$\dot{\lambda}_1 = \lambda_1 \phi + \lambda_1 \frac{q}{V} \tag{8.6.12}$$

$$\dot{\lambda}_2 = -\lambda_1 \phi + \lambda_2 \frac{q}{V} \tag{8.6.13}$$

$$\dot{\lambda}_3 = -\lambda_2 f_p - \lambda_3 \mu_x + Y \lambda_4 \mu_x + \lambda_3 \frac{q}{V} \tag{8.6.14}$$

$$\begin{aligned} \dot{\lambda}_4 &= -\lambda_1 \frac{\partial \phi}{\partial S} (P_T - P_M) - \lambda_2 \frac{\partial f_p}{\partial S} X - \lambda_3 \frac{\partial \mu_x}{\partial S} X \\ &\quad + Y \lambda_4 \frac{\partial \mu_x}{\partial S} X + \lambda_4 \frac{q}{V} \end{aligned} \tag{8.6.15}$$

$$\dot{\lambda}_5 = -\frac{q}{V^2} (\lambda_1 P_M + \lambda_2 P_T + \lambda_3 X + \lambda_4 S - \lambda_4 m) \tag{8.6.16}$$

Transversality Conditions

The Transversality Conditions for this fixed final time, free end point problem are

$$\lambda_1(t_f) = -V(t_f) \tag{8.6.17}$$

$$\lambda_2(t_f) = 0 \tag{8.6.18}$$

$$\lambda_3(t_f) = 0 \tag{8.6.19}$$

$$\lambda_4(t_f) = 0 \tag{8.6.20}$$

$$\lambda_5(t_f) = -P_M(t_f) \tag{8.6.21}$$

Optimal Control

The Hamiltonian is linear with respect to the control variable and can be represented as

$$H = \psi(\boldsymbol{x}, \boldsymbol{\lambda}) + q H_q(\boldsymbol{x}, \boldsymbol{\lambda}) \tag{8.6.22}$$

where

$$\psi = \lambda_1 \phi (P_T - P_M) + \lambda_2 f_p X + \lambda_3 \mu_x X - \lambda_4 Y \mu_x X \tag{8.6.23}$$

$$H_q = \frac{1}{V} \Big(-\lambda_1 P_M - \lambda_2 P_T - \lambda_3 X + \lambda_4 (m - S) + \lambda_5 V \Big) \tag{8.6.24}$$

The minimum principle for the optimal control requires the following:

$$\text{if } H_q \geq 0 \quad \text{then } q = q_{\min} \tag{8.6.25}$$

$$\text{if } H_q \leq 0 \quad \text{then } q = q_{\max} \tag{8.6.26}$$

$$\text{if } H_q = 0 \quad \text{and stays at zero, then } q = q_{\text{sing}} \tag{8.6.27}$$

We need to investigate the existence of the singular arc. For H_q to be zero and stay zero, then

$$\frac{dH_q}{dt} = 0 \tag{8.6.28}$$

Using the expressions for the state and costate variables, we can, after algebraic manipulations, evaluate the time derivative of the function, H_q, as

$$\frac{dH_q}{dt} = -\frac{1}{V} \left(\lambda_1 \frac{\partial \phi}{\partial S} (P_T - P_M) + \lambda_2 \frac{\partial f_p}{\partial S} X + (\lambda_3 - Y \lambda_4) \frac{\partial \mu_x}{\partial S} X \right) (m - S) \tag{8.6.29}$$

Since $(m-s) \neq 0$, the singular arc will be the solution of the following algebraic equation:

$$\lambda_1 \frac{\partial \phi}{\partial S} (P_T - P_M) + \lambda_2 \frac{\partial f_p}{\partial S} X + (\lambda_3 - Y\lambda_4) \frac{\partial \mu_x}{\partial S} X = 0 \qquad (8.6.30)$$

Park and Ramirez have investigated the roots of this equation and have shown that each one of the terms of Equation (8.6.30) dominate the solution over a specific time interval of operation. This means that there are actually three solutions to Equation (8.6.30) or multiple singular arcs. These three singular arcs are shown in Figure 8.7. The last term of Equation (8.6.30) dominates initially and maximizes the growth rate of the yeast. A high sugar concentration is required in order to realize this maximum growth rate. The middle term dominates during the next time period. This term maximizes the protein production rate and requires a low sugar concentration. Finally, close to the end of operation, the first term of Equation (8.6.30) controls. This term maximizes the protein secretion rate. Since protein secretion is a function of metabolic activity, it requires a high sugar concentration. This means we have multiple singular arcs and that we have to move from one to the other. To stay on the singular arc requires a sudden change in the sugar concentration in the fermentor. Due to the bounded permissible control action in the range, $[q_{\min}, q_{\max}]$, abrupt transitions in the fermentor sugar concentration are not physically possible. For an abrupt change in sugar concentration to occur requires infinite flow rate or selective removal of sugar from the fermentor. The only thing that can be done is to go to extremal control.

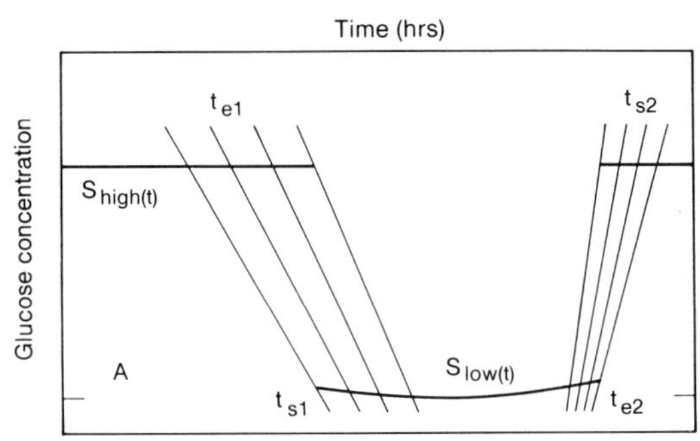

Figure 8.7 Singular arc glucose concentration (heavy lines) and trajectories constituted by the bang–bang control during the transitions (light lines). From Park and Ramirez (1988).

The boundedness of the permissible control action forces the optimal trajectory to deviate from the singular arc during transitions in culture sugar concentration. Therefore, H_q will deviate from its position at zero. There are an infinite number of possible transitions between the singular arcs through the bang–bang control mode as shown in Figure 8.7. The optimal trajectory will be the one that strictly observes the minimum principle during the transition. Because $H_q(t) = 0$ along all singular arcs, the optimal transition between a transition, $[t_1, t_2]$, will have the following properties:

$$\int_{t_1}^{t_2} \frac{dH_q}{dt}\, dt = 0 \tag{8.6.31}$$

and for any time between $[t_1, t_2]$

$$q = q_{\max} \quad \text{if} \quad \int_{t}^{t_2} \frac{dH_q}{dt}\, dt \geq 0 \tag{8.6.32}$$

$$q = q_{\min} \quad \text{if} \quad \int_{t}^{t_2} \frac{dH_q}{dt}\, dt \leq 0 \tag{8.6.33}$$

The following algorithm can be used to determine the optimal control policy.

- For an initial assumed control policy, the state equations are integrated forward in time until the final time is reached.

- The costate equations are then integrated backward in time from t_f, and the singular arc sugar concentration that is a solution to Equation (8.6.30) is calculated. When multiple solutions are obtained, the one that minimizes the Hamiltonian of Equation (8.6.22) is chosen. The control action is calculated from Equation (8.6.4). When the singular arc control action exceeds the control constraints, the actual control is constrained at the constraint value. Bang–bang control is also required when the singular trajectory requires a sudden transition in sugar concentration.

- During backward integration, the integral of dH_q/dt is calculated and used to find the optimal transition between singular arcs. To locate the optimal transition times, t_{s1} and t_{s2} in Figure 8.8, a gradient algorithm is used

$$t_{s2,\text{new}} = t_{s2,\text{old}} - K_2 \int_{t_{e2}}^{t_{s2}} \frac{dH_q}{dt}\, dt \qquad K_2 \geq 0 \tag{8.6.34}$$

$$t_{s1,\text{new}} = t_{s1,\text{new}} + K_1 \int_{t_{e1}}^{t_{s1}} \frac{dH_q}{dt}\, dt \qquad K_1 \geq 0 \tag{8.6.35}$$

- If t_{s2} exceeds t_f, bang–bang control is active at the final time. However, due to the transversality conditions, H_q is still zero at the final time. To find the optimal final sugar concentration at the final time, we use the following gradient method:

$$S(t_f)_{\text{new}} = S(t_f)_{\text{old}} + K_S \int_{t_{e2}}^{t_f} \frac{dH_q}{dt}\, dt \qquad K_S \geq 0 \tag{8.6.36}$$

- With an updated control strategy, the first four steps are repeated until convergence is obtained.

Park and Ramirez consider the system with the following parameters:

$$
\begin{aligned}
P_M(t_0) &= 0 & \qquad P_t(t_0) &= 0 \\
X(t_0) &= 1.0 \text{ g/L} & \qquad S(t_0) &= 5.0 \text{ g/L} \\
V(t_0) &= 1.0 \text{ L} & \qquad m &= 20.0 \text{ g/L} \\
0 < q &< 10.0 \text{ L/hr} & \qquad t_f &= 7.5 \text{ hr}
\end{aligned}
$$

The optimal control results for this case are given in Figure 8.8. For this system, the optimal control divides the fed-batch operation into three consecutive phases which are clearly seen in the optimal sugar concentration profile of Figure 8.8C. The first phase, during which the glucose level is maintained at the value supporting maximum specific growth rate, is dedicated for growing the host cell mass in the culture. The next phase is for expression of the foreign protein. During expression, secretion is also taking place as seen in Figure 8.8A. Close to the final time, the last singular arc phase, the optimal secretion rate, activates. For a given reactor volume, the final time, for which optimal operation can utilize the full capacity of the reactor, can be uniquely determined.

References

Bellman, R., Glicksburg, I., and Gross, I., "On the Bang–Bang Control Problem," *Quart. Appl. Math.* **14**, 11–18 (1956).

Denn, M. M., *Optimization by Variational Methods*, McGraw–Hill, New York (1969).

Engasser, J. M., Marc, I., Moll., and Duteertre, B., *EBC Proceedings*, 579 (1981).

Gee, D. A. and Ramirez, W. F., "Optimal Temperature Control for Batch Beer Fermentation," *Biotech. & Bioengr.* **31**, 224–234 (1988).

Hong, J., "Optimal Substrate Feeding Policy for a Fed Batch Fermentation with Substrate and Product Inhibition Kinetics," *Biotech. and Bioengr.* **28**, 1421–1431 (1986).

Jackson, R., *Chem. Eng. Sci.* **21**, No. 3, 241 (1966).

Kelley, H. J., *J. SIAM Control* **2**, 234–240 (1965).

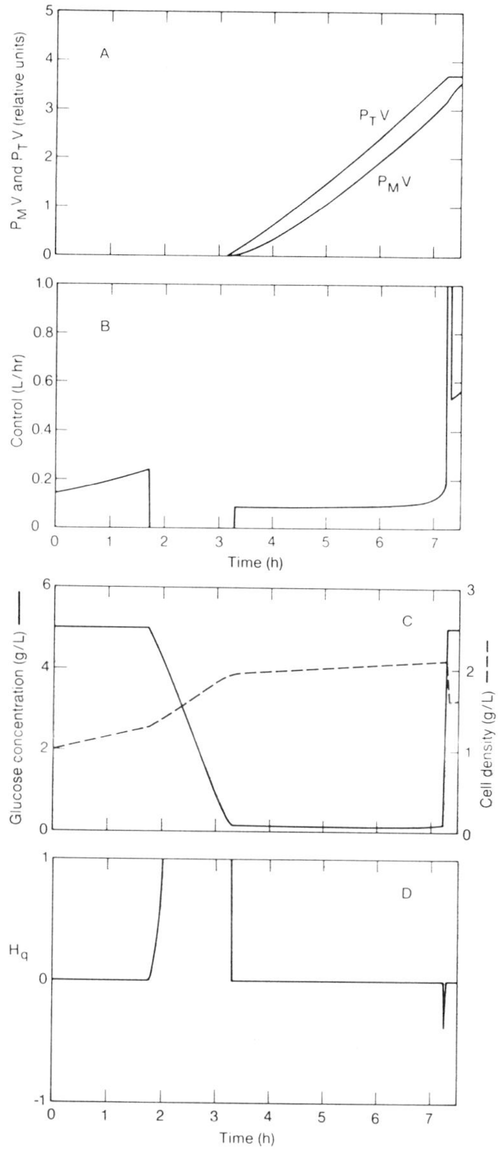

A. Optimal total expressed SUC2–s2 level, $P_T(t)$, and optimal secreted SUC2–s2 level, $P_M(t)$.

B. Optimal control function (feed rate) with $q_{\max} = 10.0$ L/h and $q_{\min} = 0.0$ L/h.

C. Optimal glucose concentration and cell density.

D. $H_q(t)$ vs. time.

Figure 8.8 Optimal trajectories with final time of 7.5. From Park and Ramirez (1988).

LaSalle, J. P., "The Time Optimal Control Problem," in *Contributions to Differential Equations, Vol. 5*, Princeton University Press, Lawrenceville, NJ, pp. 1–24 (1960).

Novik, P., Field, C., and Schekman, R., "Identification of 23 Complementation Groups Required for Post–translational Events in the Yeast Secretory Path," *Cell* **21**, 205 (1980).

Nyquist, J. K. and Ramirez, W. F., "Time Optimal Control of an Experimental Continuous Stirred Tank Reactor," *Chem. Eng. Sci.* **26**, No. 10, 1673–1688 (1971).

Park, S. and Ramirez, W. F., "Optimal Production of Secreted Protein in Fed-Batch Reactors," *AIChE Journal* **39**, 1550–1558 (1988).

Park, S. and Ramirez, W. F., "Dynamics of Foreign Protein Secretion from *Saccharomyces cerevisiae*, *Biotech. & Bioengr.* **33**, 272–281 (1989).

Pontryagin, L. S., Boltyanskii, V. G., Gamkrelidze, R. V., and Mishchenko, E. F., *The Mathematical Theory of Optimal Processes*, Wiley and Sons, New York (1962).

Sage, A. P. and White, C. C., III, *Optimmum Systems Control*, Second Edition, p. 108, Prentice–Hall, Englewood Cliffs, NJ (1977).

Exercises

1. Optimum start-up procedure for an autothermic reactor system has been treated by Jackson (1966):

Reaction $A \rightleftarrows B$

System

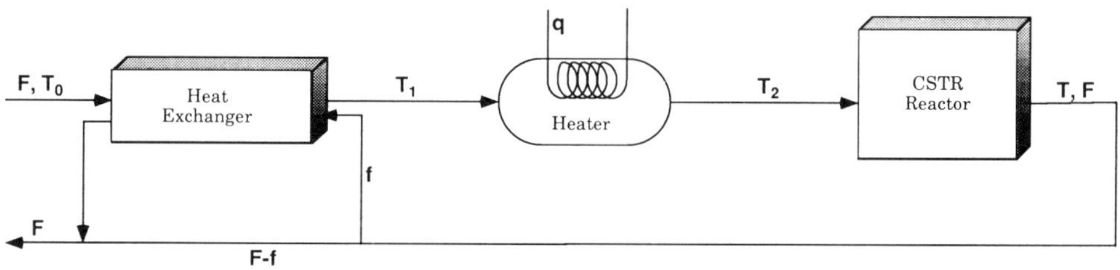

Mass Balance (Product)

$$\frac{dy}{dt} = R(y, T) - y$$

y = concentration
R = reaction rate

Energy Balance

$$\gamma \frac{dT}{dt} = q - u(T - T_o) + (-\Delta H)\ R(y, T)$$

γ = constant
T = reactor temperature
q = heat rate supplied by heater
$u = \frac{f}{F}$ = fraction of total flow through heat exchanger
$(T - T_o)$ is always greater than zero

Control Variable Constraints

$$u_{min} \leq u \leq 1$$
$$0 \leq q \leq q \max$$

The starting or initial condition is $T = 300°\text{K}$ and $y = 0$. This is a steady-state condition with $u = 1$ and $q = 0$.

The desired operating condition is the steady-state condition that is a maximum in the product composition. From steady-state optimization (function optimization), this condition is shown to be $y_s = .7$ and $T_s =$ 800°K (see the figure and the locus of steady states).

- We want to determine a start-up control policy that will take the system from its initial state to the final desired steady state so that the loss in profit by not operating at the optimal steady-state condition is minimized. This results in the following functional to be minimized:

$$J = \int_0^{t_f} [(y_s - y) - c(q_s - q)] \; dt$$

- What are the necessary conditions for optimal control?
- Study the possibility for singular or intermediate control. Compute the intermediate control law if it can exist. Forming a ratio of the state equations might aid in the calculation.
- How must you approach the optimum steady state? Note in the figure that the region above the curve $R = 0$ is not physically realizable and that various extremal controls which come into the desired steady state come from the unattainable region.
- What is the maximum length of time that you can stay on the singular arc?
- Discuss the computational strategy for determining the exact control policy to move from the given initial state to the desired final state minimizing this performance index.

2. In this problem we want to study the time optimal control of a stirred tank reactor with both cooling and feed flow controls.

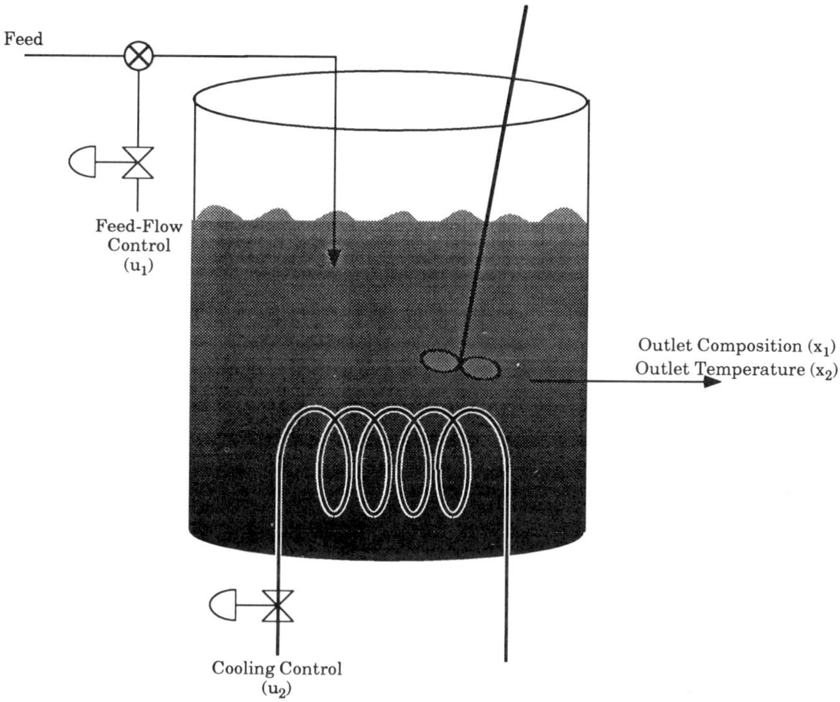

A linearized model can be expressed as

$$\dot{x}_1 = a_{11}\; x_1 + a_{12}\; x_2 + b_{11}\; u_1$$
$$\dot{x}_2 = a_{21}\; x_1 + a_{22}\; x_2 + b_{21}\; u_1 + b_{22}\; u_2$$

After a linear transformation, this problem can be expressed as

$$\dot{y}_1 = S_1\; y_1 - M_{11}\; u_1 - M_{12}\; u_2$$
$$\dot{y}_2 = S_2\; y_2 + M_{21}\; u_1 + M_{12}\; u_2$$

The following constraints exist for the controls:

$$u_{1*} \leq u_1 \leq u_1^*$$
$$u_{2*} \leq u_2 \leq u_2^*$$

- Develop the necessary conditions for the time optimal control of this system.
- Give analytical solutions to the costate equations.

- Is singular control possible?

3. The system

$$\ddot{x} + a\dot{x} + bx = u$$

where u = control variable and has an upper limit of u^* and a lower limit of u_*, is to be controlled to minimize an integral of time times error criterion:

$$J = \frac{1}{2}\int_0^{t_f} t(c_1\, x^2 + c_2\, \dot{x}^2)\, dt$$

- Develop the optimal control for this system.
- Consider whether intermediate (not bang–bang) control is ever possible.
- Can a feedback law be formulated on a singular arc?

4. The dynamics and control of pH in stirred vessels have been treated extensively in the literature. The state equation is

$$V\dot{\gamma} = F_1 C_1 - \left(1 + \frac{10^{-7}}{Ka}\right) F_2 C_2 - (F_1 + F_2)\,\gamma$$

where γ is a combined concentration and F_1 is the inlet flow which is the control variable.

Develop the time optimal control policy when

$$F_{min} \leq F_1 \leq F_{max}$$

Examine the possibility of intermediate control.

5. A first order, reversible, exothermic chemical reaction takes place in a plug–flow tubular reactor. The state equations for concentration, x_1, and temperature, x_2, are

$$\begin{aligned} \dot{x}_1 &= R(x_1, x_2) & x_1(t_o) &= x_1^0 \\ \dot{x}_2 &= JR - u(x_2 - T_c) & x_2(t_o) &= x_2^0 \end{aligned}$$

The reaction rate expression is

$$R(x_1, x_2) = (1 - x_1)k_{10} \exp\left(-\frac{\gamma_1}{x_2}\right) - x_1 k_{20} \exp\left(-\frac{\gamma_2}{x_2}\right)$$

t_f is given for this problem (the reactor length is known).

The control is bounded by $0 = u_{min} \leq u \leq u_{max}$.

We want to obtain the control strategy so that $x_1(t_f)$ is maximized.

Be sure to consider both extremal and intermediate control possibilities. You should be able to show that on the singular arc $\frac{\partial R}{\partial x_2} = 0$.

6. A transformation due to Kelley (1965) is useful for calculating singular arcs for problems with a single control variable. For problems of the form

$$x_i = p_i + g_i u \qquad i = 1, \cdots n$$

we can transform the problem into a phase state that does not explicitly involve the control variable. The phase variables, z_j, are such that

$$z_j = r_j \qquad j = 1, \cdots m$$

We choose the functions, r_j, so that

$$\dot{z}_j = -\sum_{i=1}^{n} \frac{\partial r_j}{\partial x_i} p_i + \left(\sum_{i=1}^{n} \frac{\partial r_j}{\partial x_i} g_i \right) u + \frac{\partial r_j}{\partial t} \qquad j = 1, \cdots m$$

do not contain the control u. To do this, we set

$$\sum_{i=1}^{n} \frac{\partial r_j}{\partial x_i} g_i = 0 \qquad j = 1, \cdots m \tag{1}$$

Solving this equation by the method of characteristics gives

$$\frac{dx_i}{ds} = g_i \qquad i = 1, \cdots n$$

If one of the quantities, x_i, is adopted instead of s as the independent variable, the general solution is represented in terms of the $(n-1)$ parameters

$$C_k = \phi_k \qquad k = 1, \cdots n-1$$

where C_k are constants of integration and ϕ_k are the mutually independent integrals of the solution to (1).

The first $(n-1)$ variables, z_j, are defined according to

$$r_j = \phi_j$$

and the nth is $z_n = x_\ell$.

Hong (1986) shows how Kelley's transformation can be applied to the control of a fed-batch fermenter. The original model is

$$\begin{aligned}\frac{dx}{dt} &= x\mu - \frac{ux}{V}\\ \frac{dP}{dt} &= x\epsilon - \frac{uP}{V}\\ \frac{dV}{dt} &= u\end{aligned}$$

Show that Kelley's transformation yields

$$\begin{aligned}z_1 &= \ln(xV)\\ z_2 &= \ln(PV)\\ z_3 &= V\end{aligned}$$

Also show that the singular arc is given by

$$\frac{\partial}{\partial z_3}\left(\frac{\mu}{\epsilon}\right) = 0$$

Differentiate this expression with respect to time to obtain the control policy along the singular arc.

Chapter 9

THE DISCRETE MAXIMUM PRINCIPLE

9.1 Introduction

In implementing optimal control problems using digital computers, we keep the control constant over a period of time. Problems that were originally described by differential equations defined over a continuous time domain are transformed to problems that are described by a set of discrete algebraic equations that are of the explicit form

$$\boldsymbol{x}_{n+1} = \boldsymbol{f}_n(\boldsymbol{x}_n, \boldsymbol{u}_n, n) \tag{9.1.1}$$

In solving distributed parameter control problems by finite difference methods, problems that were originally described by partial differential equations continuous in the time and spatial domains are transformed to problems described either by algebraic equations of the explicit form or the implicit form

$$\boldsymbol{f}_n(\boldsymbol{x}_{n+1}, \boldsymbol{x}_n, \boldsymbol{u}_n, n) = \boldsymbol{0} \tag{9.1.2}$$

We will develop the necessary conditions for extrema of a discrete performance function of the form

$$J = \sum_{n=0}^{N-1} F_n(\boldsymbol{x}_n, \boldsymbol{u}_n, n) \tag{9.1.3}$$

subject to the equality constraints of Equations (9.1.1) or (9.1.2) using the discrete maximum principle of Pontryagin (Denn, 1969; Sage and White, 1977).

9.2 Necessary Conditions for Explicit Models

We will consider the discrete state system to be defined by

$$\boldsymbol{x}_{n+1} = \boldsymbol{f}_n(\boldsymbol{x}_n, \boldsymbol{u}_n, n) \qquad n = 0 \cdots N-1 \tag{9.2.1}$$

with $\boldsymbol{u} \in U$ which is a subset of R^m.

We want to find the admissible sequence of controls, $\boldsymbol{u}_n$, which will extremize the performance function

$$J = \sum_{n=0}^{N-1} F_n(\boldsymbol{x}_n, \boldsymbol{u}_n, n) \tag{9.2.2}$$

subject to the equality constraints of Equation (9.2.1).

We will use a variational method to develop the necessary conditions for an extremum. First, we form an augmented performance function that includes the equality state equation constraints

$$J_A = \sum_{n=0}^{N-1} \left(F_n\left(\boldsymbol{x}_n, \boldsymbol{u}_n, n\right) - \boldsymbol{\lambda}_{n+1}^T \left[\boldsymbol{x}_{n+1} - \boldsymbol{f}_n\left(\boldsymbol{x}_n, \boldsymbol{u}_n, n\right)\right]\right) \tag{9.2.3}$$

where $\boldsymbol{\lambda}_{n+1}$ are the costate variables or Lagrange multipliers. The performance function of Equation (9.2.3) has the same extrema as that of Equation (9.2.2) when the state equality constraint relations are satisfied.

To develop the necessary conditions for the extrema of the function J_A, we formulate the first variation of the function, δJ_A. The first variation is the linear part of the increment which is defined as

$$\begin{aligned} \Delta J_A &= J_A\left(\boldsymbol{x}_{n+1} + \boldsymbol{\delta x}_{n+1},\ \boldsymbol{x}_n + \boldsymbol{\delta x}_n,\ \boldsymbol{u}_n + \boldsymbol{\delta u}_n,\ \boldsymbol{\lambda}_{n+1} + \boldsymbol{\delta\lambda}_{n+1}\right) \\ &\quad - J_A\left(\boldsymbol{x}_{n+1},\ \boldsymbol{x}_n,\ \boldsymbol{u}_n,\ \boldsymbol{\lambda}_{n+1}\right) \end{aligned} \tag{9.2.4}$$

where $\boldsymbol{\delta x}_{n+1}$, $\boldsymbol{\delta x}_n$, $\boldsymbol{\delta u}_n$, $\boldsymbol{\delta\lambda}_{n+1}$ are variations in the appropriate state, control, and costate variables.

Expanding $J_A(\boldsymbol{x}_{n+1} + \boldsymbol{\delta x}_{n+1},\ \boldsymbol{x}_n + \boldsymbol{\delta x}_n,\ \boldsymbol{u}_n + \boldsymbol{\delta u}_n,\ \boldsymbol{\lambda}_{n+1} + \boldsymbol{\delta\lambda}_{n+1})$ in a Taylor series about $J_A(\boldsymbol{x}_{n+1},\ \boldsymbol{x}_n,\ \boldsymbol{u}_n,\ \boldsymbol{\lambda}_{n+1})$ and keeping only the linear terms gives the first variation as

$$\begin{aligned} \delta J_A &= \left(\frac{\partial J_A}{\partial \boldsymbol{x}_{n+1}}\right)^T \boldsymbol{\delta x}_{n+1} + \left(\frac{\partial J_A}{\partial \boldsymbol{x}_n}\right)^T \boldsymbol{\delta x}_n \\ &\quad + \left(\frac{\partial J_A}{\partial \boldsymbol{u}_n}\right)^T \boldsymbol{\delta u}_n + \left(\frac{\partial J_A}{\partial \boldsymbol{\lambda}_{n+1}}\right)^T \boldsymbol{\delta\lambda}_{n+1} \end{aligned} \tag{9.2.5}$$

From calculus, the necessary condition for an extremal of the function J_A when the control is within the admissible set U is that the first variation be zero,

$$\delta J_A = 0 \tag{9.2.6}$$

If the control is on the constraint boundary of the admissible set U, then the necessary condition for a maximum is

$$\delta J_A \leq 0 \tag{9.2.7}$$

and the condition for a minimum is

$$\delta J_A \;\geq\; 0 \tag{9.2.8}$$

We can simplify the augmented performance function of Equation (9.2.3) by introducing the Hamiltonian, H, as

$$H_n \;=\; F_n(\boldsymbol{x}_n, \boldsymbol{u}_n, n) + \boldsymbol{\lambda}_{n+1}^T\, \boldsymbol{f}_n(\boldsymbol{x}_n \boldsymbol{u}_n, n) \tag{9.2.9}$$

Therefore, the augmented performance function can be written as

$$J_A = \sum_{n=0}^{N-1} \left[H_n - \boldsymbol{\lambda}_{n+1}^T\, \boldsymbol{x}_{n+1}\right] \tag{9.2.10}$$

and the first variation of this performance function is

$$\begin{aligned}\delta J_A \;=\; & \sum_{n=0}^{N-1} \left(\frac{\partial H_n}{\partial \boldsymbol{x}_n}\right)^T \boldsymbol{\delta x}_n - \sum_{n=0}^{N-1} \boldsymbol{\lambda}_{n+1}^T\, \boldsymbol{\delta x}_{n+1} + \sum_{n=0}^{N-1} \left(\frac{\partial H_n}{\partial \boldsymbol{u}_n}\right)^T \boldsymbol{\delta u}_n \\ & + \sum_{n=0}^{N-1} \left[\frac{\partial H_n}{\partial \boldsymbol{\lambda}_{n+1}} - \boldsymbol{x}_{n+1}\right]^T \boldsymbol{\delta \lambda}_{n+1}\end{aligned} \tag{9.2.11}$$

The variations $\boldsymbol{\delta x}_{n+1}$ and $\boldsymbol{\delta x}_n$ are not independent but can be related by a discrete version of the integration by parts as

$$-\sum_{n=0}^{N-1} \boldsymbol{\lambda}_{n+1}^T\, \boldsymbol{\delta x}_{n+1} = -\sum_{n=1}^{N} \boldsymbol{\lambda}_n^T\, \boldsymbol{\delta x}_n \tag{9.2.12}$$

or, rearranging the RHS, we have

$$-\sum_{n=0}^{N-1} \boldsymbol{\lambda}_{n+1}^T\, \boldsymbol{\delta x}_{n+1} = -\boldsymbol{\lambda}_N^T\, \boldsymbol{\delta x}_N + \boldsymbol{\lambda}_0\, \boldsymbol{\delta x}_0 - \sum_{n=0}^{N-1} \boldsymbol{\lambda}_n^T\, \boldsymbol{\delta x}_n \tag{9.2.13}$$

The first variation therefore becomes

$$\begin{aligned}\delta J_A \;=\; & -\boldsymbol{\lambda}_n^T \boldsymbol{\delta x}_N + \boldsymbol{\lambda}_0^T \boldsymbol{\delta x}_0 + \sum_{n=0}^{N-1} \left[\frac{\partial H_n}{\partial \boldsymbol{x}_n} - \boldsymbol{\lambda}_n\right]^T \boldsymbol{\delta x}_n \\ & + \sum_{n=0}^{N-1} \left(\frac{\partial H_n}{\partial \boldsymbol{u}_n}\right)^T \boldsymbol{\delta u}_n + \sum_{n=0}^{N-1} \left(\frac{\partial H_n}{\partial \boldsymbol{\lambda}_{n+1}} - \boldsymbol{x}_{n+1}\right)^T \boldsymbol{\delta \lambda}_{n+1}\end{aligned} \tag{9.2.14}$$

The variations $\boldsymbol{\delta x}_n$ and $\boldsymbol{\delta \lambda}_{n+1}$ are free and unrestricted; however, the variation $\boldsymbol{\delta u}_n$ is restricted at the boundary of the set U. Applying the necessary condition for an extremal of the performance function (Equation (9.2.12), (9.2.13), or (9.2.14)) gives the necessary conditions of the discrete maximum principle of optimal control theory or the Kuhn–Tucker (1951) necessary conditions for inequality constraints:

Costate Model

Since the variation $\boldsymbol{\delta x}_n$ is free, we have

$$\boldsymbol{\lambda}_n = \frac{\partial H_n}{\partial \boldsymbol{x}_n} \tag{9.2.15}$$

State Model

Since the variation $\boldsymbol{\delta \lambda}_{n+1}$ is free, we have

$$\frac{\partial H_n}{\partial \boldsymbol{\lambda}_{n+1}} - \boldsymbol{x}_{n+1} = \mathbf{0} \tag{9.2.16}$$

or

$$\boldsymbol{x}_{n+1} = \boldsymbol{f}_n(\boldsymbol{x}_n, \boldsymbol{u}_n, n) \tag{9.2.17}$$

Transversality Conditions

Since the variations $\boldsymbol{\delta x}_N$ and $\boldsymbol{\delta x}_0$ are independent, we have

$$\boldsymbol{\lambda}_N^T \, \boldsymbol{\delta x}_N = 0 \tag{9.2.18}$$

and

$$\boldsymbol{\lambda}_0^T \, \boldsymbol{\delta x}_0 = 0 \tag{9.2.19}$$

Optimal Control

To find a relative maximum in the performance function

$$\sum_{n=0}^{N-1} \left(\frac{\partial H_n}{\partial \boldsymbol{u}_n}\right)^T \boldsymbol{\delta u}_n \leq 0 \tag{9.2.20}$$

or from arguments similar to Section 4.2, we have

$$H_n(\boldsymbol{x}_n, \boldsymbol{u}_n, \boldsymbol{\lambda}_{n+1}, n) = \sup_{\boldsymbol{v}_n \in U} H_n(\boldsymbol{x}_n, \boldsymbol{v}_n, \boldsymbol{\lambda}_{n+1}, n) \tag{9.2.21}$$

To find a relative minimum,

$$\sum_{n=0}^{N-1} \left(\frac{\partial H_n}{\partial \boldsymbol{u}_n}\right)^T \boldsymbol{\delta u}_n \geq 0 \tag{9.2.22}$$

or

$$H_n(\boldsymbol{u}_n, \boldsymbol{u}_n, \boldsymbol{\lambda}_{n+1}, n) = \inf_{\boldsymbol{v}_n \in U} H_n(\boldsymbol{x}_n, \boldsymbol{v}_n, \boldsymbol{\lambda}_{n+1}, n) \tag{9.2.23}$$

When the control does not lie on the boundary of the admissible set U, then the variation $\boldsymbol{\delta u}_n$ is free; therefore,

$$\frac{\partial H_n}{\partial \boldsymbol{u}_n} = \mathbf{0} \tag{9.2.24}$$

9.3 Comparison of the Discrete and Continuous Maximum Principles

Having presented both the continuous and discrete maximum principles, we can now investigate the interconnections between the two. It is natural to expect that the two will yield similar and perhaps identical solutions. We will show that the discrete and continuous versions of the necessary conditions are different. However, for small time intervals, the computational results are essentially the same. Consider the optimal control problem of Section 4.1, where the performance functional to be minimized is

$$J = \int_{t_0}^{t_f} F(\boldsymbol{x}, \boldsymbol{u}, t)\, dt \tag{9.3.1}$$

subject to the state model equality constraints

$$\dot{\boldsymbol{x}} = \boldsymbol{f}(\boldsymbol{x}, \boldsymbol{u}, t) \qquad \boldsymbol{x}(t_0) = \boldsymbol{x}_0 \tag{9.3.2}$$

With the Hamiltonian defined as

$$\boldsymbol{H} = F(\boldsymbol{x}, \boldsymbol{u}, t) + \boldsymbol{\lambda}^T \boldsymbol{f}(\boldsymbol{x}, \boldsymbol{u}, t) \tag{9.3.3}$$

the optimal control is

$$\frac{\partial H}{\partial \boldsymbol{u}} = \boldsymbol{0} \tag{9.3.4}$$

or

$$\frac{\partial F}{\partial \boldsymbol{u}} + \left(\frac{\partial \boldsymbol{f}}{\partial \boldsymbol{u}}\right)^T \boldsymbol{\lambda} = \boldsymbol{0} \tag{9.3.5}$$

and the Euler–Lagrange equation is

$$\dot{\boldsymbol{\lambda}} = -\frac{\partial F}{\partial \boldsymbol{x}} - \left(\frac{\partial \boldsymbol{f}}{\partial \boldsymbol{x}}\right)^T \boldsymbol{\lambda} \tag{9.3.6}$$

with the transversality condition of

$$\boldsymbol{\lambda}(t_f) = \boldsymbol{0} \tag{9.3.7}$$

To discretize the state differential equation, we will use a first-order, forward-difference approximation

$$\dot{\boldsymbol{x}}_n = \frac{\boldsymbol{x}_{n+1} - \boldsymbol{x}_n}{T} \tag{9.3.8}$$

We use a first-order, backward-difference approximation to develop the discrete form of the costate differential equation

$$\dot{\boldsymbol{\lambda}}_n = \frac{\boldsymbol{\lambda}_n - \boldsymbol{\lambda}_{n-1}}{T} \tag{9.3.9}$$

Substituting these expressions into the necessary conditions gives the following discrete equation set which describes the optimal control problem:

$$\boldsymbol{x}_{n+1} = \boldsymbol{x}_n + T\,\boldsymbol{f}_n(\boldsymbol{x}_n, \boldsymbol{u}_n, n) \tag{9.3.10}$$

$$\boldsymbol{\lambda}_n = T\,\frac{\partial F_{n+1}}{\partial \boldsymbol{x}_{n+1}} + \left[\boldsymbol{I} + T\,\frac{\partial \boldsymbol{f}_{n+1}}{\partial \boldsymbol{x}_{n+1}}\right]\boldsymbol{\lambda}_{n+1} \tag{9.3.11}$$

$$\boldsymbol{x}_0 = \boldsymbol{x}_0 \tag{9.3.12}$$

$$\boldsymbol{\lambda}_N = \mathbf{0} \tag{9.3.13}$$

$$T\,\frac{\partial F_n}{\partial \boldsymbol{u}_n} + T\left(\frac{\partial \boldsymbol{f}_n}{\partial \boldsymbol{u}_n}\right)^T \boldsymbol{\lambda}_n = \mathbf{0} \tag{9.3.14}$$

Using the discrete maximum principle, we want to find the minimum of the performance function

$$J = T\sum_{n=0}^{N-1} F_n(\boldsymbol{x}_n, \boldsymbol{u}_n, n) \tag{9.3.15}$$

with the state equality constraints of

$$\boldsymbol{x}_{n+1} = \boldsymbol{x}_n + T\,\boldsymbol{f}_n(\boldsymbol{x}_n, \boldsymbol{u}_n, n) \tag{9.3.16}$$

The necessary conditions are

$$\boldsymbol{\lambda}_n = T\left(\frac{\partial F_n}{\partial \boldsymbol{x}_n}\right) + \left[\boldsymbol{I} + T\,\frac{\partial \boldsymbol{f}_n}{\partial \boldsymbol{x}_n}\right]\boldsymbol{\lambda}_{n+1} \tag{9.3.17}$$

$$\boldsymbol{x}_0 = \boldsymbol{x}_0 \tag{9.3.18}$$

$$\boldsymbol{\lambda}_N = \mathbf{0} \tag{9.3.19}$$

$$T\,\frac{\partial F_n}{\partial \boldsymbol{u}_n} + T\left(\frac{\partial \boldsymbol{f}_n}{\partial \boldsymbol{u}_n}\right)^T \boldsymbol{\lambda}_{n+1} = \mathbf{0} \tag{9.3.20}$$

In comparing Equations (9.3.10) through (9.3.14) with Equations (9.3.16) through (9.3.20), we observe the following equivalences. The state equations are the same, as well as the boundary information. There are slight differences in the optimal control and costate relations. The necessary control conditions arising from the continuous maximum principle has $\boldsymbol{\lambda}_n$ in Equation (9.3.14), while the discrete version is the same equation but with $\boldsymbol{\lambda}_{n+1}$ (Equation (9.3.20)). As the time interval, T, gets small, the difference between the two equations gets small.

There is also a small difference in the costate equations resulting from the two formulations. The form of the equations is the same (Equation (9.4.11) and (9.3.17)), but the continuous version has the coefficients evaluated at the time level $n+1$ while the coefficients in the discrete version are evaluated at time level n. Again, as the time interval, T, gets small, the difference between the two equations gets small.

Therefore, computational results for the two approaches are essentially the same for small enough time intervals, T, and the proper finite differencing of the differential equations. Care must be used in determining a sequence of discrete optimal controls that converge to the associated continuous optimal control problem (Cullum, 1969).

9.4 Necessary Conditions for Implicit Models

Often in finite differencing partial differential equations, the explicit form of Equation (9.2.1) does not arise. Instead, an implicit form appears of

$$\boldsymbol{f}_n(\boldsymbol{x}_{n+1}, \boldsymbol{x}_n, \boldsymbol{u}_n, n) = \mathbf{0} \tag{9.4.1}$$

We will now develop the necessary conditions for the extrema of the performance function

$$J = \sum_{n=0}^{N-1} F_n(\boldsymbol{x}_n, \boldsymbol{u}_n, n) \tag{9.4.2}$$

subject to the equality state constraints of Equation (9.4.1). We want to find the admissible sequence of controls, $\boldsymbol{u}_n$, in the admissible set, U, which will extremize the performance function. As in the previous section, we will use a variational method to develop the necessary conditions. First, we form the augmented performance function that includes the implicit equality state constraints

$$J_A = \sum_{n=0}^{N-1} F_n(\boldsymbol{x}_n, \boldsymbol{u}_n, n) + \boldsymbol{\lambda}_{n+1}^T \boldsymbol{f}_n(\boldsymbol{x}_{n+1}, \boldsymbol{x}_n, \boldsymbol{u}_n, n) \tag{9.4.3}$$

The necessary condition for an extremal of the function J_A, when the control is within the admissible set U, is that the first variation be zero,

$$\delta J_A = 0 \tag{9.4.4}$$

If the control is on the constraint boundary of the admissible set, U, then the necessary condition for a maximum is

$$\delta J_A \leq 0 \tag{9.4.5}$$

and the condition for a minimum is

$$\delta J_A \geq 0 \tag{9.4.6}$$

We will define $\phi_n(\boldsymbol{x}_{n+1},\ \boldsymbol{x}_n,\ \boldsymbol{\lambda}_{n+1},\ \boldsymbol{u}_n)$ to be

$$\phi_n(\boldsymbol{x}_{n+1}, \boldsymbol{x}_n, \boldsymbol{\lambda}_{n+1}, \boldsymbol{u}_n) = F_n + \boldsymbol{\lambda}_{n+1}^T \boldsymbol{f}_n \tag{9.4.7}$$

The augmented performance function is therefore

$$J_A = \sum_{n=0}^{N-1} \phi_n(\boldsymbol{x}_{n+1}, \boldsymbol{x}_n, \boldsymbol{\lambda}_{n+1}, \boldsymbol{u}_n) \tag{9.4.8}$$

and the first variation of this performance function is

$$\begin{aligned} \delta J_A &= \sum_{n=0}^{N-1} \left(\frac{\partial \phi_n}{\partial \boldsymbol{x}_{n+1}}\right)^T \boldsymbol{\delta x}_{n+1} + \sum_{n=0}^{N-1} \left(\frac{\partial \phi_n}{\partial \boldsymbol{x}_n}\right)^T \boldsymbol{\delta x}_n \\ &+ \sum_{n=0}^{N-1} \left(\frac{\partial \phi_n}{\partial \boldsymbol{\lambda}_{n+1}}\right)^T \boldsymbol{\delta\lambda}_{n+1} + \sum_{n=0}^{N-1} \left(\frac{\partial \phi_n}{\partial \boldsymbol{u}_n}\right)^T \boldsymbol{\delta u}_n \end{aligned} \tag{9.4.9}$$

Using the discrete version of the integration by parts, the variations $\boldsymbol{\delta x}_{n+1}$ and $\boldsymbol{\delta x}_n$ are related by

$$\begin{aligned} \sum_{n=0}^{N-1} \left(\frac{\partial \phi_n}{\partial \boldsymbol{x}_{n+1}}\right)^T \boldsymbol{\delta x}_{n+1} &= \left(\frac{\partial \phi_{N-1}}{\partial \boldsymbol{x}_N}\right)^T \boldsymbol{\delta x}_N - \left(\frac{\partial \phi_{-1}}{\partial \boldsymbol{x}_0}\right)^T \boldsymbol{\delta x}_0 \\ &+ \sum_{n=0}^{N-1} \left(\frac{\partial \phi_{n-1}}{\partial \boldsymbol{x}_n}\right)^T \boldsymbol{\delta x}_n \end{aligned} \tag{9.4.10}$$

The first variation therefore becomes

$$\begin{aligned} \delta J_A &= \sum_{n=0}^{N-1} \left(\frac{\partial \phi_{n-1}}{\partial \boldsymbol{x}_n} + \frac{\partial \phi_n}{\partial \boldsymbol{x}_n}\right)^T \boldsymbol{\delta x}_n + \left(\frac{\partial \phi_{N-1}}{\partial \boldsymbol{x}_N}\right)^T \boldsymbol{\delta x}_N - \left(\frac{\partial \phi_{-1}}{\partial \boldsymbol{x}_0}\right)^T \boldsymbol{\delta x}_0 \\ &+ \sum_{n=0}^{N-1} \left(\frac{\partial \phi_n}{\partial \boldsymbol{\lambda}_{n+1}}\right)^T \boldsymbol{\delta\lambda}_{n+1} + \sum_{n=0}^{N-1} \left(\frac{\partial \phi_n}{\partial \boldsymbol{u}_n}\right)^T \boldsymbol{\delta u}_n \end{aligned} \tag{9.4.11}$$

The discrete maximum principle gives the following necessary conditions for an extremal of the performance function:

Costate Model

Since the variation $\boldsymbol{\delta x}_n$ is free, we have

$$\frac{\partial \phi_{n-1}}{\partial \boldsymbol{x}_n} + \frac{\partial \phi_n}{\partial \boldsymbol{x}_n} = \boldsymbol{0} \tag{9.4.12}$$

State Model

Since the variation $\boldsymbol{\delta\lambda}_{n+1}$ is free, we have

$$\boldsymbol{f}_n(\boldsymbol{x}_{n+1}, \boldsymbol{x}_n, \boldsymbol{u}_n, n) = \boldsymbol{0} \tag{9.4.13}$$

Transversality Conditions

Since the variations $\boldsymbol{\delta x}_N$ and $\boldsymbol{\delta x}_0$ are independent, we have

$$\left(\frac{\partial \phi_{N-1}}{\partial \boldsymbol{x}_N}\right)^T \boldsymbol{\delta x}_N = 0 \tag{9.4.14}$$

and

$$\left(\frac{\partial \phi_{-1}}{\partial \boldsymbol{x}_0}\right)^T \boldsymbol{\delta x}_0 = 0 \tag{9.4.15}$$

Optimal Control

To find a relative maximum in the performance index

$$\sum_{n=0}^{N-1} \left(\frac{\partial \phi_n}{\partial \boldsymbol{u}_n}\right)^T \boldsymbol{\delta u}_n \leq 0 \tag{9.4.16}$$

or

$$\phi_n(\boldsymbol{x}_{n+1},\ \boldsymbol{x}_n,\ \boldsymbol{\lambda}_{n+1},\ \boldsymbol{u}_n) = \sup_{\boldsymbol{v}_n \epsilon U} \phi_n(\boldsymbol{x}_{n+1},\ \boldsymbol{x}_n,\ \boldsymbol{\lambda}_{n+1},\ \boldsymbol{v}_n) \tag{9.4.17}$$

To find a relative minimum,

$$\sum_{n=0}^{N-1} \left(\frac{\partial \phi_n}{\partial \boldsymbol{u}_n}\right)^T \boldsymbol{\delta u}_n \geq 0 \tag{9.4.18}$$

or

$$\phi_n(\boldsymbol{x}_{n+1},\ \boldsymbol{x}_n,\ \boldsymbol{\lambda}_{n+1},\ \boldsymbol{u}_n) = \inf_{\boldsymbol{v}_n \epsilon U} (\boldsymbol{x}_{n+1},\ \boldsymbol{x}_n,\ \boldsymbol{\lambda}_{n+1},\ \boldsymbol{v}_n) \tag{9.4.19}$$

When the control does not lie on the boundary of the admissible set, U, then the variation $\boldsymbol{\delta u}_n$ is free; therefore,

$$\frac{\partial \phi_n}{\partial \boldsymbol{u}_n} = \boldsymbol{0} \tag{9.4.20}$$

The formulation of the implicit discrete necessary conditions is found in the thesis of Mehos (1986). Application of the implicit discrete necessary conditions for a finite differenced two–dimensional oil recovery problem is discussed by Mehos and Ramirez (1989). In this work they consider the computation of optimal control policies that maximize the profitability of operating a carbon dioxide enhanced oil recovery project.

9.5 The Discrete Linear Regulator

A linear system with constant control action over the control interval can be expressed as shown in Section 1.7 as

$$\boldsymbol{x}_{k+1} = \boldsymbol{\phi}\boldsymbol{x}_k + \boldsymbol{\Delta}\boldsymbol{u}_k \tag{9.5.1}$$

where $\boldsymbol{\phi}$ is the state transition matrix and $\boldsymbol{\Delta}$ is the control transition matrix. The initial condition is

$$\boldsymbol{x}(0) = \boldsymbol{x}_0 \tag{9.5.2}$$

Let us consider a general quadratic performance function of

$$J = \frac{1}{2}\boldsymbol{x}_f^T \boldsymbol{S}\boldsymbol{x}_f + \frac{1}{2}\sum_{k=0}^{k_f-1}\left(\boldsymbol{x}_k^T\boldsymbol{Q}\boldsymbol{x}_k + \boldsymbol{u}_k^T\boldsymbol{R}\boldsymbol{u}_k\right) \tag{9.5.3}$$

The Hamiltonian for this system is

$$H_k = \frac{1}{2}\boldsymbol{x}_k^T\boldsymbol{Q}\boldsymbol{x}_k + \frac{1}{2}\boldsymbol{u}_k^T\boldsymbol{R}\boldsymbol{u}_k + \boldsymbol{\lambda}_{k+1}^T\left(\boldsymbol{\phi}\boldsymbol{x}_k + \boldsymbol{\Delta}\boldsymbol{u}_k\right) \tag{9.5.4}$$

The definition of the costate variables is given by

$$\boldsymbol{\lambda}_k = \frac{\partial H_k}{\partial \boldsymbol{x}_k} \tag{9.5.5}$$

or

$$\boldsymbol{\lambda}_k = \boldsymbol{Q}\boldsymbol{x}_k + \boldsymbol{\phi}^T\boldsymbol{\lambda}_{k+1} \tag{9.5.6}$$

The transversality condition becomes

$$\left(\frac{\partial h}{\partial \boldsymbol{x}_f} - \boldsymbol{\lambda}_f\right)^T \delta\boldsymbol{x}_f = 0 \tag{9.5.7}$$

which is

$$\boldsymbol{\lambda}(k_f) = \boldsymbol{S}\boldsymbol{x}(k_f) \tag{9.5.8}$$

The optimal control condition is

$$\frac{\partial H_k}{\partial \boldsymbol{u}_k} = \boldsymbol{0} \tag{9.5.9}$$

which is evaluated to be

$$\boldsymbol{0} = \boldsymbol{R}\boldsymbol{u}_k + \boldsymbol{\Delta}^T\boldsymbol{\lambda}_{k+1} \tag{9.5.10}$$

or

$$\boldsymbol{u}_k = -\boldsymbol{R}^{-1}\boldsymbol{\Delta}^T\boldsymbol{\lambda}_{k+1} \tag{9.5.11}$$

Invoking the Riccati transformation,

$$\boldsymbol{\lambda}_k = \boldsymbol{P}_k \boldsymbol{x}_k \tag{9.5.12}$$

the state response under optimal control is

$$\boldsymbol{x}_{k+1} = \boldsymbol{\phi x}_k - \boldsymbol{\Delta R}^{-1}\boldsymbol{\Delta}^T \boldsymbol{P}_{k+1}\boldsymbol{x}_{k+1} \tag{9.5.13}$$

or

$$\boldsymbol{x}_{k+1} = \left(\boldsymbol{I} + \boldsymbol{\Delta R}^{-1}\boldsymbol{\Delta}^T \boldsymbol{P}_{k+1}\right)^{-1} \boldsymbol{\phi x}_k \tag{9.5.14}$$

Now we need to develop the recursive formula for calculating the Riccati matrix, $\boldsymbol{P}_k$. Using the definition of the costate variables (9.5.6) and the Riccati transformation (9.5.12), we have

$$\boldsymbol{P}_k \boldsymbol{x}_k = \boldsymbol{Q x}_k + \boldsymbol{\phi}^T \boldsymbol{P}_{k+1} \left(\boldsymbol{I} + \boldsymbol{\Delta R}^{-1}\boldsymbol{\Delta}^T \boldsymbol{P}_{k+1}\right)^{-1} \boldsymbol{\phi x}_k \tag{9.5.15}$$

or

$$\boldsymbol{P}_k = \boldsymbol{Q} + \boldsymbol{\phi}^T \boldsymbol{P}_{k+1} \left[\boldsymbol{I} + \boldsymbol{\Delta R}^{-1}\boldsymbol{\Delta}^T \boldsymbol{P}_{k+1}\right]^{-1} \boldsymbol{\phi} \tag{9.5.16}$$

which can also be expressed as

$$\boldsymbol{P}_k = \boldsymbol{Q} + \boldsymbol{\phi}^T \left[\boldsymbol{P}_{k+1}^{-1} + \boldsymbol{\Delta R}^{-1}\boldsymbol{\Delta}^T\right]^{-1} \boldsymbol{\phi} \tag{9.5.17}$$

with the final condition

$$\boldsymbol{P}(k_f) = \boldsymbol{S} \tag{9.5.18}$$

This is a recursive relation that is equivalent to the Riccati matrix differential equation for the continuous regulator problem of Section 5.1. The discrete Riccati equation can also be expressed as

$$\boldsymbol{P}_k = \boldsymbol{\phi}^T \boldsymbol{P}_{k+1}\boldsymbol{\phi} - \boldsymbol{\phi}^T \boldsymbol{P}_{k+1}\boldsymbol{\Delta} \left[\boldsymbol{\Delta}^T \boldsymbol{P}_{k+1}\boldsymbol{\Delta} + \boldsymbol{R}\right]^{-1} \boldsymbol{\Delta}^T \boldsymbol{P}_{k+1}\boldsymbol{\phi} + \boldsymbol{Q} \tag{9.5.19}$$

This relation comes from use of the matrix inversion lemma that states that (Grenander and Rosenblatt, 1957)

$$\left[\boldsymbol{P}_{k+1}^{-1} + \boldsymbol{\Delta R}^{-1}\boldsymbol{\Delta}^T\right]^{-1} = \boldsymbol{P}_{k+1} - \boldsymbol{P}_{k+1}\boldsymbol{\Delta} \left[\boldsymbol{\Delta}^T \boldsymbol{P}_{k+1}\boldsymbol{\Delta} + \boldsymbol{R}\right]^{-1} \boldsymbol{\Delta}^T \boldsymbol{P}_{k+1} \tag{9.5.20}$$

The optimal control law is

$$\boldsymbol{u}_k = -\boldsymbol{R}^{-1}\boldsymbol{\Delta}^T \boldsymbol{P}_{k+1}\boldsymbol{x}_{k+1} \tag{9.5.21}$$

which can be expressed in terms of $\boldsymbol{x}_k$ by

$$\boldsymbol{u}_k = -\boldsymbol{R}^{-1}\boldsymbol{\Delta}^T \boldsymbol{P}_{k+1} \left[\boldsymbol{I} + \boldsymbol{\Delta R}^{-1}\boldsymbol{\Delta}^T \boldsymbol{P}_{k+1}\right]^{-1} \boldsymbol{\phi x}_k \tag{9.5.22}$$

or

$$\boldsymbol{u}_k = -\left[\boldsymbol{R} + \boldsymbol{\Delta}^T \boldsymbol{P}_{k+1} \boldsymbol{\Delta}\right]^{-1} \boldsymbol{\Delta}^T \boldsymbol{P}_{k+1} \boldsymbol{\phi} \boldsymbol{x}_k \tag{9.5.23}$$

We can define the feedback gain matrix as

$$\boldsymbol{K}_k = \left[\boldsymbol{R} + \boldsymbol{\Delta}^T \boldsymbol{P}_{k+1} \boldsymbol{\Delta}\right]^{-1} \boldsymbol{\Delta}^T \boldsymbol{P}_{k+1} \boldsymbol{\phi} \tag{9.5.24}$$

and the optimal control is a proportional feedback controller:

$$\boldsymbol{u}_k = -\boldsymbol{K}_k \boldsymbol{x}_k \tag{9.5.25}$$

There is a steady-state version of the discrete regulator problem. By applying the discrete Riccati equation (9.5.19) backwards in time, the covariance matrix will converge to a steady-state value $\boldsymbol{P}_{ss}$. Using this steady-state value of the covariance matrix, a steady-state feedback gain matrix can be computed from Equation (9.5.24). The associated optimal control expression is

$$\boldsymbol{u}_k = -\boldsymbol{K}_{ss} \boldsymbol{x}_k \tag{9.5.26}$$

9.5.1 Nyquist–Ramirez Reactor

Figure 9.1 gives the MATLAB program that implements the discrete regulator for the Nyquist–Ramirez reactor described in Section 5.4. The program first converts the continuous system to the discrete time model for a given sample frequency, T. The MATLAB routine C2D is used to perform the operation. Next the steady-state discrete Kalman feedback controller gain matrix is calculated using the routine DLQR. Finally the impulse response of the system is obtained using the routine DIMPULSE. This routine gives the values of the state at the discrete time values. The response of the actual system is continuous, but the controller action is updated every sample frequency, T. For the plotting of the results, a linear interpolation between the discrete time state values is used as an approximation to the response of the actual system. The state response for $T = 1$ min and $T = 10$ min are given in Figure 9.2. The response with a sample frequency of one minute is nearly equivalent to the continuous response given in Figure 5.7. It does show a slight delay of one minute, which is not significant in light of the 100 minute response time. The response with a sample frequency of 10 minutes starts to deviate from the continuous case, although in reality the response is quite reasonable. A sample frequency of 10 minutes for a response time of 100 minutes does result in a significant delay in the response dynamics. The normal rule of thumb is that the sample frequency should be at least 1/20 of the response time.

```
>> % discrete regulator for Nyquist-Ramirez reactor
>>
>> % input the data
>> a = [-0.05156  -0.05877;
         0.01503  -0.005887];
>> b = [0  -0.03384]';
>>
>> % convert to discrete form
>> T = 1;
>> [P,D] = c2d(a,b,T)

P =

    0.9493   -0.0571
    0.0146    0.9937
D =

    0.0010
   -0.0337

>> % calculate the discrete Kalman feedback gain
>> q = eye(2);
>> r = 1;
>> k = dlqr(P,D,q,r)

k =

    0.0417   -0.9084

>> % compute the impulse response to the closed loop system
>> aa = P - D*k;
>> bb = [1 1]';
>> c = eye(2);
>> d = [0 0]';
>> n = 100;
>> y = dimpulse(aa,bb,c,d,1,n);
>>
>> % compute for a larger time increment
>> T = 10;
>> [P,D] = c2d(a,b,T)

P =

    0.5665   -0.4383
    0.1121    0.9071
D =

    0.0821
   -0.3244

>> k = dlqr(P,D,q,r)

k =

   -0.0209   -0.7678

>> aa = P - D*k;
>> n = 10;
>> y1 = dimpulse (aa,bb,c,d,1,n);
>> nnn = [0:10:90];
>> nn = [0:1:90];
>> % prepare a plot of the discrete states
>>
>> plot ( nn(2:91),y(2:91,1:2),nnn(2:10),y1(2:10))
>> plot (nn(2:91),y(2:91,1:2),nnn(2:10),y1(2:10,1:2))
>> meta dlqr2
>> title ('Plot of optimal states vs time');
>> xlabel ('Time, min');
>> ylabel('State vector components');
>> meta dlqr2
>> diary off
```

Figure 9.1 MATLAB program for Nyquist–Ramirez reactor discrete LQR.

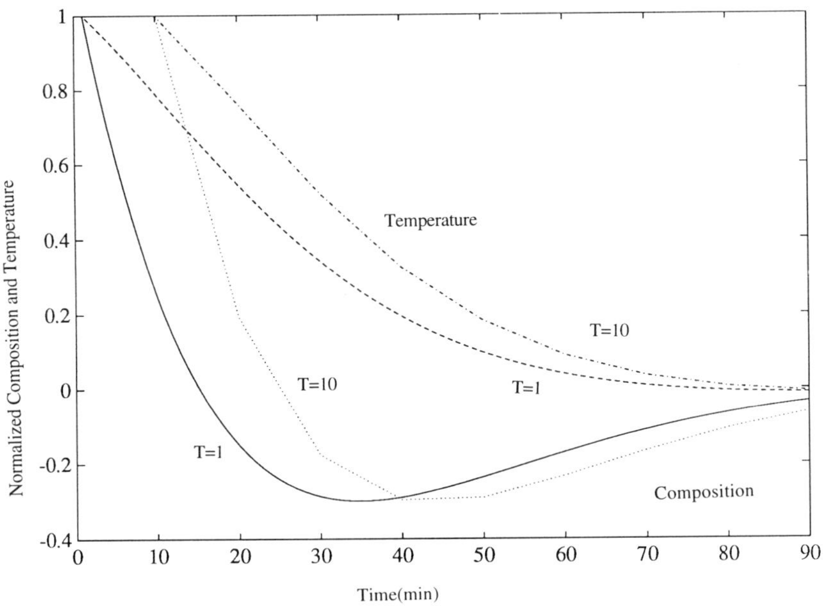

Figure 9.2 Discrete LQR state response.

9.6 Discrete Johnson P–I Regulator

There is a discrete version of the Johnson controller (Section 5.6) which compensates for unmeasurable load disturbances to the system. The actual system is modeled as

$$\dot{\boldsymbol{x}} = \boldsymbol{A}\boldsymbol{x} + \boldsymbol{B}\boldsymbol{u} + \boldsymbol{W}\boldsymbol{d}_u \tag{9.6.1}$$

Applying the Johnson transformation,

$$\boldsymbol{y} = \boldsymbol{B}\boldsymbol{u} + \boldsymbol{W}\boldsymbol{d} \tag{9.6.2}$$

for piecewise constant disturbances, we have

$$\dot{\boldsymbol{y}} = \boldsymbol{B}\dot{\boldsymbol{u}} \tag{9.6.3}$$

Therefore, the original problem of (9.6.1) can be expressed as

$$\begin{pmatrix} \dot{\boldsymbol{x}} \\ \dot{\boldsymbol{y}} \end{pmatrix} = \begin{bmatrix} \boldsymbol{A} & \boldsymbol{I} \\ \boldsymbol{0} & \boldsymbol{0} \end{bmatrix} \begin{pmatrix} \boldsymbol{x} \\ \boldsymbol{y} \end{pmatrix} + \begin{pmatrix} \boldsymbol{0} \\ \boldsymbol{B} \end{pmatrix} \dot{\boldsymbol{u}} \tag{9.6.4}$$

This is rewritten as

$$\dot{\tilde{\boldsymbol{x}}} = \tilde{\boldsymbol{A}}\tilde{\boldsymbol{x}} + \tilde{\boldsymbol{B}}\tilde{\boldsymbol{u}} \tag{9.6.5}$$

The discrete version of the transformed system is

$$\tilde{\boldsymbol{x}}_{k+1} = \tilde{\boldsymbol{\phi}}\tilde{\boldsymbol{x}}_k + \tilde{\boldsymbol{\Delta}}\tilde{\boldsymbol{u}}_k \tag{9.6.6}$$

This is the discrete Kalman control problem for the quadratic objective function

$$J = \frac{1}{2}\sum_{k=0}^{k_f-1} \left(\tilde{\boldsymbol{x}}_k^T \tilde{\boldsymbol{Q}} \tilde{\boldsymbol{x}}_k + \tilde{\boldsymbol{u}}_k^T \tilde{\boldsymbol{R}} \tilde{\boldsymbol{u}}_k\right) \tag{9.6.7}$$

where

$$\tilde{\boldsymbol{Q}} = \begin{bmatrix} \boldsymbol{Q} & \boldsymbol{0} \\ \boldsymbol{0} & \boldsymbol{0} \end{bmatrix} \tag{9.6.8}$$

$$\tilde{\boldsymbol{R}} = \boldsymbol{R} \tag{9.6.9}$$

The optimal control for $k_f \to \infty$ is

$$\tilde{\boldsymbol{u}}_k = -\tilde{\boldsymbol{K}}_{ss}\tilde{\boldsymbol{x}}_k \tag{9.6.10}$$

or

$$\tilde{\boldsymbol{u}}_k = -\begin{bmatrix} \tilde{\boldsymbol{K}}_1 & \tilde{\boldsymbol{K}}_2 \end{bmatrix} \begin{pmatrix} \boldsymbol{x}_k \\ \boldsymbol{y}_k \end{pmatrix} \tag{9.6.11}$$

The discrete time model of Equation (9.6.6) can be written as

$$\begin{pmatrix} \boldsymbol{x}_{k+1} \\ \boldsymbol{y}_{k+1} \end{pmatrix} = \begin{bmatrix} \tilde{\boldsymbol{\phi}}_{11} & \tilde{\boldsymbol{\phi}}_{12} \\ \tilde{\boldsymbol{\phi}}_{21} & \tilde{\boldsymbol{\phi}}_{22} \end{bmatrix} \begin{pmatrix} \boldsymbol{x}_k \\ \boldsymbol{y}_k \end{pmatrix} + \begin{bmatrix} \tilde{\boldsymbol{\Delta}}_1 \\ \tilde{\boldsymbol{\Delta}}_2 \end{bmatrix} \dot{\boldsymbol{u}}_k \tag{9.6.12}$$

This gives the following relation for $\boldsymbol{y}_k$:

$$\boldsymbol{y}_k = \tilde{\boldsymbol{\phi}}_{21}\boldsymbol{x}_{k-1} + \tilde{\boldsymbol{\phi}}_{22}\boldsymbol{y}_{k-1} + \tilde{\boldsymbol{\Delta}}_2\dot{\boldsymbol{u}}_{k-1} \tag{9.6.13}$$

Thus, the actual optimal control law is

$$\dot{\boldsymbol{u}}_k = -\tilde{\boldsymbol{K}}_1\boldsymbol{x}_k - \tilde{\boldsymbol{K}}_2\left(\tilde{\boldsymbol{\phi}}_{21}\boldsymbol{x}_{k-1} + \tilde{\boldsymbol{\phi}}_{22}\boldsymbol{y}_{k-1} + \tilde{\boldsymbol{\Delta}}_2\dot{\boldsymbol{u}}_{k-1}\right) \tag{9.6.14}$$

At the initial time $k = 0$; $\boldsymbol{x}_{-1}$, $\boldsymbol{y}_{-1}$, and $\dot{\boldsymbol{u}}_{-1}$ are all zero since the system starts from the original equilibrium condition without any unmeasured disturbances. It should be noted that the optimal control law is given in terms of a constant rate of change in the control variables and not a constant value over the sampling period. This is a velocity algorithm and not a position algorithm.

9.6.1 Nyquist–Ramirez Reactor

Figure 9.3 gives the MATLAB program for implementing the discrete Johnson PI controller for the Ramirez–Nyquist reactor. This program is an M file which can be run in MATLAB. It starts with the entry of the system data and the introduction of the Johnson transformation given by Equation (9.6.4). The appropriate quadratic performance index weighting matrices of Equations (9.6.8) and (9.6.9) are then inputted. The continuous transformed system is then converted into the discrete form of Equation (9.6.12), using the routine C2D. The steady-state feedback gain matrix is computed using the recursive algorithm of (9.5.19). The DLQR algorithm of MATLAB cannot be used since the transformed equation set of (9.6.12) is uncontrollable, although the actual system is controllable. The convergence of the Riccati matrix elements is given in Figure 9.4. The gain elements are nearly those of the continuous controller (Figure 5.12). The dynamic response of the system is shown in Figure 9.5. It shows that the composition is regulated about the original equilibrium state but that the temperature reaches a new equilibrium. The response is nearly equivalent to that of the continuous problem given in Figure 5.14.

9.7 Discrete Dynamic Predictor Regulator

The model predictive controller for measurable disturbances of Section 6.3 can be placed in a discrete form. The original problem model is

$$\dot{\boldsymbol{x}} = \boldsymbol{A}\boldsymbol{x} + \boldsymbol{B}\boldsymbol{u} + \boldsymbol{W}\boldsymbol{d}_m \tag{9.7.1}$$

We define the predictive state $\boldsymbol{x}_p$ and the predictive control $\boldsymbol{u}_p$ to satisfy

$$\begin{pmatrix} \boldsymbol{0} \\ \dot{\boldsymbol{x}}_{fp} \end{pmatrix} = \begin{bmatrix} \boldsymbol{A}_{11} & \boldsymbol{A}_{12} \\ \boldsymbol{A}_{21} & \boldsymbol{A}_{22} \end{bmatrix} \begin{pmatrix} \boldsymbol{x}_{sp} \\ \boldsymbol{x}_{fp} \end{pmatrix} + \begin{bmatrix} \boldsymbol{B}_1 \\ \boldsymbol{B}_2 \end{bmatrix} \boldsymbol{u}_p + \begin{bmatrix} \boldsymbol{W}_1 \\ \boldsymbol{W}_2 \end{bmatrix} d_m \tag{9.7.2}$$

where $\boldsymbol{x}_{sp}$ are the m predicted state variables that are maintained at specified values and $\boldsymbol{x}_{fp}$ are free or unspecified predicted state variables that are free to obtain new equilibrium states. Introducing the transformation

$$\overline{\boldsymbol{x}} = \boldsymbol{x} - \boldsymbol{x}_p \tag{9.7.3}$$

$$\overline{\boldsymbol{u}} = \boldsymbol{u} - \boldsymbol{u}_p \tag{9.7.4}$$

yields the state dynamic model

$$\dot{\overline{\boldsymbol{x}}} = \boldsymbol{A}\,\overline{\boldsymbol{x}} + \boldsymbol{B}\,\overline{\boldsymbol{u}} \tag{9.7.5}$$

The discrete form of this equation set is

$$\overline{\boldsymbol{x}}_{k+1} = \boldsymbol{\phi}\,\overline{\boldsymbol{x}}_k + \boldsymbol{\Delta}\,\overline{\boldsymbol{u}}_k \tag{9.7.6}$$

```
% input the data
a = [-0.05156  -0.05877;
      0.01503  -0.005887];
b = [0  -0.03384]';
w = [0.0125  0;
     0       0.0125];
% define the Johnson transformation
aa = [a      eye(2);
      zeros(2) zeros(2)];
bb = [0;
      0;
      b];
% define the weighting matrices
q = [1  0;
     0  0];
r = 1;
% Johnson transformation weighting matrices
qq = [q          zeros(2);
      zeros(2)   zeros(2)];
% convert transformed system to discrete form
T = 1;

[p,d] = c2d(aa,bb,T)

% since the transformed system is uncontrollable, use the recursive
% formula to calculate the Kalman feedback gain matrix
tmax = 100;
pt = zeros(4);
for t = tmax:-1:1,
        pt1 = p'*pt*p -p'*pt*d*inv(d'*pt*d +r)*d'*pt*p +qq;
        kt1 = inv(r +d'*pt1*d)*d'*pt1*p;
%   transform the matrix into a vector
        for col = 1:4,
            k(t,col) = kt1(1,col);
        end
        pt = pt1;
end
% plot k vs time
t = 0:1:(tmax-1);
plot (t,k)
title ('Gain Matrix vs Time')
ylabel ('Gain Components'),xlabel('Time, min')
meta pigain

% note that a step response in the states is an impulse response in y
api = p -d*kt1;
wi = [0 0 .0125 .0125]';
wi = .2* wi;
dd = [0 0 0 0]';
c = eye(4);
xpi = dimpulse(api,wi,c,dd,1,101);
t = [0:1:100];
t = t';
plot (t(1:100,1),xpi(2:101,1:2));
title('Discrete Johnson Algorithm State Response');
xlabel('Time, min');
ylabel('Normalized Composition and Temperature');
meta dxpi;
```

Figure 9.3 MATLAB program for Nyquist reactor using discrete Johnson algorithm.

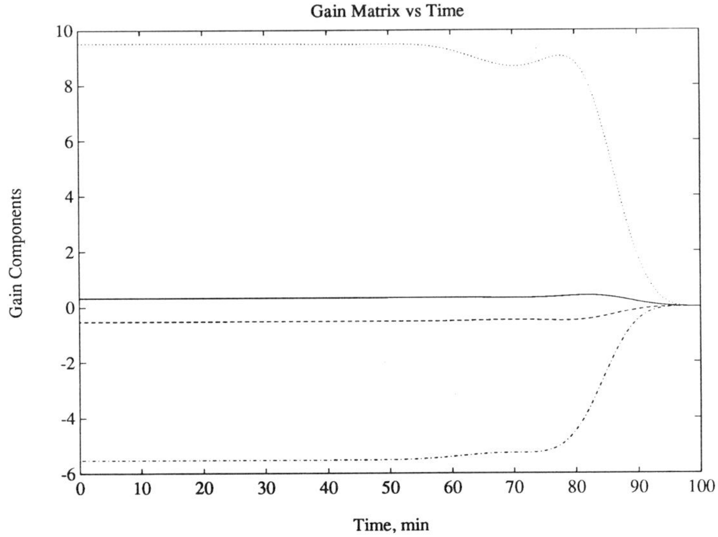

Figure 9.4 Convergence of Riccati matrix elements. K_{13} (···), K_{11} (—), K_{12} (- - - -), K_{14} (— · —·).

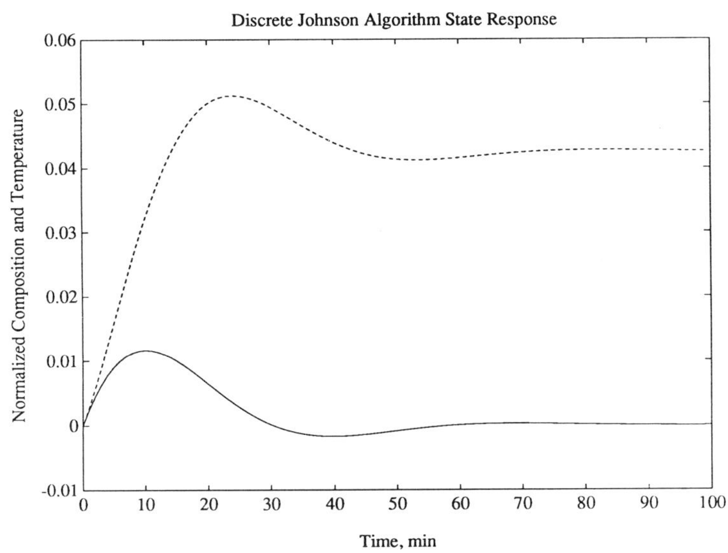

Figure 9.5 Discrete Johnson algorithm controlled state response. Temperature (···), composition (—).

This is a discrete Kalman regulator control problem with the quadratic performance index

$$J = \frac{1}{2}\sum_{k=0}^{k_f-1}\left(\overline{\boldsymbol{x}}_k^T\,\boldsymbol{Q}\,\overline{\boldsymbol{x}}_k + \overline{\boldsymbol{u}}_k^T\,\boldsymbol{R}\,\overline{\boldsymbol{u}}_k\right) \tag{9.7.7}$$

When $k_f \rightarrow \infty$, the optimal controller is

$$\boldsymbol{u}_k = -\boldsymbol{K}_{ss}\left(\boldsymbol{x}_k - \boldsymbol{x}_k^p\right) + \boldsymbol{u}_k^p \tag{9.7.8}$$

which has the model predictive control structure of the continuous problem of Section 6.3. The final point that needs to be considered is how to compute the predicted controls $\boldsymbol{u}_k^p$ and the free predicted states $\boldsymbol{x}_k^{fp}$. As in Section 6.3, one possible solution strategy is to solve the m $\boldsymbol{u}_p$ variables using the m algebraic equations of (9.7.2) so that

$$\boldsymbol{u}_k^p = \boldsymbol{B}_1^{-1}\left[-\boldsymbol{A}_{11}\boldsymbol{x}^{sp} - \boldsymbol{A}_{12}\boldsymbol{x}_k^{fp} - \boldsymbol{W}_1\boldsymbol{d}_k^m\right] \tag{9.7.9}$$

This solution depends upon the existence of $\boldsymbol{B}_1^{-1}$. The $(n-m)$ free predictive states are given by the $(n-m)$ differential equations

$$\begin{aligned}\dot{\boldsymbol{x}}_{fp} \;=\; & \left[\boldsymbol{A}_{22} - \boldsymbol{B}_2\boldsymbol{B}_1^{-1}\boldsymbol{A}_{12}\right]\boldsymbol{x}_{fp} + \left[\boldsymbol{A}_{21} - \boldsymbol{B}_2\boldsymbol{B}_1^{-1}\boldsymbol{A}_{11}\right]\boldsymbol{x}_{sp} \\ & + \left[-\boldsymbol{B}_2\boldsymbol{B}_1^{-1}\boldsymbol{W}_1 + \boldsymbol{W}_2\right]\,\boldsymbol{d}_m\end{aligned} \tag{9.7.10}$$

which has the discrete solution of

$$\boldsymbol{x}_{k+1}^{fp} = \boldsymbol{\phi}^p\boldsymbol{x}_k^{fp} + \boldsymbol{\Delta}_1^p\boldsymbol{x}^{sp} + \boldsymbol{\Delta}_2^p\boldsymbol{d}_k^m \tag{9.7.11}$$

The matrix $\left[\boldsymbol{A}_{22} - \boldsymbol{B}_2\boldsymbol{B}_1^{-1}\boldsymbol{A}_{12}\right]$ must be negative semi-definite for the $\boldsymbol{x}_{fp}$ solution to be stable. When $\boldsymbol{B}_1^{-1}$ does not exist, the partitioning given in Equation (6.3.13) is necessary. As described in Section 6.3, the m algebraic equations are used to solve for the singular part of the differential equations of (6.3.13), $\boldsymbol{x}_{s_k}^{fp}$, and for the controls not associated with the singular part of (6.3.13), $\boldsymbol{u}_{n_k}^p$,

$$\begin{pmatrix}\boldsymbol{x}_{s_k}^{fp}\\ \boldsymbol{u}_{n_k}^p\end{pmatrix} = \begin{bmatrix}\overline{\boldsymbol{A}}_1\\ \overline{\boldsymbol{A}}_2\end{bmatrix}\boldsymbol{x}^{sp} + \begin{bmatrix}\overline{\boldsymbol{B}}_{11} & \overline{\boldsymbol{B}}_{12}\\ \overline{\boldsymbol{B}}_{21} & \overline{\boldsymbol{B}}_{22}\end{bmatrix}\begin{pmatrix}\boldsymbol{x}_{n_k}^{fp}\\ \boldsymbol{u}_{s_k}^p\end{pmatrix} + \begin{bmatrix}\overline{\boldsymbol{W}}_1\\ \overline{\boldsymbol{W}}_2\end{bmatrix}\boldsymbol{d}_k^m \tag{9.7.12}$$

The $(n-m)$ differential equations of (6.3.13) are given by Equation (6.3.23) so that

$$\boldsymbol{y}_{2_k} = -\boldsymbol{D}_{22}^{-1}\boldsymbol{D}_{21}\boldsymbol{y}_{1_k} \tag{9.7.13}$$

and

$$\dot{\boldsymbol{y}}_1 = \boldsymbol{S}_d^{-1}\left[\boldsymbol{D}_{11} + \boldsymbol{D}_{12}\boldsymbol{D}_{22}^{-1}\boldsymbol{D}_{21}\right]\,\boldsymbol{y}_1 \tag{9.7.14}$$

which has a discrete form solution

$$\boldsymbol{y}_{1_{k+1}} = \boldsymbol{\phi}^D\boldsymbol{y}_{1_k} \tag{9.7.15}$$

The necessary condition for this solution is that $\boldsymbol{D}_{22}^{-1}$ must exist and $\boldsymbol{S}_d^{-1}[\boldsymbol{D}_{11} + \boldsymbol{D}_{12}\boldsymbol{D}_{22}^{-1}\boldsymbol{D}_{21}]$ must be negative semidefinite.

9.8 Robustness of Discrete Time Problems

Molander (1979) has demonstrated that there are no universally valid robustness results that can be given for discrete time systems. However, as the sample time is reduced and the discrete time system approaches the continuous time system, the robustness results of the continuous time regulator of Chapter 7 apply.

References

Cullum, J., "Discrete Approximations to Continuous Optimal Control Problems," *SIAM J. Control* **7**, 32 (1969).

Denn, M. M., *Optimization by Variational Methods*, McGraw–Hill, New York (1969).

Grenander, U. and Rosenblatt, M., *Statistical Analysis of Stationary Time Series*, Wiley and Sons, New York (1957).

Kuhn, H. W. and Tucker, A. W., *Second Berkeley Symposium on Mathematical Statistics and Probability*, California (1951).

Mehos, G., *Application of Optimal Control Theory to Carbon Dioxide Flooding*, Ph.D. Thesis, University of Colorado (1986).

Mehos, G. J. and Ramirez, W. F., "Use of Optimal Control Theory to Optimize Carbon Dioxide Miscible Flooding Enhanced Oil Recovery," *J. Pet. Sci. & Eng.* **2**, 247–260 (1989).

Molander, P., "Stabilization of Uncertain Systems," Research Report LUTFD2/CTFRT–1020, 1–111, Department of Automatic Control, Lund Inst. of Tech., Sweden (1979).

Sage, A. P. and White, C. C., III, *Optimal Systems Control*, Second Edition, Prentice–Hall, Englewood Cliffs, NJ (1977).

Exercise

1. Develop the discrete regulator results for three quite different sampling times for the following problems:

    1. Douglas reactor of problem 5.1.
    2. Newell *et al.* evaporator of problem 5.2.
    3. Ramirez *et al.* dialyzer of problem 5.3.
    4. Distillation column of Section 6.5.

Chapter 10

STATE AND PARAMETER IDENTIFICATION

10.1 Fundamental Concepts

As we have seen in the previous chapters, optimal control laws are a function of the entire state. This means that we need to have information on all the state variables in order to implement optimal control policies. However, the entire state is seldom measurable. We therefore need a way to estimate the state from the available measurement information. In order to do this, additional information about the system will have to be used. The new information will be characterizations about the uncertainties in the measurements and the uncertainties in the state model. All measurement devices have certain noise which is random in character. We can also consider the uncertainty in the process model to have a probabilistic character.

Using uncertainty concepts, an optimal estimate of the state for a linear system is obtained based upon system measurements. The result is the Kalman filter which is one of the major practical results of optimal estimation theory. We show how the Kalman filter can be extended for nonlinear systems and discrete-time models. Kalman filtering is also shown to be effective for the estimation of model parameters. Due to duality, the Kalman filter has the same excellent robustness characteristics as does the LQR problem.

The linear uncertain process model can be mathematically expressed as

$$\dot{\boldsymbol{x}} = \boldsymbol{F}(t)\boldsymbol{x}(t) + \boldsymbol{w}(t) \tag{10.1.1}$$

where $\boldsymbol{w}(t)$ is a random variable. We usually assume that the random variable $\boldsymbol{w}(t)$ is normally distributed so that

$$\mathcal{E}\left\{\boldsymbol{w}(t)\right\} = \boldsymbol{0} \tag{10.1.2}$$

where $\mathcal{E}\{\ \}$ is the expected value operator, and

$$cov\ \left\{\boldsymbol{w}(t)\right\} = \boldsymbol{Q}(t) \tag{10.1.3}$$

where *cov* is the covariance operator.

The expected value operator is defined as

$$\mathcal{E}\ \left\{\boldsymbol{w}(t)\right\} = \int_{-\infty}^{\infty} \boldsymbol{w}(t) f(\boldsymbol{w})\, d\boldsymbol{w} \tag{10.1.4}$$

where $f(\boldsymbol{w})$ is the probability density function. The covariance is defined by

$$cov\ \left\{\boldsymbol{w}(t)\right\} = \mathcal{E}\ \left\{\boldsymbol{w}(t)\boldsymbol{w}^T(t)\right\} \tag{10.1.5}$$

For a scalar random variable, w, the probability density function for a Gaussian or normal distribution is given by

$$f(w) = \frac{1}{\sqrt{2\pi\sigma}} \exp\left(-\frac{(w-\overline{w})^2}{2\sigma^2}\right) \tag{10.1.6}$$

where σ^2 = the variance of the distribution. The square root of the variance σ^2 is σ and is known as the standard deviation. The variable $\overline{w}$ is the mean value and is the expected value of the normal distribution. A plot of a normal distribution is shown in Figure 10.1. It illustrates the shape of the distribution as well as the mean value and the standard deviation.

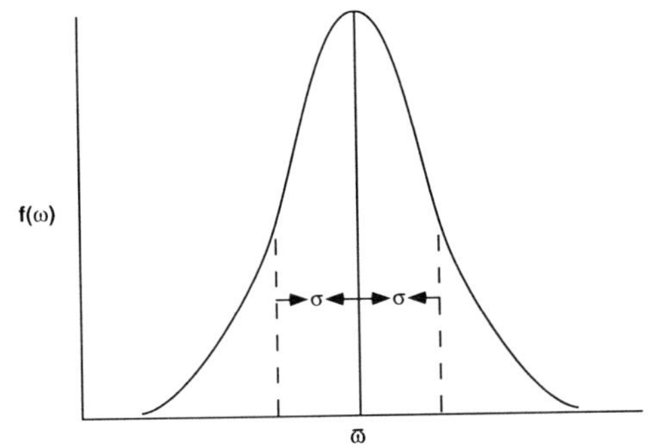

Figure 10.1 Gaussian probability density function.

For vector variables, the Gaussian distribution is given by

$$f(\boldsymbol{w}) = \frac{1}{\left((2\pi)^n \det \boldsymbol{V}\right)^{1/2}} \exp\left(-\frac{1}{2}(\boldsymbol{w}-\overline{\boldsymbol{w}})^T \boldsymbol{V}^{-1}(\boldsymbol{w}-\overline{\boldsymbol{w}})\right) \tag{10.1.7}$$

where $\overline{\boldsymbol{w}}$ is the mean value given by

$$\overline{\boldsymbol{w}} = \mathcal{E}(\boldsymbol{w}) \tag{10.1.8}$$

and $\boldsymbol{V}$ is the covariance matrix given by

$$\boldsymbol{V} = \mathcal{E}\left\{(\boldsymbol{w} - \overline{\boldsymbol{w}})(\boldsymbol{w} - \overline{\boldsymbol{w}})^T\right\} \tag{10.1.9}$$

The initial conditions for the process are

$$\mathcal{E}\left(\boldsymbol{x}(t_0)\right) = \overline{\boldsymbol{x}}_0 \tag{10.1.10}$$

and

$$\operatorname{cov}\left\{\boldsymbol{x}(t_0)\right\} = \boldsymbol{P}_0 \tag{10.1.11}$$

A linear measurement system is modeled as

$$\boldsymbol{y}(t) = \boldsymbol{H}(t)\boldsymbol{x}(t) + \boldsymbol{v}(t) \tag{10.1.12}$$

where $\boldsymbol{v}(t)$ is a random noise that is assumed normally distributed so that

$$\mathcal{E}\left\{\boldsymbol{v}(t)\right\} = \boldsymbol{0} \tag{10.1.13}$$

$$\operatorname{cov}\left\{\boldsymbol{v}(t)\right\} = \boldsymbol{R}(t) \tag{10.1.14}$$

There are several ways that the probabilistic model and measurement system can be used. These are by smoothing, filtering, and prediction.

Smoothing is the use of data before and after a time t_1 to provide an estimate of the state at time t_1. This is a nonsequential process and is illustrated in Figure 10.2.

Filtering is the estimation of the state at a time t_1 when all of the data up to time t_1 are used. This is a sequential process and is illustrated in Figure 10.3. Prediction is obtaining estimates of the state at a time t_2 from data up to a time t_1 where t_1 is less than the time $t_2, (t_1 < t_2)$. This often occurs because of time delays in the measuring elements and is illustrated in Figure 10.4.

10.2 The Kalman Filter

We now move to the development of one of the truly important results of modern control and estimation theory. This is the Kalman filtering problem. Kalman considered obtaining an optimal estimate of the state of a linear process described by

$$\dot{\boldsymbol{x}} = \boldsymbol{F}(t)\boldsymbol{x}(t) + \boldsymbol{w}(t) \tag{10.2.1}$$

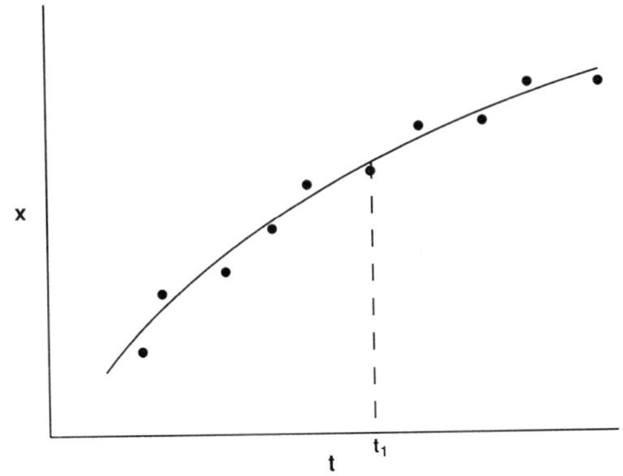

Figure 10.2 Smoothing process.

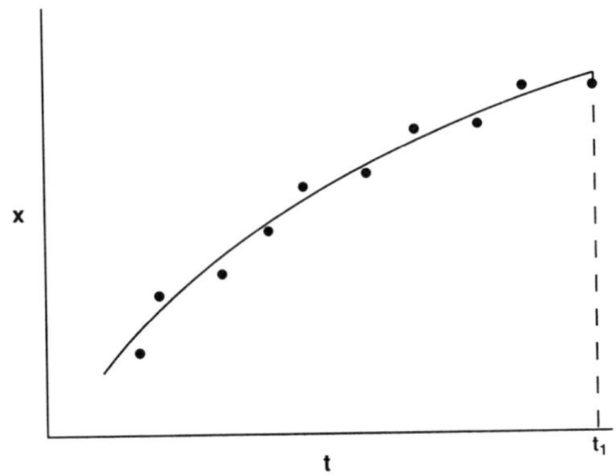

Figure 10.3 Filtering process.

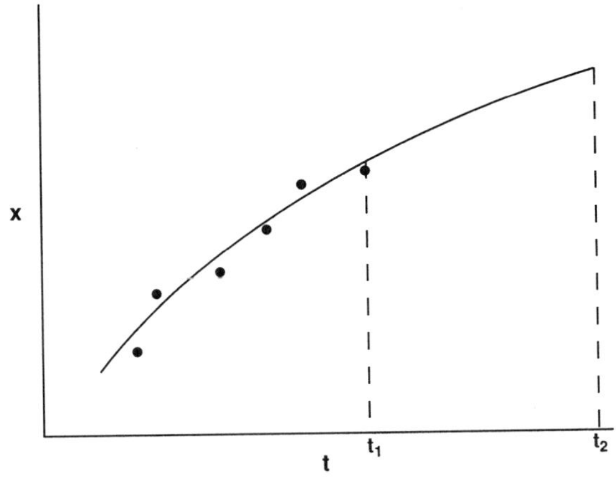

Figure 10.4 Prediction process.

with

$$\mathcal{E}\left\{\boldsymbol{w}(t)\right\} = \mathbf{0} \tag{10.2.2}$$

$$\text{cov}\left\{\boldsymbol{w}(t)\right\} = \boldsymbol{Q} \tag{10.2.3}$$

and the initial condition

$$\boldsymbol{x}(t_0) = \overline{\boldsymbol{x}}_0 + \boldsymbol{w}_0 \tag{10.2.4}$$

where

$$\mathcal{E}\left\{\boldsymbol{x}(t_0)\right\} = \overline{\boldsymbol{x}}_0 \tag{10.2.5}$$

$$\text{cov}\left\{\boldsymbol{x}(t_0)\right\} = \boldsymbol{P}_0 \tag{10.2.6}$$

The measurement model is

$$\boldsymbol{y}(t) = \boldsymbol{H}\boldsymbol{x}(t) + \boldsymbol{v}(t) \tag{10.2.7}$$

with

$$\mathcal{E}\left\{\boldsymbol{v}(t)\right\} = \mathbf{0} \tag{10.2.8}$$

$$\text{cov}\left\{\boldsymbol{v}(t)\right\} = \boldsymbol{R} \tag{10.2.9}$$

We want to find the optimal estimate of the state that minimizes the quadratic performance functional

$$\begin{aligned} J = {} & \frac{1}{2}\left(\boldsymbol{x}(t_0) - \overline{\boldsymbol{x}}_0\right)^T \boldsymbol{P}_0^{-1}\left(\boldsymbol{x}(t_0) - \overline{\boldsymbol{x}}_0\right) \\ & + \frac{1}{2}\int_{t_0}^{t_f}\left\{\left(\dot{\boldsymbol{x}} - \boldsymbol{F}(t)\boldsymbol{x}\right)^T \boldsymbol{Q}^{-1}\left(\dot{\boldsymbol{x}} - \boldsymbol{F}(t)\boldsymbol{x}\right)\right. \\ & \left. + (\boldsymbol{y} - \boldsymbol{H}\boldsymbol{x})^T \boldsymbol{R}^{-1}(\boldsymbol{y} - \boldsymbol{H}\boldsymbol{x})\right\} dt \end{aligned} \tag{10.2.10}$$

By using the inverse of the covariance matrices in the performance functional, we are weighting more heavily the more accurate states and measurements in the optimization. This also is a consistent means of combining both state and measurement uncertainties into one performance measure of an optimal estimate. This problem can be cased as an optimal control problem by defining a control vector as

$$\boldsymbol{u} = \dot{\boldsymbol{x}} - \boldsymbol{F}\boldsymbol{x} \tag{10.2.11}$$

The state equations become

$$\dot{\boldsymbol{x}} = \boldsymbol{F}\boldsymbol{x} + \boldsymbol{u} \tag{10.2.12}$$

and the performance functional

$$\begin{aligned} J &= \frac{1}{2}\left(\boldsymbol{x}(t_0) - \overline{\boldsymbol{x}}_0\right)^T \boldsymbol{P}_0^{-1}\left(\boldsymbol{x}(t_0) - \overline{\boldsymbol{x}}_0\right) \\ &+ \frac{1}{2}\int_{t_0}^{t_f}\left\{\boldsymbol{u}^T\boldsymbol{Q}^{-1}\boldsymbol{u} + (\boldsymbol{y} - \boldsymbol{H}\boldsymbol{x})^T \boldsymbol{R}^{-1}(\boldsymbol{y} - \boldsymbol{H}\boldsymbol{x})\right\} dt \end{aligned} \tag{10.2.13}$$

The Hamiltonian for this problem is

$$H = \frac{1}{2}\left[\boldsymbol{u}^T\boldsymbol{Q}^{-1}\boldsymbol{u} + (\boldsymbol{y} - \boldsymbol{H}\boldsymbol{x})^T \boldsymbol{R}^{-1}(\boldsymbol{y} - \boldsymbol{H}\boldsymbol{x})\right] + \boldsymbol{\lambda}^T(\boldsymbol{F}\boldsymbol{x} + \boldsymbol{u}) \tag{10.2.14}$$

and the necessary conditions for minimization of the performance functional are as follows:

Costate Equations

$$\dot{\boldsymbol{\lambda}} = -\frac{\partial H}{\partial \boldsymbol{x}} = -\boldsymbol{H}^T\boldsymbol{R}^{-1}\boldsymbol{H}\boldsymbol{x} + \boldsymbol{H}^T\boldsymbol{R}^{-1}\boldsymbol{y} - \boldsymbol{F}^T\boldsymbol{\lambda} \tag{10.2.15}$$

Transversality

$$\left(\frac{\partial h}{\partial \boldsymbol{x}} - \boldsymbol{\lambda}\right)^T_{t_f} \delta\boldsymbol{x}_f + \left(\frac{\partial h}{\partial \boldsymbol{x}} + \boldsymbol{\lambda}\right)^T_{t_0} \delta\boldsymbol{x}_0 = 0 \tag{10.2.16}$$

Since $\boldsymbol{x}(0)$ is free, we have

$$\left.\frac{\partial h}{\partial \boldsymbol{x}}\right|_{t_0} + \boldsymbol{\lambda}(t_0) = 0 \tag{10.2.17}$$

which is evaluated as

$$\boldsymbol{P}_0^{-1}\left(\boldsymbol{x}(t_0) - \overline{\boldsymbol{x}}_0\right) + \boldsymbol{\lambda}(t_0) = 0 \tag{10.2.18}$$

or

$$\boldsymbol{x}(t_0) = \overline{\boldsymbol{x}}_0 + \boldsymbol{P}_0\boldsymbol{\lambda}(t_0) \tag{10.2.19}$$

The final state is also free, which leads to

$$\boldsymbol{\lambda}(t_f) = \boldsymbol{0} \tag{10.2.20}$$

Optimal Control

$$\frac{\partial H}{\partial \boldsymbol{u}} = \boldsymbol{0} = \boldsymbol{Q}^{-1}\boldsymbol{u} + \boldsymbol{\lambda} \tag{10.2.21}$$

which is

$$\boldsymbol{u} = -\boldsymbol{Q}\boldsymbol{\lambda} \tag{10.2.22}$$

Let us now restrict our attention to the filtering problem which is determining optimal estimates at the final time given data up to the final time. We denote optimal filtering estimates as

$$\hat{\boldsymbol{x}}\,(t_f/t_f) \qquad \hat{\boldsymbol{u}}\,(t_f/t_f) \tag{10.2.23}$$

which are optimal estimates at time t_f with data $\boldsymbol{y}(t)$ up to time t_f.

We introduce a Riccati transformation,

$$\hat{\boldsymbol{x}}(t/t_f) = \boldsymbol{z}(t) - \boldsymbol{P}(t)\boldsymbol{\lambda}(t) \tag{10.2.24}$$

We want to determine the functions $\boldsymbol{z}(t)$ and $\boldsymbol{P}(t)$ that are consistent with the necessary conditions for optimality. The optimal estimates are given by the state model with the optimal control as

$$\dot{\hat{\boldsymbol{x}}}(t/t_f) = \boldsymbol{F}(t)\hat{\boldsymbol{x}}(t/t_f) - \boldsymbol{Q}\boldsymbol{\lambda}(t) \tag{10.2.25}$$

Using the Riccati transformation to evaluate both $\hat{\boldsymbol{x}}(t/t_f)$ and $\dot{\hat{\boldsymbol{x}}}(t/t_f)$, we have

$$\dot{\boldsymbol{z}}(t) - \dot{\boldsymbol{P}}(t)\boldsymbol{\lambda}(t) - \boldsymbol{P}(t)\dot{\boldsymbol{\lambda}}(t) = \boldsymbol{F}(t)\left[\boldsymbol{z}(t) - \boldsymbol{P}(t)\boldsymbol{\lambda}(t)\right] - \boldsymbol{Q}\boldsymbol{\lambda}(t) \tag{10.2.26}$$

or using the definition of the costates (Equation (10.2.15)), we have

$$\begin{aligned}\dot{\boldsymbol{z}}(t) &- \dot{\boldsymbol{P}}(t)\boldsymbol{\lambda}(t) - \boldsymbol{P}(t)\left[-\boldsymbol{H}^T\boldsymbol{R}^{-1}\boldsymbol{H}\left(\boldsymbol{z}(t) - \boldsymbol{P}(t)\boldsymbol{\lambda}(t)\right) + \boldsymbol{H}^T\boldsymbol{R}^{-1}\boldsymbol{y}(t)\right.\\ &\left. - \boldsymbol{F}^T(t)\boldsymbol{\lambda}(t)\right] = \boldsymbol{F}(t)\left[\boldsymbol{z}(t) - \boldsymbol{P}(t)\boldsymbol{\lambda}(t)\right] - \boldsymbol{Q}\boldsymbol{\lambda}(t)\end{aligned} \tag{10.2.27}$$

Putting all the $\boldsymbol{z}$ terms on the left-hand side and the $\boldsymbol{\lambda}$ terms on the right-hand side gives

$$\begin{aligned}\dot{\boldsymbol{z}}(t) &- \boldsymbol{P}(t)\boldsymbol{H}^T\boldsymbol{R}^{-1}\left(\boldsymbol{y}(t) - \boldsymbol{H}\boldsymbol{z}(t)\right) - \boldsymbol{F}(t)\boldsymbol{z}\\ &= \left[\dot{\boldsymbol{P}}(t) + \boldsymbol{P}(t)\boldsymbol{H}^T\boldsymbol{R}^{-1}\boldsymbol{H}\boldsymbol{P}(t) - \boldsymbol{P}(t)\boldsymbol{F}^T - \boldsymbol{F}(t)\boldsymbol{P}(t) - \boldsymbol{Q}\right]\boldsymbol{\lambda}(t)\end{aligned} \tag{10.2.28}$$

The only way that this relation can always be true is if both sides of Equation (10.2.28) are zero. This leads to

$$\dot{\boldsymbol{z}} = \boldsymbol{F}(t)\boldsymbol{z}(t) + \boldsymbol{P}(t)\boldsymbol{H}^T\boldsymbol{R}^{-1}(\boldsymbol{y} - \boldsymbol{H}\boldsymbol{z}) \tag{10.2.29}$$

$$\dot{\boldsymbol{P}}(t) = \boldsymbol{P}(t)\boldsymbol{F}^T + \boldsymbol{F}(t)\boldsymbol{P}(t) + \boldsymbol{Q} - \boldsymbol{P}(t)\boldsymbol{H}^T\boldsymbol{R}^{-1}\boldsymbol{H}\boldsymbol{P}(t) \tag{10.2.30}$$

From the initial transversality condition of Equation (10.2.19) and the Riccati transformation, we have

$$\boldsymbol{z}(0) = \overline{\boldsymbol{x}}_0 \tag{10.2.31}$$

$$\boldsymbol{P}(0) = \boldsymbol{P}_0 \tag{10.2.32}$$

Both boundary conditions are at the initial time; therefore, both Equation (10.2.31) and Equation (10.2.32) can be integrated forward in time.

For the filtering problem, we are trying to evaluate $\hat{\boldsymbol{x}}(t_f/t_f)$. This is the least squares estimate at time t_f conditional on all the data up to the final time t_f. The final transversality condition is

$$\boldsymbol{\lambda}(t_f) = \boldsymbol{0} \tag{10.2.33}$$

which means that from the Riccati transformation, we obtain

$$\hat{\boldsymbol{x}}(t_f/t_f) = \boldsymbol{z}(t_f) \tag{10.2.34}$$

Therefore, for filtering, we have

$$\dot{\hat{\boldsymbol{x}}}(t/t) = \boldsymbol{F}(t)\hat{\boldsymbol{x}}(t/t) + \boldsymbol{P}(t)\boldsymbol{H}^T\boldsymbol{R}^{-1}\left(\boldsymbol{y}(t) - \boldsymbol{H}\hat{\boldsymbol{x}}(t/t)\right) \tag{10.2.35}$$

$$\dot{\boldsymbol{P}} = \boldsymbol{P}(t)\boldsymbol{F}^T(t) + \boldsymbol{F}(t)\boldsymbol{P}(t) + \boldsymbol{Q} - \boldsymbol{P}(t)\boldsymbol{H}^T\boldsymbol{R}^{-1}\boldsymbol{H}\boldsymbol{P}(t) \tag{10.2.36}$$

with the initial conditions

$$\hat{\boldsymbol{x}}(t_0) = \overline{\boldsymbol{x}}_0 \tag{10.2.37}$$

$$\boldsymbol{P}(t_0) = \boldsymbol{P}_0 \tag{10.2.38}$$

Equations (10.2.35) and (10.2.36) are known as the Kalman filter and were first derived by Kalman and Bucy (1961). We can define the filter gain as

$$\boldsymbol{K}(t) = \boldsymbol{P}(t)\boldsymbol{H}^T\boldsymbol{R}^{-1} \tag{10.2.39}$$

As the final time progresses, Equation (10.2.36) will approach a steady state so that the steady-state filter gain is

$$\boldsymbol{K}_{ss} = \boldsymbol{P}_{ss}\boldsymbol{H}^T\boldsymbol{R}^{-1} \tag{10.2.40}$$

and the optimal state estimates are

$$\dot{\hat{\boldsymbol{x}}}(t/t) = \boldsymbol{F}(t)\hat{\boldsymbol{x}}(t/t) + \boldsymbol{K}_{ss}\left(\boldsymbol{y}(t) - \boldsymbol{H}\hat{\boldsymbol{x}}(t/t)\right) \tag{10.2.41}$$

The Kalman filter has several important properties. First, it is an unbiased estimate of the state. If we define the error in the state estimate to be

$$\boldsymbol{e}(t) = \boldsymbol{x}(t) - \hat{\boldsymbol{x}}(t/t) \tag{10.2.42}$$

then the expected value of that error is

$$\mathcal{E}\left\{e(t)\right\} = \mathbf{0} \tag{10.2.43}$$

Therefore, the estimate is not biased. The covariance of the error is

$$cov\left\{\boldsymbol{e}(t)\right\} = \boldsymbol{P}(t) \tag{10.2.44}$$

Therefore, the $\boldsymbol{P}(t)$ matrix is the covariance of the current error and is called the minimum error variance.

10.3 The Smoothing Problem

For the smoothing problem, we need to evaluate the estimate $\hat{\boldsymbol{x}}(t/t_f)$ where t is less than t_f. To do this, we first solve the filtering problem

$$\boldsymbol{z}(t) = \hat{\boldsymbol{x}}(t/t) \tag{10.3.1}$$

for both $\boldsymbol{x}(t/t)$ and $\boldsymbol{P}(t)$ forward in time up to the final time t_f:

$$\hat{\boldsymbol{x}}(t/t) = \boldsymbol{F}(t)\hat{\boldsymbol{x}}(t/t) + \boldsymbol{P}(t)\boldsymbol{H}^T\boldsymbol{R}^{-1}\left(\boldsymbol{y}(t) - \boldsymbol{H}\hat{\boldsymbol{x}}(t/t)\right) \tag{10.3.2}$$

$$\dot{\boldsymbol{P}}(t) = \boldsymbol{P}(t)\boldsymbol{F}^T(t) + \boldsymbol{F}(t)\boldsymbol{P}(t) + \boldsymbol{Q} - \boldsymbol{P}(t)\boldsymbol{H}^T\boldsymbol{R}^{-1}\boldsymbol{H}\boldsymbol{P}(t) \tag{10.3.3}$$

Next we solve the model backwards in time from the final time to the time for which we desire the smoothed value

$$\dot{\boldsymbol{x}} = \boldsymbol{F}\boldsymbol{x} + \boldsymbol{u} \tag{10.3.4}$$

which is for the optimal estimates

$$\dot{\boldsymbol{x}} = \boldsymbol{F}\boldsymbol{x} - \boldsymbol{Q}\boldsymbol{\lambda} \tag{10.3.5}$$

Using the Riccati equation, we have

$$\dot{\hat{\boldsymbol{x}}}(t/t_f) = \boldsymbol{F}(t)\hat{x}(t/t_f) + \boldsymbol{Q}\boldsymbol{P}^{-1}(t)\left(\left(\hat{\boldsymbol{x}}(t/t_f) - \hat{\boldsymbol{x}}(t/t)\right)\right) \tag{10.3.6}$$

This equation can be integrated backwards since the $\hat{\boldsymbol{x}}(t_f/t_f)$ is known from the filtering problem. Therefore the smoothing problem can be approached as first a forward-in-time integration of the filtering problem followed by a backwards-in-time integration of Equation (10.3.6).

10.4 The Prediction Problem

For the prediction problem, we want to evaluate the estimate $\hat{\boldsymbol{x}}(t/t_0)$ where $t > t_0$. The time t_0 is the last time for which data are available. To implement this procedure, we can use the filter results directly with $\boldsymbol{R}^{-1} = \boldsymbol{0}$ for $t > t_0$. The optimal estimate is therefore given by

$$\dot{\hat{\boldsymbol{x}}}(t/t_0) = \boldsymbol{F}(t)\hat{\boldsymbol{x}}(t/t_0)$$

with the initial condition

$$\hat{\boldsymbol{x}}(t_0/t_0) \tag{10.4.1}$$

where $\hat{\boldsymbol{x}}(t_0/t_0)$ is the filter estimate at time t_0. The prediction covariance matrix is given by the solution of the equation

$$\dot{\boldsymbol{P}}(t/t_0) = \boldsymbol{P}(t/t_0)\boldsymbol{F}^T + \boldsymbol{F}\boldsymbol{P}(t/t_0) + \boldsymbol{Q} \tag{10.4.2}$$

with

$$\boldsymbol{P}(t_0/t_0) = \boldsymbol{P}(t_0) \tag{10.4.3}$$

where $\boldsymbol{P}(t_0)$ is the filter covariance at t_0. Note that $\boldsymbol{P}(t/t_0)$ will tend to increase with time. This means that the estimates without measurements get more uncertain as the time between measurements increases.

10.5 The Extended Kalman Filter

We now want to consider nonlinear systems. The nonlinear system model can be expressed as

$$\dot{\boldsymbol{x}} = \boldsymbol{f}(\boldsymbol{x}, t) + \boldsymbol{w}(t) \tag{10.5.1}$$

with

$$\mathcal{E}(\boldsymbol{w}) = \boldsymbol{0} \tag{10.5.2}$$

$$\text{cov}\,(\boldsymbol{w}) = \boldsymbol{Q} \tag{10.5.3}$$

and the initial state

$$\boldsymbol{x}(0) = \overline{\boldsymbol{x}}_0 + \overline{\boldsymbol{w}}_0 \tag{10.5.4}$$

with

$$\mathcal{E}\,(\boldsymbol{w}_0) = \boldsymbol{0} \tag{10.5.5}$$

$$\text{cov}\,(\boldsymbol{w}_0) = \boldsymbol{P}_0 \tag{10.5.6}$$

The measurement model is

$$\boldsymbol{y}(t) = \boldsymbol{h}(\boldsymbol{x}, t) + \boldsymbol{v}(t) \tag{10.5.7}$$

with

$$\mathcal{E}(\boldsymbol{v}) = \boldsymbol{0} \tag{10.5.8}$$

$$\operatorname{cov}(\boldsymbol{v}) = \boldsymbol{R} \tag{10.5.9}$$

Unfortunately for nonlinear systems the conditional probability distributions for the state evolving in time are not Gaussian even when $\boldsymbol{w}(t)$, $\boldsymbol{w}_0$, and $\boldsymbol{v}(t)$ are assumed to have Gaussian distributions. This means that an infinite number of expected values (moments) are required to determine the exact distribution and that these moments are coupled. That is, the first moment depends upon the second, the second upon the third, etc. A rigorous treatment involves Ito calculus (Jazwinski, 1970). One useful approximate solution is that of the extended Kalman filter approach. In this approach, we define the performance functional to be minimized as

$$\begin{aligned} J &= \frac{1}{2}\left(\boldsymbol{x}(t_0) - \overline{\boldsymbol{x}}_0\right)^T \boldsymbol{P}_0^{-1}\left(\boldsymbol{x}(t_0) - \overline{\boldsymbol{x}}_0\right) \\ &\quad + \frac{1}{2}\int_{t_0}^{t_f} \left\{(\dot{\boldsymbol{x}} - \boldsymbol{f})^T \boldsymbol{Q}^{-1}(\dot{\boldsymbol{x}} - \boldsymbol{f}) + (\boldsymbol{y} - \boldsymbol{h})^T \boldsymbol{R}^{-1}(\boldsymbol{y} - \boldsymbol{h})\right\} dt \end{aligned} \tag{10.5.10}$$

Again we let

$$\boldsymbol{u} = \dot{\boldsymbol{x}} - \boldsymbol{f} \tag{10.5.11}$$

The optimal control problem is formulated in terms of the Hamiltonian,

$$H = \frac{1}{2}\left[\boldsymbol{u}^T \boldsymbol{Q}^{-1}\boldsymbol{u} + (\boldsymbol{y} - \boldsymbol{h})^T \boldsymbol{R}^{-1}(\boldsymbol{y} - \boldsymbol{h})\right] + \boldsymbol{\lambda}^T(\boldsymbol{f} + \boldsymbol{u}) \tag{10.5.12}$$

and the necessary conditions for a minimum of the performance functional are as follows:

Costate Equations

$$\dot{\boldsymbol{\lambda}} = \left(\frac{\partial \boldsymbol{h}}{\partial \boldsymbol{x}}\right)^T \boldsymbol{R}^{-1}(\boldsymbol{y} - \boldsymbol{h}) - \left(\frac{\partial \boldsymbol{f}}{\partial \boldsymbol{x}}\right)^T \boldsymbol{\lambda} \tag{10.5.13}$$

Transversality

$$\boldsymbol{\lambda}(t_f) = \boldsymbol{0} \tag{10.5.14}$$

$$\boldsymbol{x}(t_0) = \overline{\boldsymbol{x}}_0 - \boldsymbol{P}_0\boldsymbol{\lambda}(t_0) \tag{10.5.15}$$

Optimal Control

$$\boldsymbol{u} = -\boldsymbol{Q}\boldsymbol{\lambda} \tag{10.5.16}$$

If we assume the Riccati transformation

$$\hat{\boldsymbol{x}}(t/t_f) = \boldsymbol{z}(t) - \boldsymbol{P}(t)\boldsymbol{\lambda}(t) \tag{10.5.17}$$

then the optimal state response is

$$\dot{\hat{\boldsymbol{x}}}(t/t_f) = \boldsymbol{f}(\hat{\boldsymbol{x}}) - \boldsymbol{Q}\boldsymbol{\lambda}(t) \tag{10.5.18}$$

which can be evaluated as

$$\dot{\boldsymbol{z}}(t) - \dot{\boldsymbol{P}}(t)\boldsymbol{\lambda}(t) - \boldsymbol{P}(t)\dot{\boldsymbol{\lambda}}(t) = \boldsymbol{f}(\hat{\boldsymbol{x}}) - \boldsymbol{Q}\boldsymbol{\lambda}(t) \tag{10.5.19}$$

We now linearize the $\boldsymbol{f}(\hat{\boldsymbol{x}})$ function to obtain the first-order linearized filter

$$\dot{\boldsymbol{z}}(t) - \dot{\boldsymbol{P}}(t)\boldsymbol{\lambda}(t) - \boldsymbol{P}(t)\left[\left(\frac{\partial \boldsymbol{h}}{\partial \boldsymbol{x}}\right)^T \boldsymbol{R}^{-1}\left(\boldsymbol{y} - \left(\frac{\partial \boldsymbol{h}}{\partial \boldsymbol{x}}\right)\left(\boldsymbol{z} - \boldsymbol{P}(t)\boldsymbol{\lambda}\right)\right) - \left(\frac{\partial \boldsymbol{f}}{\partial \boldsymbol{x}}\right)^T \boldsymbol{\lambda}\right] = \frac{\partial \boldsymbol{f}}{\partial \boldsymbol{x}}\left[\boldsymbol{z} - \boldsymbol{P}(t)\boldsymbol{\lambda}(t)\right] - \boldsymbol{Q}\boldsymbol{\lambda}(t) \tag{10.5.20}$$

Equation (10.5.20) can only hold in general if both

$$\dot{\boldsymbol{z}}(t) = \frac{\partial \boldsymbol{f}}{\partial \boldsymbol{x}}\boldsymbol{z} + \boldsymbol{P}(t)\left(\frac{\partial \boldsymbol{h}}{\partial \boldsymbol{x}}\right)^T \boldsymbol{R}^{-1}\left[\boldsymbol{y} - \left(\frac{\partial \boldsymbol{h}}{\partial \boldsymbol{x}}\right)\boldsymbol{z}\right] \tag{10.5.21}$$

and

$$\dot{\boldsymbol{P}}(t) = \boldsymbol{P}(t)\left(\frac{\partial \boldsymbol{f}}{\partial \boldsymbol{x}}\right)^T + \left(\frac{\partial \boldsymbol{f}}{\partial \boldsymbol{x}}\right)\boldsymbol{P}(t) - \boldsymbol{P}(t)\left[\left(\frac{\partial \boldsymbol{h}}{\partial \boldsymbol{x}}\right)^T \boldsymbol{R}^{-1}\left(\frac{\partial \boldsymbol{h}}{\partial \boldsymbol{x}}\right)\right]\boldsymbol{P}(t) + \boldsymbol{Q} \tag{10.5.22}$$

Using the transversality conditions, the filtering problem is

$$\hat{\boldsymbol{x}}(t_f/t_f) = \boldsymbol{z}(t_f) - \boldsymbol{P}(t_f)\boldsymbol{\lambda}(t_f) \tag{10.5.23}$$

with

$$\boldsymbol{\lambda}(t_f) = 0 \tag{10.5.24}$$

Therefore

$$\hat{\boldsymbol{x}}(t_f/t_f) = \boldsymbol{z}(t_f) \tag{10.5.25}$$

The extended Kalman filter is therefore

$$\dot{\hat{\boldsymbol{x}}}(t/t) = \boldsymbol{f}\left(\hat{\boldsymbol{x}}(t/t)\right) + \boldsymbol{P}(t)\left(\frac{\partial \boldsymbol{h}}{\partial \boldsymbol{x}}\right)^T \boldsymbol{R}^{-1}\left(\boldsymbol{y}(t) - \boldsymbol{h}\right) \tag{10.5.26}$$

$$\dot{\boldsymbol{P}}(t) = \boldsymbol{P}(t)\left(\frac{\partial \boldsymbol{f}}{\partial \boldsymbol{x}}\right)^T + \left(\frac{\partial \boldsymbol{f}}{\partial \boldsymbol{x}}\right)\boldsymbol{P}(t) - \boldsymbol{P}(t)\left[\left(\frac{\partial \boldsymbol{h}}{\partial \boldsymbol{x}}\right)^T \boldsymbol{R}^{-1}\left(\frac{\partial \boldsymbol{h}}{\partial \boldsymbol{x}}\right)\right]\boldsymbol{P}(t) + \boldsymbol{Q} \tag{10.5.27}$$

with

$$\hat{\boldsymbol{x}}(t_0) = \overline{\boldsymbol{x}}_0 \qquad \boldsymbol{P}(0) = \boldsymbol{P}_0 \tag{10.5.28}$$

The optimal state estimate $\hat{\boldsymbol{x}}(t)$ and the sensitivity matrix $\boldsymbol{P}(t)$ have to be solved simultaneously due to the partial derivative terms that appear in the $\boldsymbol{P}(t)$ differential equation set. Nonetheless, the extended Kalman filter defines a reasonable procedure for obtaining optimal estimates of the states for nonlinear systems.

10.6 Discrete Kalman Filter

There are a number of approaches to developing the discrete Kalman filter. One that gives good insight to the problem is that of Bryson and Ho (1975).

10.6.1 Weighted Least Squares

We start by considering a static system. This will be the development of a sequential least squares algorithm.

Suppose we want to estimate the state vector $\boldsymbol{x}$ of a static system using a p component measurement vector $\boldsymbol{y}$ containing random errors $\boldsymbol{v}$ which are independent of the state $\boldsymbol{x}$. Therefore,

$$\boldsymbol{y} = \boldsymbol{H}\boldsymbol{x} + \boldsymbol{v} \tag{10.6.1}$$

where $\boldsymbol{H}$ is the measurement matrix (p x n) in size. The random errors $\boldsymbol{v}$ will be described by a zero mean and known covariance matrix $\boldsymbol{R}$ such that

$$\mathcal{E}(\boldsymbol{v}) = \boldsymbol{0} \tag{10.6.2}$$

$$cov(\boldsymbol{v}) = \boldsymbol{R} \tag{10.6.3}$$

Let us also suppose we had an estimate of the state before measurements were taken, $\overline{\boldsymbol{x}}$. Due to an imperfect model, we know that there will be a certain level of uncertainty in our previous estimate, which we can describe by a known (n x n) covariance matrix, $\boldsymbol{M}$,

$$\mathcal{E}\left[(\boldsymbol{x}-\overline{\boldsymbol{x}})(\boldsymbol{x}-\overline{\boldsymbol{x}})^T\right] = \boldsymbol{M} \qquad \text{or since } \boldsymbol{x} = \overline{\boldsymbol{x}} + \boldsymbol{w} \tag{10.6.4}$$

$$\text{cov}(\boldsymbol{w}) = \boldsymbol{M} \qquad \text{and } \mathcal{E}(\boldsymbol{w}) = \boldsymbol{0} \tag{10.6.5}$$

A good estimate for the state $\boldsymbol{x}$ would result from the minimization of a weighted-least-squares performance index,

$$J = \frac{1}{2}\left\{(\boldsymbol{x}-\overline{\boldsymbol{x}})^T \boldsymbol{M}^{-1}(\boldsymbol{x}-\overline{\boldsymbol{x}}) + (\boldsymbol{y}-\boldsymbol{H}\boldsymbol{x})^T \boldsymbol{R}^{-1}(\boldsymbol{y}-\boldsymbol{H}\boldsymbol{x})\right\} \tag{10.6.6}$$

Note that the weighting matrices used are $\boldsymbol{M}^{-1}$ and $\boldsymbol{R}^{-1}$ which give most weight to the more precise, and therefore less random, states. The first term of this performance index minimizes deviation between the new state estimate and the estimate before the new measurements are taken, while the second term minimizes deviations between the actual measurements and their computed values.

The differential of the function J is dJ and is given by

$$dJ = \boldsymbol{dx}^T \left[\boldsymbol{M}^{-1}(\boldsymbol{x} - \overline{\boldsymbol{x}}) - \boldsymbol{H}^T\boldsymbol{R}^{-1}(\boldsymbol{y} - \boldsymbol{Hx})\right] \tag{10.6.7}$$

The necessary condition for the extremum in J is that $dJ = 0$. Since $\boldsymbol{dx}$ is arbitrary, this means

$$\boldsymbol{0} = \left[\boldsymbol{M}^{-1}(\boldsymbol{x} - \overline{\boldsymbol{x}}) - \boldsymbol{H}^T\boldsymbol{R}^{-1}(\boldsymbol{y} - \boldsymbol{Hx})\right] \tag{10.6.8}$$

Solving for the optimal estimate of $\boldsymbol{x} = \hat{\boldsymbol{x}}$ gives

$$\hat{\boldsymbol{x}} = \overline{\boldsymbol{x}} + \boldsymbol{PH}^T\boldsymbol{R}^{-1}(\boldsymbol{y} - \boldsymbol{H}\overline{\boldsymbol{x}}) \tag{10.6.9}$$

where

$$\boldsymbol{P}^{-1} = \boldsymbol{M}^{-1} + \boldsymbol{H}^T\boldsymbol{R}^{-1}\boldsymbol{H} \tag{10.6.10}$$

The quantity $\boldsymbol{P}$ is the covariance matrix of the error in the estimate of $\hat{\boldsymbol{x}}$. This follows because

$$\mathcal{E}\left((\hat{\boldsymbol{x}} - \boldsymbol{x})(\hat{\boldsymbol{x}} - \boldsymbol{x})^T\right) = \boldsymbol{P} \tag{10.6.11}$$

The matrix $\boldsymbol{M}$ is the error covariance matrix before measurement, while $\boldsymbol{P}$ is the error covariance matrix after measurement. From Equation (10.6.10) we can see that $\boldsymbol{P}$ is never larger than $\boldsymbol{M}$ since $\boldsymbol{H}^T\boldsymbol{R}^{-1}\boldsymbol{H}$ is positive semi-definite.

The expected value of the objective function is $\dfrac{(n+p)}{2}$. This can be shown since the objective function can be written

$$J = \frac{1}{2}\text{tr}\left\{\boldsymbol{M}^{-1}(\boldsymbol{x} - \overline{\boldsymbol{x}})(\boldsymbol{x} - \overline{\boldsymbol{x}})^T\right\} + \frac{1}{2}\text{tr}\left\{\boldsymbol{R}^{-1}(\boldsymbol{y} - \boldsymbol{Hx})(\boldsymbol{y} - \boldsymbol{Hx})^T\right\} \tag{10.6.12}$$

where tr{} means the trace, which is the sum of the elements on the principle diagonal. Therefore,

$$\begin{aligned}\mathcal{E}(J) &= \frac{1}{2}\,\text{tr}\left\{\boldsymbol{M}^{-1}\Sigma\left((\boldsymbol{x} = \overline{\boldsymbol{x}})(\boldsymbol{x} - \overline{\boldsymbol{x}})^T\right) + \boldsymbol{R}^{-1}\Sigma\left((\boldsymbol{y} - \boldsymbol{Hx})(\boldsymbol{y} - \boldsymbol{Hx})^T\right)\right\} \\ &= \frac{1}{2}\,\text{tr}\left\{\boldsymbol{M}^{-1}\boldsymbol{M} + \boldsymbol{R}^{-1}\boldsymbol{R}\right\}\end{aligned} \tag{10.6.13}$$

or

$$\mathcal{E}(J) = \frac{n+p}{2} \tag{10.6.14}$$

Prior knowledge of $\boldsymbol{M}$ and $\boldsymbol{R}$ can sometimes be vague and uncertain. Therefore, as a check, after the estimation process has been completed, the actual value of J should be computed and it should be close to $(n+p)/2$, its prior expected value. If the actual value is not close to $(n+p)/2$, it should be adjusted by a scale factor to be so. This scale factor also affects the covariance matrix after measurement $\boldsymbol{P}$. This means all that was really needed was the relative magnitude of the elements of $\boldsymbol{M}$ and $\boldsymbol{R}$ since the final values of $\boldsymbol{M}$, $\boldsymbol{R}$, and $\boldsymbol{P}$ can be changed in order to be consistent with the data.

10.6.2 Optimal Filtering and Estimation for Single-Stage Linear System

Now consider a one-stage process

$$\boldsymbol{x}_1 = \boldsymbol{\phi}_0\boldsymbol{x}_0 + \boldsymbol{\Gamma}_0\boldsymbol{w}_0 \tag{10.6.15}$$

where $\boldsymbol{\phi}_0$ is a known state transition matrix, $\boldsymbol{\Gamma}_0$ is a known matrix, and the random noise is described by

$$\mathcal{E}(\boldsymbol{w}_0) = \boldsymbol{0} \tag{10.6.16}$$

$$\operatorname{cov}(\boldsymbol{w}_0) = \boldsymbol{Q}_0 \tag{10.6.17}$$

Thus, $\boldsymbol{w}_0$ is a forcing vector with zero mean and covariance $\boldsymbol{Q}_0$. The initial state is also described by its mean $\boldsymbol{x}_0$ and covariance $\boldsymbol{P}_0$.

$$\mathcal{E}(\boldsymbol{x}_0) = \boldsymbol{x}_0 \tag{10.6.18}$$

$$\operatorname{cov}(\boldsymbol{x}_0 - \overline{\boldsymbol{x}}_0) = \boldsymbol{P}_0 \tag{10.6.19}$$

From this information, it follows that $\boldsymbol{x}_1$ is also a random variable, and its mean and covariance are given by

$$\mathcal{E}(\boldsymbol{x}_1) = \overline{\boldsymbol{x}}_1 = \boldsymbol{\phi}_0\overline{\boldsymbol{x}}_0 + \boldsymbol{\Gamma}_0\overline{\boldsymbol{w}}_0 \tag{10.6.20}$$

$$\operatorname{cov}(\boldsymbol{x}_1 - \overline{\boldsymbol{x}}_1) = \boldsymbol{M}_1 = \boldsymbol{\phi}_0\boldsymbol{P}_0\boldsymbol{\phi}_0^T + \boldsymbol{\Gamma}_0\boldsymbol{Q}_0\boldsymbol{\Gamma}_0^T \tag{10.6.21}$$

Note that since $\boldsymbol{\Gamma}_0\boldsymbol{Q}_0\boldsymbol{\Gamma}_0^T$ is nonnegative, the effect of $\boldsymbol{w}_0$ is to increase the uncertainty in our knowledge of the state $\boldsymbol{x}_1$.

Now let us take measurements, $\boldsymbol{y}_1$, after the transition to state one so that

$$\boldsymbol{y}_1 = \boldsymbol{H}\boldsymbol{x}_1 + \boldsymbol{v}_1 \tag{10.6.22}$$

where

$$\mathcal{E}\,(\boldsymbol{v}_1) = \mathbf{0} \tag{10.6.23}$$

$$\mathcal{E}\left(\boldsymbol{v}_1\boldsymbol{v}_1^T\right) = \boldsymbol{R} \tag{10.6.24}$$

Now the optimal estimate for $\boldsymbol{x}_1$ is obtained using the quadratic performance index

$$J=\frac{1}{2}\left\{(\boldsymbol{x}_1-\overline{\boldsymbol{x}})^T\boldsymbol{M}_1^{-1}(\boldsymbol{x}_1-\overline{\boldsymbol{x}}_1)+(\boldsymbol{y}_1-\boldsymbol{H}\boldsymbol{x}_1)^T\boldsymbol{R}^{-1}(\boldsymbol{y}_1-\boldsymbol{H}\boldsymbol{x}_1)\right\} \tag{10.6.25}$$

This is approached in the same manner as that in the previous section and leads to the optimal estimate

$$\hat{\boldsymbol{x}}_1 = \overline{\boldsymbol{x}}_1 + \boldsymbol{P}_1\boldsymbol{H}^T\boldsymbol{R}^{-1}(\boldsymbol{y}_1 - \boldsymbol{H}\overline{\boldsymbol{x}}_1) \tag{10.6.26}$$

where

$$\boldsymbol{P}_1 = \left(\boldsymbol{M}_1^{-1} + \boldsymbol{H}^T\boldsymbol{R}^{-1}\boldsymbol{H}\right)^{-1} \tag{10.6.27}$$

Using the matrix inversion lemma, we have

$$\boldsymbol{P}_1 = \boldsymbol{M}_1 - \boldsymbol{M}_1\boldsymbol{H}^T\left(\boldsymbol{H}\boldsymbol{M}_1\boldsymbol{H}^T + \boldsymbol{R}\right)^{-1}\boldsymbol{H}\boldsymbol{M}_1 \tag{10.6.28}$$

$\boldsymbol{P}_1$ is the state error covariance matrix after measurement while $\boldsymbol{M}_1$ is the state error covariance matrix before measurement. From Equation (10.6.27) we see that $\boldsymbol{P}_1$ is never larger than $\boldsymbol{M}_1$ so that the effect of measurement is to reduce the uncertainty of the estimate for $\boldsymbol{x}_1$.

10.6.3 Optimal Filtering and Estimation for Multistage Systems

Consider the linear multistage system

$$\boldsymbol{x}_{i+1} = \boldsymbol{\phi}_i\boldsymbol{x}_i + \boldsymbol{\Gamma}_i\boldsymbol{w}_i \tag{10.6.29}$$

with the measurement model

$$\boldsymbol{y}_i = \boldsymbol{H}_i\boldsymbol{x}_i + \boldsymbol{v}_i \tag{10.6.30}$$

where

$$\mathcal{E}\,(\boldsymbol{x}_0) = \overline{\boldsymbol{x}}_0 \tag{10.6.31}$$

$$\mathrm{cov}\,(\boldsymbol{x}_0 - \overline{\boldsymbol{x}}_0) = \boldsymbol{P}_0 \tag{10.6.32}$$

$$\mathcal{E}\,(\boldsymbol{w}_i) = \mathbf{0} \tag{10.6.33}$$

$$\mathrm{cov}\,(\boldsymbol{w}_i) = \boldsymbol{Q}_i \tag{10.6.34}$$

$$\mathcal{E}\,(\boldsymbol{v}_i) = \mathbf{0} \tag{10.6.35}$$

$$\mathrm{cov}\,(\boldsymbol{v}_i) = \boldsymbol{R}_i \tag{10.6.36}$$

The optimal filter has the same form as the filter for the single stage process so that

$$\hat{\boldsymbol{x}}_i = \overline{\boldsymbol{x}}_i + \boldsymbol{K}_i(\boldsymbol{y}_i - \boldsymbol{H}_i\overline{\boldsymbol{x}}_i) \tag{10.6.37}$$

where

$$\boldsymbol{K}_i = \boldsymbol{P}_i\boldsymbol{H}_i^T\boldsymbol{R}_i^{-1} \tag{10.6.38}$$

$$\boldsymbol{P}_i = \boldsymbol{M}_i - \boldsymbol{M}_i\boldsymbol{H}_i^T\left(\boldsymbol{H}_i\boldsymbol{M}_i\boldsymbol{H}_i^T + \boldsymbol{R}_i\right)^{-1}\boldsymbol{H}_i\boldsymbol{M}_i \tag{10.6.39}$$

$$\boldsymbol{M}_{i+1} = \boldsymbol{\phi}_i\boldsymbol{P}_i\boldsymbol{\phi}_i^T + \boldsymbol{\Gamma}_i\boldsymbol{Q}_i\boldsymbol{\Gamma}_i^T \tag{10.6.40}$$

This is the Kalman filter for a linear multistage system. The form of the filter, $\boldsymbol{K}_i$, is a proportional correction to the state computed by the model before measurement. The correction term is the error between the actual measurement, $\boldsymbol{y}_i$, and model predicted values for those measurements, $\boldsymbol{H}_i\overline{\boldsymbol{x}}_i$.

Also, the propagation of state error covariance (Equations (10.6.39) and (10.6.40)) is independent from the measurements, $\boldsymbol{y}_i$. Therefore, the covariance matrices can be computed beforehand and stored for on-line use. For stationary processes (when $\boldsymbol{Q}$, $\boldsymbol{R}$, $\boldsymbol{\phi}$, $\boldsymbol{\Gamma}$, and $\boldsymbol{H}$ are constants and not a function of time), the filter covariance matrices $\boldsymbol{P}_i$ and $\boldsymbol{M}_i$ will approach a steady-state constant value ($\boldsymbol{P}_{ss}$ and $\boldsymbol{M}_{ss}$). The steady state is usually reached after the application of the filter for several process time constants. Use of the steady-state filter is attractive since only a constant filter gain of $\boldsymbol{K}_{ss}$ need be stored for on-line application.

10.6.4 The Extended Kalman Filter

For a nonlinear system, we have

$$\boldsymbol{x}_{i+1} = \boldsymbol{f}_i(\boldsymbol{x}_i) + \boldsymbol{\Gamma}_i\boldsymbol{w}_i \tag{10.6.41}$$

with the measurement model

$$\boldsymbol{y}_i = \boldsymbol{h}_i(\boldsymbol{x}_i) + \boldsymbol{v}_i \tag{10.6.42}$$

Where $\boldsymbol{f}_i$ and $\boldsymbol{h}_i$ are nonlinear functions, we use the following extended Kalman filter algorithm:

$$\hat{\boldsymbol{x}}_i = \overline{\boldsymbol{x}}_i + \boldsymbol{K}_i(\boldsymbol{y}_i - \boldsymbol{h}_i) \tag{10.6.43}$$

where

$$\overline{\boldsymbol{x}}_i = \boldsymbol{f}_{i-1}(\hat{\boldsymbol{x}}_{i-1}) + \boldsymbol{\Gamma}_{i-1}\overline{\boldsymbol{w}}_{i-1} \tag{10.6.44}$$

$$\boldsymbol{K}_i = \boldsymbol{P}_i\left(\frac{\partial \boldsymbol{h}_i}{\partial \boldsymbol{x}}\right)^T\boldsymbol{R}^{-1} \tag{10.6.45}$$

$$\boldsymbol{P} = \boldsymbol{M}_i - \boldsymbol{M}_i \left(\frac{\partial \boldsymbol{h}_i}{\partial \boldsymbol{x}_i}\right)^T \left[\left(\frac{\partial \boldsymbol{h}_i}{\partial \boldsymbol{x}_i}\right) \boldsymbol{M}_i \left(\frac{\partial \boldsymbol{h}_i}{\partial \boldsymbol{x}_i}\right)^T + \boldsymbol{R}_i\right]^{-1} \left(\frac{\partial \boldsymbol{h}_i}{\partial \boldsymbol{x}_i}\right) \boldsymbol{M}_i \tag{10.6.46}$$

$$\boldsymbol{M}_{i+1} = \left(\frac{\partial \boldsymbol{f}_i}{\partial \boldsymbol{x}_i}\right) \boldsymbol{P}_i \left(\frac{\partial \boldsymbol{f}_i}{\partial \boldsymbol{x}_i}\right)^T + \boldsymbol{\Gamma}_i \boldsymbol{Q}_i \boldsymbol{\Gamma}_i^T \tag{10.6.47}$$

Here we have to linearize the process in order to compute the propagation of the state error covariance, but we use the actual nonlinear model for computation of the state $\overline{\boldsymbol{x}}_i$ and the computation of the measurement values $\boldsymbol{h}_i$.

10.7 Kalman Filtering for a Stirred Tank Reactor

Lynch and Ramirez (1975) extended the Nyquist–Ramirez reactor (Nyquist and Ramirez, 1971) to include control on both cooling water flow and the flow rate of the potassium iodide catalyst concentration to the reactor. The model for the system now includes the material balance for the catalyst concentration since it is no longer constant. A schematic of the system is given in Figure 10.5.

The mathematical model for the process is as follows:

Peroxide Material Balance

$$\frac{dC}{dt} = \frac{F_p C_o}{V_x} - \frac{FC}{V_x} - R \tag{10.7.1}$$

Catalyst Material Balance

$$\frac{dC_{KI}}{dt} = \frac{F_{KI}}{V_x} C_{KI_o} - \frac{C_{KI} F}{V_x} \tag{10.7.2}$$

Energy Balance

$$\frac{dT}{dt} = \frac{F(T_o - T)}{V_x} + \frac{Q_s}{V_x \rho C_p} + \frac{(\Delta H) R V_x}{V_H \rho C_p} - \frac{\beta (T - T_c)}{V_H \rho C_p} \tag{10.7.3}$$

with the reaction rate given as

$$R = k_o C_{KI} C e^{-E/R'(T+273)} \tag{10.7.4}$$

Table 10.1 gives typical steady-state parameter values. The model was shown to predict the actual performance with a 3% maximum error in the process variables.

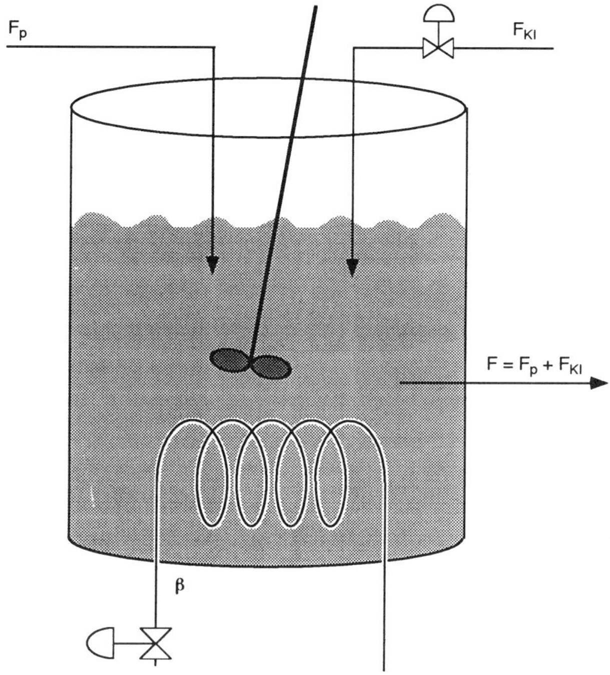

Figure 10.5. CSTR reactor with cooling rate and catalyst concentration control.

We introduce normalized dimensionless state and control variables of

$$x_1 = \frac{C - C_s}{C_s} \qquad x_2 = \frac{C_{KI} - C_{KI_s}}{C_{KI_s}} \qquad x_3 = \frac{T - T_s}{T_s}$$
$$u_1 = \frac{\beta - \beta_s}{\beta_s} \qquad u_2 = \frac{F_{KI} - F_{KI_s}}{F_{KI_s}} \tag{10.7.5}$$

For this system, two measurements are available. One is the system temperature and the other is the reaction rate, R. The reaction rate is measurable due to the fact that we can measure the rate of evolution of the produced oxygen gas from the reaction

$$2H_2O_2 \rightarrow O_2 + 2H_2O \tag{10.7.6}$$

The normalized dimensionless measurement variables are therefore

$$m_1 = x_3 \qquad m_2 = \frac{R - R_s}{R_s} \tag{10.7.7}$$

The temperature is measurable with a noise level characterized by a normal distribution with a standard deviation of 1%, but the reaction rate has a 10% standard deviation error associated with its measurement. We wish to apply Kalman filtering in order to estimate the entire state using the available

Table 10.1.
Steady-State and Parameter Values

C_{KI_o}	=	.01615 g moles/liter	Q_s	=	1284.6 − 23.61 T (cal/min)
C_{KI_s}	=	.0019 g moles/liter	T_c	=	19.2°C
C_0	=	2.25 g moles/liter	T_0	=	25°C
C_p	=	1 cal/g °C	T_s	=	30°C
C_s	=	1.718 g moles/liter	V_H	=	17.36 liters
E/R'	=	6524 cal/g mole	V_x	=	15.78 liters
F_{KI_s}	=	100 ml/min	β_s	=	122.1 cal/min °C
F_p	=	750 ml/min	ρ	=	1000 g/liter
k_0	=	9.775×10^9	$(-\Delta H)$	=	22,600 cal/g mole

measurements. The linearized system is observable. This means that the two measurements of temperature and reaction rate allow for the estimation of the entire state vector.

Since the system is nonlinear, the extended Kalman filter approach of Section 10.5 is used. In order to calculate the filter gain, $\boldsymbol{K}$, we need to linearize the system about an operating condition which was chosen as the desired steady-state condition. The linearized model is characterized by the state matrix $\boldsymbol{A}$ and the measurement matrix $\boldsymbol{H}$ given below,

$$\boldsymbol{A} = \begin{bmatrix} -.0621 & -.00826 & .0188 \\ 0 & -.0479 & 0 \\ .0097 & .0097 & -.0359 \end{bmatrix} \tag{10.7.8}$$

$$\boldsymbol{H} = \begin{bmatrix} 0 & 0 & 1 \\ 1 & 1 & .1741 \end{bmatrix} \tag{10.7.9}$$

Rather than apply the time-varying filter gain of Equations (10.5.26) and (10.5.27), we will assume that a steady-state gain based upon the constant matrices of Equations (10.7.8) and (10.7.9) can be used. This leads to the computation, using MATLAB program LQE, of the continuous filter gain of

$$\boldsymbol{K}_{ss} = \begin{bmatrix} .0129 & .1476 \\ -.0047 & .1628 \\ .9638 & .0117 \end{bmatrix} \tag{10.7.10}$$

with

$$\boldsymbol{Q} = \begin{bmatrix} .0001 & 0 & 0 \\ 0 & .0001 & 0 \\ 0 & 0 & .0001 \end{bmatrix} \quad \boldsymbol{R} = \begin{bmatrix} .0001 & 0 \\ 0 & .01 \end{bmatrix} \tag{10.7.11}$$

The covariance matrix $\boldsymbol{Q}$ reflects the fact that the standard deviation of the model uncertainty of each state is 1%. The covariance matrix $\boldsymbol{R}$ reflects that the temperature measurement is characterized by a Gaussian noise with a standard deviation of 1%. The accuracy of the reaction rate measurement is characterized by a noise standard deviation of 10%.

It is possible to use Equation (10.5.26) to calculate the estimated states; however, due to the fact that the RHS contains random fluctuations due to the noise associated with the measurements $\boldsymbol{y}(t)$, numerical integration techniques have a difficult time actually performing the computations. The high frequency noise on the RHS of a differential equation forces the integration step size to become very small, and the amount of computational effort becomes excessive. Linear problems do not suffer this drawback since analytical solutions are available to the differential equations.

A better computational choice for this nonlinear problem is to use the discrete extended Kalman filter of Section 10.6.4. Note that the optimal estimate of Equation (10.6.43) is an algebraic equation and not a differential equation. Again, for simplicity, we will use a constant steady-state filter gain based upon linearization at the desired operating condition rather than the actual time-varying gain based upon linearization along the operational trajectory. Figure 10.6 shows the comparison between the linear and nonlinear response of the system. There is quite good agreement between the temperature and catalyst concentration profiles. There is more discrepancy in the reactant concentration linear and nonlinear profiles.

Figure 10.7 gives the MATLAB program that was used for testing this application of extended Kalman filtering. The M files DER1.M and LIN.M that are called in the program are given in Figure 10.8. The program starts by calculating the nonlinear and linear responses of the system and develops the comparison plot of Figure 10.6. The transition matrix for a sampling time of .5 minutes is

$$\boldsymbol{\phi} = \begin{bmatrix} .9694 & -.00826 & .0092 \\ 0 & .9758 & 0 \\ .0047 & .0047 & .9822 \end{bmatrix} \tag{10.7.12}$$

Next the program calculates the steady-state discrete Kalman filter (Equations (10.6.38) to (10.6.40)) using the routine DLQE. The result is

$$\boldsymbol{K}_{ss} = \begin{bmatrix} .0069 & .0522 \\ .0042 & .0579 \\ .6142 & .0012 \end{bmatrix} \tag{10.7.13}$$

By examining the second column of this matrix, we observe that the weighting put on the rate measurement is small. This is the result of the fact that the rate measurement is quite noisy and the model is quite accurate. By examining the first column, we see that the temperature state estimation is affected

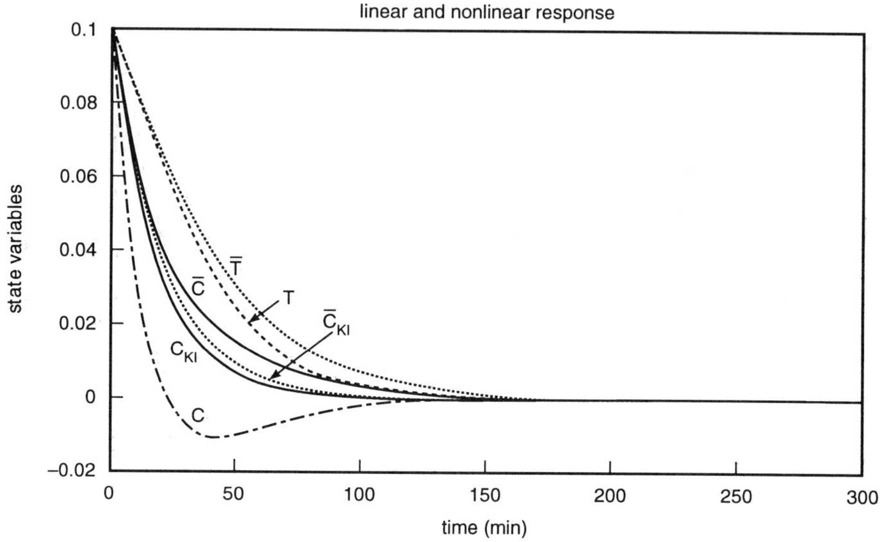

C_{KI}	=	actual catalyst conc.	$\overline{C}$	=	model reactant conc.
$\overline{C}_{KI}$	=	model catalyst conc.	T	=	actual temp.
C	=	actual reactant conc.	$\overline{T}$	=	model temp.

Figure 10.6 Comparison of linear model and actual nonlinear states.

strongly by the accurate temperature measurement, but neither concentration is influenced greatly by the accurate temperature measurement. In fact, their estimation is more influenced by the less accurate rate measurement.

Next the program runs a simulation where there is an initial error of .1 in each state variable, but the model is assumed to start from a condition with each state variable off-set by .07. Figure 10.9 gives the results of the actual and filtered measurements. The results clearly show the fact that there is much more noise associated with the reaction rate measurement. It also shows the smoothing effect of the Kalman filter on both measurements. There are less fluctuations associated with the filtered values than the raw data.

Figure 10.10 shows the actual but unknown temperature response of the system, the best estimated temperature, and the uncorrected model prediction. These results illustrate the ability of the extended Kalman filter to track the correct system temperature in the face of incorrect initial model conditions and model inaccuracy by the use of an approximate linear model. The excellent tracking is largely due to the fact that the state variable is measured and that measurement is accurate.

Figure 10.11 gives results for the actual, estimated, and model compositions. It can be seen that the optimal estimates basically track the response of the inaccurate model. This is due to the fact that the estimator has knowledge that the model is accurate and that the rate measurement is not. Also, the coupling effect of the system temperature on the system concentration is also

```
>> % look at the nonlinear response
>> % x1 is the actual state vector
>>
>> x0 = [.1 .1 .1];
>> [t1 x1] = ode45('der1',0,300,x0);
>> plot(t1,x1)
>> meta cstr
>> title('nonlinear response')
>> xlabel('time (min)')
>> ylabel('state variables')
>> meta cst
>>
>> % look at a linear model
>> % compute the transition matrix
>>
>> a = [-.0621 -.00826 .0188;
            0   -.049     0;
        .0097  .0097 -.0359];
>> h = [0 0 1;
       1 1 .1741];
>> b = [0 0 0]';
>> [phi,del] = c2d(a,b,.5)

phi =

    0.9694   -0.0040    0.0092
         0    0.9758         0
    0.0047    0.0047    0.9822


del =

     0
     0
     0

>> % xl is the linear model response
>>
>> d = [0 0]';
>> [yl,xl] = dinitial(phi,del,h,d,x0,600);
>> for n = 1:600,
   nn(n) = n*.5;
   end
>> xla = [x0;
   xl];
>> nnn = [0 nn];
>> plot(nnn',xla,t1,x1)
>> title('linear and nonlinear response')
>> xlabel('time (min)')
>> ylabel('state variables')
>> meta comp
>>
>> % develop the discrete Kalman Filter
>>
>> q = .0001*eye(3);
>> r = [.0001 0;
        0   .01];
>> g = eye(3);
>> k = dlqe(phi,g,h,q,r)

k =

    0.0069    0.0522
    0.0042    0.0579
    0.6142    0.0012

>> % response of the nonlinear system using Kalman filtering
>> % lin is an m file that computes the actual and estimated
>> %   states and measurements
>>
```

Figure 10.7 MATLAB program for implementing extended Kalman filter.

```
>> x00 = [.07 .07 .07];

>> t0 = 0;
>> tf = .5;
>> x0 = [.1 .1 .1];
>> x00 = [.07 .07 .07];
>> lin
>> plot various outputs
>> plot(nn',me,nn',yh)
>> title('actual and filtered measurements')
>> xlabel('time (min)')
>> ylabel('measured variables')
>> meta meas
>> plot(nn',xa(:,1),nn',xh(:,1))
>> plot(nn',xa(:,1),nn',xh(:,1),nn',xa(:,2),nn',xh(:,2))
>> title('actual and filtered concentrations')
>> xlabel('time (min)')
>> ylabel('concnetration states')
>> meta conc
>> plot(nn',xa(:,3),nn',xh(:,3),nn',me(:,1))
>> title('acutal, measured, filtered temperature')
>> xlabel('time (min)')
>> ylabel('temperature state')
>> meta temp
>> plot(nn',xa,nn',xh)
>> title('actual and estimalted states')
>> xlabel('time (min)')
>> ylabel('state variables')
>> meta states
>>
>> % compute the linear model response without Kalman filtering
>>
>> yl = [];
>> xl = [];
>> x0 = [.07 .07 .07];
>> [yl,xl] = dinitial(phi,del,h,d,x0,50);
>> nn = [];
>> for n = 1:50,
     nn(n) = n*.5;
   end
>> xla = [];
>> xla = [x0;
         xl];
>> nnn = [];
>> nnn = [0 nn];
>>
>> % prepare some final plots
>>
>> plot(nnn',xla)
>> plot(nn',xa,nn',xh,nnn',xla)
>> plot(nn',xa(:,3),nn',xh(:,3),nnn',xla(:,3))
>> title('estimated,actual,model temperatures')
>> xlabel('time (min)')
>> ylabel('temperature state variable')
>> meta temp
>> plot(nn',xa(:,1:2),nn',xh(:,1:2),nnn',xla(:,1:2))
>> title('actual, filtered, model concentrations')
>> xlabel('time (min)')
>> ylabel('concentration state variables')
>> meta conc
>> diary off
```

Figure 10.7 continued.

```
M File DER1.M
function dx = der (t,x)
%
% this function defines the nonlinear differential equations
%
% the actual dimensional variables are computed
%
y(1) = x(1)*1.737 + 1.737;
y(2) = x(2)*.0019 + .0019;
y(3) = x(3)*30.23 + 30.23;
%
% the dimensionless derivatives are computed
%
dx(1) = (.759*2.25/15.78 - .85*y(1)/15.78 - 9.775e9*y(1)*y(2)*...
        exp(-6524/(y(3)+273)))/1.737;
dx(2) = (.1*.01615/15.78 -.85*y(2)/15.78)/.0019;

dx(3) = (.85*(25-y(3))/17.36 +(1284.6-23.61*y(3))/(17.36*1000)+...
        22600*9.775e9*y(1)*y(2)*exp(-6524/(y(3)+273))*15.78/(17.36*1000)...
        -122.1*(y(3)-19.2)/(17.36*1000))/30.23;
```

```
M File LIN.M


% m file to perform simulation for extended Kalman filtering
%
for n = 1:50,
%
% calculate the linear model response
%
 xm = phi*x00';
%
% calculate the nonlinear actual response
%
 [t,x] = ode45('der1',t0,tf,x0);
%
% update the differential equation solving routine
%
 [m,mm] = size(t);
 t0 = tf;
 tf = tf+.5;
 x0(1,:) = x(m,:);
%
% calculate the actual measurements
%
 me(n,1) = x(m,3) + sqrt(.0001)*rand;
 me(n,2) = (1 + x(m,1) + x(m,2) + x(m,1)*x(m,2))*exp(-6524/(30*x(m,3)..
          +303))/exp(-6524/303) - 1 + sqrt(.01)*rand;
%
% calculate the predicted measurements
%
 yh(n,1) = xm(3);
 yh(n,2) = (1 + xm(1)+xm(2) + xm(1)*xm(2))*exp(-6524/(30*xm(3)+303))...
          /exp(-6524/303) -1;
%
% calculate the estimated states
%
 xh(n,1) = xm(1) + k(1,1)*(me(n,1)-yh(n,1)) + k(1,2)*(me(n,2)-yh(n,2));
 xh(n,2) = xm(2) + k(2,1)*(me(n,1)-yh(n,1)) + k(2,2)*(me(n,2)-yh(n,2));
 xh(n,3) = xm(3) + k(3,1)*(me(n,1)-yh(n,1)) + k(3,2)*(me(n,2)-yh(n,2));
%
% save the time, actual state, and update the linear model
%
 xa(n,:) = x(m,:);
 x00 = xh(n,:);
 nn(1,n) = n*.5;
end
```

Figure 10.8. M Files DER1.M and LIN.M to compute nonlinear differential equations and logic to implement extended Kalman filter.

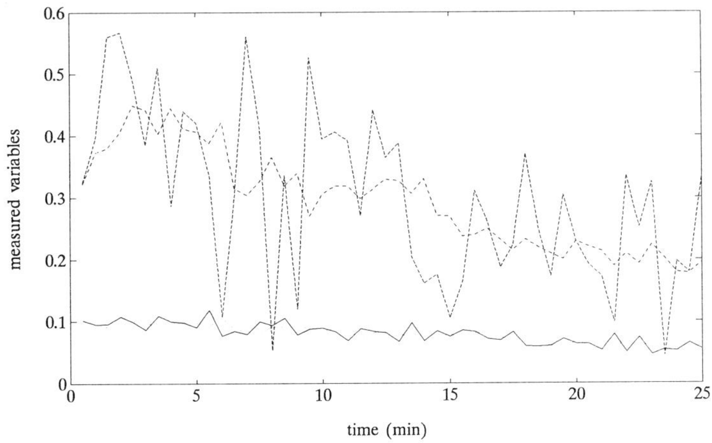

Figure 10.9 Comparison of actual and filtered measurements.

T_m = meas. temperature state (—)
R_m = meas. normalized reaction rate (- - - -)
$\hat{T}$ = filtered temperature state (···)
$\hat{R}$ = filtered normalized reaction rate (— · —·)

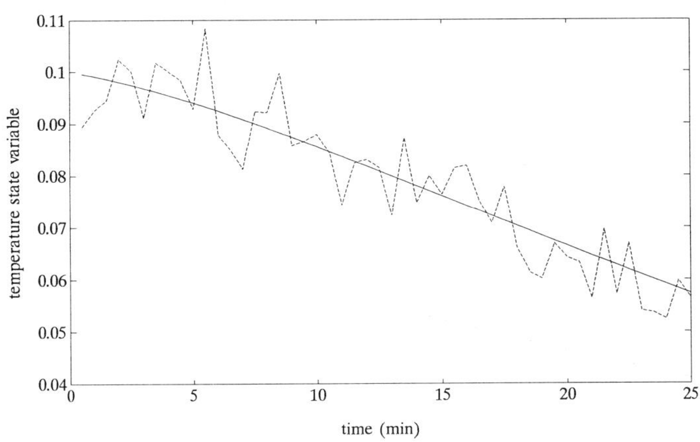

Figure 10.10 Comparison of actual, filtered, and model temperature profiles.

T = actual temperature state (—)
$\hat{T}$ = filtered temperature state (- - - -)
$\overline{T}$ = model temperature state (···)

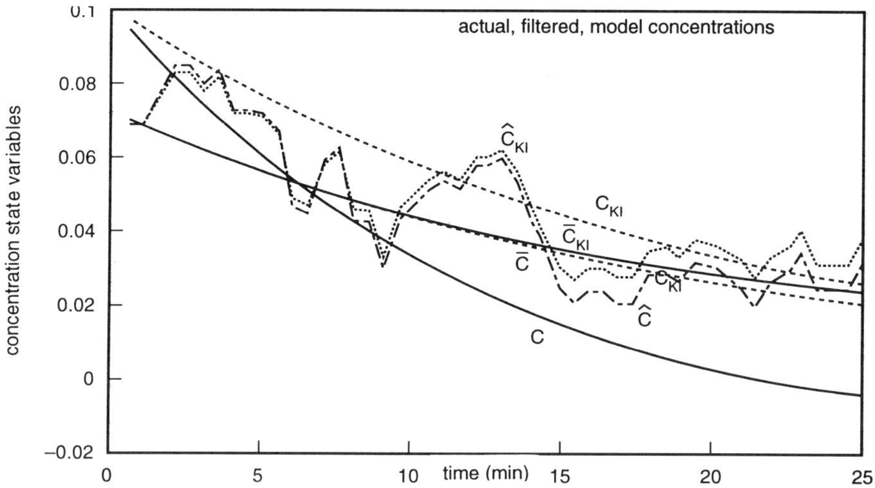

Figure 10.11 Comparison of actual, filtered, and model concentrations.

C	$=$	actual reactant conc. state	C_{KI}	$=$	actual catalyst conc. state
$\hat{C}$	$=$	filtered reactant conc. state	$\hat{C}$	$=$	filtered catalyst conc. state
$\overline{C}$	$=$	model reactant conc. state	$\overline{C}$	$=$	model catalyst conc. state

very weak. Finally there is more discrepancy in the reactant concentration due to the fact that the linearized model did not well represent that state's dynamic response (Figure 10.6). The tracking ability for the reactant concentration can be improved by performing the step by step updating of Equations (10.6.45) to (10.6.47).

10.8 Estimation of Model Parameters

Kalman filtering can be used effectively for the sequential estimation of unknown or uncertain model parameters. This is accomplished by introducing new state variables corresponding to the model parameters to be identified. If the parameters are expected to be constants or slowly varying over the process time domain, then the following state dynamic model is appropriate:

$$\dot{\boldsymbol{a}} = \boldsymbol{0} + \boldsymbol{w}_1 \tag{10.8.1}$$

This model simply states that we expect the parameters to be constant and that they have an uncertainty that is characterized by the random variable $\boldsymbol{w}_1$. The random variable vector $\boldsymbol{w}_1$ has an expected value of zero and a covariance

$$\text{cov}\ (\boldsymbol{w}_1) = \boldsymbol{Q}_1 \tag{10.8.2}$$

If we expect the parameters to vary quickly with various loads, controls, or states, then more complex models of the following form are introduced,

$$\dot{\boldsymbol{a}} = \boldsymbol{b} + \boldsymbol{w}_2 \tag{10.8.3}$$

$$\dot{\boldsymbol{b}} = \boldsymbol{F}(\boldsymbol{u},\ \boldsymbol{d},\ \boldsymbol{x})\boldsymbol{b} + \boldsymbol{w}_3 \tag{10.8.4}$$

with cov $(\boldsymbol{w}_2) = \boldsymbol{Q}_2$ and cov $(\boldsymbol{w}_3) = \boldsymbol{Q}_3$.

This model allows for the parameters $\boldsymbol{a}$ to vary quickly as changes in either the controls $(\boldsymbol{u})$, the disturbances $(\boldsymbol{d})$, or the states $(\boldsymbol{x})$ are detected. Of course, the matrix function $\boldsymbol{F}$ must be specified based upon some physical insight into the behavior of the system.

10.8.1 Simultaneous Kinetic Determinations

Wentzell *et al.* (1989) have used the Kalman filter to effectively process kinetic data from simultaneous reaction rate data. A typical system is one in which two analyte species X and Y react with a common reagent R to produce products P_1 and P_2 which are similar but not identical:

$$\begin{aligned} X + R &\rightarrow P_1 \\ Y + R &\rightarrow P_2 \end{aligned} \tag{10.8.5}$$

The measurable total absorbance A_t is given by

$$A_t = A_1\left[1 - \exp(-k_1, t)\right] + A_2\left[1 - \exp(-k_2 t)\right] + B \tag{10.8.6}$$

The reaction rates k_1 and k_2 can be determined from pure species experiments. We wish to determine the best values for A_1, A_2, and B from the observed measurement $A_t(t)$. For this system, there is one measurement,

$$y = A_t \tag{10.8.7}$$

and three state variables,

$$\boldsymbol{x} = \begin{pmatrix} A_1 \\ A_2 \\ B \end{pmatrix} \tag{10.8.8}$$

with the state model

$$\dot{\boldsymbol{x}} = \boldsymbol{0} + \boldsymbol{w} \tag{10.8.9}$$

This expresses the fact that the states or model parameters are constants. The measurement is related to the states by

$$y = \boldsymbol{H}(t)\ \boldsymbol{x} + \boldsymbol{v} \tag{10.8.10}$$

with

$$\boldsymbol{H}(t) = \begin{pmatrix} 1 - \exp(-k_1 t) \\ 1 - \exp(-k_2 t) \\ 1 \end{pmatrix}^T \tag{10.8.11}$$

and

$$\text{cov}\ (\boldsymbol{v}) = \boldsymbol{R} \tag{10.8.12}$$

We will use the discrete Kalman filter for data processing. Therefore, the integrated form of Equation (10.8.9) is needed:

$$\boldsymbol{x}_{i+1} = \boldsymbol{\phi x}_i + \Gamma w_i \tag{10.8.13}$$

with

$$\boldsymbol{\Phi} = \boldsymbol{\Gamma} = \boldsymbol{I} \tag{10.8.14}$$

For this problem, the nonstationary form of the Kalman filter is applied,

$$\hat{\boldsymbol{x}}_i = \bar{\boldsymbol{x}}_i + \boldsymbol{K}_i(\boldsymbol{y}_i - \boldsymbol{H}_i\bar{\boldsymbol{x}}_i) \tag{10.8.15}$$

with $\boldsymbol{K}_i$ defined by Equations (10.6.38) to (10.6.40).

Wentzell *et al.* (1989) use this Kalman filtering method for the kinetic analysis of the reaction of amino acid mixtures of glycine and asparagine with the reagent trinitrobenzenesulfonic acid (TNBS). Simulations were first run to test the expected effectiveness of the technique. The ratio of rate constants k_2/k_1 varied from 1.2 to 10. Data were always generated for times equal to $2/k_1$. Actual values of A_1 and A_2 were 1 and 0.1, and the background absorbance value, B, was 0.1. A Gaussian noise with standard deviation of 1% was used to simulate the measurement noise and model uncertainty. The initial value of the covariance matrix $\boldsymbol{P}_0$ must also be provided. The diagonal elements of this matrix are set to a large value to ensure an unbiased estimate, while the off-diagonal terms are set to zero. For their work,

$$\boldsymbol{P}_0 = \text{diag}\ (10^8 \times \text{ measurement variance}) \tag{10.8.16}$$

There is considerable flexibility in this value. The initial states or model parameters were set to zero although this value is not critical when diag ($\boldsymbol{P}_0$) is large.

A typical absorbance response is shown in Figure 10.12, and it is clear that the two individual reaction effects are not easily distinguishable from this response curve. The simulation results for the parameters A_1 and A_2 over the range of k_2/k_1 values are given in Figure 10.13. The estimates are good; however, larger errors are apparent when the k_2/k_1 ratio is small.

Since $\boldsymbol{P}$ is the covariance of the optimal estimate, precise information is available on the accuracy of the states (the model parameters). As an example, the predicted error associated with A_2 (when $A_2 = .1$) is shown in

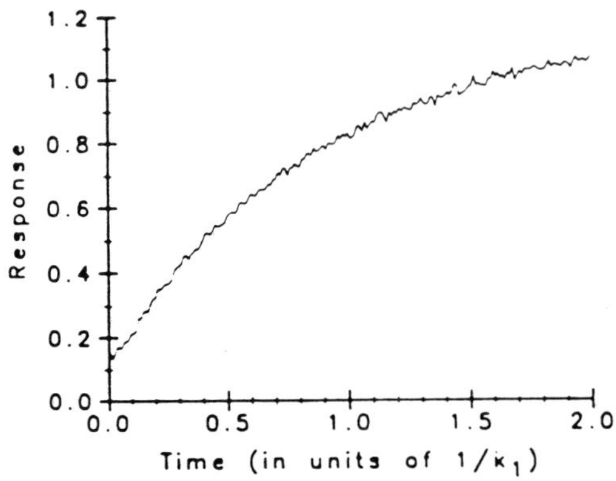

Figure 10.12 Typical synthetic biexponential data used to test the Kalman filter. From Wentzell *et al.* (1989), with permission.

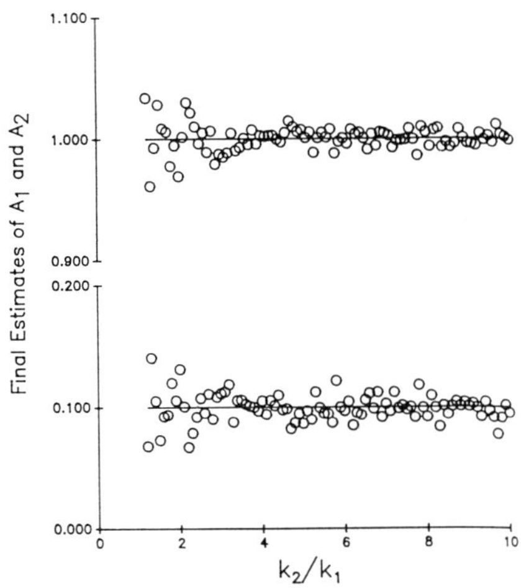

Figure 10.13 Results of the application of the Kalman filter to synthetic data as a function of the ratio of rate constants. Solid lines are the actual values; points are the Kalman filter estimates. From Wentzell *et al.* (1989), with permission.

Figure 10.14. The predicted value is twice the standard deviation $(2\sqrt{P_{22}})$. The figure also shows actual errors computed by the simulations. The plot indicates the limitations of differential kinetic methods when the k_2/k_1 ratio gets small.

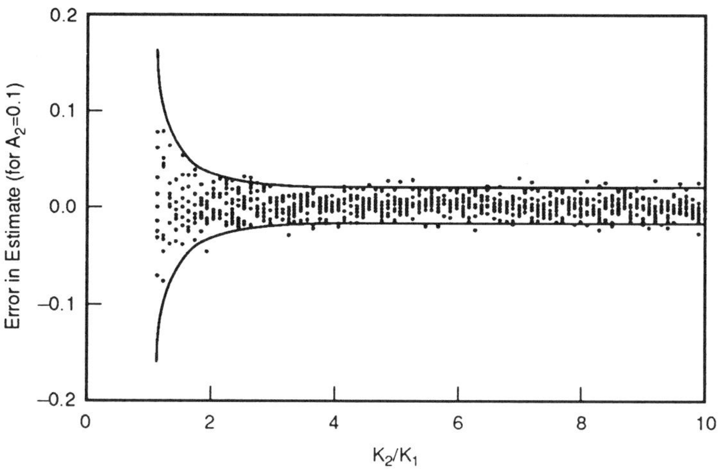

Figure 10.14. Comparison of actual (points) and predicted (curved solid lines) errors for the filter analysis of synthetic data as a function of the ratio of rate constants. From Wentzell *et al.* (1989), with permission.

Wentzell *et al.* (1989) also applied the method to actual absorbance data for two mixtures of glycine and asparagine. These results, given in Table 10.2, indicate that there is a 10% agreement between the actual and computed values. It is believed that the major source of error is related to the chemical system itself rather than the method of data analysis.

10.8.2 Bioreactor Identification

Stephanopoulos and San (1984) and San and Stephanopoulos (1984) present a good example of the use of extended Kalman filtering for the determination of state and model parameters for a bioreactor. The system that they consider is the growth of a microorganism on a limiting substrate sugar. The two state variables for the system are the biomass concentration (X) and the sugar concentration (S). The mathematical model is a material balance on the biomass,

$$\dot{X} = \mu X - DX \tag{10.8.17}$$

where D is a design parameter associated with the reactor type. It is zero for a batch reactor and reflects the outlet flow in a stirred tank reactor. For a fed-batch reactor, it is related to the change in the reactor volume ($\dot{V}/V$).

Table 10.2
Glycine and Asparagine Mixture Identifications

Mixture	[Gly](μm)		% Error	[Asp](μm)		% Error
	Actual	computed		Actual	Computed	
1	14.3	12.8	-10	5.7	5.0	-12
2	12.0	10.3	-14	24.0	24.0	0

We also have the material balance for the sugar,

$$\dot{S} = D(S_f - S) - \frac{1}{Y_s}\,\mu X \tag{10.8.18}$$

where S_f is the feed sugar concentration. There are two model parameters, the specific growth rate μ and the yield coefficient Y_s. These two model parameters need to be identified. It is expected that the yield coefficient either remains constant or is a slowly varying function. Therefore, the appropriate state model for this parameter is

$$\dot{Y}_s = 0 + w_y \tag{10.8.19}$$

where w_y is a Gaussian random variable. On the other hand, we expect the growth rate to be proportional to the dilution rate. Since we need a rapid change in this parameter with dilution rate, D, the following model was proposed:

$$\begin{aligned}\dot{\mu} &= C + w_\mu \\ \dot{C} &= -DC + w_\mu\end{aligned} \tag{10.8.20}$$

where w_μ is a Gaussian random variable. As formulated, the problem can be solved via extended Kalman filtering. Stephanopoulos and San have assumed that the only significant model uncertainties are those associated with the identification of the model parameters. Therefore, they use a state covariance matrix,

$$\boldsymbol{Q} = \begin{bmatrix} 0 & 0 & 0 & 0 & 0 \\ 0 & 0 & 0 & 0 & 0 \\ 0 & 0 & q_\mu & q_\mu & 0 \\ 0 & 0 & q_\mu & q_\mu & 0 \\ 0 & 0 & 0 & 0 & q_y \end{bmatrix} \tag{10.8.21}$$

Stephanopoulos and San have used the adaptive algorithm of Jazwinski (1970) to determine model parameter covariances. Instead of looking at residuals after the fact, Jazwinski developed a real-time algorithm to let the residuals themselves determine appropriate noise covariance levels. Using actual residuals, the adaptive covariances are

$$q_\mu = \begin{cases} \frac{1}{k_1}\Big[(R - \hat{\mu}\hat{X})^2_{t+1} - \left(P_\mu \hat{X}^2 + 2P_{\mu X}\hat{\mu}\hat{X} + P_X\hat{\mu}^2\right)_{t+1} \\ \quad -q_R\Big] = \Phi_1 \qquad \text{for } \Phi_1 > 0 \\ 0 \text{ for } \Phi_2 \leq 0 \end{cases} \tag{10.8.22}$$

$$q_y = \begin{cases} \frac{1}{k_2}\left[\left(Y_s - \hat{Y}_s\right)^2_{t+1} - (P_{Y_s})^2_{t+1} - q_y\right] = \phi_2 \qquad \text{for } \Phi_2 > 0 \\ 0 \text{ for } \Phi_2 \leq 0 \end{cases} \tag{10.8.23}$$

The $\boldsymbol{P}$ values are the appropriate elements of the recursive covariance matrix of the extended Kalman filter (Equation (10.5.27)). The response rate constants k_1 and k_2 need to be determined through simulation. Values of 1/3 were shown to work well.

The measurements on the system that are available are the total rate of microbial growth, R, and the yield, Y_s. The total rate of microbial growth is

$$R = \mu X \tag{10.8.24}$$

The measurement covariance matrix is

$$\boldsymbol{R} = \begin{bmatrix} q_R & 0 \\ 0 & q_Y \end{bmatrix} \tag{10.8.25}$$

where q_R is the covariance associated with the growth measurement and q_Y the covariance associated with the yield measurement. A 15% measurement error in both measurements was assumed.

Figure 10.15 shows the response of the actual and estimated values of biomass, X, and the specific growth rate, μ, to a series of step changes in the dilution rate of a continuous stirred fermenter (chemostat). Very good agreement was obtained, implying that the method can be applied to an actual chemostat in order to study the properties of the specific growth rate under steady-state and dynamic conditions. It is observed that μ converges to the value of the maximum specific growth rate used in the Monod model that generated the response measurements,

$$\mu = \frac{\mu_{\max} X}{K + X} \tag{10.8.26}$$

following a step increase in dilution rate to a value larger than $\mu_{\max}$. This suggests that $\mu_{\max}$ can be estimated by a washout chemostat experiment.

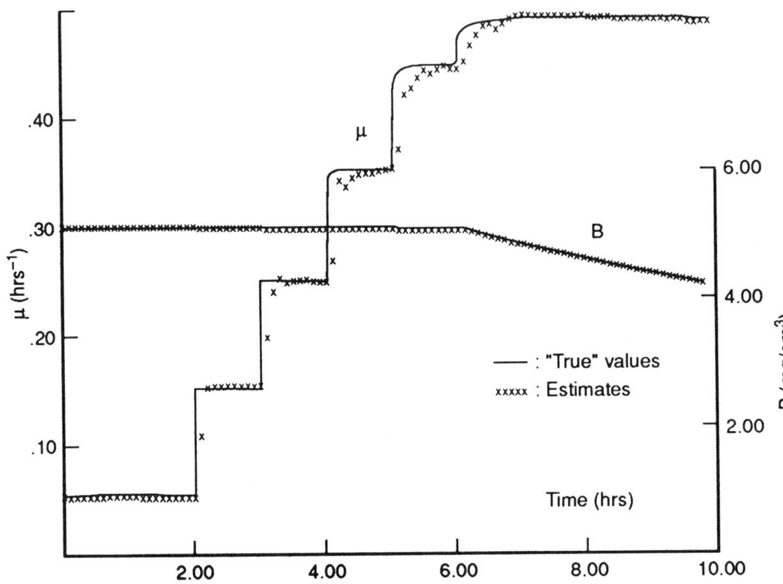

Figure 10.15 Response of estimator to a series of step changes of the dilution rate for a continuous reactor; constant yield: (——) true values and (X) estimates. Reprinted from San and Stephanopoulos, *Biotech. & Bioeng.* **26**, 1189–1197, ©1984, John Wiley and Sons.

10.9 Robustness of the Kalman Filter

By duality, the Kalman filter has equivalent robustness properties to the state regulator (LQR) problem of Chapter 7. This means that to test the stability of the Kalman filter, all that is needed is to substitute $\boldsymbol{A}^T$ for $\boldsymbol{A}$ and $\boldsymbol{H}^T$ for $\boldsymbol{B}$ in the Safonov (1980) stability development of Equation (7.2.11) and the Wong *et al.* (1978) integrity test of Equation (7.3.1). It can be concluded that the Kalman filter has the same excellent robustness characteristics as the LQR problem.

If we consider the Nyquist–Ramirez reactor where the state dynamics are described by the matrix

$$\boldsymbol{A} = \begin{bmatrix} -.05156 & -.05877 \\ .01503 & -.005887 \end{bmatrix} \tag{10.9.1}$$

and we assume that both the measurement and state models have a 1% level of uncertainty, then for the filter problem the state covariance matrix is

$$\boldsymbol{Q} = \text{diag}\,(.0001) \tag{10.9.2}$$

The covariances associated with the measurements are

$$\boldsymbol{R} = \text{diag}\ (.0001) \tag{10.9.3}$$

The optimal stationary filter gain is computed by the MATLAB program LQE. Three different measurement situations are considered. The first is that both states are measured,

$$\boldsymbol{H}_1 = \begin{bmatrix} 1 & 0 \\ 0 & 1 \end{bmatrix} \tag{10.9.4}$$

This gives a filter gain of

$$\boldsymbol{K}_1 = \begin{bmatrix} .9508 & -.0220 \\ -.0220 & .9936 \end{bmatrix} \tag{10.9.5}$$

If only the composition is measured, then

$$\boldsymbol{H}_2 = \begin{bmatrix} 1 & 0 \end{bmatrix} \tag{10.9.6}$$

and the filter gain is

$$\boldsymbol{K}_2 = \begin{bmatrix} 1.0004 \\ -.8848 \end{bmatrix} \tag{10.9.7}$$

Finally, if only the temperature is measured, then

$$\boldsymbol{H}_3 = \begin{bmatrix} 0 & 1 \end{bmatrix} \tag{10.9.8}$$

and

$$\boldsymbol{K}_3 = \begin{bmatrix} .0809 \\ .9953 \end{bmatrix} \tag{10.9.9}$$

Applying the Wong robust stability criteria of Equation (7.3.1),

$$n > \frac{1}{2}\,(1 - \lambda_0) \tag{10.9.10}$$

and where λ_0 is the minimum eigenvalue of

$$\left(\boldsymbol{R}^{1/2}\boldsymbol{K}^T\boldsymbol{Q}^{-1}\boldsymbol{K}\boldsymbol{R}^{1/2}\right)^{-1} \tag{10.9.11}$$

we have the minimum eigenvalues for the three cases as

$$\begin{aligned} \lambda_1 &= .9941 \\ \lambda_2 &= .5606 \\ \lambda_3 &= 1.0028 \end{aligned} \tag{10.9.12}$$

This yields guaranteed robust stability for measurement perturbation slopes greater than

$$\begin{aligned} n_1 &> .002525 \\ n_2 &> .2197 \\ n_3 &> -.0028 \end{aligned} \tag{10.9.13}$$

These results show that the stability region with only the concentration being measurable is fairly close to the minimum value of 1/2 that is guaranteed by the general result of Safonov. However, once the temperature is measured, the region of robust stability is greatly enhanced. It is therefore important that any measurement scheme for this system include measurements on the temperature. Fortunately, this is the easiest of the states to actually measure, using thermocouple instrumentation.

10.10 Kalman Filtering with Multiple Time Domains

Systems demonstrating multiple time scales occur frequently in practice. Examples from the chemical industry include heat exchangers, catalytic reactors, distillation columns, and fluidized beds. Heat exchangers have separate time constants for each flowing fluid as well as longer time scales associated with fouling. Catalytic reactors can have time scales associated with the residence time of the reactants, a thermal time constant due to the interaction with the catalyst, and a long-term time constant due to catalyst degradation. Distillation columns have tray time constants as well as time constants associated with tray fouling. Fluidized beds have process time constants as well as long-term time constants associated with the elutriation of particles.

Kalman filtering does not normally use the knowledge that multiple time constants exist. Ignoring this information leads to stiff differential equations for the calculation of the covariance matrices used to compute filter gains. The use of perturbation theory for improving the computational efficiency of filter gains has been shown to be effective. Perturbation analysis is used to approximate solutions to differential equations by defining an arbitrary expansion scalar variable. The solution of the differential equations is studied as a series expansion around this scalar variable. If the order of the equations is reduced when the perturbation variables are set to zero, we have singular perturbation. If the equations become decoupled, then we have nonsingular perturbation.

Perturbation analysis in modern control and estimation theory began with the work of Sannuti and Kokotovic (1969). They demonstrated that a fifth-order model for an electrical power generator could be approximated by the singular perturbation technique. The theory for systems characterized by two time scales has been studied by Kokotovic *et al.* (1976) and by Soliman and Ray (1979) among others. Multiple time scales have been treated by Gaafar and Ramirez (1985).

Let us consider a system with three time scales modeled as

$$\dot{\boldsymbol{x}}_1 = \boldsymbol{A}_{11}\boldsymbol{x}_1 + \boldsymbol{A}_{12}\boldsymbol{x}_2 + \boldsymbol{A}_{13}\boldsymbol{x}_3 + \boldsymbol{B}_1\nu$$

$$
\begin{aligned}
\epsilon_1 \dot{\boldsymbol{x}}_2 &= \boldsymbol{A}_{21}\boldsymbol{x}_1 + \boldsymbol{A}_{22}\boldsymbol{x}_2 + \boldsymbol{A}_{23}\boldsymbol{x}_3 + \boldsymbol{B}_2\nu \qquad (10.10.1)\\
\epsilon_2 \dot{\boldsymbol{x}}_3 &= \boldsymbol{A}_{31}\boldsymbol{x}_1 + \boldsymbol{A}_{32}\boldsymbol{x}_2 + \boldsymbol{A}_{33}\boldsymbol{x}_3 + \boldsymbol{B}_3\nu\\
\boldsymbol{y} &= \boldsymbol{C}_1\boldsymbol{x}_1 + \boldsymbol{C}_2\boldsymbol{x}_2 + \boldsymbol{C}_3\boldsymbol{x}_3 + \omega \qquad (10.10.2)
\end{aligned}
$$

where $\boldsymbol{x}_1$, $\boldsymbol{x}_2$, and $\boldsymbol{x}_3$ are states associated with the three separate time scales of order one, order ϵ_1 and order ϵ_2. The vector $\boldsymbol{y}$ is the measurement vector. The vectors $\boldsymbol{\nu}$ and $\boldsymbol{\omega}$ are zero mean uncorrelated Gaussian random vectors with

$$
\begin{aligned}
\boldsymbol{E}(\nu) &= \mathbf{0} \qquad (10.10.3)\\
\boldsymbol{E}(\omega) &= \mathbf{0} \qquad (10.10.4)\\
\text{cov}\left(\nu(t)\nu^T(t+\tau)\right) &= \boldsymbol{Q}\delta(\tau) \qquad (10.10.5)\\
\text{cov}\left(\omega(t)\omega^T(t+\tau)\right) &= \boldsymbol{R}\delta(\tau) \qquad (10.10.6)\\
\text{cov}\left(\omega(t)\nu^T(\tau)\right) &= \mathbf{0} \qquad (10.10.7)
\end{aligned}
$$

We can cast this problem into a classical Kalman filtering problem by the introduction of the following definitions,

$$
x = \begin{bmatrix} \boldsymbol{x}_1 \\ \boldsymbol{x}_2 \\ \boldsymbol{x}_3 \end{bmatrix} \qquad (10.10.8)
$$

$$
\boldsymbol{A} = \begin{bmatrix} \boldsymbol{A}_{11} & \boldsymbol{A}_{12} & \boldsymbol{A}_{13} \\ \epsilon_1^{-1}\boldsymbol{A}_{21} & \epsilon_1^{-1}\boldsymbol{A}_{22} & \epsilon_1^{-1}\boldsymbol{A}_{23} \\ \epsilon_2^{-1}\boldsymbol{A}_{31} & \epsilon_2^{-1}\boldsymbol{A}_{32} & \epsilon_2^{-1}\boldsymbol{A}_{33} \end{bmatrix} \qquad (10.10.9)
$$

$$
\boldsymbol{B} = \begin{bmatrix} \boldsymbol{B}_1 \\ \epsilon_1^{-1}\boldsymbol{B}_2 \\ \epsilon_2^{-1}\boldsymbol{B}_3 \end{bmatrix} \qquad (10.10.10)
$$

$$
\boldsymbol{C} = \begin{bmatrix} \boldsymbol{C}_1 & \boldsymbol{C}_2 & \boldsymbol{C}_3 \end{bmatrix} \qquad (10.10.11)
$$

This gives

$$
\begin{aligned}
\dot{x} &= \boldsymbol{A}x + \boldsymbol{B}\nu \qquad (10.10.12)\\
y &= \boldsymbol{C}x + \omega \qquad (10.10.13)
\end{aligned}
$$

From the Kalman filtering development of Section 10.2, we have

$$
\dot{\hat{\boldsymbol{x}}} = \boldsymbol{A}\hat{\boldsymbol{x}} + \boldsymbol{K}[\boldsymbol{y} - \boldsymbol{C}\hat{\boldsymbol{x}}] \qquad (10.10.14)
$$

where $\boldsymbol{K}$ is the Kalman gain matrix defined by the covariance matrix $\boldsymbol{P}$

$$
\boldsymbol{K} = \boldsymbol{P}\boldsymbol{C}^T\boldsymbol{R}^{-1} \qquad (10.10.15)
$$

$$\dot{\boldsymbol{P}} = \boldsymbol{AP} + \boldsymbol{PA}^T - \boldsymbol{PC}^T\boldsymbol{R}^{-1}\boldsymbol{CP} + \boldsymbol{BQB}^T \tag{10.10.16}$$

For problems with discrete measurements, the optimal estimates between measurements are

$$\dot{\hat{\boldsymbol{x}}} = \boldsymbol{A}\hat{\boldsymbol{x}} \tag{10.10.17}$$

$$\dot{\boldsymbol{P}} = \boldsymbol{AP} + \boldsymbol{PA}^T + \boldsymbol{BQB}^T \tag{10.10.18}$$

and at the measurement times we have

$$\hat{\boldsymbol{x}}^+ = \hat{\boldsymbol{x}}^- + \boldsymbol{K}(\boldsymbol{y} - \boldsymbol{C}\hat{\boldsymbol{x}}^-) \tag{10.10.19}$$

where

$$\boldsymbol{K} = \boldsymbol{P}^-\boldsymbol{C}^T(\boldsymbol{CP}^-\boldsymbol{C}^T + \boldsymbol{R})^{-1} \tag{10.10.20}$$

$$\boldsymbol{P}^+ = \boldsymbol{P}^- - \boldsymbol{KCP}^- \tag{10.10.21}$$

Gaafar and Ramirez (1985) have shown that the covariance matrix calculation simplifies when the covariance matrix is partitioned as

$$\boldsymbol{P} = \begin{bmatrix} \boldsymbol{P}_{11} & \boldsymbol{P}_{12} & \boldsymbol{P}_{13} \\ \boldsymbol{P}_{21} & \epsilon_1^{-1}\boldsymbol{P}_{22} & \epsilon_1^{-1}\boldsymbol{P}_{23} \\ \boldsymbol{P}_{31} & \epsilon_1^{-1}\boldsymbol{P}_{32} & \epsilon_2^{-1}\boldsymbol{P}_{33} \end{bmatrix} \tag{10.10.22}$$

In terms of the submatrices of (10.10.22), the covariance matrix equation of (10.10.18) becomes

$$\begin{aligned} \dot{\boldsymbol{P}}_{11} &= \boldsymbol{F}_{11}(\boldsymbol{P}_{ij}, \epsilon_1, \epsilon_2) && (10.10.23) \\ \epsilon_1\dot{\boldsymbol{P}}_{12} &= \boldsymbol{F}_{12}(\boldsymbol{P}_{ij}, \epsilon_1, \epsilon_2) && (10.10.24) \\ \epsilon_2\dot{\boldsymbol{P}}_{13} &= \boldsymbol{F}_{13}(\boldsymbol{P}_{ij}, \epsilon_1, \epsilon_2) && (10.10.25) \\ \epsilon_1\dot{\boldsymbol{P}}_{22} &= \boldsymbol{F}_{22}(\boldsymbol{P}_{ij}, \epsilon_1, \epsilon_2) && (10.10.26) \\ \epsilon_2\dot{\boldsymbol{P}}_{23} &= \boldsymbol{F}_{23}(\boldsymbol{P}_{ij}, \epsilon_1, \epsilon_2) && (10.10.27) \\ \epsilon_2\dot{\boldsymbol{P}}_{33} &= \boldsymbol{F}_{33}(\boldsymbol{P}_{ij}, \epsilon_1, \epsilon_2) && (10.10.28) \end{aligned}$$

These equations are the exact solution to the linear filtering problem. When the perturbation variables which multiply these equations are small, the equations become stiff and difficult to solve efficiently. To simplify, the covariance submatrices are decomposed in slow, medium, and fast parts,

$$\boldsymbol{P}_{ij} = \boldsymbol{G}_{ij}(T_2, \epsilon_1, \epsilon_2) + \boldsymbol{H}_{ij}(T_1, \epsilon_1, \epsilon_2) + \boldsymbol{I}_{ij}(t, \epsilon_1, \epsilon_2) \tag{10.10.29}$$

where $T_i = t/\epsilon_i$ and $\boldsymbol{G}$, $\boldsymbol{H}$, and $\boldsymbol{I}$ are the fast, medium, and slow decompositions. The derivative of $\boldsymbol{P}$ with respect to time is

$$\dot{\boldsymbol{P}}_{ij} = \epsilon_2^{-1}\dot{\boldsymbol{G}}_{ij} + \epsilon_1^{-1}\dot{\boldsymbol{H}}_{ij} + \dot{\boldsymbol{I}}_{ij} \tag{10.10.30}$$

Using this decomposition, the system of covariance equations becomes

$$\dot{\boldsymbol{I}}_{11} + \epsilon_1^{-1}\dot{\boldsymbol{H}}_{11} + \epsilon_2^{-1}\dot{\boldsymbol{G}}_{11} = \boldsymbol{F}_{11}(\boldsymbol{G}_{ij} + \boldsymbol{H}_{ij} + \boldsymbol{I}_{ij}) \tag{10.10.31}$$

$$\epsilon_1\dot{\boldsymbol{I}}_{\ell 2} + \dot{\boldsymbol{H}}_{\ell 2} + \frac{\epsilon_1}{\epsilon_2}\dot{\boldsymbol{G}}_{\ell 2} = \boldsymbol{F}_{\ell 2}(\boldsymbol{G}_{ij} + \boldsymbol{H}_{ij} + \boldsymbol{I}_{ij}) \qquad \ell = 1, 2 \tag{10.10.32}$$

$$\epsilon_2\dot{\boldsymbol{I}}_{\ell 3} + \frac{\epsilon_2}{\epsilon_1}\dot{\boldsymbol{H}}_{\ell 3} + \dot{\boldsymbol{G}}_{\ell 3} = \boldsymbol{F}_{\ell 3}(\boldsymbol{G}_{ij} + \boldsymbol{H}_{ij} + \boldsymbol{I}_{ij}) \qquad \ell = 1, \cdots, 3 \tag{10.10.33}$$

In Equation (10.10.31), the quantities ϵ_1^{-1} and ϵ_2^{-1} are large; therefore, to satisfy the equation, the derivatives which we multiply must approach zero. This is also true for the $\boldsymbol{G}_{\ell 2}$ derivatives described in (10.10.32) since $\epsilon_1 >> \epsilon_2$. This leads to

$$\dot{\boldsymbol{G}}_{11} = \boldsymbol{0} \tag{10.10.34}$$

$$\dot{\boldsymbol{H}}_{11} = \boldsymbol{0} \tag{10.10.35}$$

$$\dot{\boldsymbol{G}}_{12} = \boldsymbol{0} \tag{10.10.36}$$

$$\dot{\boldsymbol{G}}_{22} = \boldsymbol{0} \tag{10.10.37}$$

Since these variables are constants and must die out with time, they must be zero for all time under singular decomposition. If we keep only first-order terms in, then the submatrices become

$$\boldsymbol{0} = \boldsymbol{F}_{k\ell}(\boldsymbol{I}_{ij}) \qquad k = 1, \cdots, \ell \quad \ell = 2, 3 \tag{10.10.38}$$

$$\dot{\boldsymbol{I}}_{11} = \boldsymbol{F}_{11}(\boldsymbol{I}_{ij}) \tag{10.10.39}$$

$$\boldsymbol{0} = \boldsymbol{F}_{\ell 3}(\boldsymbol{H}_{ij} + \boldsymbol{I}_{ij}) \qquad \ell = 1, \cdots, 3 \tag{10.10.40}$$

$$\dot{\boldsymbol{H}}_{\ell 2} = \boldsymbol{F}_{\ell 2}(\boldsymbol{H}_{ij} + \boldsymbol{I}_{ij}) \qquad \ell = 1, 2 \tag{10.10.41}$$

$$\dot{\boldsymbol{G}}_{\ell 3} = \boldsymbol{F}_{\ell 3}(\boldsymbol{G}_{ij} + \boldsymbol{H}_{ij} + \boldsymbol{I}_{ij}) \qquad \ell = 1, \cdots, 3 \tag{10.10.42}$$

This set of equations is solved in a direct series fashion. Equations (10.10.38) and (10.10.39) specify the slow terms $\boldsymbol{I}$. With those known, then the medium time scale equations (10.10.40) and (10.10.41) are solved. Finally, with both $\boldsymbol{I}$ and $\boldsymbol{H}$ known, the fast terms $\boldsymbol{G}$ are computed from Equation (10.10.42).

10.10.1 Slurry Reactor with Catalyst Deactivation

Gaafar and Ramirez (1985) applied the singular decomposition algorithm to a continuous stirred tank slurry reactor used for the dehydrogenation of ethylbenzene to form styrene monomer. Rate equations for the reaction are given by Clough and Ramirez (1976). The model for the system is described by a slow catalyst deactivation relation

$$\tau\dot{\alpha} = -\alpha \tag{10.10.43}$$

a medium time scale thermal response given by the energy balance,

$$\begin{aligned}&[(1-\gamma)\rho_c C_{pc} + \gamma\rho C_p]V\dot{T} = \\ &\qquad \rho C_p q(T_0 - T) - \Delta H k_0 \alpha W_c C T e^{-E/T}\end{aligned} \tag{10.10.44}$$

and a fast composition response given by the material balance,

$$\gamma V\dot{C} = q(C_0 - C) - k_0\alpha W_c C T e^{-E/T} \tag{10.10.45}$$

In order to apply the extended Kalman filter to this problem, the covariance matrix differential equation must be solved. To do this, the system equations are linearized about the current operating condition. This results in a system of linearized equations

$$\dot{\boldsymbol{x}} = \begin{bmatrix} a_{11} & a_{12} & a_{13} \\ a_{21} & a_{22} & a_{23} \\ a_{31} & a_{32} & a_{33} \end{bmatrix} \boldsymbol{x} \tag{10.10.46}$$

The partitioned covariance matrix differential equation becomes for this problem

$$\dot{I}_{11} = 2a_{11}I_{11} + 2a_{12}I_{12} + 2a_{13}I_{13} + b_1^2 q_{11} \tag{10.10.47}$$

$$\begin{aligned}0 &= 2_{12}I_{22} + a_{13}I_{23} + a_{21}I_{11} + a_{22}I_{12} + a_{23}I_{13} \\ 0 &= a_{13}I_{33} + a_{31}I_{11} + a_{32}I_{12} + a_{33}I_{13} \\ 0 &= 2a_{22}I_{22} + 2a_{23}I_{23} + b_2^2 q_{22} \\ 0 &= a_{23}I_{33} + a_{32}I_{22} + a_{33}I_{23} \\ 0 &= 2a_{33}I_{33} + b_3^2 q_{33}\end{aligned} \tag{10.10.48}$$

$$\begin{aligned}\dot{H}_{12} &= a_{12}H_{22} + a_{13}H_{23} + a_{21}H_{11} + a_{22}H_{12} + a_{23}H_{13} \\ \dot{H}_{22} &= 2a_{22}H_{22} + 2a_{23}H_{23} + b_2^2 q_{22}\end{aligned} \tag{10.10.49}$$

$$\begin{aligned}0 &= a_{13}H_{13} + a_{31}H_{11} + a_{32}H_{12} + a_{33}H_{13} \\ 0 &= a_{23}H_{33} + a_{32}H_{22} + a_{33}H_{23} \\ 0 &= H_{33} \\ 0 &= H_{11}\end{aligned} \tag{10.10.50}$$

$$\begin{aligned}\dot{G}_{13} &= a_{13}G_{33} + a_{31}G_{11} + a_{32}G_{12} + a_{33}G_{13} \\ \dot{G}_{23} &= a_{23}G_{33} + a_{32}G_{22} + a_{33}G_{23} \\ \dot{G}_{33} &= 2a_{33}G_{33} + b_3^2 q_{33}\end{aligned} \tag{10.10.51}$$

$$\begin{aligned}G_{11} &= 0 \\ G_{12} &= 0 \\ G_{22} &= 0\end{aligned} \tag{10.10.52}$$

These equations illustrate the fact that each time scale can be solved independently. The original set of six stiff differential equations has been partitioned into six non-stiff differential equations and twelve algebraic relations. Each of the various time modes of the covariance matrix differential equation can be solved independently of the others. This leads to computational efficiency. For this problem, Gaafar and Ramirez report a 50% savings in computational effort. Figures 10.16, 10.17, and 10.18 give the results of estimating the three state variables using both the efficient singular perturbation algorithm and the less efficient complete solution using a stiff differential equation solver. Since the number of differential equations for this example is small (six), it is still feasible to carry out the stiff equation solution using appropriate software. In these figures, the thin, solid line is the actual unknown state, the thick, solid line is the singular perturbation estimate, and the dashed line is the complete filter estimate. Figure 10.16 shows that the singular and complete filter results are nearly equivalent. There is essentially no loss of information for the slow dynamic state by using the singular perturbation algorithm. Figures 10.17 and 10.18 show that there is some loss of accuracy in the estimation of the other two modes by the use of the more efficient singular perturbation algorithm. The estimation of the faster concentration state is less accurate than the estimation of the intermediate time mode of temperature. The singular perturbation technique increases in usefulness as the stiffness of the equation set increases or as the size of the problem increases.

10.11 Kalman Filtering for Dynamic and Static Systems with Uncertain Parameters

Fathi *et al.* (1992) have considered modifications that are necessary to Kalman filtering in order to incorporate differential dynamic states and equilibrium static states. Figure 10.19 depicts a general stochastic system. The system inputs are divided into

1. process noise, $\boldsymbol{w}$, which is a random model uncertainty;
2. known inputs which include the control variables, $\boldsymbol{u}$, and the measured disturbances, $\boldsymbol{d}_m$; and
3. unknown system inputs which include unmeasured disturbances, $\boldsymbol{d}_u$, and changing physical coefficients, $\boldsymbol{c}$.

The process states are partitioned into

1. slowly varying dynamic states, $\boldsymbol{x}_d$; and

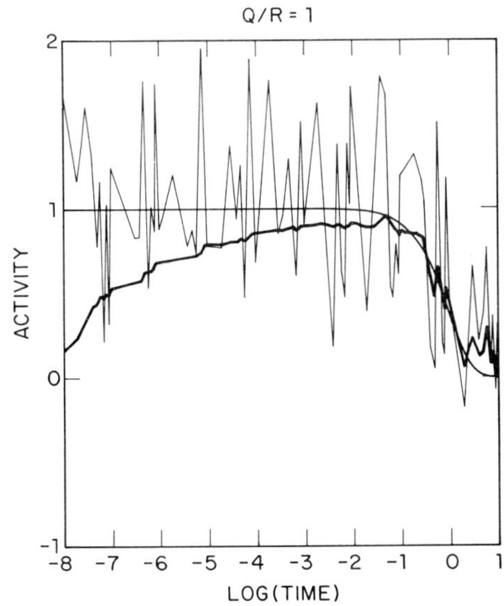

Figure 10.16 Comparison of complete and singular perturbation filters. Reprinted with permission from Gaafar and Ramirez, *Ind. Eng. Process Des. Dev.* **24**, 1055–1062, copyright ©1985 American Chemical Society.

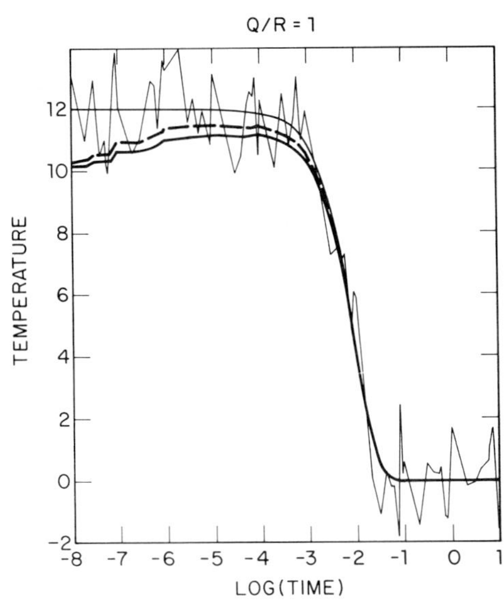

Figure 10.17 Comparison of complete and singular perturbation filters. Reprinted with permission from Gaafar and Ramirez, *Ind. Eng. Process Des. Dev.* **24**, 1055–1062, copyright ©1985 American Chemical Society.

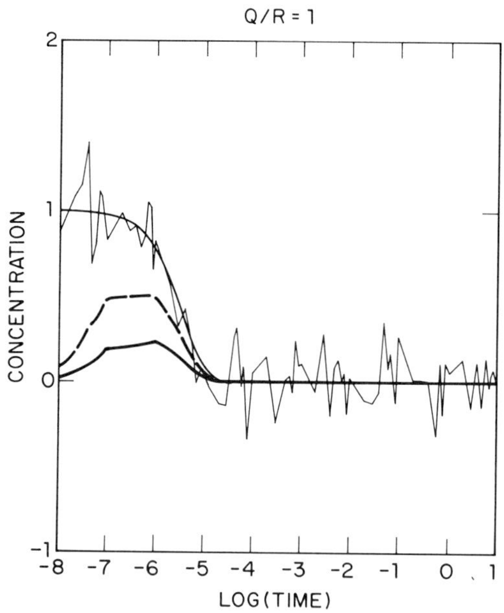

Figure 10.18 Comparison of complete and singular perturbation filters. Reprinted with permission from Gaafar and Ramirez, *Ind. Eng. Process Des. Dev.* **24**, 1055–1062, copyright ©1985 American Chemical Society.

2. rapidly responding steady-state or equilibrium states, $\boldsymbol{x}_s$.

The measurement system has measurement noise, $\boldsymbol{v}$. The discrete time model of the system is

Dynamic Model

$$\boldsymbol{x}_d(t) = \boldsymbol{f}_d\Big(t, \boldsymbol{x}_d(t-1), \boldsymbol{x}_s(t-1), \boldsymbol{\theta}(t-1)\Big) + \boldsymbol{w}_d \tag{10.11.1}$$

Steady-State Model

$$\boldsymbol{0} = \boldsymbol{f}_s\Big(t, \boldsymbol{x}_d(t), \boldsymbol{x}_s(t), \boldsymbol{\theta}(t)\Big) + \boldsymbol{w}_s \tag{10.11.2}$$

Measurement Model

$$\boldsymbol{y}(t) = \boldsymbol{h}\Big(t, \boldsymbol{x}_d(t), \boldsymbol{x}_s(t), \boldsymbol{\theta}(t)\Big) + \boldsymbol{v} \tag{10.11.3}$$

with

$$\boldsymbol{\theta}(t) = \begin{pmatrix} \boldsymbol{c}(t) \\ \boldsymbol{d}_u(t) \end{pmatrix} \tag{10.11.4}$$

To cope with time-varying parameters, the parameter vector is postulated to vary as

$$\boldsymbol{\theta}(t) = \boldsymbol{\theta}(t-1) + \boldsymbol{w}_\theta \tag{10.11.5}$$

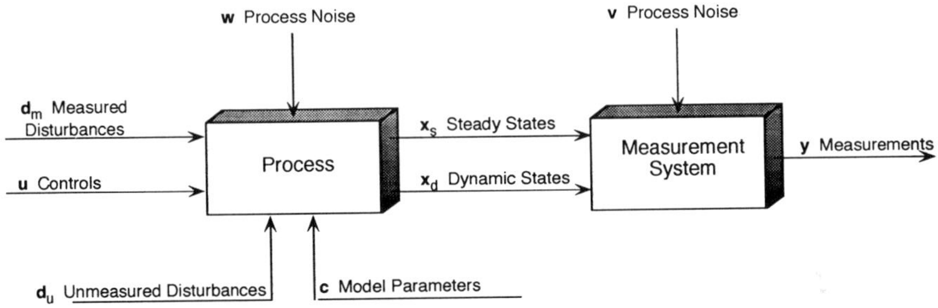

Figure 10.19 General stochastic system.

In order to transform this system of equations into a standard Kalman filtering problem, the augmented state $\boldsymbol{z}$ is defined as

$$\boldsymbol{z}(t) = \begin{pmatrix} \boldsymbol{x}_d(t) \\ \boldsymbol{x}_s(t) \\ \boldsymbol{\theta}(t) \end{pmatrix} \tag{10.11.6}$$

Furthermore, we have the following statistical information:

$$\begin{aligned} &\mathcal{E}\Big(\boldsymbol{z}(o)\Big) = \overline{z}_0 &\quad \text{cov}\Big(\boldsymbol{z}(0) - \overline{z}_0\Big) = \boldsymbol{M}_0 \\ &\mathcal{E}\Big(\boldsymbol{w}(t)\Big) = \boldsymbol{0} &\quad \text{cov}\Big(\boldsymbol{w}(t)\Big) = \boldsymbol{Q}_w \\ &\mathcal{E}\Big(\boldsymbol{v}(t)\Big) = \boldsymbol{0} &\quad \text{cov}\Big(\boldsymbol{v}(t)\Big) = \boldsymbol{Q}_v \end{aligned} \tag{10.11.7}$$

The Kalman filter that minimizes the quadratic performance index

$$\begin{aligned} J = {} & \frac{1}{2}\sum_{t=0}^{T}\Big(\boldsymbol{z}(t) - \overline{\boldsymbol{z}}(t)\Big)^T \boldsymbol{M}^{-1}(t)\Big(\boldsymbol{z}(t) - \overline{\boldsymbol{z}}(t)\Big) \\ & + \Big(\boldsymbol{y}(t) - \boldsymbol{h}(t)\Big)^T \boldsymbol{Q}_v^{-1}\Big(\boldsymbol{y}(t) - \boldsymbol{h}(t)\Big) \end{aligned} \tag{10.11.8}$$

where $\overline{\boldsymbol{z}}(t)$ is the estimate prior to measurement at time t, is

$$\hat{\boldsymbol{z}}(t) = \overline{\boldsymbol{z}}(t) + \boldsymbol{K}(t)\Big(\boldsymbol{y}(t) - \boldsymbol{h}(t)\Big) \tag{10.11.9}$$

with

$$\begin{aligned} \boldsymbol{K}(t) &= \boldsymbol{P}(t)\boldsymbol{H}^T(t)\boldsymbol{Q}_v^{-1} \\ \boldsymbol{P}(t) &= \boldsymbol{M}(t) - \boldsymbol{M}(t)\boldsymbol{H}^T(t)\Big[\boldsymbol{H}(t)\boldsymbol{M}(t)\boldsymbol{H}^T(t) + \boldsymbol{Q}_v\Big]^{-1}\boldsymbol{H}(t)\boldsymbol{M}(t) \\ \boldsymbol{M}(t) &= \boldsymbol{F}(t-1)\boldsymbol{P}(t-1)\boldsymbol{F}^T(t-1) + \boldsymbol{Q}_w(t-1) \quad \text{with } \boldsymbol{M}(0) = \boldsymbol{M}_0 \\ \boldsymbol{H}(t) &= \frac{\partial \boldsymbol{h}(t)}{\partial \boldsymbol{z}} \end{aligned} \tag{10.11.10}$$

and $\boldsymbol{F}$ and $\boldsymbol{Q}_w$ are partitionable as

$$
\begin{aligned}
&\boldsymbol{F}_{11} = \frac{\partial \hat{\boldsymbol{f}}_d}{\partial \boldsymbol{x}_d} \qquad \boldsymbol{F}_{12} = \frac{\partial \hat{\boldsymbol{f}}_d}{\partial \boldsymbol{x}_s} \qquad \boldsymbol{F}_{13} = \frac{\partial \hat{\boldsymbol{f}}_d}{\partial \boldsymbol{\theta}} \\
&\boldsymbol{F}_{21} = -\left(\frac{\partial \overline{\boldsymbol{f}}_s}{\partial \boldsymbol{x}_s}\right)^{-1} \frac{\partial \overline{\boldsymbol{f}}_s}{\partial \boldsymbol{x}_d} \frac{\partial \hat{\boldsymbol{f}}_d}{\partial \boldsymbol{x}_d} \qquad \boldsymbol{F}_{22} = -\left(\frac{\partial \overline{\boldsymbol{f}}_s}{\partial \boldsymbol{x}_s}\right)^{-1} \frac{\partial \overline{\boldsymbol{f}}_s}{\partial \boldsymbol{x}_d} \frac{\partial \hat{\boldsymbol{f}}_d}{\partial \boldsymbol{x}_s} \\
&\boldsymbol{F}_{23} = -\left(\frac{\partial \overline{\boldsymbol{f}}_s}{\partial \boldsymbol{x}_s}\right)^{-1} \left(\frac{\partial \overline{\boldsymbol{f}}_s}{\partial \boldsymbol{x}_d} \frac{\partial \hat{\boldsymbol{f}}_d}{\partial \boldsymbol{\theta}} + \frac{\partial \overline{\boldsymbol{f}}_s}{\partial \boldsymbol{\theta}}\right) \\
&\boldsymbol{F}_{31} = \boldsymbol{0} \qquad \boldsymbol{F}_{32} = \boldsymbol{0} \qquad \boldsymbol{F}_{33} = \boldsymbol{I}
\end{aligned}
\tag{10.11.11}
$$

$$
\begin{aligned}
&\boldsymbol{Q}_{11} = \boldsymbol{Q}_d \qquad \boldsymbol{Q}_{12} = -\boldsymbol{Q}_d \left(\frac{\partial \overline{\boldsymbol{f}}_s}{\partial \boldsymbol{x}_d}\right)^T \left(\frac{\partial \overline{\boldsymbol{f}}_s}{\partial \boldsymbol{x}_s}\right)^{-T} \qquad \boldsymbol{Q}_{13} = \boldsymbol{0} \\
&\boldsymbol{Q}_{21} = -\left(\frac{\partial \overline{\boldsymbol{f}}_s}{\partial \boldsymbol{x}_s}\right)^{-1} \frac{\partial \overline{\boldsymbol{f}}_s}{\partial \boldsymbol{x}_d} \boldsymbol{Q}_d \qquad \boldsymbol{Q}_{23} = -\left(\frac{\partial \overline{\boldsymbol{f}}_s}{\partial \boldsymbol{x}_s}\right)^{-1} \frac{\partial \overline{\boldsymbol{f}}_s}{\partial \boldsymbol{\theta}} \boldsymbol{Q}_\theta \\
&\boldsymbol{Q}_{22} = \left(\frac{\partial \overline{\boldsymbol{f}}_s}{\partial \boldsymbol{x}_s}\right)^{-1} \left[\boldsymbol{Q}_s + \left(\frac{\partial \overline{\boldsymbol{f}}_s}{\partial \boldsymbol{x}_d}\right) \boldsymbol{Q}_d \left(\frac{\partial \overline{\boldsymbol{f}}_s}{\partial \boldsymbol{x}_d}\right)^T \right. \\
&\qquad \left. + \left(\frac{\partial \overline{\boldsymbol{f}}_s}{\partial \boldsymbol{\theta}}\right) \boldsymbol{Q}_\theta \left(\frac{\partial \overline{\boldsymbol{f}}_s}{\partial \boldsymbol{\theta}}\right)^T\right] \left(\frac{\partial \overline{\boldsymbol{f}}_s}{\partial \boldsymbol{x}_s}\right)^{-T} \\
&\boldsymbol{Q}_{31} = \boldsymbol{0} \quad \boldsymbol{Q}_{32} = -\boldsymbol{Q}_\theta \left(\frac{\partial \overline{\boldsymbol{f}}_s}{\partial \boldsymbol{\theta}}\right)^T \left(\frac{\partial \overline{\boldsymbol{f}}_s}{\partial \boldsymbol{x}_s}\right)^{-T} \qquad \boldsymbol{Q}_{33} = \boldsymbol{Q}_\theta
\end{aligned}
\tag{10.11.12}
$$

where

$$
\frac{\partial \overline{\boldsymbol{f}}_s}{\partial \boldsymbol{z}} = \left.\frac{\partial \boldsymbol{f}_s}{\partial \boldsymbol{z}}\right|_{\boldsymbol{z}=\overline{\boldsymbol{z}}(t+1)} \tag{10.11.13}
$$

$$
\frac{\partial \hat{\boldsymbol{f}}_d}{\partial \boldsymbol{z}} = \left.\frac{\partial \boldsymbol{f}_d}{\partial \boldsymbol{z}}\right|_{\boldsymbol{z}=\hat{\boldsymbol{z}}(t)} \tag{10.11.14}
$$

In general, the structure of the modified extended Kalman filter algorithm is the same as the standard extended Kalman filter for nonlinear dynamic models. The main difference is that the system matrix $\boldsymbol{F}$ and the system uncertainty covariance matrix $\boldsymbol{Q}_w$ include appropriate static–dynamic interaction terms.

To avoid filter degradation in estimating faulty parameters, the method of Yoshimura *et al.* (1979) is adopted. In this technique, the condition

$$
\left|\theta_i^n - \hat{\theta}_i(t)\right| > d\left[M_{\theta i}(t+1)\right]^{1/2} \tag{10.11.15}
$$

is checked. Here θ_i^n is the nominal parameter value, $\hat{\theta}_i(t)$ is the estimate of the parameter, $M_{\theta i}(t)$ is the error variance of θ_i, and d is a positive constant usually equal to 1.96 which corresponds to a 95% confidence level. If condition (10.11.15) is satisfied for a parameter estimate, then modified variances and parameters are taken as

$$\begin{aligned} M_{\theta i}(t+1) &= \left[\theta_i^n - \hat{\theta}_i(t)\right]^2 / d^2 \\ \theta_i^n &= \hat{\theta}_i(t) \end{aligned} \tag{10.11.16}$$

Fathi *et al.* (1993) have shown that this filter is very effective in estimating states and parameters needed for an integrated approach to fault diagnosis. The integrated approach includes analytical redundancy models using state and parameter estimation modules in a diagnostic reasoning system of a knowledge-based set of rules of an expert system. They have been successful in diagnosing faults for a subsystem of a coal-fired power plant.

References

Bryson, A. E. and Ho, Y. C., *Applied Optimal Control*, Hemisphere Pub. (1975).

Clough, D. E. and Ramirez, W. F., "Mathematical Modeling and Optimization of the Dehydrogenation of Ethylbenzene to Form Styrene," *AIChE Journal* **22**, 1097 (1976).

Corcoran, C. A. and Rutan, S. C., "Correction for Temperature Variations in Kinetic Methods of Analysis with the Extended Kalman Filter," *Anal. Chem.* **60**, 1146–1153 (1988).

Fathi, Z., Korbicz, J., and Ramirez, W. F., "Analytical and Knowledge-Based Redundancy for Fault Diagnosis in Process Plants," *AIChE Journal* (1993).

Fathi, Z., Ramirez, W. F., and Aarna, O., "Joint State and Parameter Estimation for a Nuclear Reactor with Fast and Slow Modes," *Optimal Control Appl. and Meth.* **13**, 105–120 (1992).

Gaafar, W. S. and Ramirez, W. F., "Multiple Time Scale State Estimation," *Ind. Eng. Process Des. Dev.* **24**, 1055–1062 (1985).

Jazwinski, A. H., *Stochastic Processes and Filtering*, Academic Press, Cambridge, MA (1970).

Kalman, R. E., "A New Approach to Linear Filtering and Prediction Problems," *J. Basic Engr.* **82**, 35–45, March (1960).

Kalman, R. E. and Bucy, R. S., "New Results in Linear Filtering and Prediction Theory," *J. Basic Engr.* **83**, 95–100, March (1961).

Kokotovic, P. V., O'Malley, R. E., and Sannuti, P., *Automatica*, **11**, 123 (1976).

Lynch, E. B. and Ramirez, W. F., "Real-time Time-optimal Control of a Stirred Tank Reactor Using Kalman Filtering for State Estimation," *AIChE Journal* **21**, 799–804 (1975).

Nahlik, J. and Barianec, Z., "On-Line Parameter and State Estimation of Continuous Cultivation by Extended Kalman Filter," *Applied Microbiol. & Biotech.* **28**, 128–134 (1988).

Nyquist, J. K. and Ramirez, W. F., "Time Optimal Control of an Experimental Continuous Stirred Tank Reactor," *Chem. Engr. Sci.* **26**, 1673–1688 (1971).

Safonov, M. G., *Stability and Robustness of Multivariable Feedback Systems*, MIT Press, Cambridge, MA (1980).

San, K. and Stephanopoulos, G., "Studies on On-line Bioreactor Identification, II. Numerical and Experimental Results," *Biotech. and Bioengr.* **26**, 1189–1197 (1984).

Sannuti, P. and Kokotovic, P. V., *IEEE Trans. Auto. Control* **AC–14**, 15 (1969).

Soliman, M. A. and Ray, W. H., *Int. J. Control* **30**, 757 (1979).

Stephanopoulos, G. and San, K., "Studies on On-line Bioreactor Identification, I. Theory," *Biotech. and Bioengr.* **26**, 1176–1188 (1984).

Wentzell, P. D., Karayannis, M. I., and Crouch, S. R., "Simultaneous Kinetic Determinations with the Kalman Filter," *Analytica Chimica Acta* **224**, 263–274 (1989).

Wong, P. K., Stein, G., and Athans, M., "Structural Reliability and Robustness Properties of Optimal Linear Quadratic Multivariable Regulators," *IFAC Preprints Helsinki*, 1797–1805 (1978).

Yoshimura, T., Watanabe, K., Konish, K., and Soeda, T., "A Sequential Failure Detection Approach and the Identification of Failure Parameters," *Int. J. Syst. Sci.* **10**, 827–836 (1979).

Exercises

1. The following reaction sequence is being carried out in an isothermal batch chemical reactor. The scheme is characteristic of partial oxidation.

$$\begin{array}{l} A \xrightarrow{k_1} B \xrightarrow{k_2} C \\ \downarrow k_3 \\ C \end{array}$$

The mathematical model for the process is

$$\begin{aligned} \frac{dx_1}{dt} &= -(k_1 + k_3)x_1 \\ \frac{dx_2}{dt} &= k_1 x_1 - k_2 x_2 \\ \frac{dx_3}{dt} &= k_3 x_1 + k_2 x_2 \end{aligned}$$

Assume that x_2 and x_3 are measurable.

a) Find the optimal steady-state Kalman filter gain for this system when $k_1 = 1$, $k_2 = 2$, $k_3 = .5$, $\boldsymbol{Q} = \text{diag}\ (.0001)$, and $\boldsymbol{R} = \boldsymbol{Q}$, $\boldsymbol{R} = 10\ \boldsymbol{Q}$, and $\boldsymbol{R} = .1\ \boldsymbol{Q}$.

b) Simulate the response of the system (MATLAB program LSIM is useful for this) when

$$\begin{array}{ll} x_1(0) = 1 & \hat{x}_1(0) = .8 \\ x_2(0) = 0 & \hat{x}_2(0) = 0 \\ x_3(0) = 0 & \hat{x}_3(0) = 0 \end{array}$$

and $y_1 = x_2 + v$; $y_2 = x_3 + v$, where v is Gaussian with covariance $\boldsymbol{R}$.

Be sure to plot $\boldsymbol{x}(t)$, $\hat{\boldsymbol{x}}(t)$, and $\boldsymbol{y}(t)$.

2. Investigate the effects of more frequent measurements on the results of discrete Kalman filtering for the stirred tank problem of Section 10.7. In particular, look at sampling times of $\Delta T = .1$ min. and .2 min. What conclusions can you draw from your results?

3. Investigate the improvement achievable if the complete extended discrete Kalman filter with time-varying covariance matrices and the nonlinear predictive model are used for the stirred tank problem of Section 10.7. Which algorithm would you implement and why?

4. Often physical processes are continuous but measurements are discrete due to sampling for processing by digital control systems. In this case, the covariance matrix calculations become predictive during periods without measurements and filtering at the measurement time. Calculate the covariance matrix for this sequence for the linearized stirred tank reactor problem of Section 10.7 when $\Delta T = .5$ min. Perform the calculation for several sampling periods until a periodic behavior is obtained.

5. Implement an extended Kalman filter for the on–line state and parameter estimation of the continuous cultivation of yeast. You are to estimate the state concentrations of biomass (X) and glucose (S) in a stirred tank reactor. The mathematical model is

$$\begin{aligned}\frac{dX}{dt} &= \mu X - DX \\ \frac{dS}{dt} &= D(S_i - S) - \frac{1}{Y_{XS}}\,\mu X \\ \mu &= \mu_m \frac{S}{K_S + S}\end{aligned}$$

where μ is the specific growth rate (1/hr), Y_{XS} the yield coefficient, μ_m the maximum specific growth rate (1/hr), K_S the saturation constant (kg/m^3), and S_i the inlet glucose concentration, which is 15 kg/m^3.

You are to also identify the three constant model parameters μ_{max}, K_S, and Y_{XS}. The augmented state vector is therefore

$$\boldsymbol{x} = \begin{pmatrix} X \\ S \\ \mu_m \\ K_S \\ Y_{XS} \end{pmatrix}$$

The measurement on the system is the glucose concentration.

Obtain optimal estimates using extended Kalman filtering when the system is run for 40 hours and measurements available every 10 minutes. The state uncertainties are given by

$$\boldsymbol{Q} = \text{diag}\begin{pmatrix} 2.0 & .03 & .003 & .001 & .01 \end{pmatrix}$$

The measurement covariance is

$$R = .0064$$

The initial augmented state is

$$\boldsymbol{x}(t_0) = \begin{pmatrix} 2.5 \\ .64 \\ .36 \\ .4 \\ .3 \end{pmatrix}$$

and the initial state uncertainty is given by

$$\boldsymbol{P}(t_0) = \text{diag}\begin{pmatrix} 1.6 & .1 & .06 & .18 & .022 \end{pmatrix}$$

Nahlik and Barianec (1988) have considered this problem.

10.6 Corcoran and Rutan (1988) have shown the usefulness of extended Kalman filtering for the estimation of kinetic parameters in the absence of precise temperature control provided that the reaction temperatures are measurable. This system is similar to the problem discussed in Section 10.8. A single reaction is being investigated by measuring absorbance (A_t):

$$A_t = A\left(1 - e^{-kt}\right)$$

where A is a constant, t the reaction time, and k the reaction rate which is modeled as a function of temperature as

$$k = k_0 e^{-E/RT} \qquad \text{or} \qquad \ln k = \ln k_0 - E/RT$$

Investigate the effectiveness of two different estimation strategies. The first is to estimate the parameter A and the rate constant k,

$$\boldsymbol{x} = \begin{pmatrix} A \\ k \end{pmatrix}$$

and the second is to identify A and the parameters k_0 and E with

$$\boldsymbol{x} = \begin{pmatrix} A \\ \ln k_0 \\ E \end{pmatrix}$$

To test the effectiveness of each strategy, assume that the measurements have a 1% accuracy and the model is 10% accurate. Nominal values are $k = 2.5 \times 10^{-3}$ sec^{-1}, $E = 28$ kJ/mol, and $A = 3.1 \times 10^{-3}$. The temperature rises exponentially with a time constant of 50 hrs and has a noise level of 1%. The initial temperature is 342°K, and the final value is 347°K.

When identifying the reaction rate k, use a model of the form

$$k_{j+1} = g\ k_j + \nu$$

where

$$g = \exp\left(\frac{E}{R(T_{j+1} - T_j)}\right)$$

since the parameter is known to be a function of temperature.

In your simulations, start with initial estimates of

$$\begin{aligned} A &= 0 \\ k &= 1 \times 10^{-3}\ \mathrm{s}^{-1} \\ K &= 10\ \mathrm{kJ/mol} \end{aligned}$$

The final time is 400 sec., and the sampling rate is $\Delta t = 1$ sec.

10.7 Derive an optimal Kalman filter for systems with dynamic measurements. The measurement model is

$$\dot{\boldsymbol{y}} = \boldsymbol{M}\boldsymbol{y} + \boldsymbol{\nu}$$

where $\boldsymbol{\nu}$ is Gaussian.

A transformation of the form

$$\tilde{\boldsymbol{x}} = \begin{pmatrix} \boldsymbol{x} \\ \boldsymbol{y} \end{pmatrix}$$

should be considered.

Chapter 11

SEQUENTIAL LEAST SQUARES PARAMETER ESTIMATION

11.1 Sequential Stochastic Approximation Methods

This chapter considers the use of sequential least squares techniques for the recessive estimation of uncertain model parameters. There is a statistical advantage in taking this approach to model parameter identification over that of incorporating model parameter estimation directly into Kalman filtering as presented in Sections 10.8 and 10.12. This is due to the fact that by using Kalman filtering for state identification and recessive least squares for parameter identification, the innovations error is the difference between the actual measurements and the current measurement estimates, based upon current state estimates, rather than model predictions of the measurements. The concept of stochastic approximation has been developed in statistics for certain parameter estimation problems. Let us consider the simple difference model

$$y(t) \quad = \quad \boldsymbol{\theta}^T \boldsymbol{x}(t) + v(t) \tag{11.1.1}$$

where $y(t)$ are the scalar measurements, $\boldsymbol{x}(t)$ is the known independent variable vector, and $\boldsymbol{\theta}$ are the parameters to be determined. The measurement error is given by $v(t)$. We wish to select $\boldsymbol{\theta}$ so that the variance of $V(t)$ is minimized,

$$\min_{\boldsymbol{\theta}} \; V(\boldsymbol{\theta}) \tag{11.1.2}$$

where

$$V(\boldsymbol{\theta}) \quad = \quad \frac{1}{2}\,\mathcal{E}\left(y(t) - \boldsymbol{x}^T\boldsymbol{\theta}\right)^2 \tag{11.1.3}$$

where $y(t)$ are the actual scalar measurements.

The necessary condition for a minimum is obtained by differentiating Equation (11.1.3) and setting it to zero. This gives

$$\mathcal{E}\left(\boldsymbol{x}(t)\left(y - \boldsymbol{x}^T\boldsymbol{\theta}\right)\right) = \mathbf{0} \tag{11.1.4}$$

Since the probability distributions of $y(t)$ and $\boldsymbol{x}(t)$ are not known, an exact solution is impossible. Robbins and Monro (1951) introduced the concept of stochastic approximation, which yields the algorithm

$$\hat{\boldsymbol{\theta}}(t) = \hat{\boldsymbol{\theta}}(t-1) + \gamma(t)\left[\boldsymbol{x}(t)\left(y(t) - \boldsymbol{x}^T(t)\hat{\boldsymbol{\theta}}(t-1)\right)\right] \tag{11.1.5}$$

where the weighting factor $\gamma(t)$ is defined such that $\{\gamma(t)\}$ is a sequence of positive scalars tending to zero.

The stochastic gradient method is a stochastic analog to the method of steepest descent,

$$\boldsymbol{\theta}(t) = \boldsymbol{\theta}(t-1) - \gamma(t)\left[\frac{d}{d\boldsymbol{\theta}}V(\boldsymbol{\theta})\right]^T_{\boldsymbol{\theta}=\boldsymbol{\theta}(t-1)} \tag{11.1.6}$$

Therefore, the term in the square brackets in Equation (11.1.5) is essentially the steepest descent direction, which is also the term in the square brackets in Equation (11.1.6).

Instead of using a steepest descent algorithm, a Newton method can be used. Newton methods are known to have superior quadratic convergence properties. Newton methods are based upon a Taylor series truncating the nonlinear forms

$$\mathbf{0} = \left.\frac{d\,V(\boldsymbol{\theta})}{d\boldsymbol{\theta}}\right|_{\boldsymbol{\theta}(t-1)} + \left.\frac{d^2\,V(\boldsymbol{\theta})}{d\boldsymbol{\theta}^2}\right|_{\boldsymbol{\theta}(t-1)}\left(\boldsymbol{\theta}(t) - \boldsymbol{\theta}(t-1)\right) \tag{11.1.7}$$

or

$$\boldsymbol{\theta}(t) = \boldsymbol{\theta}(t-1) - \left.\left(\frac{d^2V(\boldsymbol{\theta})}{d\boldsymbol{\theta}^2}\right)^{-1}\right|_{\boldsymbol{\theta}(t-1)} \left.\frac{d}{d\boldsymbol{\theta}}V(\boldsymbol{\theta})\right|_{\boldsymbol{\theta}(t-1)} \tag{11.1.8}$$

Here the second derivative term is the Hessian and is

$$\frac{d^2\,V(\boldsymbol{\theta})}{d\boldsymbol{\theta}^2} = \mathcal{E}\left(\boldsymbol{x}(t)\boldsymbol{x}^T(t)\right) = \boldsymbol{R} \tag{11.1.9}$$

or

$$\mathcal{E}\left(\boldsymbol{x}(t)\boldsymbol{x}^T(t) - \boldsymbol{R}\right) = \mathbf{0} \tag{11.1.10}$$

We can apply the Robbins–Monro steepest descent procedure for the calculation of the Hessian or covariance to obtain the algorithm

$$\boldsymbol{R}(t) = \boldsymbol{R}(t-1) + \gamma(t)\left(\boldsymbol{x}(t)\boldsymbol{x}^T(t) - \boldsymbol{R}(t-1)\right) \tag{11.1.11}$$

and the stochastic Newton algorithm for the calculation of the parameters is

$$\boldsymbol{\theta}(t) \;=\; \boldsymbol{\theta}(t-1) + \gamma(t)\boldsymbol{R}^{-1}(t)\boldsymbol{x}(t)\left(y(t) - \boldsymbol{x}^T(t)\boldsymbol{\theta}(t-1)\right) \tag{11.1.12}$$

Notice that Equations (11.1.11) and (11.1.12) are sequential in that they update the covariance and the parameters as the data are obtained.

11.2 Sequential Linear Regression Example

Let us consider the application of sequential stochastic approximation methods for the determination of parameters in the linear model

$$y(t) = \theta_1 + \theta_2 z(t) + v(t) \tag{11.2.1}$$

This model is cast in the form of Equation (11.1.1) with

$$\boldsymbol{\theta} = \begin{pmatrix} \theta_1 \\ \theta_2 \end{pmatrix} \qquad \text{and} \qquad \boldsymbol{x} = \begin{pmatrix} 1 \\ z(t) \end{pmatrix} \tag{11.2.2}$$

As a specific example, consider the data set collected to determine the relationship between a rotameter reading $z(t)$ and the flow rate $y(t)$ given in Table 11.1.

We will use the second order Newton algorithm of Equations (11.1.11) and (11.1.12) with

$$\gamma = \frac{1}{t} \qquad \text{where } \; t = 1, 2, \cdots n \tag{11.2.3}$$

when the covariance matrix $\boldsymbol{R}$ converges

$$\mathcal{E}\left(\boldsymbol{x}\boldsymbol{x}^T\right) = \boldsymbol{R} \tag{11.2.4}$$

which is for this example

$$\boldsymbol{R} = \begin{bmatrix} 1 & \overline{z} \\ \overline{z} & \overline{z^2} \end{bmatrix} \tag{11.2.5}$$

with $\overline{z}$ the average value of $z(t)$ and $\overline{z^2}$ the average value of $z^2(t)$.

An initial guess must be provided for θ_1 and θ_2 as well as the initial values for the covariance matrix $\boldsymbol{R}$. Table 11.2 gives the results of applying the second order Newton algorithm for

$$\boldsymbol{\theta}_0 \;=\; \begin{pmatrix} 10 \\ 100 \end{pmatrix} \tag{11.2.6}$$

$$\boldsymbol{R}_0 \;=\; \begin{bmatrix} 1 & 3 \\ 3 & 9 \end{bmatrix} \tag{11.2.7}$$

Table 11.1
Rotameter Calibration Data

Data Set, t	y (cc/min)	z (in)
1	112	1.14
2	115	1.37
3	152	1.89
4	199	2.09
5	161	2.45
6	209	2.04
7	237	2.73
8	231	3.04
9	233	3.19
10	259	3.09
11	287	3.05
12	240	3.10
13	281	3.34
14	311	3.75
15	392	4.19
16	357	4.59

Note that reasonable values for the elements of $\boldsymbol{R}_0$ are chosen consistent with expected mean values and the square of the mean value.

Table 11.3 gives the results with completely different initial values for $\boldsymbol{\theta}$ as

$$\boldsymbol{\theta}_0 = \begin{pmatrix} 0 \\ 0 \end{pmatrix} \tag{11.2.8}$$

The results show that the system has converged after two iterations since the sequential parameters identified, $\boldsymbol{\theta}$, and the covariance, $\boldsymbol{R}$, are identical starting with the second data set. The same convergence is obtained when both $\boldsymbol{\theta}_0$ and $\boldsymbol{R}_0$ are changed as shown in Table 11.4.

It is obvious that there is significant noise in the data and that the confidence in θ_1 is low since its value fluctuates greatly with each additional data point processed. The confidence in θ_2 is much better. Also, the values obtained for $\boldsymbol{\theta}$ by off-line regression are identical to those obtained using the on-line sequential algorithm.

Although the convergence of the second-order algorithm was very rapid, the use of the first-order algorithm given by Equation (11.1.5) has poor convergence properties. This is shown in Table 11.5. The final values of θ_1 and θ_2 after 16 data sets are in significant error in comparison to the off-line and

Table 11.2
Convergence of Second-Order Newton Algorithm

Data Set	θ_1	θ_2	R_{11}	R_{12}	R_{21}	R_{22}
0	10	100	1	3	3	9
1	-2	100	1	1.14	1.14	1.3
2	97.13	13.04	1	1.26	1.26	1.59
3	43.66	56.37	1	1.47	1.47	2.25
4	4.52	86.27	1	1.62	1.62	2.78
5	50.81	54.24	1	1.79	1.79	3.42
6	43.35	62.65	1	1.83	1.83	3.55
7	28.42	71.92	1	1.96	1.96	4.11
8	37.37	66.69	1	2.09	2.09	4.75
9	44.77	62.49	1	2.22	2.22	5.35
10	38.66	66.06	1	2.30	2.30	5.77
11	28.33	72.22	1	2.37	2.37	6.09
12	30.82	70.78	1	2.43	2.43	6.38
13	27.37	72.58	1	2.50	2.50	6.75
14	23.81	74.27	1	2.59	2.59	7.27
15	5.22	82.57	1	2.70	2.70	7.96
16	13.51	79.02	1	2.82	2.82	8.78

Table 11.3
Convergence of Second-Order Newton Algorithm

Data Set	θ_1	θ_2	R_{11}	R_{12}	R_{21}	R_{22}
0	0	0	1	3	3	9
1	112	0	1	1.14	1.14	1.3
2	97.13	13.04	1	1.26	1.26	1.59
3	43.66	56.37	1	1.47	1.47	2.25
4	4.52	86.27	1	1.62	1.62	2.78
5	50.81	54.24	1	1.79	1.79	3.42
6	43.35	62.65	1	1.83	1.83	3.55
7	28.42	71.92	1	1.96	1.96	4.11
8	37.37	66.69	1	2.09	2.09	4.75
9	44.77	62.49	1	2.22	2.22	5.35
10	38.66	66.06	1	2.30	2.30	5.77
11	28.33	72.22	1	2.37	2.37	6.09
12	30.82	70.78	1	2.43	2.43	6.38
13	27.37	72.58	1	2.50	2.50	6.75
14	23.81	74.27	1	2.59	2.59	7.27
15	5.22	82.57	1	2.70	2.70	7.96
16	13.51	79.02	1	2.82	2.82	8.78

Table 11.4
Convergence of Second-Order Newton Algorithm

Data Set	θ_1	θ_2	R_{11}	R_{12}	R_{21}	R_{22}
0	1000	-100	1	0	0	100
1	226	-100	1	1.14	1.14	1.3
2	97.13	13.04	1	1.26	1.26	1.59
3	43.66	56.37	1	1.47	1.47	2.25
4	4.52	86.27	1	1.62	1.62	2.78
5	50.81	54.24	1	1.79	1.79	3.42
6	43.35	62.65	1	1.83	1.83	3.55
7	28.42	71.92	1	1.96	1.96	4.11
8	37.37	66.69	1	2.09	2.09	4.75
9	44.77	62.49	1	2.22	2.22	5.35
10	38.66	66.06	1	2.30	2.30	5.77
11	28.33	72.22	1	2.37	2.37	6.09
12	30.82	70.78	1	2.43	2.43	6.38
13	27.37	72.58	1	2.50	2.50	6.75
14	23.81	74.27	1	2.59	2.59	7.27
15	5.22	82.57	1	2.70	2.70	7.96
16	13.51	79.02	1	2.82	2.82	8.78

Table 11.5
Convergence of First-Order Algorithm

Data Set	θ_1	θ_2
0	10	100
1	4	93.16
2	-1.54	85.57
3	-3.59	81.70
4	2.78	95.01
5	-9.65	64.57
6	2.77	89.90
7	1.37	86.08
8	-2.19	75.25
9	-2.68	73.70
10	.41	83.23
11	3.14	91.55
12	- .47	80.36
13	.46	83.48
14	.29	82.85
15	3.08	94.52
16	-1.62	72.94

second-order algorithm results. This is especially true of parameter θ_1, which has poor confidence limits. The first-order algorithm can be used effectively when the parameters to be identified are quite certain and have tight confidence limits.

11.3 Model Parameter Estimation

Often there are uncertain model parameters in process models. It is therefore important to be able to update selected model parameters sequentially based upon actual response data. Ljung and Soderstrom (1983) have considered the problem of sequential parameter identification and have proposed a method of coupling parameter identification with state variable identification based upon sequential stochastic approximation concepts. We will consider a general linear discrete prediction model of the form

$$\boldsymbol{x}(t+1,\boldsymbol{\theta}) = \boldsymbol{F}(\boldsymbol{\theta})\boldsymbol{x}(t,\boldsymbol{\theta}) + \boldsymbol{G}(\boldsymbol{\theta})\boldsymbol{y}(t) \tag{11.3.1}$$

where $\boldsymbol{x}$ = the state vector
$\boldsymbol{y}$ = the measurement vector
$\boldsymbol{\theta}$ = the unknown parameter vector

The measurement model is

$$\overline{\boldsymbol{y}}(t|\boldsymbol{\theta}) = \boldsymbol{H}(\boldsymbol{\theta})\boldsymbol{x}(t,\boldsymbol{\theta}) \tag{11.3.2}$$

where $\overline{\boldsymbol{y}}$ = the predicted measurements.

We can relate the predicted measurements, $\overline{\boldsymbol{y}}(t|\boldsymbol{\theta})$, to the unknown parameters by the matrix

$$\boldsymbol{\psi}(t,\boldsymbol{\theta}) = \frac{d}{d\boldsymbol{\theta}}\,\overline{\boldsymbol{y}}(t|\boldsymbol{\theta}) \tag{11.3.3}$$

The matrix $\boldsymbol{\psi}(t,\boldsymbol{\theta})$ can be formed from knowledge of the measurements $\boldsymbol{y}(t)$ by a finite dimensional filter by introducing various differentiations with respect to the unknown parameters,

$$\boldsymbol{W}(t,\boldsymbol{\theta}) = \frac{\partial \boldsymbol{x}}{\partial \boldsymbol{\theta}}\,(t,\boldsymbol{\theta}) \tag{11.3.4}$$

$$\boldsymbol{M}(\boldsymbol{\theta},\boldsymbol{x},\boldsymbol{y}) = \frac{\partial}{\partial \boldsymbol{\theta}}\left(\boldsymbol{F}(\boldsymbol{\theta})\boldsymbol{x}(t,\boldsymbol{\theta}) + \boldsymbol{G}(\boldsymbol{\theta})\boldsymbol{y}(\boldsymbol{t})\right) \tag{11.3.5}$$

$$\boldsymbol{D}(\boldsymbol{\theta},\boldsymbol{x}) = \frac{\partial}{\partial \boldsymbol{\theta}}\left(\boldsymbol{H}(\boldsymbol{\theta})\boldsymbol{x}(t,\boldsymbol{\theta})\right) \tag{11.3.6}$$

Therefore, from Equations (11.3.1) and (11.3.2), we have the gradient model

$$\boldsymbol{W}(t+1,\boldsymbol{\theta}) = \boldsymbol{F}(\boldsymbol{\theta})\boldsymbol{W}(t,\boldsymbol{\theta}) + \boldsymbol{M}(\boldsymbol{\theta},\boldsymbol{x},\boldsymbol{y}) \tag{11.3.7}$$

$$\boldsymbol{\psi}^T(t,\boldsymbol{\theta}) = \boldsymbol{H}(\boldsymbol{\theta})\boldsymbol{W}(t,\boldsymbol{\theta}) + \boldsymbol{D}(\boldsymbol{\theta},\boldsymbol{x}) \tag{11.3.8}$$

It is convenient to combine the state space model of Equations (11.3.1) and (11.3.2) with the gradient model of Equations (11.3.7) and (11.3.8) into one formulation. We define an augmented state vector to be the combination of the state vector and the gradient vector,

$$\boldsymbol{\xi}(t,\boldsymbol{\theta}) = \begin{pmatrix} \boldsymbol{x}(t,\boldsymbol{\theta}) \\ \text{col}\,\boldsymbol{W}(t,\boldsymbol{\theta}) \end{pmatrix} \tag{11.3.9}$$

which gives the augmented state and gradient model of

$$\boldsymbol{\xi}(t+1,\boldsymbol{\theta}) = \boldsymbol{A}(\boldsymbol{\theta})\xi(t,\boldsymbol{\theta}) + \boldsymbol{B}(\boldsymbol{\theta})\boldsymbol{y}(t) \tag{11.3.10}$$

with

$$\boldsymbol{A}(\boldsymbol{\theta}) = \begin{bmatrix} \boldsymbol{F} & & & \\ & \boldsymbol{F} & & \\ & & \boldsymbol{F} & \\ & & & \ddots \end{bmatrix} \tag{11.3.11}$$

$$\boldsymbol{B}(\boldsymbol{\theta}) = \begin{bmatrix} \boldsymbol{G} & & & \\ & \dfrac{\partial \boldsymbol{M}}{\partial \boldsymbol{y}} & & \\ & & \dfrac{\partial \boldsymbol{M}}{\partial \boldsymbol{y}} & \\ & & & \ddots \end{bmatrix} \tag{11.3.12}$$

The augmented measurement model is

$$\begin{pmatrix} \overline{\boldsymbol{y}}(t|\boldsymbol{\theta}) \\ \text{col}\,\boldsymbol{\psi}(t,\boldsymbol{\theta}) \end{pmatrix} = \boldsymbol{C}(\boldsymbol{\theta})\boldsymbol{\xi}(t,\boldsymbol{\theta}) \tag{11.3.13}$$

with

$$\boldsymbol{C}(\boldsymbol{\theta}) = \begin{bmatrix} \boldsymbol{H} & & & \\ & \boldsymbol{H} + \dfrac{\partial \boldsymbol{D}}{\partial \boldsymbol{W}} & & \\ & & \boldsymbol{H} + \dfrac{\partial \boldsymbol{D}}{\partial \boldsymbol{W}} & \\ & & & \ddots \end{bmatrix} \tag{11.3.14}$$

We will now develop a recursive Gauss–Newton algorithm for minimization of a quadratic error criterion. We want to select so that the quadratic function ℓ is minimized where

$$\ell = \frac{1}{2}\,\boldsymbol{\epsilon}^T \boldsymbol{\Lambda}^{-1} \boldsymbol{\epsilon} \tag{11.3.15}$$

and where $\boldsymbol{\Lambda}$ is a weighting matrix and $\boldsymbol{\epsilon}$ is the error defined as

$$\boldsymbol{\epsilon} = \boldsymbol{y}(t) - \overline{\boldsymbol{y}}(t|\boldsymbol{\theta}) \tag{11.3.16}$$

The necessary conditions for the minimization of ℓ are

$$\left(\frac{d}{d\boldsymbol{\theta}}\,\ell\right)^T = -\boldsymbol{\psi}(t,\boldsymbol{\theta})\boldsymbol{\Lambda}^{-1}\boldsymbol{\epsilon}(t,\boldsymbol{\theta}) = \mathbf{0} \tag{11.3.17}$$

Equation (11.3.17) follows from the fact that

$$\frac{d\boldsymbol{\epsilon}}{d\boldsymbol{\theta}} = \frac{d}{d\boldsymbol{\theta}}\Big(\boldsymbol{y}(t) - \overline{\boldsymbol{y}}(t|\theta)\Big) = -\boldsymbol{\psi}^T(t,\boldsymbol{\theta}) \tag{11.3.18}$$

11.4 Stochastic Approximation Methods Using Models

For the problem of estimating the coefficients of the model of Equation (11.3.10), and minimizing the quadratic error criterion of Equation (11.3.15), the necessary conditions for minimization are given by Equation (11.3.17). We will use the stochastic Newton algorithm of (11.1.8) to find the roots of the necessary conditions of Equation (11.3.17). This gives

$$\hat{\boldsymbol{\theta}}(t) = \hat{\boldsymbol{\theta}}(t-1) + \gamma(t)\boldsymbol{R}^{-1}(t)\boldsymbol{\psi}\Big(t,\hat{\boldsymbol{\theta}}(t-1)\Big)\boldsymbol{\Lambda}^{-1}\boldsymbol{\epsilon}\Big(t,\hat{\boldsymbol{\theta}}(t-1)\Big) \tag{11.4.1}$$

We apply the Robbins–Monro steepest descent method for the updating of the Hessian, $\boldsymbol{R}(t)$,

$$\boldsymbol{R}(t) = \boldsymbol{R}(t-1) + \gamma(t)\Big(\boldsymbol{\psi}(t)\boldsymbol{\Lambda}^{-1}\boldsymbol{\psi}^T(t) - \boldsymbol{R}(t-1)\Big) \tag{11.4.2}$$

We also have the augmented model

$$\boldsymbol{\xi}\Big(k+1,\hat{\boldsymbol{\theta}}(t-1)\Big) = \boldsymbol{A}\Big(\hat{\boldsymbol{\theta}}(t+1)\Big)\,\boldsymbol{\xi}\Big(k,\hat{\boldsymbol{\theta}}(t-1)\Big) + \boldsymbol{B}\Big(\hat{\boldsymbol{\theta}}(t-1)\Big)\,\boldsymbol{y}(k) \tag{11.4.3}$$

$$k = 0, 1\cdots t-1$$

$$\begin{pmatrix} \overline{\boldsymbol{y}}\left(t|\hat{\boldsymbol{\theta}}(t-1)\right) \\ \operatorname{col}\boldsymbol{\psi}\left(t,\ \hat{\boldsymbol{\theta}}(t-1)\right) \end{pmatrix} = \boldsymbol{C}\Big(\hat{\boldsymbol{\theta}}(t-1)\Big)\,\boldsymbol{\xi}\Big(t,\ \hat{\boldsymbol{\theta}}(t-1)\Big) \tag{11.4.4}$$

where the error is defined as

$$\boldsymbol{\epsilon}\left(t,\ \hat{\boldsymbol{\theta}}(t-1) = \boldsymbol{y}(t) - \overline{\boldsymbol{y}}\left(t|\hat{\boldsymbol{\theta}}(t-1)\right)\right) \tag{11.4.5}$$

The only problem with this algorithm is that it is not recursive since we need the whole data record, $\boldsymbol{y}(k)$ for $k = 0, 1, \cdots, t-1$, in order to update the model parameters. We can make the procedure recursive with the following approximation:

$$\hat{\boldsymbol{\theta}}(t) = \hat{\boldsymbol{\theta}}(t-1) + \gamma(t)\boldsymbol{R}^{-1}(t)\boldsymbol{\psi}(t)\boldsymbol{\Lambda}^{-1}\boldsymbol{\epsilon}(t) \tag{11.4.6}$$

$$\boldsymbol{R}(t) = \boldsymbol{R}(t-1) + \gamma(t)\left(\boldsymbol{\psi}(t)\boldsymbol{\Lambda}^{-1}\boldsymbol{\psi}(t) - \boldsymbol{R}(t-1)\right) \tag{11.4.7}$$

$$\boldsymbol{\xi}(t) = \boldsymbol{A}\left(\hat{\boldsymbol{\theta}}(t-1)\right)\boldsymbol{\xi}(t-1) + \boldsymbol{B}\left(\hat{\boldsymbol{\theta}}(t-1)\right)\boldsymbol{y}(t-1) \tag{11.4.8}$$

$$\begin{pmatrix} \overline{\boldsymbol{y}}(t) \\ \text{col}\,\boldsymbol{\psi}(t) \end{pmatrix} = \boldsymbol{C}\left(\hat{\boldsymbol{\theta}}(t-1)\right)\boldsymbol{\xi}(t) \tag{11.4.9}$$

$$\boldsymbol{\epsilon}(t) = \boldsymbol{y}(t) - \overline{\boldsymbol{y}}(t) \tag{11.4.10}$$

This is a recursive algorithm since we only need to store $\boldsymbol{\xi}(t-1)$, $\boldsymbol{\theta}(t-1)$, and $\boldsymbol{y}(t-1)$ at time $(t-1)$.

The best choice of $\boldsymbol{\Lambda}$ is the covariance matrix of the true prediction error

$$\boldsymbol{\Lambda} = \mathcal{E}\left\{\boldsymbol{\epsilon}\left(t,\, \boldsymbol{\theta})\boldsymbol{\epsilon}^T(t,\, \boldsymbol{\theta}\right)\right\} \tag{11.4.11}$$

Since $\boldsymbol{\Lambda}$ is typically unknown, a reasonable choice is based upon stochastic approximation

$$\boldsymbol{\Lambda}(t) = \boldsymbol{\Lambda}(t-1) + \gamma(t)\left(\boldsymbol{\epsilon}(t)\boldsymbol{\epsilon}^T(t) - \boldsymbol{\Lambda}(t-1)\right) \tag{11.4.12}$$

Ljung and Soderstrom (1983) also present an algebraic rearrangement that eliminates $\boldsymbol{R}^{-1}(t)$. We introduce

$$\boldsymbol{P}(t) = \gamma(t)\boldsymbol{R}^{-1}(t) \tag{11.4.13}$$

From the recursive formula for $\boldsymbol{R}$, we can derive the relation

$$\begin{aligned}\boldsymbol{P}(t) = \frac{1}{\lambda(t)}\Big(\boldsymbol{P}(t-1) - \boldsymbol{P}(t-1)\boldsymbol{\psi}(t)\Big[\boldsymbol{\psi}^T(t)\boldsymbol{P}(t-1)\boldsymbol{\psi}(t) \\ +\lambda(t)\boldsymbol{\Lambda}(t)\Big]^{-1}\boldsymbol{\psi}^T(t)\boldsymbol{P}(t-1)\Big)\end{aligned} \tag{11.4.14}$$

where $\lambda(t)$ is the forgetting factor as is given by

$$\lambda(t) = \gamma(t-1)\left(1 - \gamma(t)\right)\Big/\gamma(t) \tag{11.4.15}$$

Equation (11.4.14) can be simplified by the introduction of

$$\boldsymbol{L}(t) = \boldsymbol{P}(t-1)\boldsymbol{\psi}(t)\boldsymbol{S}^{-1}(t) \qquad (11.4.16)$$

where

$$\boldsymbol{S}(t) = \boldsymbol{\psi}^T(t)\boldsymbol{P}(t-1)\boldsymbol{\psi}(t) + \lambda(t)\boldsymbol{\Lambda}(t) \qquad (11.4.17)$$

Then Equation (11.4.14) becomes

$$\boldsymbol{P}(t) = \frac{1}{\lambda(t)}\left[\boldsymbol{P}(t-1) - \boldsymbol{L}(t)\boldsymbol{S}(t)\boldsymbol{L}^T(t)\right] \qquad (11.4.18)$$

The matrix $\boldsymbol{P}$ is the covariance of the model uncertainty.

11.5 A Recursive Parameter Identification Algorithm for State Space Models

We can now summarize the Ljung–Soderstrom algorithm for the state space model

$$\boldsymbol{x}(t+1|\boldsymbol{\theta}) = \boldsymbol{F}(\boldsymbol{\theta})\boldsymbol{x}(t|\boldsymbol{\theta}) + \boldsymbol{G}\,\boldsymbol{y}(t) \qquad (11.5.1)$$

with a measurement model

$$\overline{\boldsymbol{y}}(t|\boldsymbol{\theta}) = \boldsymbol{H}(\boldsymbol{\theta})\boldsymbol{x}(t|\boldsymbol{\theta}) \qquad (11.5.2)$$

We define the Jacobian matrices

$$\boldsymbol{M}(\boldsymbol{\theta}) = \frac{\partial}{\partial\boldsymbol{\theta}}\left(\boldsymbol{F}(\boldsymbol{\theta})\boldsymbol{x}\right) \qquad (11.5.3)$$

and

$$\boldsymbol{D}(\boldsymbol{\theta}) = \frac{\partial}{\partial\boldsymbol{\theta}}\left(\boldsymbol{H}(\boldsymbol{\theta})\boldsymbol{x}\right) \qquad (11.5.4)$$

which can be computed for specific $\boldsymbol{H}(\boldsymbol{\theta})$ and $\boldsymbol{F}(\boldsymbol{\theta})$ matrices. We also define the Jacobian gradient matrices

$$\boldsymbol{W}(t) = \frac{d}{d\boldsymbol{\theta}}\boldsymbol{x}(t|\boldsymbol{\theta}) \qquad (11.5.5)$$

and

$$\boldsymbol{\psi}(t) = \frac{d}{d\boldsymbol{\theta}}\overline{\boldsymbol{y}}(t|\boldsymbol{\theta}) \qquad (11.5.6)$$

These gradient matrices can be updated using the state and measurement model relations by

$$\boldsymbol{W}(t+1) = \boldsymbol{F}(\boldsymbol{\theta})\boldsymbol{W}(t) + \boldsymbol{M}(\boldsymbol{\theta}) \tag{11.5.7}$$

and

$$\boldsymbol{\psi}(t) = \boldsymbol{H}(\boldsymbol{\theta})\boldsymbol{W}(t) + \boldsymbol{D}(\boldsymbol{\theta}) \tag{11.5.8}$$

The optimal estimate of the parameters $\boldsymbol{\theta}(t)$ is obtained by minimization of a quadratic error criterion, where the error is

$$\boldsymbol{\epsilon}(t) = \boldsymbol{y}(t) - \boldsymbol{H}\,\boldsymbol{x}\left(t,\ \boldsymbol{\theta}(t-1)\right) \tag{11.5.9}$$

where $\boldsymbol{y}(t)$ are the actual measurements. The weighting matrix $\boldsymbol{\Lambda}$ of the quadratic error function of Equation (11.3.15) is updated using stochastic approximation as

$$\boldsymbol{\Lambda}(t) = \boldsymbol{\Lambda}(t-1) + \gamma(t)\left(\boldsymbol{\epsilon}(t)\boldsymbol{\epsilon}^T(t) - \boldsymbol{\Lambda}(t-1)\right) \tag{11.5.10}$$

The recursive algorithm for the model parameters is therefore

$$\boldsymbol{\theta}(t) = \boldsymbol{\theta}(t-1) + \boldsymbol{L}(t)\boldsymbol{\epsilon}(t) \tag{11.5.11}$$

where the gain matrix $\boldsymbol{L}(t)$ is defined as

$$\boldsymbol{L}(t) = \boldsymbol{P}(t-1)\boldsymbol{\psi}(t)\boldsymbol{S}^{-1}(t) \tag{11.5.12}$$

with $\boldsymbol{S}(t)$ defined as

$$\boldsymbol{S}(t) = \boldsymbol{\psi}^T(t)\boldsymbol{P}(t-1)\boldsymbol{\psi}(t) + \lambda(t)\boldsymbol{\Lambda}(t) \tag{11.5.13}$$

and the covariance of the model uncertainty is recursively computed from

$$\boldsymbol{P}(t) = \left(\boldsymbol{P}(t-1) - \boldsymbol{L}(t)\boldsymbol{S}(t)\boldsymbol{L}^T(t)\right) \Big/ \lambda(t) \tag{11.5.14}$$

The scalar gain sequence $\gamma(t)$ is a sequence that tends to a small number as time increases. It is usually related to the forgetting factor, λ, by

$$\lambda(t) = \frac{\gamma(t-1)}{\gamma(t)}\left(1 - \gamma(t)\right) \tag{11.5.15}$$

A constant forgetting factor, λ, yields a constant gain, γ. A constant forgetting factor is also related to the exponential decay time constant through

$$T_o = \frac{1}{1-\lambda} \tag{11.5.16}$$

Hence, a prediction error older than T_o time units has a weight less than 36% of that of the most recent data. If a constant forgetting factor is 0.9, then the exponential decay times constant, T_o, is 10 time units.

In order to initiate this algorithmic procedure, we need to specify the initial parameter estimates,

$$\boldsymbol{\theta}(0) = \boldsymbol{\theta}_0 \tag{11.5.17}$$

the initial state Jacobian matrix, which is usually assumed to be zero,

$$\boldsymbol{W}(0) = \mathbf{0} \tag{11.5.18}$$

the initial covariance of the model uncertainty,

$$\boldsymbol{P}(0) = \boldsymbol{P}_0 \tag{11.5.19}$$

the initial weighting matrix, $\boldsymbol{\Lambda}(0)$, which is the initial error covariance estimate,

$$\boldsymbol{\Lambda}(0) = \mathcal{E}\left(\boldsymbol{\epsilon}_0\boldsymbol{\epsilon}_0^T\right) \tag{11.5.20}$$

and the initial forgetting factor,

$$\lambda(0) = \lambda_0 \tag{11.5.21}$$

11.6 Simultaneous State and Parameter Identification

Recursive least squares parameter identification can be coupled with Kalman filtering to give an algorithm for simultaneous state and parameter identification. For state identification, we have

$$\hat{\boldsymbol{x}}(t) = \overline{\boldsymbol{x}}(t) + \boldsymbol{K}_{ss}\left(\boldsymbol{y}(t) - \boldsymbol{H}\,\overline{\boldsymbol{x}}(t)\right) \tag{11.6.1}$$

where $\boldsymbol{K}_{ss}$ is the steady-state Kalman filter gain, which is computed from

$$\boldsymbol{K}_{ss} = \boldsymbol{P}_{ss}\boldsymbol{H}^T\boldsymbol{R}^{-1} \tag{11.6.2}$$

The steady-state covariance matrix $\boldsymbol{P}_{ss}$ comes from the recursive algorithm

$$\boldsymbol{P}_i = \boldsymbol{M}_i - \boldsymbol{M}_i\,\boldsymbol{H}^T\left(\boldsymbol{H}\,\boldsymbol{M}_i\boldsymbol{H}^T + \boldsymbol{R}\right)^{-1}\boldsymbol{H}\,\boldsymbol{M}_i \tag{11.6.3}$$

$$\boldsymbol{M}_{i+1} = \boldsymbol{\phi}\,\boldsymbol{P}_i\boldsymbol{\phi}^T + \boldsymbol{Q} \tag{11.6.4}$$

The state model is

$$\boldsymbol{x}(t) = \boldsymbol{\phi}\,\boldsymbol{x}(t-1) + \boldsymbol{w}(t) \tag{11.6.5}$$

and the measurement model is

$$\boldsymbol{y} = \boldsymbol{H}\,\boldsymbol{x}(t) + \boldsymbol{v}(t) \tag{11.6.6}$$

The general state model for parameter identification using Kalman filtering is given by

$$\begin{aligned}\hat{\boldsymbol{x}}\left(t|\boldsymbol{\theta}(t-1)\right) &= \boldsymbol{\phi}\left(\boldsymbol{\theta}(t-1)\right)\hat{\boldsymbol{x}}\left(t-1|\boldsymbol{\theta}(t-1)\right) + \boldsymbol{K}_{ss}\Big[\boldsymbol{y}(t) \\ &- \boldsymbol{H}\boldsymbol{\phi}\left(\boldsymbol{\theta}(t-1)\right)\hat{\boldsymbol{x}}\left(t-1|\boldsymbol{\theta}(t-1)\right)\Big]\end{aligned} \tag{11.6.7}$$

or

$$\begin{aligned}\hat{\boldsymbol{x}}\left(t|\boldsymbol{\theta}(t-1)\right) &= \left(\boldsymbol{I}-\boldsymbol{K}\boldsymbol{H}\right)\boldsymbol{\phi}\left(\boldsymbol{\theta}(t-1)\right)\hat{\boldsymbol{x}}\left(t-1|\boldsymbol{\theta}(t-1)\right) \\ &+ \boldsymbol{K}_{ss}\boldsymbol{y}(t)\end{aligned} \tag{11.6.8}$$

which can be expressed as

$$\begin{aligned}\hat{\boldsymbol{x}}\left(t|\boldsymbol{\theta}(t-1)\right) &= \boldsymbol{F}\left(\boldsymbol{\theta}(t-1)\right)\hat{\boldsymbol{x}}\left(t-1|\boldsymbol{\theta}(t-1)\right) \\ &+ \boldsymbol{G}\boldsymbol{y}(t)\end{aligned} \tag{11.6.9}$$

where

$$\boldsymbol{F}(\boldsymbol{\theta}) = \left(\boldsymbol{I}-\boldsymbol{K}\boldsymbol{H}\right)\boldsymbol{\phi}\left(\boldsymbol{\theta}(t-1)\right) \tag{11.6.10}$$

$$\boldsymbol{G} = \boldsymbol{K}_{ss} \tag{11.6.11}$$

and the measurement model for the estimation problem is

$$\hat{\boldsymbol{y}}\left(t|\boldsymbol{\theta}(t-1)\right) = \boldsymbol{H}\hat{\boldsymbol{x}}\left(t|\boldsymbol{\theta}(t-1)\right) \tag{11.6.12}$$

This is an estimate of the measurements based upon the best state estimate available using Kalman filtering. The Ljung and Soderstrom sequential least squares algorithm can be modified to accept data at the time of estimation.

Prediction Error (Innovation)

This gives

$$\boldsymbol{\epsilon}(t) = \boldsymbol{y}(t) - \hat{\boldsymbol{y}}\left(t|\boldsymbol{\theta}(t-1)\right) \tag{11.6.13}$$

where $\hat{\boldsymbol{y}}(t/\boldsymbol{\theta}(t-1))$ is computed from Equation (11.6.12). It is important to recognize that this innovations error for updating the model parameters is based upon a prediction of the measurements $\hat{\boldsymbol{y}}(t)$ and not $\overline{\boldsymbol{y}}(t)$. This means that we are using a prediction based upon the current state estimates $\hat{\boldsymbol{x}}(t/\boldsymbol{\theta}(t-1))$ rather than one based upon the predicted states $\overline{\boldsymbol{x}}(t/\boldsymbol{\theta}(t-1))$.

Gradient Update (State and Measurement)

$$\boldsymbol{W}(t) = \boldsymbol{F}\left(\boldsymbol{\theta}(t-1)\right)\boldsymbol{W}(t-1) + \boldsymbol{M}\left(\boldsymbol{\theta}(t-1)\right) \tag{11.6.14}$$

$$\boldsymbol{\psi}(t) = \left(\boldsymbol{H}\boldsymbol{W}(t)\right)^T \tag{11.6.15}$$

with

$$\boldsymbol{M}\left(\boldsymbol{\theta}(t-1)\right) = \frac{\partial}{\partial \boldsymbol{\theta}}\left[\boldsymbol{F}\left(\boldsymbol{\theta}(t-1)\right)\hat{\boldsymbol{x}}\left((t-1)|\boldsymbol{\theta}(t-1)\right)\right] \tag{11.6.16}$$

Covariance Matrix Update

$$\begin{aligned}
\boldsymbol{\Lambda}(t) &= \boldsymbol{\Lambda}(t-1) + \gamma(t)\left(\boldsymbol{\epsilon}(t)\boldsymbol{\epsilon}^T(t) - \boldsymbol{\Lambda}(t-1)\right) && (11.6.17)\\
\boldsymbol{S}(t) &= \boldsymbol{\psi}^T(t)\boldsymbol{P}(t-1)\boldsymbol{\psi}(t) + \lambda(t)\boldsymbol{\Lambda}(t) && (11.6.18)\\
\boldsymbol{P}(t) &= \left(\boldsymbol{P}(t-1) - \boldsymbol{L}(t)\boldsymbol{S}(t)\boldsymbol{L}^T(t)\right)\Big/\lambda(t) && (11.6.19)
\end{aligned}$$

with

$$\boldsymbol{L}(t) = \boldsymbol{P}(t-1)\boldsymbol{\psi}(t)\boldsymbol{S}^{-1}(t) \tag{11.6.20}$$

Parameter Estimate Update

$$\boldsymbol{\theta}(t) = \boldsymbol{\theta}(t-1) + \boldsymbol{L}(t)\boldsymbol{\epsilon}(t) \tag{11.6.21}$$

Initial Conditions

$$\begin{aligned}
\boldsymbol{\theta}(0) &= \boldsymbol{\theta}_0 && (11.6.22)\\
\boldsymbol{W}(0) &= \boldsymbol{0} && (11.6.23)\\
\boldsymbol{P}(0) &= \boldsymbol{P}_0 && (11.6.24)\\
\boldsymbol{\Lambda}(0) &= \mathcal{E}\left(\boldsymbol{\epsilon}_0\boldsymbol{\epsilon}_0^T\right) && (11.6.25)\\
\lambda(0) &= \lambda_0 && (11.6.26)
\end{aligned}$$

11.7 State and Parameter Identification for Batch Fermentation

A specific system that requires simultaneous identification of process states and model parameters is that of batch beer fermentation. This problem was considered by Ramirez (1987). A mathematical model of batch beer fermentation has been developed by Gee and Ramirez (1988) based upon the work of Engasser *et al.* (1981). The fermenting medium is assumed to contain three sugars—glucose, maltose, and maltrotriose—as the limiting nutrients. The rates of uptake of the three sugars are given by the following material balances based upon Monod kinetics:

Glucose

$$\frac{dG}{dt} = -\frac{V_G G}{K_G + G} X \tag{11.7.1}$$

Maltose

$$\frac{dM}{dt} = -\frac{V_M M}{K_M + M} \frac{K'_g}{K'_G + G} X \tag{11.7.2}$$

Maltotriose

$$\frac{dN}{dt} = -\frac{V_N N}{K_N + N} \frac{K'_G}{K'_G + G} \frac{K'_M}{K'_M + M} X \tag{11.7.3}$$

where G = glucose concentration, mol/m^3
M = maltose concentration, mol/m^3
N = maltotriose concentration, mol/m^3
X = biomass concentration, mol/m^3
V_i = velocity rate constant for species i, hr^{-1}
K_i = Michaelis constant for species i, mol/m^3
K'_i = inhibition rate constant for species i, mol/m^3

The rates of biomass and ethanol production are related to the individual sugar concentrations by

$$X = X_O - R_{XG}(G - G_O)) - R_{XM}(M - M_O) \tag{11.7.4}$$

$$E = -R_{EG}(G - G_O) - R_{EM}(M - M_O) - R_{EN}(N - N_O) \tag{11.7.5}$$

where R_{Xi} = the stoichiometric yield of biomass per mole of sugar reacted, mol biomass/mol sugar
R_{Ei} = the stoichiometric yield of ethanol per mole of sugar reacted, mol ethanol/mol sugar

Since biomass and ethanol are algebraic functions of the sugar concentrations, they are not state variables needed for the dynamic response of the system.

The rate of change of temperature in the fermenting medium is given by

$$\begin{aligned}\frac{dT}{dt} &= \frac{1}{\rho C_p}\left[\Delta H_{FG}\frac{dG}{dt} + \Delta H_{FM}\frac{dM}{dt}\right. \\ &\qquad \left. + \Delta H_{FN}\frac{dN}{dt} - u(T - T_c)\right] \end{aligned} \tag{11.7.6}$$

with the cooling rate control, u, being

$$u = \frac{UA}{V} \tag{11.7.7}$$

where C_p = mixture heat capacity, kJ/kg°C
ρ = mixture density, kg/m^3
T = temperature, °C
ΔH_{Fi} = heat of fermentation, kJ/mol
u = fermentor cooling rate, kJ/h m$^{3\circ}$C
U = overall heat transfer coefficient, kJ/h m$^{2\circ}$
A = heat transfer area, m^2
V = volume of fermentation mixture, m^3
T = coolant temperature, °C

The kinetic rate constants are all assumed to follow an Arrhenius temperature dependency as

$$V_i = V_{i0}\exp\left[-E_{vi}/R(T + 273.15)\right] \tag{11.7.8}$$

$$K_i = k_{i0}\exp\left[-E_{ki}/R(T + 273.15)\right] \tag{11.7.9}$$

$$K_i' = k_{i0}'\exp\left[-E_{ki}'/R(T + 273.15)\right] \tag{11.7.10}$$

where V_{i0} = velocity rate frequency factor, hr^{-1}
k_{i0} = Michaelis constant frequency factor, mol/m^3
k_{i0}' = inhibition constant frequency factor, mol/m^3
E_{vi} = velocity rate activation energy, cal/mol
E_{ki} = Michaelis constant activation energy, cal/mol
E_{ki}' = inhibition constant activation energy, cal/mol

Gee and Ramirez (1988) have determined model parameters for batch beer fermentations. They have also presented optimal cooling policies that extremize a performance functional that is a weighted sum of maximizing ethanol concentration within a minimum batch time. In order to implement this or any optimal control policy, estimates of the state variables must be available. A state identification algorithm or Kalman filter should therefore be used.

Also, the velocity rate frequency factors, V_{i0}, can change from batch to batch or during a fermentation due to the genetic changes that occur in different generations of the yeast organisms. For this reason, we expect that the velocity rate parameters will change from batch to batch and during a fermentation, but in an unpredicted manner. It therefore becomes important to sequentially update these model parameters as well as perform a state identification. In order to achieve these objectives, we will couple the Ljung parameter identification algorithm with the Kalman filter state identification algorithm.

Table 11.6 gives the physical parameters used in simulating a typical beer fermentation. The optimal controller will be approximated by the proportional feedback control law

$$u = K_p(T - T_{set}) \tag{11.7.11}$$

Results of a typical simulation are shown in Figures 11.1 to 11.4. Figure 11.1 gives the transient response of the sugar uptake, showing the initial consumption of glucose followed by maltose and then maltotriose. Figure 11.2 gives the temperature response of the system. There is a gradual increase in the temperature from its initial value of 8°C to the set-point value of 12°C. Figure 11.3 shows the control that produced the dynamic response. No cooling is used until the system reaches its set-point temperature of 12°C. At that point the heat of fermentation is offset by varying the cooling capacity of the system. Figure 11.4 shows the biomass (yeast) production as well as the ethanol production profiles. The transition in the ethanol production rate after glucose is consumed is obvious.

11.7.1 State Identification

The process model for batch beer fermentation can be put in state variable form as

$$\dot{\boldsymbol{x}} = \boldsymbol{f}(\boldsymbol{x},\ u) \tag{11.7.12}$$

with

$$\boldsymbol{x} = \begin{bmatrix} G \\ M \\ N \\ T \end{bmatrix}$$

where $\boldsymbol{x}$ is the state vector.

Since process control and identification will be implemented using an on-line computer, the discrete time version of the Kalman filter will be used. This assumes the following state model with random state uncertainty given by the Gaussian random vector, $\boldsymbol{w}_i$:

$$\boldsymbol{x}_{i+1} = \boldsymbol{\phi}_i(\Delta t)\boldsymbol{x}_i + \boldsymbol{w}_i \tag{11.7.13}$$

Table 11.6

Physical Parameters

Sugar	Reaction Parameters R_{Fi}	R_{Xi}	ΔH_{Ri}(kJ/mol)
Glucose	1.92	0.134	−91.2
Maltose	3.84	0.268	−226.3
Maltotriose	5.76	0.402	−361.3

Parameter	Arrhenius Constants Activation Energy (kcal/mol)	Natural Log of Frequency Factor
V_G	22.6	35.77
V_M	11.3	16.4
V_N	7.16	10.59
K_G	−68.6	−121.3
K_M	−14.4	−19.15
K_N	−19.9	−26.78
K'_G	10.2	23.33
K'_M	26.3	55.61

Initial Conditions and Equipment Parameters	
$G(t_0) = 70$ g mol/m^3	$X(t_0) = 175$ g mol/m^3
$M(t_0) = 220$ g mol/m^3	$T(t_0) = 8$°C
$N(t_0) = 40$ g mol/m^3	
$A = 0.188$ m^2	$u_{\min} = 0$ kJ/m^3 h°C
$C_p = 4.106$ kJ/kg°C	$u_{\max} = 40$ kJ/m^3 h°C
$\rho = 1040$ kg/m^3	$V = 0.1$ m^3
$K_p = 1000$ kJ/h°C m^3	$T_c = 0$°C

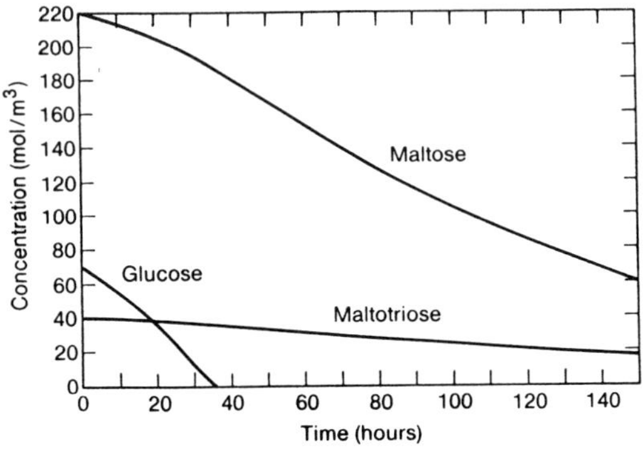

Figure 11.1 Sugar transient response. From Ramirez (1987).

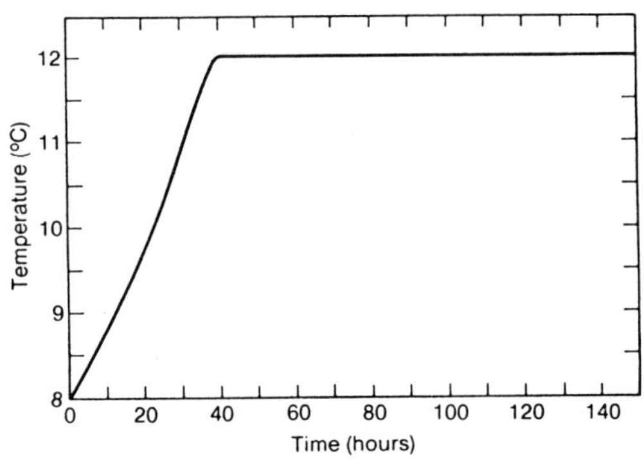

Figure 11.2 Temperature response. From Ramirez (1987).

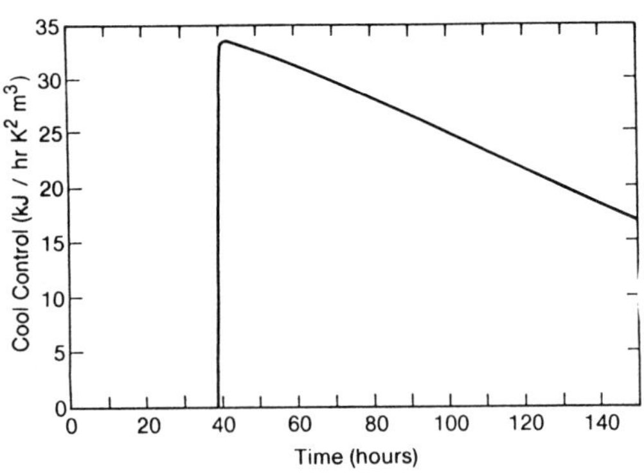

Figure 11.3 Control response. From Ramirez (1987).

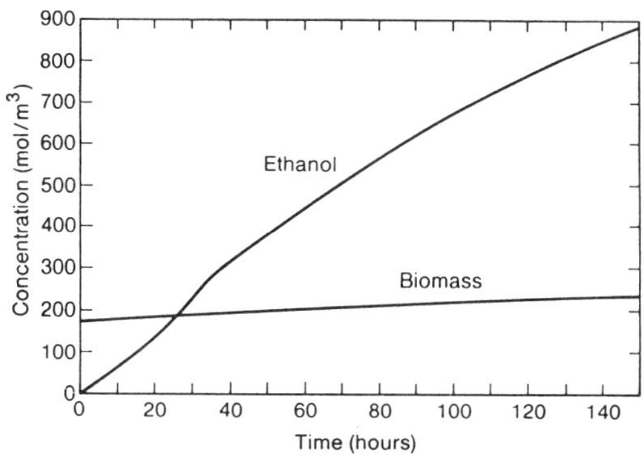

Figure 11.4 Production response. From Ramirez (1987).

with the expected value of $\boldsymbol{w}_i$ being zero and its covariance equal to $\boldsymbol{Q}$,

$$\mathcal{E}(\boldsymbol{w}_i) = \boldsymbol{0} \tag{11.7.14}$$

$$\mathcal{E}(\boldsymbol{w}_i\boldsymbol{w}_i^T) = \boldsymbol{Q} \tag{11.7.15}$$

The measurements are assumed to be linearly related to the states and corrupted with noise $\boldsymbol{v}_i$:

$$\boldsymbol{y}_i = \boldsymbol{H}\boldsymbol{x}_i + \boldsymbol{v}_i \tag{11.7.16}$$

The expected value of the measurement noise is zero, and its covariance is equal to $\boldsymbol{R}$:

$$\mathcal{E}(\boldsymbol{v}_i) = \boldsymbol{0} \tag{11.7.17}$$

$$\mathcal{E}(\boldsymbol{v}_i\boldsymbol{v}_i^T) = \boldsymbol{R} \tag{11.7.18}$$

The optimal state filter or Kalman filter is

$$\hat{\boldsymbol{x}}_i = \overline{\boldsymbol{x}}_i + \boldsymbol{K}_i(\boldsymbol{y}_i - \boldsymbol{H}\overline{\boldsymbol{x}}_i) \tag{11.7.19}$$

$$\boldsymbol{K}_i = \boldsymbol{P}_i\boldsymbol{H}^T\boldsymbol{R}^{-1} \tag{11.7.20}$$

$$\boldsymbol{P}_i = \boldsymbol{M}_i - \boldsymbol{M}_i\boldsymbol{H}^T(\boldsymbol{H}\boldsymbol{M}_i\boldsymbol{H}^T + \boldsymbol{R})^{-1}\boldsymbol{H}\boldsymbol{M}_i \tag{11.7.21}$$

$$\boldsymbol{M}_{i+1} = \boldsymbol{\phi}_i\boldsymbol{P}_i\boldsymbol{\phi}_i^T + \boldsymbol{Q} \tag{11.7.22}$$

In these equations, $\hat{\boldsymbol{x}}_i$ are the optimal state estimates, $\overline{\boldsymbol{x}}_i$ are the model states given by Equation (11.7.24) below, $\boldsymbol{P}_i$ is the state covariance matrix, and $\boldsymbol{K}_i$ is the filter gain.

The main advantage to the discrete version of the Kalman filter is that neither state model nor covariance differential equations need to be solved numerically over very small time steps. Instead, recursive algebraic equations are solved over large sample periods.

The initial covariance between the model and the true state is given by

$$\left\langle (\boldsymbol{x}_0 - \overline{\boldsymbol{x}}_0)(\boldsymbol{x}_0 - \overline{\boldsymbol{x}}_0)^T \right\rangle = \boldsymbol{P}_0 \tag{11.7.23}$$

By finite differencing, the state model of Equation (11.7.12) can be transformed into the following discrete model:

$$\overline{\boldsymbol{x}}_i = \hat{\boldsymbol{x}}_{i-1} + \Delta t \boldsymbol{f}_{i-1}(\hat{\boldsymbol{x}}_{i-1}) \tag{11.7.24}$$

and the state transition matrix $\boldsymbol{\phi}_i$ can be approximated by

$$\boldsymbol{\phi}_i = \left[\boldsymbol{I} + \frac{\partial \boldsymbol{f}_i}{\partial \boldsymbol{x}} \Delta t\right] \tag{11.7.25}$$

If we assume that data will be processed every 0.2 h for the beer fermentation example, then there is less than a 0.1% error between the approximate state transition matrix and the analytic one based upon the eigenvalues of the Jacobian matrix $\partial \boldsymbol{f}_i / \partial \boldsymbol{x}$. The same degree of accuracy is obtained between the discrete state model of Equation (11.7.24) and the continuous model of Equation (11.7.12). These results are, of course, not general but depend upon the specific problem sample period and Jacobian matrix elements.

When the state transition matrix $\boldsymbol{\phi}_i$ remains essentially constant, the filter equations of (11.7.20) through (11.7.22) will converge to a constant steady-state filter gain, $\boldsymbol{K}_{ss}$. Implementation of the steady-state filter gain is simpler since the gain equations are uncoupled from the state estimation step (Equation (11.7.19)) and can therefore be computed off-line. In order to test the feasibility of using the steady-state filter gain for the batch beer fermentation problem, gains were computed for linearization about various conditions during a fermentation (Ramirez, 1987). The parameters used in the calculations are given in Table 11.7. All state variables are assumed measurable so that the measurement matrix, $\boldsymbol{H}$, is an identity matrix. It was assumed that there is equal uncertainty in the model and measurement values. Also, these covariance matrices reflect the condition that over 95% of all the data will fall in a two standard deviation variation. For the concentrations, the two standard deviation value is 2 g mol/m^3. A two standard deviation variation for the temperature is 0.25°C.

Results of steady-state filter gains for linearization about the initial conditions, conditions at 40 h when the glucose concentration goes to zero, and at 150 h when the fermentation is terminated are given in Table 11.8. Both the steady-state filter gain and the final state covariance matrix were diagonally dominant, so only the diagonal terms are presented. These results show that neither the final state covariance matrix nor the steady-state filter gain are

Table 11.7

Kalman Filter Parameters

Measurement matrix	Measurement covariance matrix
$\boldsymbol{H} = \begin{bmatrix} 1 & 0 & 0 & 0 \\ 0 & 1 & 0 & 0 \\ 0 & 0 & 1 & 0 \\ 0 & 0 & 0 & 1 \end{bmatrix}$	$\boldsymbol{R} = \begin{bmatrix} 4 & 0 & 0 & 0 \\ 0 & 4 & 0 & 0 \\ 0 & 0 & 4 & 0 \\ 0 & 0 & 0 & 0.0625 \end{bmatrix}$
Model covariance matrix	**Initial state covariance matrix**
$\boldsymbol{Q} = \begin{bmatrix} 4 & 0 & 0 & 0 \\ 0 & 4 & 0 & 0 \\ 0 & 0 & 4 & 0 \\ 0 & 0 & 0 & 0.0625 \end{bmatrix}$	$\boldsymbol{P} = \begin{bmatrix} 4 & 0 & 0 & 0 \\ 0 & 4 & 0 & 0 \\ 0 & 0 & 4 & 0 \\ 0 & 0 & 0 & 0.625 \end{bmatrix}$

Table 11.8

Steady-State Filter Results

Linearization about initial condition
$\boldsymbol{P}_{ss}\text{diag} = (2.47 \quad 2.47 \quad 2.47 \quad 0.0387)$
$\boldsymbol{K}_{ss}\text{diag} = (0.618 \quad 0.618 \quad 0.618 \quad 0.619)$
Linearization about t = 40 h
$\boldsymbol{P}_{ss}\text{diag} = (2.10 \quad 2.47 \quad 2.47 \quad 0.0390)$
$\boldsymbol{K}_{ss}\text{diag} = (0.505 \quad 0.618 \quad 0.618 \quad 0.620)$
Linearization about t = 150 h
$\boldsymbol{P}_{ss}\text{diag} = (2.00 \quad 2.47 \quad 2.47 \quad 0.0390)$
$\boldsymbol{K}_{ss}\text{diag} = (0.500 \quad 0.618 \quad 0.618 \quad 0.621)$

strong functions of the linearization conditions. There was a 18% lowering of the glucose diagonal gain term in moving from linearization about the initial conditions to that at either 40 or 150 h. All other gain elements remained essentially constant. We are therefore justified in using the steady-state filter gain for this system. Updating of the gain once the glucose is consumed can be considered. Even though we are justified in using the steady-state approximation for the filter gain matrix, it may not be necessary to invoke when implementing with fast real-time computers.

Figures 11.5 and 11.6 show that glucose and ethanol filtered responses follow closely their actual response profiles. In fact, all state variables can be very successfully identified using the steady-state version of the Kalman filter. Figure 11.7 shows a small time slice of the glucose response curve that illustrates the smoothing properties of the Kalman filter. The true nonlinear glucose response is shown predicting a zero glucose concentration, along with random data taken every hour. The filtered response, which uses both an approximate state model and actual data, is also shown. The filtered response has less random variations than the data and agrees well with the actual state variable.

Simulations were also run to test the sensitivity of the steady-state filter gain to changes in the reaction rate parameters which will be sequentially updated. Changing these parameters an order of magnitude had little effect on the steady-state filter gain, $\boldsymbol{K}_{ss}$, or the state covariance matrix, $\boldsymbol{P}_{ss}$. The filter parameters that had the largest effect on the filter gains were the ratio of the $\boldsymbol{Q}$ to $\boldsymbol{R}$ covariance terms. When the state covariance uncertainty, $\boldsymbol{Q}$, is increased, then the filter gain terms increase, giving more weight to the measurements. Whereas, when the state covariance uncertainty is decreased, increased weight is placed on the model. Optimal performance (rapid response to the data without excessive noise in the response) is obtained with essentially equal weighting on state and model covariances. This situation can be achieved experimentally when model parameters are either constant, so that model results are of the same accuracy as the data, or when time varying-parameters are updated using a recursive parameter identification algorithm. Without updating, the model accuracy will deteriorate due to time variations in model parameters.

Filtering results can also be affected by the set of measurements available. Ramirez (1987) tested the filter with certain practical indirect measurements, i.e., the CO_2 rate and total sugar concentration. The Kalman filter performs well with that set of measurements as well as those shown previously. Results can certainly deteriorate with incomplete measurements, especially when starting from a bad initial condition.

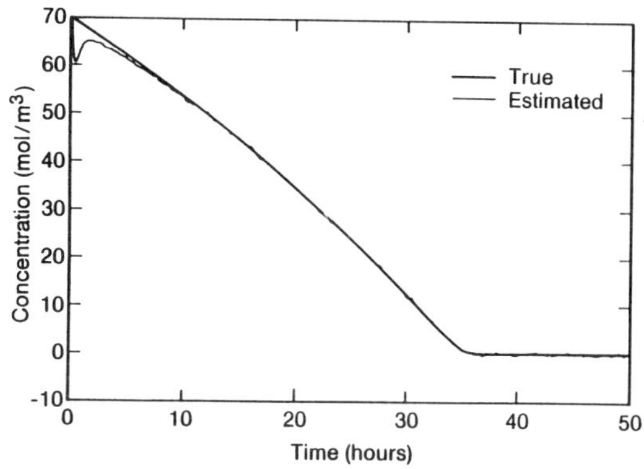

Figure 11.5 Glucose response (true and filtered responses). From Ramirez (1987).

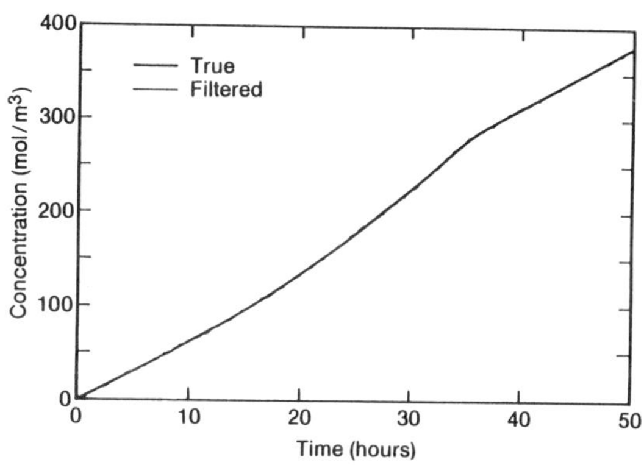

Figure 11.6 Ethanol response (true and filtered responses. From Ramirez (1987).

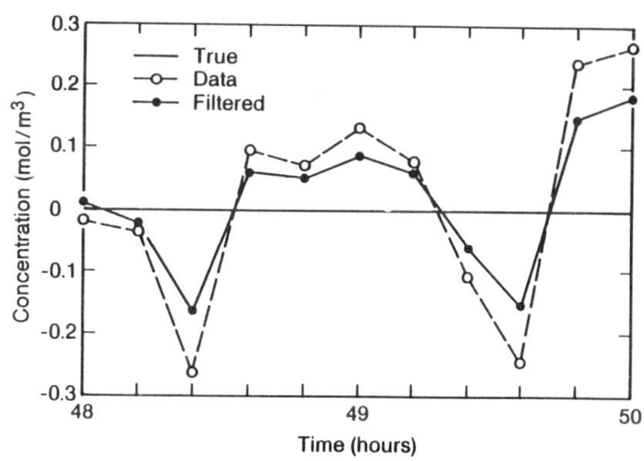

Figure 11.7 Glucose response. From Ramirez (1987).

11.7.2 Simultaneous State and Parameter Identification

The steady-state Kalman filter problem can be transformed into the simultaneous state and parameter identification problem of Section 11.6. The new state model is Equation (11.6.9) or

$$\hat{\boldsymbol{x}}(t|\boldsymbol{\theta}-1) = \boldsymbol{F}(\boldsymbol{\theta})\hat{\boldsymbol{x}}(t-1|\boldsymbol{\theta}-1)+\boldsymbol{G}\boldsymbol{y}(t) \qquad (11.7.26)$$

with

$$\boldsymbol{F}(\boldsymbol{\theta}) = (\boldsymbol{I}-\boldsymbol{K}\boldsymbol{H})\,\phi(\boldsymbol{\theta}) \qquad (11.7.27)$$

$$\boldsymbol{G} = \boldsymbol{K}_{ss} \qquad (11.7.28)$$

and the measurement model for the estimation problem is

$$\hat{\boldsymbol{y}}(t|\boldsymbol{\theta}-1) = \boldsymbol{H}\hat{\boldsymbol{x}}(t|\boldsymbol{\theta}-1) \qquad (11.7.29)$$

The sequential least squares algorithm is given by Equations (11.6.13) to (11.6.27).

In order to test the feasibility of the simultaneous Kalman filtering and parameter identification algorithm, the batch beer fermentation process is considered when the initial natural logarithm of the glucose reaction rate frequency factor is actually 35.77 (ln h^{-1}) but incorrectly estimated to be 40.0 in the state model. The states to be identified are the three sugar concentrations and temperature. The unknown model parameter is the glucose reaction rate frequency factor, V_{Go}. Figure 11.8 shows the comparison of the true, estimated, and model predicted values for the glucose concentration using classical Kalman filtering, without model parameter updating, with this large discrepancy in the reaction rate parameter. The model consumes all the glucose within two hours of fermentation, while the actual consumption time is a little over 35 h. The filtered response, which is also sensitive to process measurements, tries to correct for this discrepancy but cannot do so very well. The other state variables of temperature, maltose, and maltotriose concentration are estimated quite well. Figure 11.9 shows the maltose filtered and actual response curves. Since the glucose concentration was poorly estimated, there are significant discrepancies between the actual and estimated ethanol concentrations as shown in Figure 11.10.

The combined parameter and state identification algorithm was applied to this problem (Ramirez, 1987). Table 11.9 gives the parameters that were used in the algorithm. A constant forgetting factor, λ, was used. The initial error matrix, $\hat{\Lambda}$, reflects the fact that we expect a large error in the difference between the glucose concentrations measured and computed. All states were assumed measured, and the appropriate steady-state Kalman filter gain was used.

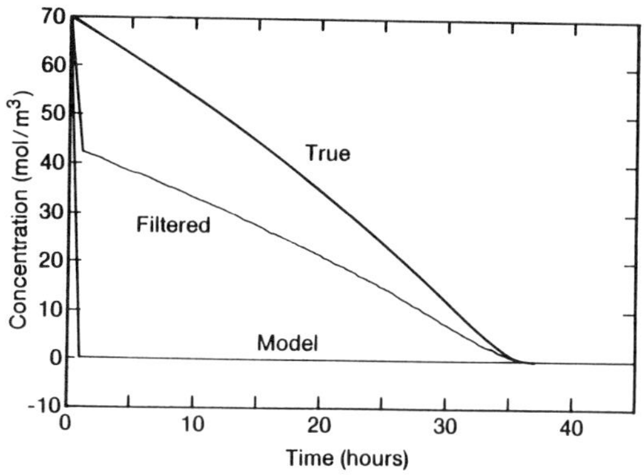

Figure 11.8 Glucose response. From Ramirez (1987).

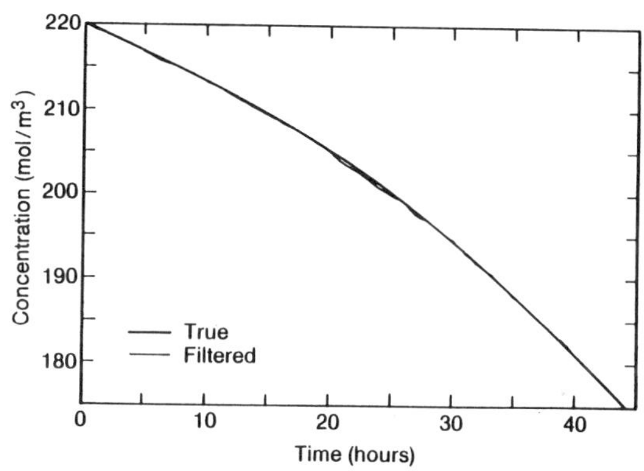

Figure 11.9 Maltose response (true and filtered responses). From Ramirez (1987).

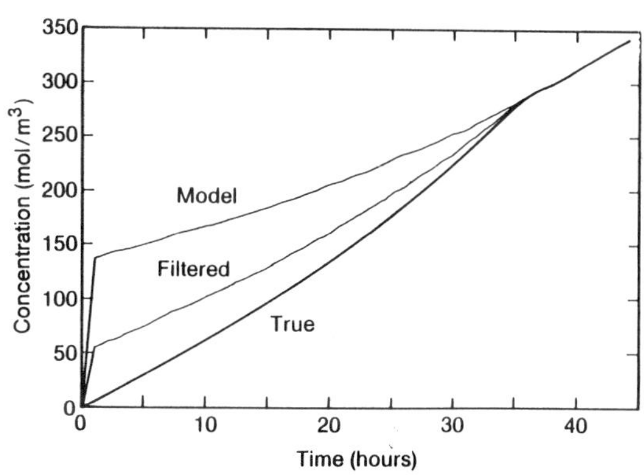

Figure 11.10 Ethanol response. From Ramirez (1987).

Table 11.9
Combined Algorithm Parameters

$\hat{\theta}(0) = 40\ (\ln\ \mathrm{h}^{-1})$
$P(0) = 1.0$
$\lambda_{\mathrm{const}} = 0.99$
$\gamma_{\mathrm{const}} = 0.01$

$$\hat{\boldsymbol{\Lambda}}(0) = \begin{bmatrix} 900 & 0 & 0 & 0 \\ 0 & 4 & 0 & 0 \\ 0 & 0 & 4 & 0 \\ 0 & 0 & 0 & 0.0625 \end{bmatrix}$$

$$\boldsymbol{H} = \begin{bmatrix} 1 & 0 & 0 & 0 \\ 0 & 1 & 0 & 0 \\ 0 & 0 & 1 & 0 \\ 0 & 0 & 0 & 1 \end{bmatrix}$$

$$\boldsymbol{K}_{ss} = \begin{bmatrix} 0.618 & 1.2 \times 10^{-2} & 8.2 \times 10^{-3} & -57.3 \times 10^{-1} \\ 1.2 \times 10^{-2} & 0.618 & 3.2 \times 10^{-3} & -2.05 \times 10^{-1} \\ 8.2 \times 10^{-3} & 3.2 \times 10^{-3} & 0.618 & -3.15 \times 10^{-2} \\ -9.0 \times 10^{-3} & -3.2 \times 10^{-3} & -4.9 \times 10^{-4} & 0.6185 \end{bmatrix}$$

The gradient matrix is

$$\boldsymbol{M}(\boldsymbol{\theta}) = \frac{\partial}{\partial \boldsymbol{\theta}} \Big((\boldsymbol{I} - \boldsymbol{K}_{ss}\boldsymbol{H})\, \boldsymbol{\phi}(\boldsymbol{\theta})\hat{\boldsymbol{x}}(t-1|\boldsymbol{\theta}-1) \Big) \tag{11.7.30}$$

which, for a four-state variable, one unknown parameter problem yields a vector with four elements. For this problem, analytical partials of $\boldsymbol{\phi}(\boldsymbol{\theta})$ were evaluated.

Figure 11.11 shows the comparison between the estimated glucose concentration profile and the actual profile. After a short delay period, due to a very bad initial parameter estimate, the estimated response recovers and follows the actual profile quite well. The estimated glucose response using this combined algorithm is significantly better than the Kalman filter results (Figure 11.8). Figure 11.12 shows that similar results are obtained for the product ethanol concentration. Again, the combined algorithm is vastly better in estimating ethanol concentration when compared to the Kalman filter results using the incorrect model parameter.

Figure 11.13 gives the results obtained for the recursive identification of

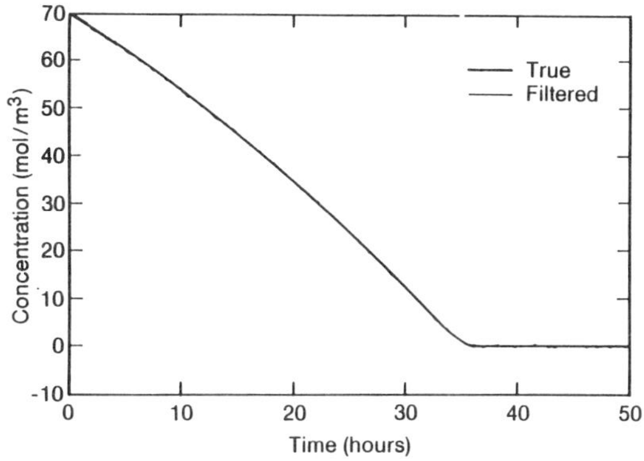

Figure 11.11 Glucose response. From Ramirez (1987).

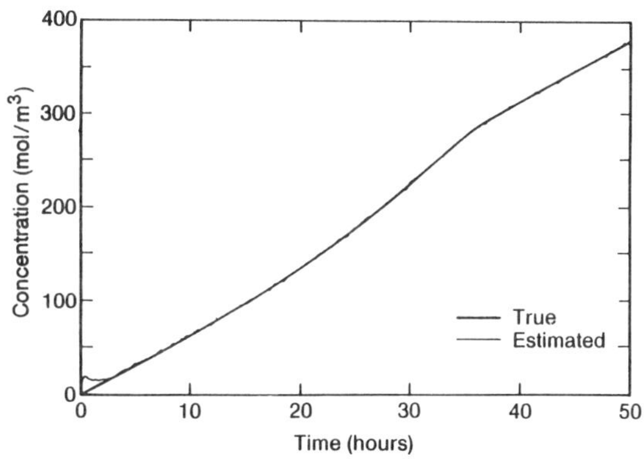

Figure 11.12 Ethanol response. From Ramirez (1987).

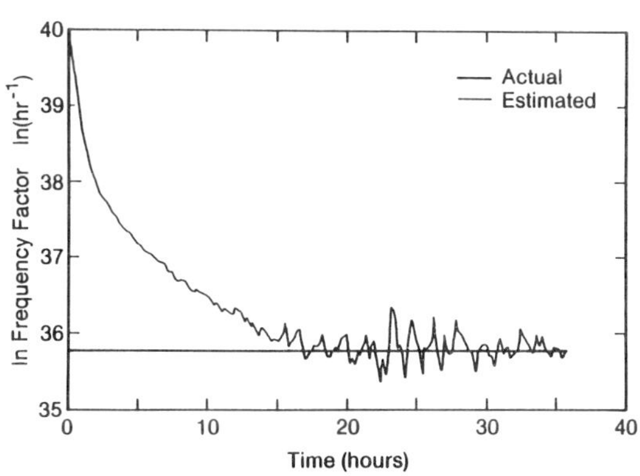

Figure 11.13 Model parameter estimation.

the reaction rate constant model parameter. The model parameter rapidly approaches the correct but unknown parameter and then oscillates around the correct value. It is this rather efficient sequential updating of the reaction rate/velocity rate parameter that allows for the improved glucose state identification using the combined algorithm.

These results clearly show the advantages that can be achieved in coupling sequential parameter identification with the Kalman filter state identification algorithm. Such a combined algorithm has the capability of accurately estimating the entire state of processes even when some model parameters are uncertain. This strategy of using the best deterministic process information coupled with sequential updating, based upon actual observations, can be applied to many chemical and biochemical processing problems.

11.8 Simultaneous State and Parameter Identification for Nonlinear Systems

The algorithm of Section 11.6 can be modified for nonlinear systems in a fashion similar to extended Kalman filtering. For nonlinear systems, the process model is

$$\overline{\boldsymbol{x}}(t|\boldsymbol{\theta}-1) = \hat{\boldsymbol{x}}\left(t-1|\boldsymbol{\theta}(t-1)\right) + \boldsymbol{f}\left(\hat{\boldsymbol{x}}(t-1),\ \boldsymbol{\theta}(t-1)\right)\Delta t \tag{11.8.1}$$

when $\overline{\boldsymbol{x}}(t)$ is the model prediction at time t, $\hat{\boldsymbol{x}}(t-1)$ is the optimal estimate at time $(t-1)$, and $\boldsymbol{f}$ is the nonlinear process model vector. Using Kalman filtering, the optimal estimate for the state at time t is

$$\begin{aligned}\hat{\boldsymbol{x}}\left(t|\boldsymbol{\theta}(t-1)\right) &= \overline{\boldsymbol{x}}\left(t|\boldsymbol{\theta}(t-1)\right) + \boldsymbol{K}_{ss}\left(\boldsymbol{\theta}(t-1)\right)\Big[\boldsymbol{y}(t) \\ &\quad - \boldsymbol{h}\left(\overline{\boldsymbol{x}}\left(t, \boldsymbol{\theta}(t-1)\right), \boldsymbol{\theta}(t-1)\right)\Big]\end{aligned} \tag{11.8.2}$$

Note that in this expression we have allowed $\boldsymbol{K}_{ss}$ to be a function of $\boldsymbol{\theta}$. As the model parameters change, there can be significant changes in the Kalman filter gain matrix.

The measurement model for parameter estimation is given by

$$\hat{\boldsymbol{y}}\left(t|\boldsymbol{\theta}(t-1)\right) = \boldsymbol{h}\left(\hat{\boldsymbol{x}}\left(t|\boldsymbol{\theta}(t-1)\right),\ \boldsymbol{\theta}(t-1)\right) \tag{11.8.3}$$

The prediction error or innovation is

$$\boldsymbol{\epsilon}(t) = \boldsymbol{y}(t) - \hat{\boldsymbol{y}}\left(t|\boldsymbol{\theta}(t-1)\right) \tag{11.8.4}$$

The gradient update for the state model is

$$\begin{aligned} \boldsymbol{W}(t) &= \boldsymbol{F}\left(\boldsymbol{\theta}(t-1)\right)\boldsymbol{W}(t-1) \\ &+ \boldsymbol{M}\left(\boldsymbol{\theta}(t-1),\ \hat{\boldsymbol{x}}(t-1)\right) \\ &- \boldsymbol{K}_{ss}\left(\boldsymbol{\theta}(t-1)\right)\boldsymbol{D}\left(\boldsymbol{\theta}(t-1),\ \hat{\boldsymbol{x}}(t-1)\right) \end{aligned} \tag{11.8.5}$$

with

$$\boldsymbol{F}\left(\boldsymbol{\theta}(t-1)\right) = (\boldsymbol{I} - \boldsymbol{K}_{ss}\boldsymbol{H})\,\boldsymbol{\phi}\left(\boldsymbol{\theta}(t-1)\right) \tag{11.8.6}$$

$$\boldsymbol{\phi}\left(\boldsymbol{\theta}(t-1)\right) = \left(\boldsymbol{I} + \frac{\partial \boldsymbol{f}}{\partial \boldsymbol{x}}\right)\Bigg|_{\substack{\boldsymbol{x} = \hat{\boldsymbol{x}}(\mathrm{t}-1) \\ \boldsymbol{\theta} = \boldsymbol{\theta}(\mathrm{t}-1)}} \tag{11.8.7}$$

$$\boldsymbol{H}\left(\boldsymbol{\theta}(t-1)\right) = \frac{\partial \boldsymbol{h}}{\partial \boldsymbol{x}}(\boldsymbol{\theta},\ \boldsymbol{x})\Bigg|_{\substack{\boldsymbol{\theta} = \boldsymbol{\theta}(\mathrm{t}-1) \\ \boldsymbol{x} = \overline{\boldsymbol{x}}(\mathrm{t}-1)}} \tag{11.8.8}$$

$$\begin{aligned} \boldsymbol{M}\left(\boldsymbol{\theta}(t-1),\ \hat{\boldsymbol{x}}(t-1)\right) &= (\boldsymbol{I} - \boldsymbol{K}\boldsymbol{H})\,\frac{\partial \boldsymbol{f}}{\partial \boldsymbol{\theta}}\Bigg|_{\substack{\boldsymbol{\theta} = \boldsymbol{\theta}(\mathrm{t}-1) \\ \boldsymbol{x} = \hat{\boldsymbol{x}}(\mathrm{t}-1)}} \\ &+ \frac{\partial \boldsymbol{K}(\boldsymbol{\theta})}{\partial \boldsymbol{\theta}}\Bigg|_{\substack{\boldsymbol{\theta} = \boldsymbol{\theta}(\mathrm{t}-1) \\ \boldsymbol{x} = \hat{\boldsymbol{x}}(\mathrm{t})}} \left(\boldsymbol{y}(t) - \boldsymbol{h}\left(\hat{x}(t)\right)\right) \end{aligned} \tag{11.8.9}$$

$$\boldsymbol{D}\left(\boldsymbol{\theta}(t-1),\ \hat{\boldsymbol{x}}(t-1)\right) = \frac{\partial \boldsymbol{H}}{\partial \boldsymbol{\theta}}\Bigg|_{\substack{\boldsymbol{\theta} = \boldsymbol{\theta}(\mathrm{t}-1) \\ \boldsymbol{x} = \hat{\boldsymbol{x}}(\mathrm{t}-1)}} \tag{11.8.10}$$

The measurement gradient system is

$$\boldsymbol{\psi}(t) = \boldsymbol{W}^T(t)\boldsymbol{H}_1^T\left(\boldsymbol{\theta}(t-1)\right) + \boldsymbol{D}_1\left(\boldsymbol{\theta}(t-1)\right) \tag{11.8.11}$$

where

$$\boldsymbol{H}_1\left(\boldsymbol{\theta}(t-1)\right) = \frac{\partial \boldsymbol{h}}{\partial \boldsymbol{x}}(\boldsymbol{\theta},\ \boldsymbol{x})\Bigg|_{\substack{\boldsymbol{\theta} = \boldsymbol{\theta}(\mathrm{t}-1) \\ \boldsymbol{x} = \hat{\boldsymbol{x}}(\mathrm{t})}} \tag{11.8.12}$$

$$\boldsymbol{D}_1\left(\boldsymbol{\theta}(t-1)\right) = \frac{\partial \boldsymbol{h}}{\partial \boldsymbol{\theta}}(\boldsymbol{\theta},\ \boldsymbol{x})\Bigg|_{\substack{\boldsymbol{\theta} = \boldsymbol{\theta}(\mathrm{t}-1) \\ \boldsymbol{x} = \hat{\boldsymbol{x}}(\mathrm{t})}} \tag{11.8.13}$$

The covariance update equations are the same as given in Section 11.6 and are

$$\boldsymbol{\Lambda}(t) = \boldsymbol{\Lambda}(t-1) + \gamma(t)\left(\boldsymbol{\epsilon}(t)\boldsymbol{\epsilon}^T(t) - \boldsymbol{\Lambda}(t-1)\right) \tag{11.8.14}$$

$$\boldsymbol{S}(t) = \boldsymbol{\psi}^T(t)\boldsymbol{P}(t-1)\boldsymbol{\psi}(t) + \lambda(t)\boldsymbol{\Lambda}(t) \tag{11.8.15}$$

$$\boldsymbol{P}(t) = \left(\boldsymbol{P}(t-1) - \boldsymbol{L}(t)\boldsymbol{S}(t)\boldsymbol{L}^T(t)\right)\Big/\lambda(t) \tag{11.8.16}$$

with

$$\boldsymbol{L}(t) = \boldsymbol{P}(t-1)\boldsymbol{\psi}(t)\boldsymbol{S}^{-1}(t) \tag{11.8.17}$$

The parameter estimate update is

$$\boldsymbol{\theta}(t) = \boldsymbol{\theta}(t-1) + \boldsymbol{L}(t)\boldsymbol{\epsilon}(t) \tag{11.8.18}$$

Ljung and Soderstrom (1983) give a detailed comparison of the extended Kalman filter approach to model parameter identification (Section 10.8) and the recursive Gauss–Newton approach. The result of that analysis is that the two are essentially identical except for a coupling term, $\boldsymbol{\kappa}_t$, which is missing from the EKF. This term is

$$\boldsymbol{\kappa}_t = \left.\frac{\partial \boldsymbol{K}(\boldsymbol{\theta})}{\partial \boldsymbol{\theta}}\right|_{\substack{\boldsymbol{\theta} = \boldsymbol{\theta}(\mathrm{t}-1) \\ \boldsymbol{x} = \hat{\boldsymbol{x}}(\mathrm{t})}} \left(\boldsymbol{y}(t) - \boldsymbol{h}\left(\hat{\boldsymbol{x}}(t)\right)\right) \tag{11.8.19}$$

11.9 State and Parameter Identification of Positive Photoresist Development

Positive optical photoresists are used in semiconductor and printed circuit board applications to mask selected areas from chemical or physical changes. A substrate is cleaned, dried, and coated with a liquid photoresist solution. The solution consists of solvent, photoactive compound, and a resin. The solvent is driven off by baking the coating, which leaves behind a film that is typically 1–20 μm thick. A patterned glass photomask is aligned over the substrate, and the resist is exposed by a light in the wavelength range of 350–450 nm (monochromatic or polychromatic). The areas exposed convert from a diazonapthoquinone to a carboxylic acid, which may later be developed using a base. The resist film may be subjected to a post-exposure bake, which alleviates a standing wave problem created by the combination of light diffraction caused by the photomask slits and the reflection of light back from the substrate. The resulting film is developed and rinsed, leaving behind the desired pattern. After a hard bake, the required process is performed on the substrate.

Finally, the resist film is completely removed using a general solvent such as acetone.

The development phase of the positive optical photoresist process has been monitored using interferometric techniques described by Willson (1983), Lauchlan *et al.* (1985), and Novembre *et al.* (1989). During development, a collimated beam of nonexposing light is allowed to shine on a substrate coated with a film of exposed photoresist. The light reflections from the resist/developer interface and the substrate are collected and read by a photodiode. The resulting signal, sinusoidal in nature, indicates the constructive and destructive interferences caused by the two reflections, thus allowing calculation of both the rate of development and the resist thickness of the developing region. Sautter *et al.* (1989) expanded the work of Lauchlan *et al.* to use the end-point detector to monitor line width variation, development uniformity, and certain process failures.

Mack (1985, 1987) has presented a mathematical model which describes the development process. This model, combined with the available process measurement, can be used to estimate the depth of penetration of the developer into the resist film and to update model parameters. By using on-line parameter and state identification, process parameters can be adjusted from one substrate to the next automatically.

Carroll and Ramirez (1990) describe state and parameter identification for a spray development process using a modified APT spray-spin developer. Simulations and experimental data are presented to show the power of combining model and real-time data to obtain a useful end-point detector which not only provides a means of development control, but also a process parameter monitoring scheme.

11.9.1 System and State Definition

The contact/proximity printing process for positive optical resists has several steps, of which exposure and development are the most important in constructing a process model. Measurements are also required if any model modification is to be implemented on-line during process operation. Development progress can be monitored by striking the resist-coated substrate with a collimated beam of nonexposing light. Interference fringes, resulting from the changes in reflection from the resist/developer interface with respect to reflection from the substrate, provide resist thickness data and rate of dissolution during development (Lauchlan *et al.*, 1985).

Figure 11.14 shows a simplified cross-sectional representation of an exposed line of resist, thickness z_0. Let z represent the depth of penetration of development into the resist film and y be the distance from the center of the line. We desire to develop the resist until a width L is cleared at $z = z_0$. Mack (1985, 1987) has derived a development model which combined the chemistry

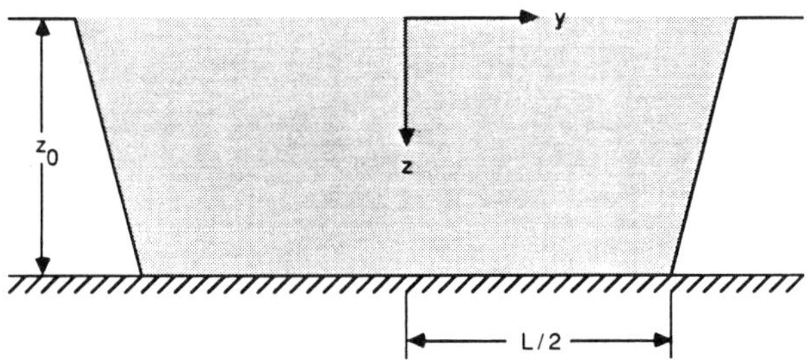

Figure 11.14 Resist cross-section. From Carroll and Ramirez (1990).

of dissolution and the mass transfer rate of developer from the bulk to the resist surface. The development rate, expressed as length per time penetration, is given by

$$r = \left(k_{\min} + k_D \frac{(1-M(y,z))^n}{a + (1-M(y,z))^n}\right)\left(1 - (1-r_0)\exp(-4z/z_t)\right) u(t) \quad (11.9.1)$$

where r is the rate of dissolution, $M(y, z)$ is the relative photoactive compound concentration after exposure, $u(t)$ is the control effort expressed as developer concentration, r_0 is the surface development rate relative to the bulk rate, y is the distance from the line-width center, z is the depth of penetration into the resist, z_t is the characteristic film depth where surface effects no longer play a significant role in development, $k_{\min}$ is the minimum development rate constant, k_D is the mass transfer coefficient, n is the order of the kinetics of dissolution, and a is a constant given by

$$a = k_D/k_R m_0^n \quad (11.9.2)$$

where m_0 is the PAC concentration before exposure and k_R is the reaction rate constant for dissolution. $M(y, z)$ is calculated from PROLITH (Mack, 1985) for given exposure parameters appropriate to the resist and the film's prebake time.

The only available on-line process measurement is the depth of penetration of developer into the resist film. During exposure, light diffraction and reflections from the substrate cause the resist in the line to be exposed unevenly. Thus, the measurement represents only the development of the central region of the line. Once development has penetrated to the substrate in this region ("breakthrough"), real-time data are no longer available.

For on-line processing, it is important to make as many simplifications to the model as possible without removing too much information about the actual physical process. The following assumptions, therefore, were made: (1) the line is symmetric about the $y = 0$ line and (2) development proceeds in the z–direction only. Assumption (2) is not strictly true, but it is a reasonable approximation in the breakthrough region.

It is convenient to define a set of depths of penetration z_i at given distances y_i from the center of the line width. The states z_i are defined as a vector $\boldsymbol{x}$, which consists of depths z_i from $y = 0$ to $y = L/2$ and beyond. The states of interest are in the breakthrough region only. The rate of dissolution at a given y_i is then

$$r_i = \frac{dz_i}{dt} \tag{11.9.3}$$

The state dynamics in the breakthrough region are given by

$$\frac{dz_i}{dt} = \left[k_{\min} + k_D \frac{(1 - M_i)^n}{a + (1 - M_i)^n}\right] \left(1 - (1 - r_0) \exp\left(-\frac{4z_i}{z_t}\right)\right) u \tag{11.9.4}$$

where M_i is the value of PAC concentration evaluated as a function of state z_i, and the subscript i denotes all states inside the linewidth ($i = 1, ..., m$). Note that these Equations (11.9.4) are uncoupled.

11.9.2 State Identification

Kalman filtering is used to balance the mathematical model of a process and available process measurements to produce an optimal state estimate. The process is assumed to have a model given by

$$\boldsymbol{x}_{k+1} = \boldsymbol{\phi}_k(\Delta t)\boldsymbol{x}_k + \boldsymbol{w}_k \tag{11.9.5}$$

where $\boldsymbol{\phi}$ is the state transition matrix and $\boldsymbol{w}$ is the uncertainty related to white Gaussian noise, with expected value $\mathcal{E}(\boldsymbol{w}) = \boldsymbol{0}$ and covariance $\mathcal{E}(\boldsymbol{w}\boldsymbol{w}^T) = \boldsymbol{Q}$. The measurements are related to the states by

$$y_k = \boldsymbol{H}\boldsymbol{x}_k + v_k \tag{11.9.6}$$

where $\boldsymbol{H}$ is the measurement matrix and v is the measurement noise, with expected value $\mathcal{E}(v) = 0$ and covariance $\mathcal{E}(v^2) = R$. Since only one measurement is available, y_k and R are scalars.

The estimated states, $\hat{\boldsymbol{x}}$, which are used to determine the control status of the development system, are calculated from Kalman filtering to be

$$\hat{\boldsymbol{x}}_k = \overline{\boldsymbol{x}}_k + \boldsymbol{K}_k(y_k - \boldsymbol{H}\overline{\boldsymbol{x}}_k) \tag{11.9.7}$$

The state transition matrix, $\boldsymbol{\phi}_k$, is given by the linear approximation

$$\boldsymbol{\phi}_k = \boldsymbol{I} + \left(\partial \boldsymbol{f}_k^T / \partial \boldsymbol{x}\right) \Delta t \tag{11.9.8}$$

where

$$\boldsymbol{f}_k = d\boldsymbol{x}_k/dt \tag{11.9.9}$$

The expected value of the states, $\overline{\boldsymbol{x}}$, is determined from the previous estimate using

$$\overline{\boldsymbol{x}}_k = \hat{\boldsymbol{x}}_{k-1} + \boldsymbol{f}_{k-1}(\hat{\boldsymbol{x}}_{k-1})\Delta t \tag{11.9.10}$$

The maxima and minima in the interference waveform, described previously, yield depth measurements given by

$$\Delta z = \lambda/4n_r \tag{11.9.11}$$

where n_r is the average refractive index of the film and λ is the wavelength of light. The measurement is thus given by

$$y_k = \frac{(k-1)\lambda}{4n_r} \tag{11.9.12}$$

where k represents each interference fringe during development.

For this situation, not all states are observable. The measurement matrix must be split into two sections:

$$\boldsymbol{H} = (1/q)[\boldsymbol{H}_1 \mid \boldsymbol{H}_2] \tag{11.9.13}$$
$$\boldsymbol{H}_1 = [1 \ \ 1 \ \dots \ 1] \tag{11.9.14}$$
$$\boldsymbol{H}_2 = [0 \ \ 0 \ \dots \ 0] \tag{11.9.15}$$

$\boldsymbol{H}_1$ is a $q \times 1$ matrix (q being the number of states in the breakthrough region) and $\boldsymbol{H}_2$ a $(m-q) \times 1$ matrix.

If $\boldsymbol{\phi}_k$ is nearly constant during the process, a steady-state gain matrix $\boldsymbol{K}_{ss}$ can be used in order to save several minutes of computational time. The time savings are extremely important in light of the small process time; on the order of one minute. Since end-point detection is desired, the values of the states are most important around the breakthrough time.

The useful information from the measurement is limited to the time from development start-up to breakthrough in the film. Only the q states defined in this limited area are observable. This implies that the Kalman filter algorithm is useful only to optimally identify these states, and it is only useful until breakthrough occurs. The remaining states can only be estimated using the mathematical model. Since the model prediction is a function of the model parameters, it is useful also to identify, on-line, actual model parameters during the time that data are available.

11.9.3 Parameter Identification

Process variations prior to development may cause significant deviations in a resist line thickness, and these are evident only after development. By using on-line parameter identification, we can fine tune the process model by the time breakthrough occurs. A simplified development model was used for this application,

$$\frac{dz_i}{dt} = \left[k_{\min} + k_D \frac{(1-M_i)^n}{a+(1-M_i)^n}\right] u(t) \tag{11.9.16}$$

where the surface induction effect described by Equation (11.9.4) has been neglected. It is not important to know the specific model dynamics at the surface since data are available until breakthrough. If we let

$$u(t) = Du_1(t) \tag{11.9.17}$$

where $u_1 = 1$ when developer fluid is dispensed and 0 when no development takes place and D is the bulk concentration of active developer in the spray, Equation (11.9.16) can be recast as

$$\frac{dz_i}{dt} = k_1 \left[k_2 + \frac{(1-M_i)^n}{a+(1-M_i)^n}\right] u_1(t) \tag{11.9.18}$$

where

$$k_1 = Dk_D \tag{11.9.19}$$

and

$$k_2 = k_{\min}/k_D \tag{11.9.20}$$

The parameters k_1, k_2, a, and n are calculated from off-line experimental results and used as a first approximation for the on-line identification.

Let $\boldsymbol{\theta}$ represent the vector of parameters to be identified. The working process model is obtained by combining the model prediction of Equation (11.9.10) with the optimal estimator of Equation (11.9.7), using the steady-state Kalman filter gain such that

$$\hat{\boldsymbol{x}}_k = (\boldsymbol{I} - \boldsymbol{K}_{ss}\boldsymbol{H})\left(\hat{\boldsymbol{x}}_{k-1} + \Delta t \boldsymbol{f}_k(\hat{\boldsymbol{x}}_{k-1})\right) + \boldsymbol{K}_{ss} y_k \tag{11.9.21}$$

The linear measurement model is for the parameter identification problem at the current time,

$$\hat{y}_k = \boldsymbol{H}\hat{\boldsymbol{x}}_k \tag{11.9.22}$$

where $\hat{y}_k$ is the estimated measurement, determined after the actual measurement has been taken and the Kalman filter implemented.

The parameters are updated using the algorithm of Section 11.8 for a nonlinear process. However, in this example, the measurements are linearly related to the states. The parameter update is given by

$$\hat{\boldsymbol{\theta}}_k = \hat{\boldsymbol{\theta}}_{k-1} + \boldsymbol{L}_k \epsilon_k \tag{11.9.23}$$

where the local error or innovation is defined as

$$\epsilon_k = y_k - \hat{y}_k \tag{11.9.24}$$

The gain $\boldsymbol{L}_k$ is calculated recursively using a Gauss–Newton search direction, based upon the covariance of the model, $\boldsymbol{P}$, given by Equations (11.8.13) to (11.8.16).

For this problem, the measurement gradient system of Equation (11.8.11) is

$$\psi_k^T = \boldsymbol{H}\boldsymbol{W}_k \tag{11.9.25}$$

where $\boldsymbol{W}$ is the gradient matrix of the states with respect to model parameters. For this problem, Equation (11.8.5) reduces to

$$\boldsymbol{W}_k = (\boldsymbol{F}_{k-1} - \boldsymbol{K}_{ss}\boldsymbol{H})\,\boldsymbol{W}_{k-1} + \boldsymbol{M}_k \tag{11.9.26}$$

where $\overline{\boldsymbol{M}}_k$ is obtained from

$$\overline{\boldsymbol{M}} = (\boldsymbol{I} - \boldsymbol{K}\boldsymbol{H})\,\frac{\partial \boldsymbol{f}_k^T}{\partial \boldsymbol{\theta}}\,\Delta t \tag{11.9.27}$$

and

$$\boldsymbol{F}_{k-1} = (\boldsymbol{I} - \boldsymbol{K}\boldsymbol{H})\phi_{k-1} \tag{11.9.28}$$

11.9.4 State Estimation

State and parameter identification was simulated for a typical exposed line of AZ1350J photoresist using the process parameters shown in Table 11.10 by Carroll and Ramirez (1990). The substrate is assumed not to reflect light for the purposes of simplicity. The values for $M(x_i)$ in the states defined in the breakthrough region are calculated from Dill's model, using parameters taken from the literature (Dill *et al.*, 1975), and stored in memory. The development parameters shown in Table 11.10 were obtained from Mack (1987). The light source is assumed to be a He–Ne laser.

The measurements, given as depth of penetration, are simulated using Equation (11.9.1) as the actual system. The development equation is solved

Table 11.10
Exposure and Development Parameters Used in Simulation

PAC absorbance, A	0.86 μm^{-1}
Inert absorbance, B	0.07 μm-1
Exposure reaction rate constant, C	0.014 cm^2/mJ
Line width, L	2.0 μm
Resist thickness, z_0	2.2 μm
Exposure dosage	300 mJ/cm^2
a	0.01353
n	5.0
$k_{\min}u$	0.005 μm/sec
$k_D u$	0.1 μm/sec
r_0	0.755
z_t	0.4 μm
Light source	632.8 nm
n_r	1.68

Table 11.11
Kalman Filter Parameters

Measurement matrix

$$\boldsymbol{H} = 0.125 \begin{bmatrix} 1 & 1 & 1 & 1 & 1 & 1 & 1 & 1 & 0 & 0 & 0 & 0 & 0 \end{bmatrix}$$

Measurement covariance

$$r = 0.04\ \mu\text{m}^2$$

Model covariance matrix

$$\boldsymbol{Q}_{diag} = .04 \begin{bmatrix} 1 & 1 & 1 & 1 & 1 & 1 & 1 & 1 & 4 & 4 & 4 & 4 & 4 \end{bmatrix} \mu\text{m}^2$$

Initial state covariance matrix

$$\boldsymbol{P}(0)_{diag} = 2.2 \begin{bmatrix} 1 & 1 & 1 & 1 & 1 & 1 & 1 & 1 & 0 & 0 & 0 & 0 & 0 \end{bmatrix} \mu\text{m}^2$$

to obtain the "true" response to the given inputs and parameters. Using Equations (11.9.11) and (11.9.12), the interference fringe extrema times may be computed. Noise is added to each data point (t, z) to simulate fluctuations in the development and to aid in the detection of interference fringes. The noise is introduced as ±0.2 seconds for each fringe occurrence.

The parameters used to evaluate the Kalman filter and state estimations are given in Table 11.11. The worst case deviation of the normalized depth is assumed to be ±0.2 μm for observable states and ± 0.4 μm for all others. The measurement is assumed to deviate ± 0.2 μm, which is much worse than the actual filtered signal. These severe penalties are used to illustrate the usefulness of Kalman filtering.

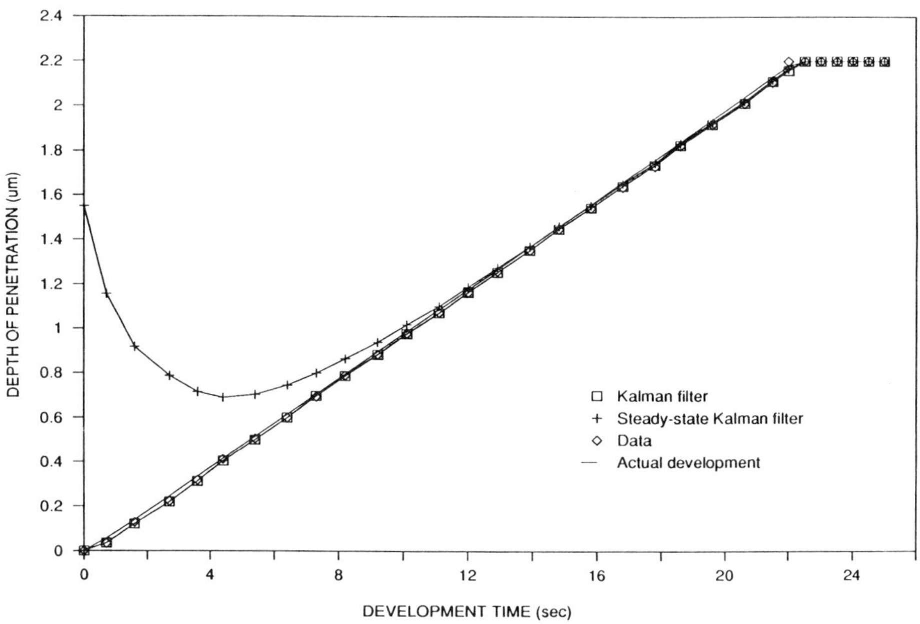

Figure 11.15 Kalman filter for AZ1350. From Carroll and Ramirez (1990).

Figure 11.15 shows the simulated estimation of the state defined at the line-width center. The simulated data differ from the actual system response because of the presence of noise in the data. Each time a data point is obtained, the Kalman filter and state are updated. Figure 11.15 illustrates three cases: (1) a continuously updated Kalman filter, which assumes an initial worst-case estimate of the depth at z_i = 2.2 μm; (2) a steady-state Kalman filter, also assuming $z_i = 2.2$ μm initially; and (3) the "actual" development assuming $z_i = 0$. The optimum filter, case (1), tracks very closely to actual state value, despite the assumption that the model's "expected" initial depth is 2.2 μm. This occurs because the Kalman filter gain is initially very large, due to the large value of $\boldsymbol{P}(0)$. The dynamic filter gain is in fact so large that a considerable undershoot would occur if constraints on z were not imposed. The states must be constrained, such that $0 \leq z \leq z_0$, in order to insure proper operation of the algorithm and a proper physical description of the phenomenon. The steady-state filter is much slower than the dynamic filter in correcting the discrepancy between measurement and model since the steady-state gain is two orders of magnitude less than the initial dynamic gain. However, the filter with the steady-state gain tracks well at the crucial time when breakthrough occurs, even for this worst-case scenario.

11.9.5 Parameter Estimation

Table 11.12 shows the parameters used in a combined parameter and state estimation simulation. Parameter k_1 is to be identified on-line. The vector $\boldsymbol{\theta}$ is simply a scalar in this case, k_1. Figure 11.16 indicates the variation of k_1 as the development proceeds, with an initial estimate of the parameter which is half its actual final value. The actual rate parameter value, $k_{1,a}$, is calculated using

$$k_{1,a} = \left(1 - (1 - r_0)\exp\left(-\frac{4z_i}{z_t}\right)\right) \tag{11.9.29}$$

Rate parameter $k_{1,a}$ is shown to illustrate the surface induction effect. Note that the estimated parameter, k_1, tracks beyond the final rate parameter value in order to compensate for the initial low guess. The state response, shown in Figure 11.17, is initially sluggish because the model parameter has not attained the value that is true to the actual system.

11.9.6 Experimental Results

Carroll and Ramirez (1990) carried out actual on-line identification experiments using a spray developer. The substrates were coated with 4.2 μm of AZ4330 resist, exposed by a Canon proximity printer with an energy dose of 206.7 mJ/cm^2. The exposed area is roughly 90%, and the lines are approximately 5 μm wide. The exposure parameters shown in Table 11.13 were determined using Dill's model and experimental technique. An approximate photoactive compound (PAC) concentration distribution of the central region is also calculated from Dill's model.

Figure 11.18 shows a monitored development experiment. The interference filter used in the fiber optic source is 630 nm with a bandwidth of 10 nm. Using Equation (11.9.11), the number of interference fringes and the development

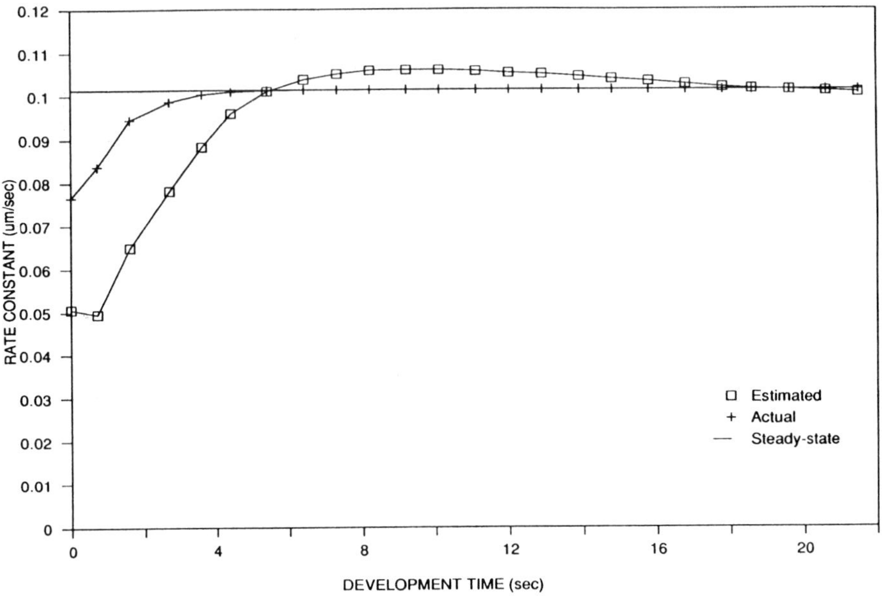

Figure 11.16 Rate constant identification for AZ1350.
From Carroll and Ramirez (1990).

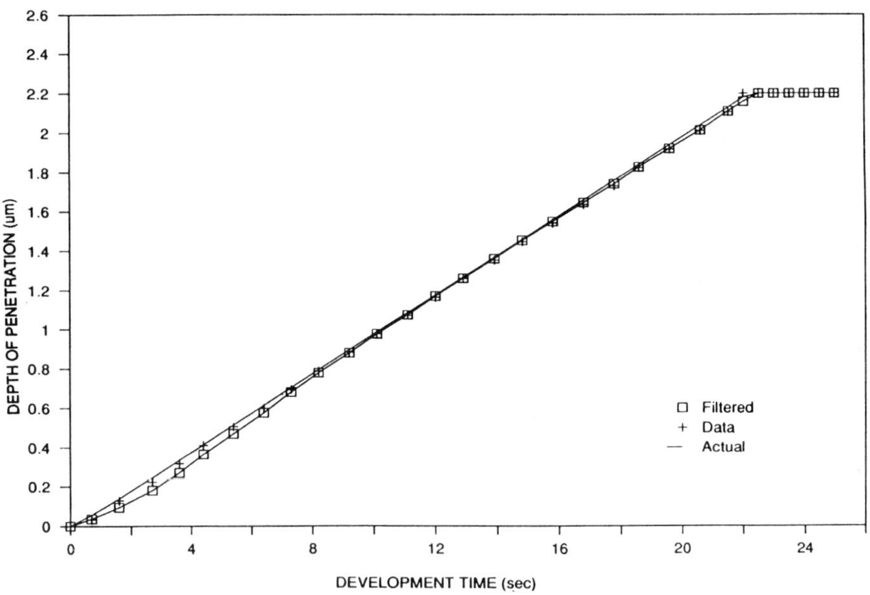

Figure 11.17 Development progress for AZ1350.
From Carroll and Ramirez (1990).

Table 11.12

Parameter/State Estimation Parameters for Simulation

$k_1 = 0.101\ \mu\text{m/sec}$
$k_2 = 0.00493$
Covariance of model uncertainty
$\boldsymbol{P}_0 = 0.1\ \mu\text{m}^2/\text{sec}^2$
Local error covariance
$\hat{\Lambda}_0 = 0.1\ \mu\text{m}^2$
Forgetting factor
$\lambda_k = 0.9$
Initial parameter estimate
$\hat{\theta}_0 = 0.101\ \mu\text{m/sec}$

rate may be calculated. The Kalman filter parameters, unchanged from the simulation, are used to calculate the steady-state Kalman filter gain matrix. The combined parameter and state identifier calculates the depth of penetration of the development, shown in Figure 11.19. The filtered state initially shows a faster development than the data indicate, but after the transients have died away, the response tracks closely with the raw data.

Figure 11.20 shows the updated parameter, which represents the development rate constant. Note that the parameter value dips initially, indicating the surface induction effect. The parameter becomes relatively constant as the development nears breakthrough.

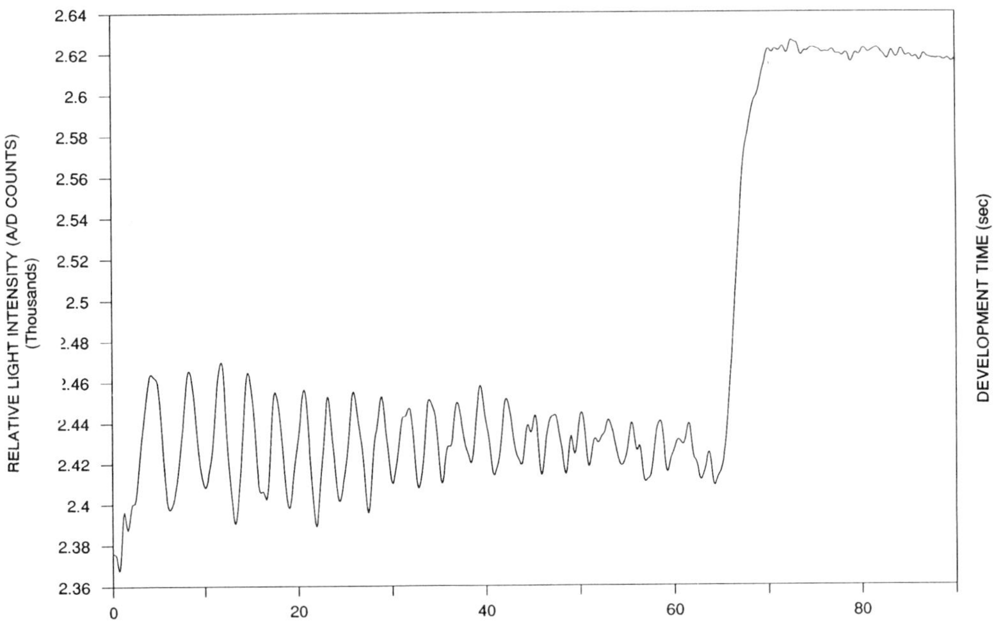

Figure 11.18 Fringe data for AZ4330. From Carroll and Ramirez (1990).

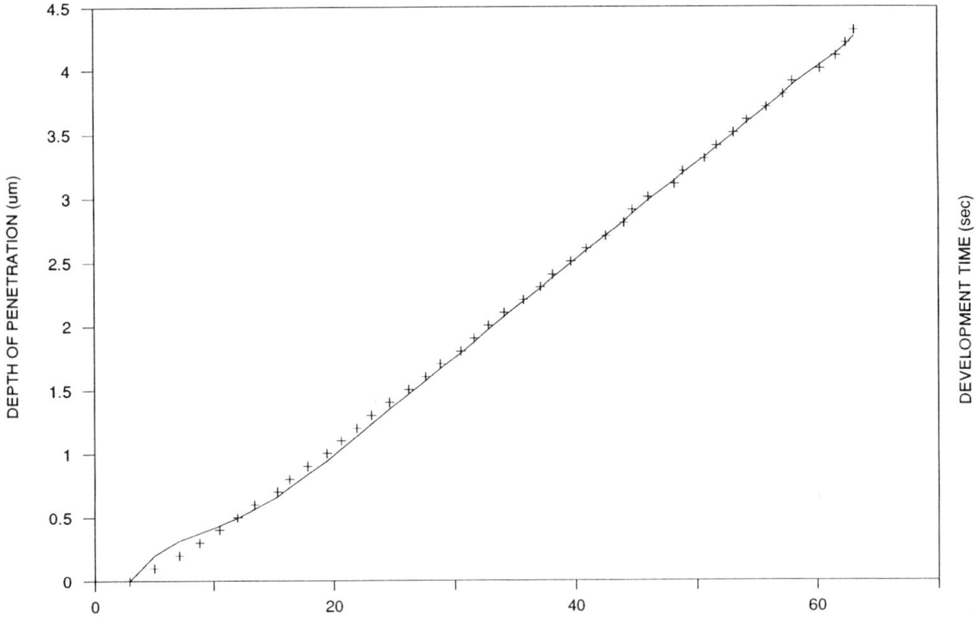

Figure 11.19 State identification for AZ4330. From Carroll and Ramirez (1990).

Table 11.13
AZ4330 Parameters Used in Experiments

Development parameters	
$a = 0.163$	$n = 3.10$
$k_1 = 0.145\ \mu\text{m/sec}$	$k_2 = 0.004$
$\boldsymbol{P}_0 = 0.1\ \mu\text{m}^2/\text{sec}^2$	$\hat{\Lambda}_0 = 0.1\ \mu\text{m}^2$
$\lambda_k = 0.9$	$\hat{\theta}_0 = 0.145\ \mu\text{m/sec}$

Exposure parameters:

Wavelength (nm)	$A(\mu\text{m}-1)$	$B(\mu\text{m}^{-1})$	$C(\text{cm}^2/\text{mJ})$	Fraction of total light
365.0	0.468	0.0850	0.0103	0.255
404.7	0.470	0.0310	0.0162	0.203
435.8	0.306	0.0273	0.0085	0.542

$n_r = 1.57$	Substrate index =2.30 + j3.96
Dose = 206.7 mJ/cm^2	$z_0 = 4.2\ \mu\text{m}$
Line width = 5 μm	Prebake time = 30 min

Steady-state Kalman Filter gain matrix:

$$\boldsymbol{K}_{ss}^T = 0.296\ [\ 1\ \ 1\ \ 1\ \ 1\ \ 1\ \ 1\ \ 1\ \ 1\ \ 0\ \ 0\ \ 0\ \ 0\ \ 0\ ]$$

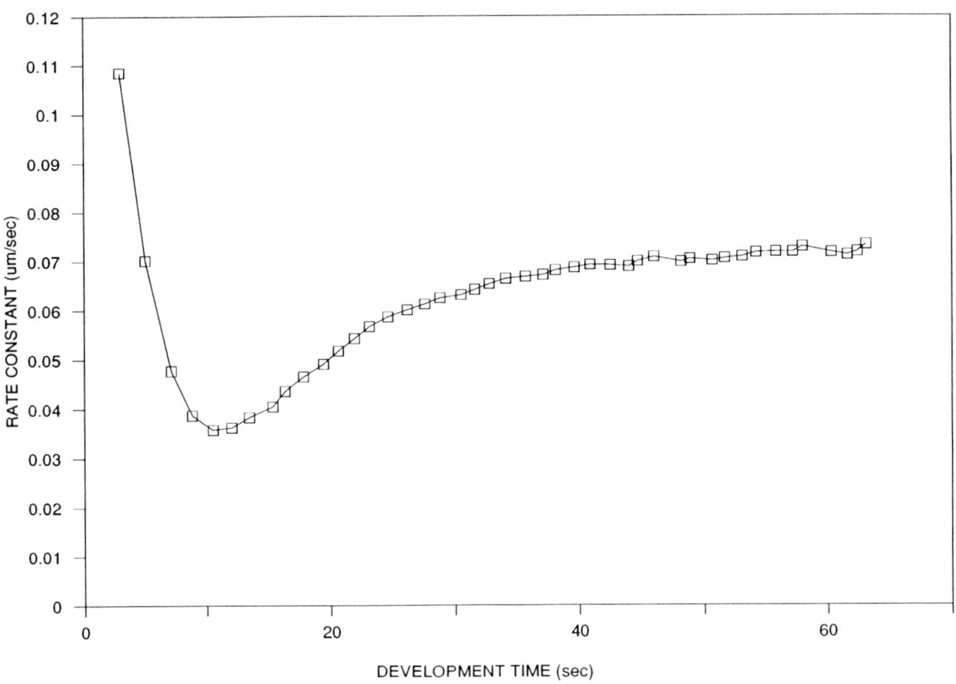

Figure 11.20 Rate constant for AZ4330 development. From Carroll and Ramirez (1990).

References

Carroll, T. A. and Ramirez, W. F., "On-Line State and Parameter Identification of Positive Photoresist Development," *AIChE Journal* (1990).

Chattaway, T. and Stephanopoulos, G., "Adaptive Estimation of Bioreactors: Monitoring Plasmid Instability," *Chem. E. Sci.* **44** 41–48 (1989).

Dill, F. H., Hornberger, W. P., Hauge, P. S., and Shaw, J. M., "Characterization of Positive Photoresist," *IEEE Trans. Elect. Dev.* **ED–22**(7), 445–452 (1975).

Engasser, J. M., Marc, I., Moll, M., and Duteurtre, B., *EBC Congress*, 579–583 (1981).

Gee, D. A. and Ramirez, W. F., "Optimal Temperature Control for Batch Beer Fermentation," *Biotech. Bioeng.* **31**, 224–234 (1988).

Lauchlan, L., Sautter, K., Batchelder, T., and Irwin, J., "On-line Automatic Photoresist Process Control," *Proc. SPIE (Advances in Resist Technology and Processing II)* **539**, 227–233 (1985).

Ljung, L. and Soderstrom, T., *Theory and Practice of Recursive Identification*, MIT Press, Cambridge, MA (1983).

Mack, C. A., "PROLITH: A Comprehensive Optical Lithography Model," *Proc. SPIE (Optical Microlithography IV)* **530**, 207–220 (1985).

Mack, C. A., "Development of Positive Photoresists," *J. Electrochem. Soc.* **134**(1), 148–152 (1987).

Novembre, A. E., Tang, W. T., and Hsieh, P., "An In Situ Interferometric Analysis of Resist Development on Photomask Substrates," *Proc. SPIE (Integrated Circuit Metrology, Inspection, and Process Control III)* **1087**, 460–468 (1989).

Ramirez, W. F., "Optimal State and Parameter Identification—An Application to Batch Fermentation," *Chem. Engr. Sci.* **42**, 2749–2756 (1987).

Robbins, H. and Monro, S., "A Stochastic Approximation Method," *Annals of Math. Stat.* **22**, 400–407 (1951).

Sautter, K. M., Ha, M., and Batchelder, T., "Development Process Control Utilizing an End Point Monitor," *Proc. SPIE (Integrated Circuit Metrology, Inspection, and Process Control III)* **1087**, 312–321 (1989).

Willson, C. G., "Organic Resist Materials—Theory and Chemistry," *Introduction to Microlithography*, ACS Symposium Series **219**, 100–105, edited by L. F. Thompson, C. G. Willson, and M. J. Bowden (1983).

Exercises

1. Raw material used in the production of a synthetic fiber is stored in a place that has no humidity control. Develop a sequential linear least squares correlation for the moisture content for the following 12 days of data.

Humidity (%)	Moisture Content (%)
42	12
35	8
50	14
43	9
48	11
62	16
31	7
36	9
44	12
39	10
55	13
48	11

2. The following data have been collected in the processing of an electronic device. Fit a linear regression model to the data sequentially. What is the expected gain when the next processing condition is 2.2 hours and 90 ohm–cm?

Diffusion Time (hrs)	Resistance (ohm–cm)	Gain
1.5	66	5.3
2.5	87	7.8
0.5	69	7.4
1.2	141	9.8
2.6	93	10.8
0.2	105	9.1
2.4	111	8.1
2.0	78	7.2
0.7	66	6.5
1.6	123	12.6

3. The amount of an additive is expected to have a parabolic effect on the drying time of a varnish,

$$t_d = a + bx + cx^2$$

Analyze the following batches of data sequentially:

Additive Amount (g)	Drying Time (hrs)
0	12.0
1	10.5
2	10.0
3	8.0
4	7.0
5	8.0
6	7.5
7	8.5
8	9.0

What is the expected drying time when 6.5 g of the additive is added?

4. Chattaway and Stephanopoulos (1989) apply combined Kalman filter state estimation and recursive least squares parameter identification to a problem of monitoring plasmid instability. For this problem there are two cultures X^+ and X^-. Recombinant cells are unstable due to unequal partition of plasmids between two daughter cells during division. Detecting the appearance of plasmid free cells (X^-) at an early stage of operation is highly desirable. The process dynamics of the plasmid carrying cells (X^+) and the plasmid free cells (X^-) in a continuous chemostat reactor is given by

$$\begin{aligned}
\frac{dX^+}{dt} &= \mu^+ X^+(1-p) - DX^+ \\
\mu^+ &= \mu^+_{\max} \frac{S}{S+K_s^+} \\
\frac{dX^-}{dt} &= \mu^- X^- + p\mu^+ X^+ - DX^- \\
\mu^- &= \mu^-_{\max} \frac{S}{S+K_s^-} \\
\frac{dS}{dt} &= -\frac{\mu^+}{Y^+} X^+ - \frac{\mu^-}{Y^-} X^- + D(S_f - S)
\end{aligned}$$

where
$$\begin{aligned} \mu^+_{\max} &= .7 \text{ h}^{-1} & \mu^-_{\max} &= 1.0 \text{ h}^{-1} \\ K^+_s &= .01 \text{ g}/\ell & K^-_s &= .01 \text{ g}/\ell \\ Y^+_s &= 1 \text{ g/g} & Y^-_s &= 1 \text{ g/g} \\ p &= .05 \end{aligned}$$

The parameters to be identified are

$$\boldsymbol{\theta} = \begin{pmatrix} \mu^+ \\ \mu^- \end{pmatrix}$$

The measurements are

$$y = \begin{pmatrix} X^+ + X^- \\ \mu^+ X^+ + \mu^- X^- \end{pmatrix}$$

which are the total biomass and the total growth rate.

Identify model parameters for a chemostat when

$$D = .5 \text{ h}^{-1},\ S_f = 20 \text{ g}/\ell,\ X^+(t_0) = 5 \text{ g}/\ell,\ X^-(t_0) = 0$$

$$\text{and } S(t_0) = 20 \text{ g}/\ell$$

The measurement noise is given by

$$\boldsymbol{R} = \text{diag}\begin{pmatrix} .15^2 & .1^2 \end{pmatrix}$$

Assume the model uncertainty is given as

$$\boldsymbol{Q} = .05^2\ \boldsymbol{I}$$

Identify the model parameters when an incorrect parameter set

$$\boldsymbol{\theta} = \begin{pmatrix} .8 \\ .9 \end{pmatrix}$$

is assumed.

Chapter 12

THE LINEAR QUADRATIC GAUSSIAN (LQG) PROBLEM

12.1 The Separation Principle for Continuous Systems

This chapter considers the combination of optimal control with state and parameter estimation. We present the important separation principle which states that the design of an optimal control problem with measurement and model uncertainty can be treated by first performing a Kalman filter estimate of the states and then developing the optimal control law based upon the estimated states. For linear regulator problems, the problem is known as the linear quadratic Gaussian (LQG) problem. The application of the separation principle allows for the practical implementation of model control concepts to systems with significant process and measurement uncertainties. It provides the modern framework for the treatment of process control problems.

This section considers the optimal regulator control of linear stochastic systems. We consider the following linear stochastic system,

$$\dot{\boldsymbol{x}} = \boldsymbol{A}\boldsymbol{x} + \boldsymbol{B}\boldsymbol{u} + \boldsymbol{w} \tag{12.1.1}$$

where $\boldsymbol{w}$ is Gaussian white mean zero and covariance $\hat{\boldsymbol{Q}}$, $\boldsymbol{x}$ is the state vector, and $\boldsymbol{u}$ is the control vector.

The measurement system is modeled as

$$\boldsymbol{y} = \boldsymbol{H}\boldsymbol{x} + \boldsymbol{v} \tag{12.1.2}$$

where the measurements, $\boldsymbol{y}$, are corrupted by white Gaussian noise with zero mean and covariance $\hat{\boldsymbol{R}}$.

Optimal regulator control is sought by minimizing a quadratic performance index. It is not possible to use the functional

$$J = \frac{1}{2}\, \boldsymbol{x}^T(t_f)\boldsymbol{S}\boldsymbol{x}(t_f) + \frac{1}{2}\int_0^{t_f} \left(\boldsymbol{x}^T\boldsymbol{Q}\boldsymbol{x} + \boldsymbol{u}^T\boldsymbol{R}\boldsymbol{u}\right) dt \tag{12.1.3}$$

since this functional is a random variable depending upon $\boldsymbol{w}$ and $\boldsymbol{v}$. In order to eliminate this difficulty, we use the expected value operator and define the following performance functional,

$$J = \mathcal{E}\left\{\frac{1}{2}\, \boldsymbol{x}^T(t_f)\boldsymbol{S}\boldsymbol{x}(t_f) + \frac{1}{2}\int_0^{t_f} \left(\boldsymbol{x}^T\boldsymbol{Q}\boldsymbol{x} + \boldsymbol{u}^T\boldsymbol{R}\boldsymbol{u}\right) dt\right\} \tag{12.1.4}$$

From Bayes's theorem on conditional probability, we have the fact that

$$\mathcal{E}\left\{\mathcal{E}(\boldsymbol{x}/\boldsymbol{y})\right\} = \mathcal{E}(\boldsymbol{x}) \tag{12.1.5}$$

From Kalman filtering we know that

$$\mathcal{E}(\boldsymbol{x}/\boldsymbol{y}) = \hat{\boldsymbol{x}} \tag{12.1.6}$$

where $\hat{\boldsymbol{x}}$ is the optimal estimate of $\boldsymbol{x}$ based upon the condition $\boldsymbol{y}$. Therefore, Equation (12.1.5) becomes

$$\mathcal{E}(\boldsymbol{x}) = \mathcal{E}(\hat{\boldsymbol{x}}) \tag{12.1.7}$$

and the performance index of Equation (12.1.4) can be written as

$$J = \mathcal{E}\left\{\frac{1}{2}\, \hat{\boldsymbol{x}}^T(t_f)\boldsymbol{S}\hat{\boldsymbol{x}}(t_f) + \int_0^{t_f} \frac{1}{2}\left(\hat{\boldsymbol{x}}^T\boldsymbol{Q}\hat{\boldsymbol{x}} + \boldsymbol{u}^T\boldsymbol{R}\boldsymbol{u}\right) dt\right\} \tag{12.1.8}$$

The system of equations that defines the optimal Kalman filter estimates is

$$\dot{\hat{\boldsymbol{x}}} = \boldsymbol{A}\hat{\boldsymbol{x}} + \boldsymbol{B}\boldsymbol{u} + \hat{\boldsymbol{K}}(\boldsymbol{y} - \boldsymbol{H}\hat{\boldsymbol{x}}) \tag{12.1.9}$$

where

$$\hat{\boldsymbol{K}} = \hat{\boldsymbol{P}}\boldsymbol{H}^T\hat{\boldsymbol{R}}^{-1} \tag{12.1.10}$$

with

$$\dot{\hat{\boldsymbol{P}}}(t) = \boldsymbol{A}\hat{\boldsymbol{P}} + \hat{\boldsymbol{P}}\boldsymbol{A}^T - \hat{\boldsymbol{P}}\boldsymbol{H}^T\hat{\boldsymbol{R}}^{-1}\boldsymbol{H}\hat{\boldsymbol{P}} + \hat{\boldsymbol{Q}} \tag{12.1.11}$$

and

$$\hat{\boldsymbol{x}}(t_o) = \overline{\boldsymbol{x}}(t_o) \tag{12.1.12}$$

$$\hat{\boldsymbol{P}}(t_o) = \hat{\boldsymbol{P}}_o \tag{12.1.13}$$

Since the expected value of the innovations process, $\boldsymbol{z}$, is zero, we have

$$\mathcal{E}(\boldsymbol{z}) = \mathcal{E}(\boldsymbol{y} - \boldsymbol{H}\hat{\boldsymbol{x}}) = \boldsymbol{0} \tag{12.1.14}$$

Therefore,

$$\mathcal{E}\left\{\dot{\hat{\boldsymbol{x}}} = \boldsymbol{A}\hat{\boldsymbol{x}} + \boldsymbol{B}\boldsymbol{u} + \hat{\boldsymbol{K}}(\boldsymbol{y} - \boldsymbol{H}\hat{\boldsymbol{x}})\right\} \tag{12.1.15}$$

is

$$\dot{\hat{\boldsymbol{x}}} = \boldsymbol{A}\hat{\boldsymbol{x}} + \boldsymbol{B}\boldsymbol{u} \tag{12.1.16}$$

and the objective functional is

$$J = \frac{1}{2}\,\hat{\boldsymbol{x}}^T(t_f)\boldsymbol{S}\hat{\boldsymbol{x}}(t_f) + \frac{1}{2}\int_0^{t_f}\left(\hat{\boldsymbol{x}}^T\boldsymbol{Q}\hat{\boldsymbol{x}} + \boldsymbol{u}^T\boldsymbol{R}\boldsymbol{u}\right)dt \tag{12.1.17}$$

Equations (12.1.16) and (12.1.17) are the statement of a classical Kalman linear quadratic regulator (LQR) problem which has the optimal regulator control solution of

$$\boldsymbol{u} = -\boldsymbol{K}\hat{\boldsymbol{x}} \tag{12.1.18}$$

where

$$\boldsymbol{K} = \boldsymbol{R}^{-1}\boldsymbol{B}^T\boldsymbol{P} \tag{12.1.19}$$

with

$$\dot{\boldsymbol{P}} = -\boldsymbol{P}\boldsymbol{A} - \boldsymbol{A}^T\boldsymbol{P} + \boldsymbol{P}\boldsymbol{B}\boldsymbol{R}^{-1}\boldsymbol{B}^T\boldsymbol{P} - \boldsymbol{Q} \tag{12.1.20}$$

and

$$\boldsymbol{P}(t_f) = \boldsymbol{S} \tag{12.1.21}$$

We have therefore obtained the well-known separation principle of linear stochastic control. The solution to the stochastic LQG control problem has been shown to use the optimal deterministic LQR problem except that the control is feedback in the optimal estimates obtained from the Kalman filter. Figure 12.1 shows a block diagram of the resulting regulator control structure for the LQG problem.

The importance of the separation principle is that the linear quadratic Gaussian regulator can be designed in two separate stages: the Kalman filter design and the Kalman linear quadratic regulator problem.

The development presented here has minimized the mathematical complexities that lead to the separation principle. The interested reader is referred to the books by Sage and White (1977) and Lewis (1986) for the Bayesian approach and that by Grimble and Johnson (1988) for the functional analysis approach. Other important references are the works by Astrom (1970), Kwakernaak and Sivan (1972) and Balakrishnan (1972).

Through a similar development, the optimal stochastic regulator with Kalman filtering state identification and sequential least squares parameter identification can be treated using the separation principle. This leads to an adaptive optimal control system. The block diagram for the adaptive LQG control problem is shown in Figure 12.2. It is important to note that the sequential least squares estimator works on the innovations error $(y - \hat{y})$ and not just the measurements y as does the Kalman filter. Also, the current estimate of the parameters is used to update the Kalman LQR controller, the Kalman filter, and the measurement prediction.

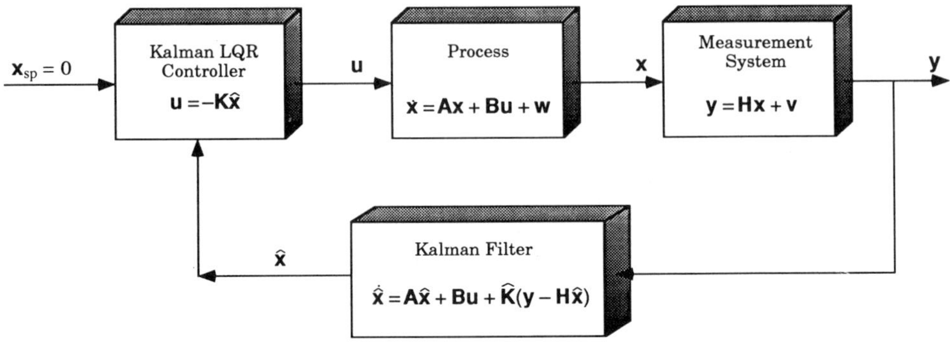

Figure 12.1 LQG block diagram.

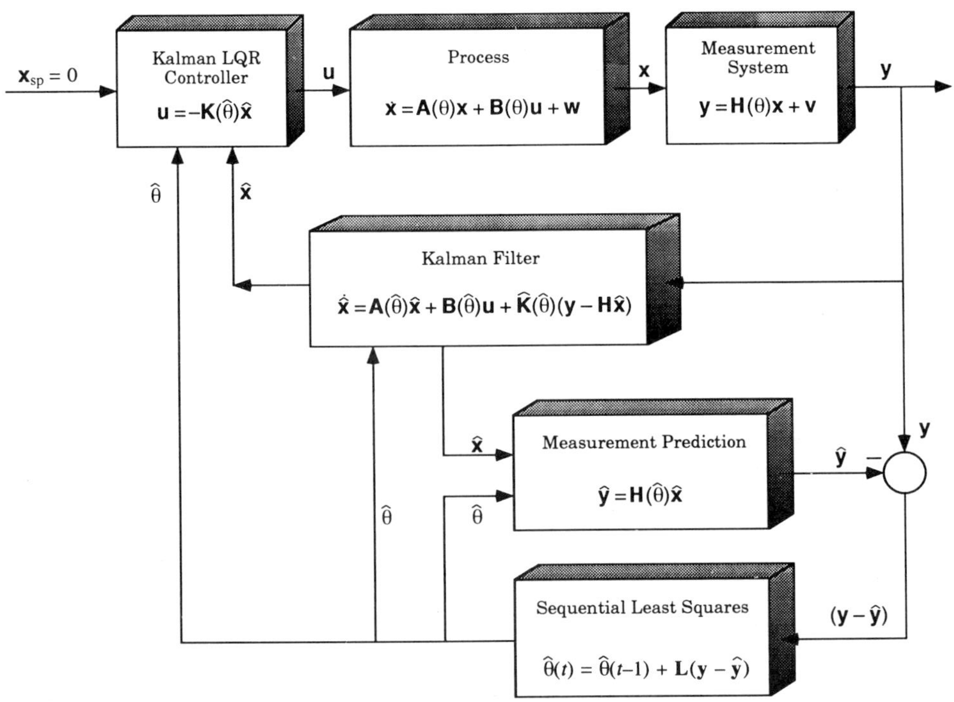

Figure 12.2 Adaptive LQG block diagram.

12.2 The Separation Principle for Discrete Systems

It is not surprising that the separation principle applies to discrete systems as well as to continuous systems. The state dynamics are given by

$$\boldsymbol{x}_{i+1} = \boldsymbol{\Phi}\boldsymbol{x}_i + \boldsymbol{\Delta}\boldsymbol{u}_i + \boldsymbol{\Gamma}\boldsymbol{w}_i \tag{12.2.1}$$

where $\boldsymbol{w}_i$ is Gaussian with mean zero and covariance $\hat{\boldsymbol{Q}}$.

The measurement system is

$$\boldsymbol{y}_i = \boldsymbol{H}\boldsymbol{x}_i + \boldsymbol{v}_i \tag{12.2.2}$$

with $\boldsymbol{v}_i$ Gaussian having covariance $\hat{\boldsymbol{R}}$.

The appropriate quadratic performance index is

$$J = \mathcal{E}\left\{\frac{1}{2}\,\boldsymbol{x}_f^T\boldsymbol{S}\boldsymbol{x}_f + \frac{1}{2}\sum_{k=0}^{k_f-1}\left(\boldsymbol{x}_k^T\boldsymbol{Q}\boldsymbol{x}_k + \boldsymbol{u}_k^T\boldsymbol{R}\boldsymbol{u}_k\right)\right\} \tag{12.2.3}$$

The separation principle leads to the solution of the following LQR problem:

$$\hat{\boldsymbol{x}}_{i+1} = \boldsymbol{\Phi}\hat{\boldsymbol{x}}_i + \boldsymbol{\Delta}\boldsymbol{u}_i \tag{12.2.4}$$

with

$$J = \frac{1}{2}\,\hat{\boldsymbol{x}}_f^T\boldsymbol{S}\hat{\boldsymbol{x}}_f + \frac{1}{2}\sum_{k=0}^{k_f-1}\left(\hat{\boldsymbol{x}}_k^T\boldsymbol{Q}\hat{\boldsymbol{x}}_k + \boldsymbol{u}_k^T\boldsymbol{R}\boldsymbol{u}_k\right) \tag{12.2.5}$$

where $\hat{\boldsymbol{x}}_i$ is the optimal estimate of the state defined by the Kalman filter,

$$\hat{\boldsymbol{x}}_{i+1} = \boldsymbol{\Phi}\hat{\boldsymbol{x}}_i + \boldsymbol{\Delta}\boldsymbol{u}_i + \hat{\boldsymbol{K}}_i\left(\boldsymbol{y}_i - \boldsymbol{H}\hat{\boldsymbol{x}}_i\right) \tag{12.2.6}$$

with

$$\hat{\boldsymbol{K}}_i = \hat{\boldsymbol{P}}_i\boldsymbol{H}^T\hat{\boldsymbol{R}}_i^{-1} \tag{12.2.7}$$

$$\hat{\boldsymbol{P}}_i = \boldsymbol{M}_i - \boldsymbol{M}_i\boldsymbol{H}^T\left(\boldsymbol{H}\boldsymbol{M}_i\boldsymbol{H}^T + \hat{\boldsymbol{R}}\right)^{-1}\boldsymbol{H}\boldsymbol{M}_i \tag{12.2.8}$$

$$\boldsymbol{M}_{i+1} = \boldsymbol{\Phi}\hat{\boldsymbol{P}}_i\boldsymbol{\Phi}^T + \boldsymbol{\Gamma}\hat{\boldsymbol{Q}}\boldsymbol{\Gamma}^T \tag{12.2.9}$$

The optimal controller is

$$\boldsymbol{u}_i = -\boldsymbol{K}_i\hat{\boldsymbol{x}}_i \tag{12.2.10}$$

with

$$\boldsymbol{K}_i = \left[\boldsymbol{R} + \boldsymbol{\Delta}^T\boldsymbol{P}_{i+1}\boldsymbol{\Delta}\right]^{-1}\boldsymbol{\Delta}^T\boldsymbol{P}_{i+1}\boldsymbol{\Phi} \tag{12.2.11}$$

$$\boldsymbol{P}_i = \boldsymbol{\Phi}^T\boldsymbol{P}_{i+1}\boldsymbol{\Phi} - \boldsymbol{\Phi}^T\boldsymbol{P}_{i+1}\boldsymbol{\Delta}\left[\boldsymbol{\Delta}^T\boldsymbol{P}_{i+1}\boldsymbol{\Delta} + \boldsymbol{R}\right]^{-1}\boldsymbol{\Delta}^T\boldsymbol{P}_{i+1}\boldsymbol{\Phi} + \boldsymbol{Q} \tag{12.2.12}$$

$$\boldsymbol{P}(k_f) = \boldsymbol{S} \tag{12.2.13}$$

It is important to keep in mind that the Kalman filter gain equations (12.2.7) to (12.2.9) are computed forward in time, while the Kalman controller gain relations (12.2.11) to (12.2.13) are computed backwards in time from the final state and final time.

12.3 Time Optimal Control of a Stirred Tank Reactor

As discussed in Section 10.7, Lynch and Ramirez (1975) extended the Nyquist–Ramirez reactor (Nyquist and Ramirez, 1971) to include control on both cooling water flow and the flow rate of the potassium iodide catalyst concentration to the reactor. The system is characterized by three state variables,

$$\boldsymbol{x} = \begin{pmatrix} x_1 \\ x_2 \\ x_3 \end{pmatrix} \tag{12.3.1}$$

where x_1 is the normalized dimensionless reactant concentration, x_2 is the normalized dimensionless catalyst concentration, and x_3 is the normalized dimensionless temperature. There are two measurements,

$$\boldsymbol{y} = \begin{pmatrix} x_3 \\ m_2 \end{pmatrix} \tag{12.3.2}$$

where m_2 is the reaction rate which is a function of all three state variables. There are two controls,

$$\boldsymbol{u} = \begin{pmatrix} u_1 \\ u_2 \end{pmatrix} \tag{12.3.3}$$

where u_1 is the normalized dimensionless cooling capacity and u_2 is the normalized dimensionless catalyst flow rate.

The system dynamics are described in Section 10.7. The equations are nonlinear in the state variables but linear in the controls and can be described by

$$\dot{\boldsymbol{x}} = \boldsymbol{f}(\boldsymbol{x}) + \boldsymbol{B}(x)\boldsymbol{u} \tag{12.3.4}$$

For time optimal control to a target set about a desired steady state, the optimal control law from Pontryagin's maximum principle (Chapter 8) is

$$\begin{aligned} u_1 &= \operatorname{sgn}(b_2\lambda_1) \\ u_2 &= \operatorname{sgn}(a_1\lambda_1 + a_2\lambda_2 + a_3\lambda_3) \end{aligned} \tag{12.3.5}$$

where λ_1, λ_2, and λ_3 are the costate variables and

$$\begin{aligned} b_2 &= -\frac{(T - T_c)}{V_H \rho c_p} \qquad & a_1 &= \frac{(T_o - T)}{V_H} \\ a_2 &= -\frac{C}{V_x} \qquad & a_3 &= \frac{(C_{KI_o} - C_{KI})}{V_x} \end{aligned} \tag{12.3.6}$$

Lynch and Ramirez (1975) chose the target set given in Figure 12.3. For a trajectory to be time-optimal, it must lie on the boundary of the target set and the costate vector is normal to a support hyperplane on the target set directed outward at the time t_f (Hermes and LaSalle, 1969). With the costate boundary condition established, the six state and costate equations are integrated backwards in time starting from the surface of the target set with the optimal control law of Equation (12.3.5). By starting at various points on the target set, optimal switching surfaces can be developed. These are given in Figures 12.4 and 12.5. These switching surfaces define the bang–bang control algorithm for time-optimal control of this stirred tank reactor.

In order to implement this control policy, all three states must be known. There are only two measurements, that of temperature and reaction rate. There is considerable noise associated with the measurement of the reaction rate as shown in Figure 12.6. In order to filter the noise in the reaction rate measurement and estimate the unmeasured states of reactant concentration and catalyst concentration, Kalman filtering is used. The actual system is uncertain so that the stochastic system model is

$$\dot{\boldsymbol{x}} = \boldsymbol{f}(\boldsymbol{x}) + \boldsymbol{B}(\boldsymbol{x})\boldsymbol{u} + \boldsymbol{w} \tag{12.3.7}$$

and the measurement model is

$$\boldsymbol{y} = \boldsymbol{h}(\boldsymbol{x}) + \boldsymbol{v} \tag{12.3.8}$$

where $\boldsymbol{w}$ is Gaussian with covariance $\hat{\boldsymbol{Q}}$ and $\boldsymbol{v}$ is Gaussian with covariance $\hat{\boldsymbol{R}}$.

From the separation principle, we can solve this problem as a standard optimal control problem using the optimal estimates for the process states,

$$\dot{\hat{\boldsymbol{x}}} = \boldsymbol{f}(\hat{\boldsymbol{x}}) + \boldsymbol{B}(\hat{\boldsymbol{x}})\boldsymbol{u} \tag{12.3.9}$$

which leads to the bang–bang policy given by Equations (12.3.5).

Since a digital computer control system was used to implement the time-optimal strategy, measurements are only available at discrete time increments, $\Delta T = 12$ sec. The discrete extended Kalman filter was therefore used to obtain estimates of the states $\hat{\boldsymbol{x}}$ based upon the discrete time version of the mathematical model of (12.3.9) and the system measurements, $\boldsymbol{y}$. In the design of the discrete extended Kalman filter, the state uncertainty covariance matrix was

$$\hat{\boldsymbol{Q}} = \text{diag}(.001\ ,\quad .001\ ,\quad .001) \tag{12.3.10}$$

and the measurement noise covariance matrix was

$$\hat{\boldsymbol{R}} = \text{diag}(.003\ ,\quad .0144) \tag{12.3.11}$$

This implies a 3% model uncertainty, a 5% temperature measurement uncertainty, and a 12% reaction rate measurement uncertainty.

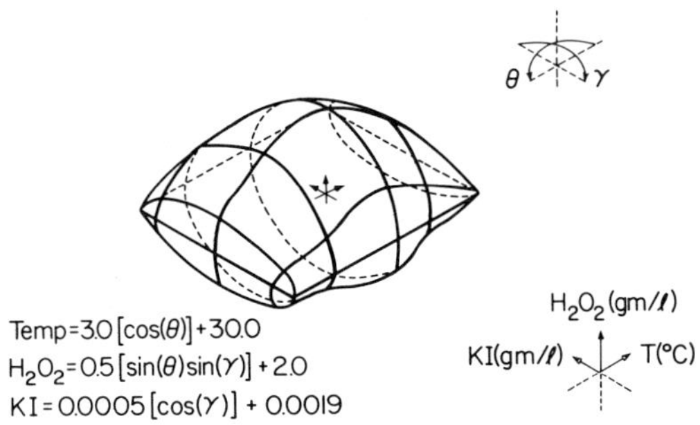

Figure 12.3 Three-dimensional view of target set region. From Lynch and Ramirez (1975).

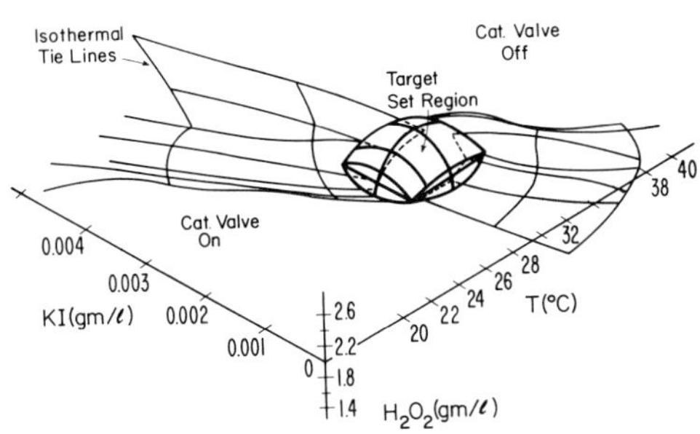

Figure 12.4 Catalyst valve switching surfaces. From Lynch and Ramirez (1975).

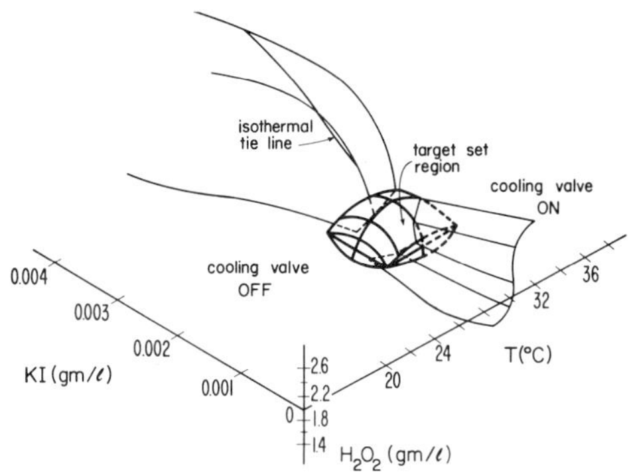

Figure 12.5 Cooling valve switching surfaces. From Lynch and Ramirez (1975).

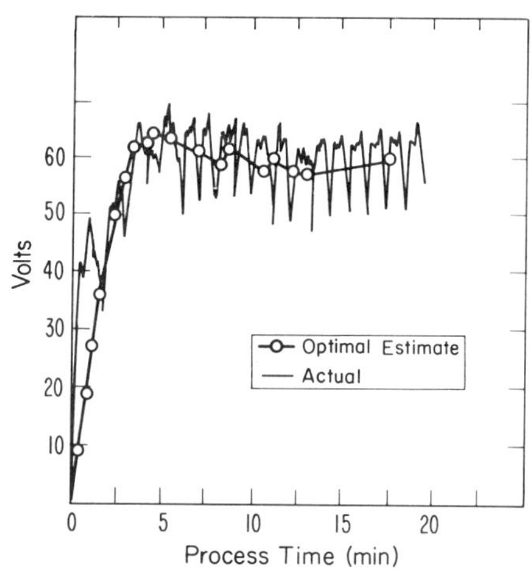

Figure 12.6 Actual vs. optimally estimated reaction rate. From Lynch and Ramirez (1975).

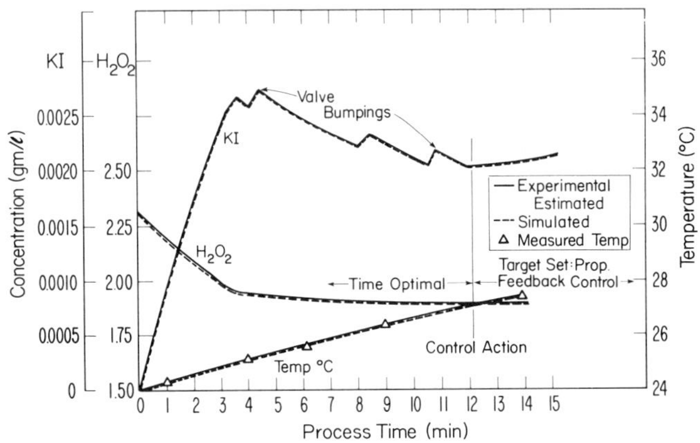

Figure 12.7 Comparison of simulated and experimental trajectories. From Lynch and Ramirez (1975).

Lynch and Ramirez (1975) applied the time-optimal control algorithm with on-line filtering and estimation to an actual experimental reactor system using a small process control computer. Typical experimental results are given in Figures 12.6 and 12.7. Figure 12.6 shows the actual noisy reaction rate signal and the optimal estimate. It is clear that the Kalman filter has eliminated significant process noise associated with this measured variable. Figure 12.7 shows the comparison of actual estimated responses with simulated expected responses. Actual measured temperatures are also shown on the figure. The results indicate that the actual experimental system behaved as expected from computer simulations. Application of the separation principle allowed the system to be operated under time-optimal conditions.

12.4 Optimal Adaptive Regulation of a Bioreactor

In Section 8.6 the work of Park and Ramirez (1988) on determining the optimal production strategy for the manufacture of a secreted foreign protein in a fed-batch reactor was discussed. The optimal strategy involves operation on singular arcs of constant glucose concentration associated with maximizing either the growth rate of the yeast cells or the synthesis rate of the protein product. In order to implement this strategy, it is necessary to regulate the glucose concentration at a specified value. In this section we will develop an optimal adaptive regulation scheme for such a bioreactor regulation problem. This problem was first considered by Park and Ramirez (1990). For this problem, combined Kalman state variable filtering and sequential least squares on-line parameter estimation are performed. This allows for the estimation of state variables and time-varying uncertain model parameters. Then, based upon the separation principle, an optimal regulatory control law is designed with both predictive feed-forward and corrective feedback elements.

The following growth model of a fed-batch bioreactor is considered,

$$\frac{dX}{dt} = \mu_x X - \frac{q}{V} X \tag{12.4.1}$$

$$\frac{dS}{dt} = \frac{q}{V}(m - S) - Y\mu_x X \tag{12.4.2}$$

$$\frac{dV}{dt} = q \tag{12.4.3}$$

where X is the cell density (g/ℓ), S is the glucose concentration (g/ℓ), μ_x is the cell's specific growth rate (1/hr), Y is the inverse of the glucose to biomass yield, V is the culture volume (1), q is the feed flow rate into the reactor (1/hr), and m is the glucose concentration of the feed medium (g/ℓ).

The model is first transformed into a discrete time model using a first-order time approximation and allowing for model uncertainties through a probabilistic law

$$\boldsymbol{x}_{t+1} = \boldsymbol{F}(\boldsymbol{\theta}, \tau)\boldsymbol{x}_t + \boldsymbol{G}(\tau)\boldsymbol{u}_t + \boldsymbol{w}_t \tag{12.4.4}$$

where

$$\boldsymbol{x}_t = \begin{pmatrix} (XV)_t \\ (SV)_t \\ V_t \end{pmatrix} \qquad \boldsymbol{u}_t = q_t \tag{12.4.5}$$

and $\boldsymbol{w}_t$ is Gaussian with covariance $\hat{\boldsymbol{Q}}$. The state transition matrix $\boldsymbol{F}$, and the control to state matrix $\boldsymbol{G}$ are

$$\boldsymbol{F} = \begin{bmatrix} \theta_1 & 0 & 0 \\ \theta_2 & 0 & 0 \\ 0 & 0 & 1 \end{bmatrix} \tag{12.4.6}$$

$$\boldsymbol{G} = \begin{bmatrix} 0 \\ \tau m \\ \tau \end{bmatrix} \tag{12.4.7}$$

where τ is the sampling frequency and the parameters θ_1 and θ_2 are

$$\theta_1 = 1 + \tau\mu_x \tag{12.4.8}$$

$$\theta_2 = -\tau\mu_x Y \tag{12.4.9}$$

For this system the state variables are all measurable; therefore, the measurement matrix $\boldsymbol{H}$ is the identity matrix

$$\boldsymbol{y}_t = \boldsymbol{H}\overline{\boldsymbol{x}}_t + \boldsymbol{v}_t \tag{12.4.10}$$

where $\boldsymbol{v}_t$ is Gaussian with covariance $\hat{\boldsymbol{R}}$ and $\overline{\boldsymbol{x}}_t$ is the model prediction,

$$\overline{\boldsymbol{x}}_t = \boldsymbol{F}(\boldsymbol{\theta})\hat{\boldsymbol{x}}_{t-1} + \boldsymbol{G}\boldsymbol{u}_{t-1} \tag{12.4.11}$$

The optimal state estimates from Kalman filtering (Section 10.6) are

$$\hat{\boldsymbol{x}}_t = \overline{\boldsymbol{x}}_t + \hat{\boldsymbol{K}}_\theta(t)(\boldsymbol{y}_t - \boldsymbol{H}\overline{\boldsymbol{x}}_t) \tag{12.4.12}$$

Typical filter performance shows an initial transient in $\hat{\boldsymbol{K}}_\theta(t)$ followed by an essentially steady-state filter operation. The steady-state filter gain $\hat{\boldsymbol{K}}_\theta$ is determined from the algebraic Riccati equation

$$\hat{\boldsymbol{K}}_\theta = \hat{\boldsymbol{P}}_\theta \boldsymbol{H}^T \left[\boldsymbol{H}\hat{\boldsymbol{P}}_\theta \boldsymbol{H}^T + \hat{\boldsymbol{Q}}\right]^{-1} \tag{12.4.13}$$

$$\hat{\boldsymbol{P}}_\theta = \boldsymbol{F}(\boldsymbol{\theta})\hat{\boldsymbol{P}}_\theta \boldsymbol{F}^T(\boldsymbol{\theta}) + \hat{\boldsymbol{R}} - \hat{\boldsymbol{K}}_\theta \left[\boldsymbol{H}\hat{\boldsymbol{P}}_\theta \boldsymbol{H}^T + \hat{\boldsymbol{Q}}\right] \hat{\boldsymbol{K}}_\theta \tag{12.4.14}$$

In order to estimate the uncertain model parameters, the sequential least squares approach of Chapter 11 is used. Here the innovations error is defined as

$$\boldsymbol{\epsilon}_t = \boldsymbol{z}_t - \boldsymbol{H}\hat{\boldsymbol{x}}_t \tag{12.4.15}$$

and the recursive parameter updating algorithm is

$$\hat{\boldsymbol{\theta}}_t = \hat{\boldsymbol{\theta}}_{t-1} + \boldsymbol{L}_t \boldsymbol{\epsilon}_t \tag{12.4.16}$$

where the gain matrix $\boldsymbol{L}_t$ is defined by Equations (11.5.13) and (11.5.14).

To design an optimal regulatory controller, we make use of the separation principle which states that the optimal control problem should be formulated in terms of the optimal state estimates $\hat{\boldsymbol{x}}$. For this adaptive algorithm, it should be noted that the estimates $\hat{\boldsymbol{x}}$ are a function of the estimated parameters $\hat{\boldsymbol{\theta}}$. The system model can be linearized and expressed as

$$\dot{\hat{\boldsymbol{x}}} = \boldsymbol{A}(\hat{\boldsymbol{\theta}}, \hat{\boldsymbol{x}})\hat{\boldsymbol{x}} + \boldsymbol{B}(\hat{\boldsymbol{\theta}}, \hat{\boldsymbol{x}})\boldsymbol{u} \tag{12.4.17}$$

The design of the regulatory control law is important since significant deviation from the glucose set point can perturb the cell's metabolic pattern which can result in different growth rates and nutrient consumption. This is especially true when operating near the bifurcating point between oxidative and fermentative metabolism. Alternatively, if the operating glucose range is very low, the failure of the control performance can threaten cell viability when the major catabolite vanishes.

First we design a predictive feed-forward control algorithm based upon the desire to maintain a known level of glucose in the fermentor. We use the dynamic predictor concept of Section 6.3 to obtain a model predictive algorithm. We partition the state into a trajectory specified state, $\boldsymbol{x}_s$, which is the glucose concentration, and nonspecified states, $\boldsymbol{x}_f$, which are the estimated biomass concentration and the estimated reactor volume

$$\boldsymbol{x}_s = S \tag{12.4.18}$$

$$\boldsymbol{x}_f = \begin{pmatrix} \hat{X} \\ \hat{V} \end{pmatrix} \tag{12.4.19}$$

Based upon this partitioning, the linearized system dynamics are

$$\begin{pmatrix} \dot{\boldsymbol{x}}_s \\ \dot{\boldsymbol{x}}_f \end{pmatrix} = \begin{bmatrix} \boldsymbol{A}_{11}(\hat{\boldsymbol{\theta}}) & \boldsymbol{A}_{12}(\hat{\boldsymbol{\theta}}) \\ \boldsymbol{A}_{21}(\hat{\boldsymbol{\theta}}) & \boldsymbol{A}_{22}(\hat{\boldsymbol{\theta}}) \end{bmatrix} \begin{pmatrix} \boldsymbol{x}_s \\ \boldsymbol{x}_f \end{pmatrix} + \begin{bmatrix} \boldsymbol{B}_1(\hat{\boldsymbol{\theta}}) \\ \boldsymbol{B}_2(\hat{\boldsymbol{\theta}}) \end{bmatrix} \boldsymbol{u}_f \tag{12.4.20}$$

The feed-forward control action, $\boldsymbol{u}_f$, can be obtained by solving the first partition equation set for the control as

$$\boldsymbol{u}_f = \boldsymbol{B}_1^{-1}(\hat{\boldsymbol{\theta}}) \left[-\boldsymbol{A}_{11}(\hat{\boldsymbol{\theta}})\boldsymbol{x}_s - \boldsymbol{A}_{12}(\hat{\boldsymbol{\theta}})\boldsymbol{x}_f \right] \tag{12.4.21}$$

Finally, a feedback element can be designed by using the information we have on the current estimated glucose concentration, $\hat{\boldsymbol{x}}_s$. There would be errors associated with the desired value, $\boldsymbol{x}_s$, and the estimated value $\hat{\boldsymbol{x}}_s$ due to modeling errors and unmeasurable disturbances affecting the system. The inclusion of unmeasurable disturbances, $\boldsymbol{d}$, leads to a model

$$\dot{\hat{\boldsymbol{x}}}_s = \boldsymbol{A}_{11}(\hat{\boldsymbol{\theta}})\hat{\boldsymbol{x}}_s + \boldsymbol{A}_{12}(\hat{\boldsymbol{\theta}})\hat{\boldsymbol{x}}_f + \boldsymbol{B}_1(\hat{\boldsymbol{\theta}})\boldsymbol{u} + \boldsymbol{W}\boldsymbol{d} \tag{12.4.22}$$

Introducing the deviation variables

$$\begin{aligned} \overline{\boldsymbol{x}} &= \hat{\boldsymbol{x}}_s - \boldsymbol{x}_s \\ \overline{\boldsymbol{u}} &= \boldsymbol{u} - \boldsymbol{u}_f \end{aligned} \tag{12.4.23}$$

yields

$$\dot{\overline{\boldsymbol{x}}}_s = \boldsymbol{A}_{11}(\hat{\boldsymbol{\theta}})\overline{\boldsymbol{x}}_s + \boldsymbol{B}_{11}(\hat{\boldsymbol{\theta}})\overline{\boldsymbol{u}} + \boldsymbol{W}\boldsymbol{d} \tag{12.4.24}$$

which is the Johnson unmeasurable problem considered in Section 5.6. Introducing the Johnson transformation leads to the proportional-integral feedback algorithm

$$\overline{\boldsymbol{u}} = -\boldsymbol{K}_2\overline{\boldsymbol{x}} - \left[\boldsymbol{K}_1 - \boldsymbol{K}_2\boldsymbol{A}_{11}(\hat{\boldsymbol{\theta}})\right] \int \overline{\boldsymbol{x}}dt \tag{12.4.25}$$

where the gain matrices $\boldsymbol{K}_1$ and $\boldsymbol{K}_2$ are functions of the state weighting matrix $\boldsymbol{Q}$ and the control weighting matrix $\boldsymbol{R}$ in a quadratic performance functional.

In terms of the original variables, the final feed-forward–feedback regulatory controller is

$$\boldsymbol{u} = \boldsymbol{u}_f - \boldsymbol{K}_2\left(\hat{\boldsymbol{x}}_s - \boldsymbol{x}_s\right) - \left[\boldsymbol{K}_1 - \boldsymbol{K}_2\boldsymbol{A}_{11}(\hat{\boldsymbol{\theta}})\right]\int(\hat{\boldsymbol{x}}_s - \boldsymbol{x}_s)\,dt \tag{12.4.26}$$

This is clearly a model predictive control law in the spirit of Chapter 6. However, it is also adaptive since the state estimates, $\hat{\boldsymbol{x}}$, the feed-forward control element, $\boldsymbol{u}_f$, and the integral feedback gain depend upon the current estimate of the model parameters $\hat{\boldsymbol{\theta}}$.

Park and Ramirez (1990) implemented and tested this regulatory control structure using a 16 liter fermentor and a genetically engineered yeast system for the production of the protein invertase.

The importance of system identification under the current control scheme is shown by comparing results when the Kalman filter is sensitive and conservative to the measurements. Figure 12.8 shows the operation when the system identification algorithm is forced to be relatively slow in utilizing the innovation signals. This is accomplished by choosing $||\hat{\boldsymbol{R}}/\hat{\boldsymbol{Q}}|| = 100$. Thus, little weight is placed upon the measurements in the identification algorithm due to the large relative uncertainty in the measurements. In the design of the controller, $||\boldsymbol{Q}/\boldsymbol{R}|| = .1$, which results in relatively large proportional-integral controller gain matrices. Except for the first few sampling periods when the parameter vector is quite sensitive to small innovation signals, there is a slow change in the estimated parameters even though there are significant differences between the measured and estimated states. Therefore, with no significant improvement in the feed-forward control tuning and unreliable state estimation, the control scheme results in relatively poor regulation of culture glucose level, as expected. However, even with unreliable measurements, the parameter identification still extracts some information and updates the parameters slowly.

When the measurements are accurate, $||\hat{R}/\hat{Q}|| = 1.0$, the state estimates are nearly identical to the raw measurements as shown in Figure 12.9. Even with an erroneous initial parameter vector, the regulation improves as more information from the process is integrated into the control algorithm. At the beginning of operation, the feed-forward control tuning is poor, which results in increased feedback action as shown in Figure 12.9(C). As the model is updated, the ratio of feedback to feed-forward control action decays, indicating that the model predictability is enhanced and the quality of regulation improved.

Although the PI control action is only supplementary, it plays an important role in the overall control scheme. With the existence of model inaccuracy, it allows for the opportunity for accurate system identification by keeping the system in the proper operating neighborhood.

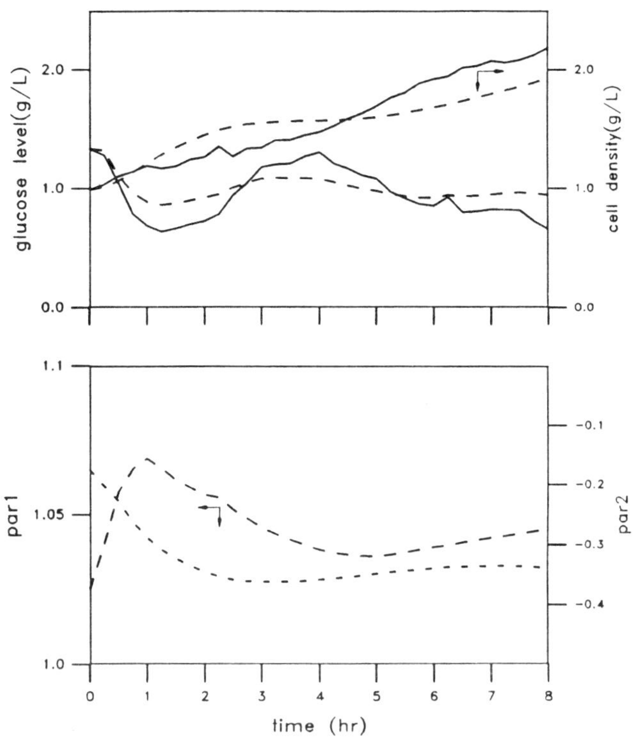

Figure 12.8. Control with $|R_2/R_1| = 100.00$ and $|Q/R| = 0.1$. Set point of culture glucose level is 1.0 g/ℓ. (A) Profiles of state variables, raw (—) and filtered (- - -). (B) Profiles of identified parameters. From Park and Ramirez (1990).

12.5 Robustness of the LQG Problem

Since both the linear quadratic regulator (LQR) and the Kalman filter have excellent robustness properties (Chapter 7 and Section 10.9), it is natural to expect, due to the separation principle, that the linear quadratic Gaussian (LQG) problem would also exhibit excellent robustness properties. However, Doyle (1978) showed that there are no guaranteed stability results for the LQG problem. The reason for this loss of robustness on the inclusion of an estimator is that when the plant differs from its model in the estimator, there is additional inaccuracy in state estimation. Doyle (1978) illustrates this with the following example:

$$\begin{pmatrix} \dot{x}_1 \\ \dot{x}_2 \end{pmatrix} = \begin{bmatrix} 1 & 1 \\ 0 & 1 \end{bmatrix} \begin{pmatrix} x_1 \\ x_2 \end{pmatrix} + \begin{bmatrix} 0 \\ 1 \end{bmatrix} u + \begin{bmatrix} 1 \\ 1 \end{bmatrix} w \tag{12.5.1}$$

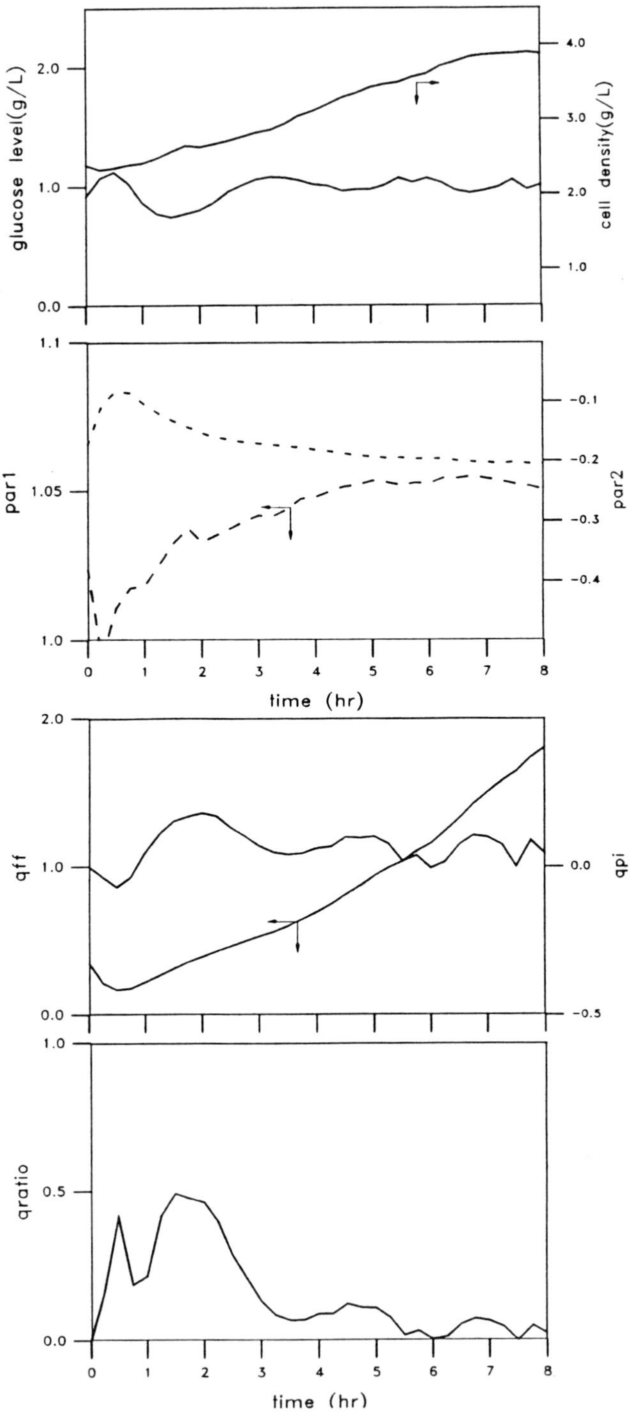

glucose level(g/L)
cell density(g/L)
par1
par2
time (hr)
qff
qpi
qratio
time (hr)

Figure 12.9

$$y = \begin{bmatrix} 1 & 0 \end{bmatrix} \begin{pmatrix} x_1 \\ x_1 \end{pmatrix} + v \tag{12.5.2}$$

with

$$\begin{matrix} \hat{\boldsymbol{Q}} = \sigma \boldsymbol{I} & \hat{\boldsymbol{R}} = 1 \\ \boldsymbol{Q} = \rho \boldsymbol{I} & \boldsymbol{R} = 1 \end{matrix} \tag{12.5.3}$$

Analytical solutions for the gain matrices are

$$\begin{aligned} \boldsymbol{K} &= -\alpha \begin{bmatrix} 1 \\ 1 \end{bmatrix} \qquad \alpha = 2 + \sqrt{4+\rho} \\ \hat{\boldsymbol{K}} &= -\beta \begin{bmatrix} 1 \\ 1 \end{bmatrix} \qquad \beta = 2 + \sqrt{4+\rho} \end{aligned} \tag{12.5.4}$$

The full system matrix is

$$\begin{bmatrix} 1 & 1 & 0 & 0 \\ 1 & 0 & -m & -m\alpha \\ \beta & 0 & 1-\beta & 1 \\ \beta & 0 & -\beta-\alpha & 1-\alpha \end{bmatrix} \tag{12.5.5}$$

where m is the plant input gain. The characteristic equation is

$$s^4 + As^3 + Bs^2 + \Big[\beta + \alpha - 4 + 2(m-1)\alpha\beta\Big] s + \Big[1 + (1-m)\alpha\beta\Big] = 0 \tag{12.5.6}$$

For stability

$$\begin{aligned} &1 + (1-m)\alpha\beta > 0 \\ &\beta + \alpha - 4 + 2(m-1)\alpha\beta > 0 \end{aligned} \tag{12.5.7}$$

For large α, β even for small m, this stability relation can be violated. Robustness properties of the LQR can be recovered by the LQG by using the loop recovery procedure of Doyle and Stein (1979). By adding fictitious noise to the process,

$$\hat{\boldsymbol{Q}} = \boldsymbol{Q}_o + q^2 \boldsymbol{B}\boldsymbol{V}\boldsymbol{B}^T \tag{12.5.8}$$

where $\boldsymbol{B}$ is the control matrix and $\boldsymbol{V}$ a positive definite matrix, robustness is improved. As q approaches infinity, the robustness of the LQR and LQG problems becomes equivalent. Robustness is improved; however, there is a performance loss since this technique detunes the model predictive capability. As q increases, the system relies more on the system measurements. If the measurements have significant noise, the filtered response will be noisy.

The inclusion of adaptive model parameter identification is another way to improve the robustness of the LQG design. An adaptive system automatically responds to innovation errors by updating model parameters and adjusting the effective level of model uncertainty. One must be sensitive to the effect of model uncertainty in applying LQG concepts.

References

Astrom, K. J., *Introduction to Stochastic Control Theory*, Academic Press, Cambridge, MA (1970).

Balakrishnan, A. V., "Stochastic Control: A Functional Space Approach," *SIAM J. Control* **10**, 285–297 (1972).

Doyle, J. C., "Guaranteed Margins for LQG Regulators," *IEEE Trans. Auto. Control* **AC–23**, 756–757 (1978).

Doyle, J. C. and Stein, G., "Robustness with Observers," *IEEE Trans. Auto. Control* **AC–24**, 607–611 (1979).

Grimble, M. J. and Johnson, M. A., *Optimal Control and Stochastic Estimation, Theory and Applications*, Vol. 2, Wiley and Sons, New York (1988).

Hermes, H. and LaSalle, J. P., *Function Analysis and Time Optimal Control*, Academic Press, New York (1979).

Kwakernaak, H. and Sivan, R., *Linear Optimal Control Systems*, Wiley and Sons, New York (1972).

Lewis, F. L., *Optimal Estimation*, Wiley Interscience, New York (1986).

Lynch, E. B. and Ramirez, W. F., "Realtime Time–Optimal Control of a Stirred Tank Reactor Using Kalman Filtering for State Estimation," *AIChE Journal* **21**, 799–804 (1975).

Nyquist, J. K. and Ramirez, W. F., "Time Optimal Control of an Experimental Continuous Stirred Tank Reactor," *Chem. Eng. Sci.* **26**, 1673–1688 (1971).

Park, S. and Ramirez, W. F., "Optimal Production of Secreted Protein in Fed-Batch Reactors," *AIChE Journal* **34**, 1550–1558 (1988).

Park, S. and Ramirez, W. F., "Optimal Regulatory Control of Bioreactor Nutrient Concentration Incorporating System Identification," *Chem. Eng. Sci.* **45**, 3467–3481 (1990).

Sage, A. P. and White, C. C., *Optimal Systems Control*, Second Edition, Prentice–Hall, Englewood Cliffs, NJ (1977).

Appendix A—MATLAB Functions

For further information regarding MATLAB, please contact the Math Works, Inc., 24 Prime Park Way, Natick, MA 01760, Telephone (508) 653-1415, FAX (508) 653-2997.

Function	**Purpose**
c2d	Converts a continuous time domain model into a discrete-time model. Appropriate discrete-time system matrices are calculated.
ctrb	Computes the controllability matrix for linear state space systems.
det	Computes the determinant of a square matrix.
dgram	Computes the controllability and observability gramians for a discrete-time system.
diary	Command used to create a copy of keyboard input and most of the resulting output.
dimpulse	Calculates the impulse response of a discrete-time linear system.
dlqe	Solves the discrete-time linear estimator problem by solving appropriate Riccati equations.
dlqr	Solves the discrete-time linear regulator problem by solving appropriate Riccati equations.
dlsim	Simulates the response of discrete-time linear systems with arbitrary inputs.
dstep	Calculates the step response of a discrete-time linear system.
d2c	Converts a discrete-time model into a continuous time model. Appropriate continuous time system matrices are calculated.
eig	Calculates the eigenvalues of a square matrix and the associated eigenvectors.
expm	Computes the exponential matrix.
eye	Generates an identity matrix.

gram	Computes the controllability and observability gramians.
help	Gives information on how to use specific MATLAB functions.
impulse	Calculates the impulse response of a linear continuous time system.
inv	Computes the inverse of a square matrix.
label	Generates a label beneath the axes of a graph.
lqe	Solves the linear quadratic estimator problem by computing the appropriate Riccati equations.
lqr	Solves the linear quadratic regulator problem by computing the appropriate Riccati equations.
lsim	Simulates a continuous time linear system with arbitrary inputs.
lu	Performs an lu factorization of a matrix into a permutated upper triangular matrix and a lower triangular matrix.
meta	Used to generate high quality graphic output.
obs	Computes the observability matrix for linear state space systems.
ode23	Integrates a system of ordinary differential equations using second and third order Runge–Kutta formulas.
ode45	Integrates a system of ordinary differential equations using fourth and fifth order Runge–Kutta formulas.
ones	Generates a matrix of all ones.
pinv	Calculates the Moore–Pinrose pseudoinverse.
plot	Generates a plot of the dependent variables verses the independent variable.
rand	Generates random numbers and matrices.
step	Calculates the step response of a linear continuous time system.
tf2ss	Converts a system in transfer function form to state space form.
title	Generates a title at the top of a graph.
zeros	Generates a matrix of all zeros.

Index